XI International Conference on Hyperons, Charm and Beauty Hadrons

(BEACH 2014)

Journal of Physics: Conference Series Volume 556

Birmingham, United Kingdom
21 – 26 July 2014

Editors:

M. Bozzo

ISBN: 978-1-63439-667-7
ISSN: 1742-6588

Printed from e-media with permission by:

Curran Associates, Inc.
57 Morehouse Lane
Red Hook, NY 12571

Some format issues inherent in the e-media version may also appear in this print version.

Printed by Curran Associates, Inc. (2015)

For permission requests, please contact the Institute of Physics
at the address below.

Institute of Physics
Dirac House, Temple Back
Bristol BS1 6BE UK

Phone: 44 1 17 929 7481
Fax: 44 1 17 920 0979

techtracking@iop.org

Additional copies of this publication are available from:

Curran Associates, Inc.
57 Morehouse Lane
Red Hook, NY 12571 USA
Phone: 845-758-0400
Fax: 845-758-2634
Email: curran@proceedings.com
Web: www.proceedings.com

TABLE OF CONTENTS

SESSION 1

CHAIR: EVGUENI GOUDZOYSKI

SESSION 2

CHAIR: ALEXANDER KHODJAMIRIAN

SESSION 3

CHAIR: NORMAN CHRIST

SESSION 4

CHAIR: KAMAL SETH

SESSION 5

CHAIR: JORGE MARTIN CAMALICH

SESSION 6

CHAIR: PATRIZIA CENCI

SESSION 7

CHAIR: SVETLJANA FAJFER

SESSION 8

CHAIR: MARCO BOZZO

SESSION 9

CHAIR: VALERIE GIBSON

SESSION 10

CHAIR: CRISTINA LAZZERONI

SESSION 11

CHAIR: ROBERT FLEISCHER

SESSION 21

CHAIR: NIGEL WATSON

Journal of Physics: Conference Series **556** (2014) 011001 doi:10.1088/1742-6596/556/1/011001

Proceedings of the

XI Conference on Beauty, Charm, Hyperons in Hadronic Interactions BEACH

Birmingham University - 21st to 26th of July 2014

Editor:

Marco Bozzo - University of Genoa and INFN Genoa, Italy

Reviewers:

Patrizia Cenci - Istituto Nazionale di Fisica Nucleare, Sezione di Perugia, Italy
Evgueni Goudzovski - School of Physics and Astronomy, University of Birmingham, UK
Martin Gorbahn - Department of Mathematical Sciences, University of Liverpool, UK
Cristina Lazzeroni - School of Physics and Astronomy, University of Birmingham, UK
Alexander Lenz - IPPP, Durham University, UK
Janis McKenna - University of British Columbia, Vancouver, Canada
Holger Meyer - Wichita University, Kansas, USA
Nick Solomey - Wichita University, Kansas, USA
Nigel Watson - School of Physics and Astronomy, University of Birmingham, UK

Preface

This volume contains the invited and contributed papers presented at the 11th*International Conference on Hyperons, Charm and Beauty Hadrons*, currently known as the BEACH Conferences. The BEACH conferences cover a broad range of physics topics in the field of Hyperon and heavy-flavor physics. This conference continues the BEACH series, which began with a meeting in Strasbourg in 1995 and since then offers a biennial opportunity for both theorists and experimentalists from the high-energy physics community to discuss all aspects of flavour physics.

The 11th Conference took place in the Lecture Theatre of the Physics West Building of the University of Birmingham (United Kingdom) from July 22nd to July 26th and was attended by 107 participants. All of the sessions were plenary sessions accommodating review talks and shorter contributions discussing both theory and recent experiments. At the end of the conference Valerie Gibson (Cavendish Laboratory, University of Cambridge, UK) and Sebastian Jaeger (School of Physics and Astronomy, University of Sussex, UK) summarized and put in context all the presentations of the conference giving two very interesting Summary talks.

These Conference Proceedings are particularly interesting since, due to the long shutdown of the LHC in Geneva(CH), most of the data presented were from the entire data set available. This volume in fact offers an interesting panorama of the present situation and allows a comparison of the experimental data and the theory in a field that is always in continuous evolution.

The conference was impeccably organized by the Local Organizing Committee chaired by Cristina Lazzeroni (Birmingham Univeristy, Birmingham, UK) that I want to thank particularly here. Many from the University Staff have contributed to the smooth running of the conference. We would like to thank the Local Scientific Secretariat for their invaluable help in making the conference a truly enjoyable and unforgettable event; a special thanks goes also to Maria Hobbs, our local secretary, who worked tirelessly in the organization of every detail. Finally we would like thank the European Organization for Nuclear Research, the European Research Council, the UK Science and Technology Facility Council, the UK Institute of Particle Physics Phenomenology and the University of Birmingham for their generous support.

The next BEACH Conference will be held at George Mason University, George Mason University Fairfax, Virginia (USA) at the beginning of summer 2016 and I hope that we will all meet again there.

Marco Bozzo,
Univeristy of Genoa and INFN Genoa (Italy)

Journal of Physics: Conference Series **556** (2014) 011001 doi:10.1088/1742-6596/556/1/011001

Advisory Committee

Marco Bozzo	Università di Genova and INFN Genova (chair)
Guido Altarelli	Università di Roma Tre
Federico Antinori	INFN Padova
Stan Brodsky	SLAC and Stanford University
Antonio Ferrer	IFIC, Valencia
Valerie Gibson	University of Cambridge
Francis Halzen	University of Wisconsin–Madison
David Hitlin	California Institute of Technology, Pasadena
Calvin Kalman	Concordia University, Montreal
Vera Luth	SLAC and Stanford University
Tatsuya Nakada	EPFL, Lausanne
Chris Quigg	Fermilab, Chicago
Gigi Rolandi	CERN, Geneva
Patrick Roudeau	Université Paris Sud, Orsay
Chris Sachrajda	University of Southampton
Marjorie Shapiro	University of California, Berkeley
Teodor Siemiarczuk	NCBJ, Warsaw
Maria Smizanska	Lancaster University
Hermine Wöhri	CERN, Geneva

Organizing Committee

Cristina Lazzeroni	University of Birmingham (chair)
Marco Bozzo	Università di Genova and INFN Genova
Patrizia Cenci	INFN Perugia
Martin Gorbahn	University of Liverpool
Evgueni Goudzovski	University of Birmingham
Chris Hawkes	University of Birmingham
Alexander Lenz	Durham University
Janis McKenna	University of British Columbia
Nick Solomey	Wichita State University
Nigel Watson	University of Birmingham

Local Scientific Secretariat

Katie Graham	James McCarthy	Francis Newson
Angela Romano	Antonino Sergi	Mark Slater
Matt Williams	Nicolas Lurkin	Maria Hobbs

Journal of Physics: Conference Series **556** (2014) 011001 doi:10.1088/1742-6596/556/1/011001

BEACH 2014 - The participants.

BEACH 2014 Highlights: photos from the conference

Journal of Physics: Conference Series **556** (2014) 011001 doi:10.1088/1742-6596/556/1/011001

BEACH 2014 – List of Participants

1 AKAR Simon
Centre de Physique des Particules de Marseille, France
akar@cppm.in2p3.fr

2 ANDERLINI Lucio
Universitá e INFN Firenze, Italy
lucio.anderlini@cern.ch

3 AZIZI Kazem
Dogus University, Turkey
kazizi@dogus.edu.tr

4 BASTID Nicole
LPC Clermont-Ferrand, France
bastid@clermont.in2p3.fr

5 BELL Guido
University of Oxford, UK
guido.bell@physics.ox.ac.uk

6 BERTRAM Iain
Lancaster University, UK
i.bertram@lancaster.ac.uk

7 BEVAN Adrian
Queen Mary University of London, UK
a.j.bevan@qmul.ac.uk

8 BOBETH Christoph
Technical University Munich, Germany
bobeth@ph.tum.de

9 BOUHOVA-THACKER Eva
Lancaster University, UK
e.bouhova-thacker@lancaster.ac.uk

10 BOZZO Marco
Universitá e INFN Genova, Italy
marco.bozzo@cern.ch

11 BRAMBILLA Nora
Technical University Munich, Germany
nora.brambilla@ph.tum.de

12 BUCCI Francesca
INFN Firenze, Italy
francesca.bucci@fi.infn.it

13 CAMALICH Jorge Martin
University of Mainz, Germany
jmartincamalich@ucsd.edu

14 CANDELISE Vieri
Universitá e INFN Trieste, Italy
vieri.candelise@ts.infn.it

15 CASAROSA Giulia
INFN Pisa, Italy
giulia.casarosa@pi.infn.it

16 CELENTANO Andrea
INFN Genova, Italy
andrea.celentano@ge.infn.it

17 CENCI Patrizia
INFN Perugia, Italy
patrizia.cenci@pg.infn.it

18 CHALUPKOVA Ina
Charles University Prague, Czech Republic
ina.chalupkova@cern.ch

19 CHEN Kai-Feng
National Taiwan University, Taiwan
kai-feng.chen@cern.ch

20 CHIARELLI Giorgio
INFN Pisa, Italy
giorgio.chiarelli@pi.infn.it

21 CHRIST Norman
Columbia University, USA
nhc@phys.columbia.edu

22 CODISPOTI Giuseppe
Universitá e INFN Bologna, Italy
giuseppe.codispoti@bo.infn.it

23 DE BRUYN Kristof
NIKHEF, The Netherlands
debkr@nikhef.nl

24 DE SIMONE Patrizia
INFN Laboratori Nazionali di Frascati, Italy
patrizia.desimone@lnf.infn.it

25 DE SOUZA Mario Everardo
Universidade Federal de Sergipe, Brazil
mariodesouza.ufs@gmail.com

26 DI LODOVICO Francesca
Queen Mary University of London, UK
f.di.lodovico@qmul.ac.uk

27 DJALALI Chaden
University of Iowa, USA
chaden-djalali@uiowa.edu

28 DOBRIGKEIT CHINELLATO David
Instituto de Física "Gleb Wataghin", Brazil
daviddc@ifi.unicamp.br

29 DOSSETT David Alasdair
University of Warwick, UK
david@ddossett.com

30 DRASAL Zbynek
Charles University Prague, Czech Republic
drasal@ipnp.troja.mff.cuni.cz

31 DUK Viacheslav
INFN Perugia, Italy
Viacheslav.Duk@cern.ch

32 EDMONDS Andrew
University College London, UK
a.edmonds@ucl.ac.uk

33 ERKOL Guray
Ozyegin University, Turkey
guray.erkol@ozyegin.edu.tr

34 FAJFER Svetlana
University of Ljubljana and Inst. J. Stefan, Slovenia
svjetlana.fajfer@ijs.si

35 FIGUEREDO Marcel A. S.
University of Liverpool, UK
Marcel.Araujo-Silva-Figueredo@liverpool.ac.uk

36 FLEISCHER Robert
Nikhef and VU University Amsterdam, The Netherlands
Robert.Fleischer@nikhef.nl

37 GHEZ Philippe
Centre National pour la Recherche Scientifique,
Annecy, France
ghez@lapp.in2p3.fr

38 GIBSON Valerie
University of Cambridge, UK
gibson@hep.phy.cam.ac.uk

39 GORBAHN Martin
University of Liverpool, UK
mgorbahn@liv.ac.uk

40 GUDZOVSKIY Evgueni
University of Birmingham, UK
eg@hep.ph.bham.ac.uk

41 HAWKES Chris
University of Birmingham, UK
c.m.hawkes@bham.ac.uk

42 HE Xionghong
Peking University, China
hexionghong@pku.edu.cn

43 HERNANDEZ Eliecer
University of Salamanca, Spain
gajatee@usal.es

44 HIROSE Minoru
Tokyo Institute of Technology, Japan
hirose@hep.phys.titech.ac.jp

45 HUBER Stephan
University of Sussex, UK
s.huber@sussex.ac.uk

46 HYUN HyoJung
Kyungpook National University, Korea
hjhyun@gmail.com

47 ILIESCU Mihail Antoniu
INFN Laboratori Nazionali di Frascati, Italy
Mihai.Iliescu@lnf.infn.it

48 JAEGER Sebastian
University of Sussex, UK
S.Jaeger@sussex.ac.uk

49 JUETTNER Andreas
University of Southampton, UK
juettner@soton.ac.uk

50 KAMENIK Jernej
Jozef Stefan Institute, Slovenia
jernej.kamenik@ijs.si

51 KARTVELISHVILI Vato
Lancaster University, UK
v.kartvelishvili@lancaster.ac.uk

52 KHODJAMIRIAN Alex
University of Siegen, Germany
khodjamirian@physik.uni-siegen.de

53 KIM Tae Jeong
Chonbuk National University, Korea
tae.jeong.kim@jbnu.ac.kr

54 KO Byeong Rok
Korea University, Korea
brko@hep.korea.ac.kr

55 KRÄTSCHMER Ilse
Institue of High Energy Physics, Vienna, Austria
ilse.kratschmer@cern.ch

56 LATHAM Thomas Edward
University of Warwick, UK
t.latham@warwick.ac.uk

57 LAZZERONI Cristina
University of Birmingham, UK
cl@hep.ph.bham.ac.uk

58 LURKIN Nicolas
University of Birmingham, UK
nicolas.lurkin@cern.ch

59 MACOLINO Carla
INFN Laboratori Nazionali del Gran Sasso, Italy
carla.macolino@lngs.infn.it

60 MARTI-GARCIA Salvador
Institut de Física Corpuscular, Spain
salvador.marti@ific.uv.es

61 MCCARTHY James
University of Birmingham, UK
james.mccarthy@cern.ch

62 MEYER Holger
Wichita State University, USA
holger.meyer@wichita.edu

63 NEWMAN Paul
University of Birmingham, UK
p.r.newman@bham.ac.uk

64 NEWSON Francis
University of Birmingham, UK
fon@hep.ph.bham.ac.uk

65 NIEVES Juan M
IFIC, Universitat de Valencia, Spain
jmnieves@ific.uv.es

66 NUMAO Toshio
TRIUMF, Canada
toshio@triumf.ca

67 OGILVY Stephen
University of Glasgow, UK
stephen.ogilvy@cern.ch

68 OWEN Patrick Haworth
Imperial College London, UK
p.owen10@imperial.ac.uk

69 PAKVASA Sandvip
University of Hawaii, USA
pakvasa@phys.hawaii.edu

70 PALAZZO Antonio
Max-Planck-Institut für Physik, Germany
palazzo@mpp.mpg.de

71 PAPA Angela
Paul Scherrer Institute, Switzerland
angela.papa@psi.ch

72 PARADISI Paride
Universitá di Padova, Italy
paride.paradisi@cern.ch

73 PECJAK Benjamin
Institute for Particle Physics Phenomenology, Durham, UK
ben.pecjak@durham.ac.uk

74 PETERS Yvonne
University of Manchester, UK
reinhild.peters@cern.ch

75 PETROV Alexey
Wayne State University, USA
apetrov@wayne.edu

76 PICH Antonio
Institut de Física Corpuscular, Universitat de Valencia, Spain
Antonio.Pich@ific.uv.es

77 PLACAKYTE Ringaile
DESY, Germany
ringaile@mail.desy.de

78 RAGGI Mauro
INFN Laboratori Nazionali di Frascati, Italy
mauro.raggi@lnf.infn.it

79 RAMACHERS Yorck
University of Warwick, UK
y.a.ramachers@warwick.ac.uk

80 ROMANO Angela
University of Birmingham, UK
angela.romano@cern.ch

81 RUBIN Philip
George Mason University, USA
prubin@gmu.edu

82 SAPRONOV Andrey
Joint Institute for Nuclear Research, Russia
andrey.a.sapronov@gmail.com

83 SCHÖNNING Karin
Uppsala University, Sweden
karin.schonning@physics.uu.se

84 SCHWALLER Pedro
CERN , Switzerland
pedro.schwaller@cern.ch

85 SERGI Antonino
University of Birmingham, UK
a.sergi@bham.ac.uk

86 SETH Kamal
Northwestern University, USA
kseth@northwestern.edu

87 SHEN Chengpin
Beihang University, P.R.China
shencp@ihep.ac.cn

88 SIBIDANOV Alexei
University of Sydney, Australia
sibid@physics.usyd.edu.au

89 SILARSKI Michal
Institute of Physics, Jagiellonian University, Poland
michal.silarski@uj.edu.pl

90 SMITH Jackson
Cambridge University, UK
jackson.smith@cern.ch

91 SOLOMEY Nick
Wichita State University, USA
Nick.solomey@wichita.edu

92 STEINHAUSER Matthias
Karlsruhe Institut für Technologie, Germany
matthias.steinhauser@kit.edu

93 STEINHEIMER Jan
Frankfurt Institute for Advanced Studies, Germany
steinheimer@fias.uni-frankfurt.de

94 TAMSETT Matthew
University of Sussex, UK
m.tamsett@sussex.ac.uk

95 TASDAN Ismail Ufuk
Middle East Technical University, Turkey
tasdan@metu.edu.tr

96 TELLARINI Giulia
Universitá di Ferrara e INFN, Italy
giulia.tellarini@cern.ch

97 TOGAWA Manabu
Osaka university, Japan
togawa@champ.hep.sci.osaka-u.ac.jp

98 TISHCHENKO Vladimir
Brookhaven National Laboratory, USA
vtishchenko@bnl.gov

99 UKLEJA Artur
National Centre for Nuclear Research, Poland
artur.ukleja@fuw.edu.pl

100 VALLIER Alexis Roger Louis
Laboratoire de l'Accélrateur Linéaire, France
vallier@lal.in2p3.fr

101 VECCHIO Alberto
University of Birmingham, UK
vecchioa@adf.bham.ac.uk

102 VELASCO SEVILLA Liliana
University of Bergen, Norway
liliana.velascosevilla
@gmail.com

103 WANG Yuming
Technische Universität München, Germany
yuming.wang@tum.de

104 WATSON Miriam
University of Birmingham, UK
Miriam.Watson@cern.ch

105 WATSON Nigel
University of Birmingham, UK
nigel.watson@cern.ch

106 WHITEHEAD Lisa
University of Houston, USA
lawhitehead@uh.edu

107 XU Zhirui
Ecole Polytechnique Federale de Lausanne, Switzerland
zhirui.xu@cern.ch

108 YUAN Xuhao
Tsinghua University, China
xuhao.yuan@cern.ch

109 ZHANG Cen
Universitè Catholique de Louvain, Belgium
cen.zhang@uclouvain.be

Journal of Physics: Conference Series **556** (2014) 011002 doi:10.1088/1742-6596/556/1/011002

Peer review statement

All papers published in this volume of *Journal of Physics: Conference Series* have been peer reviewed through processes administered by the proceedings Editors. Reviews were conducted by expert referees to the professional and scientific standards expected of a proceedings journal published by IOP Publishing.

The Flavourful Road to New Physics

Jernej F. Kamenik

J. Stefan Institute, Jamova 39, P. O. Box 3000, 1001 Ljubljana, Slovenia
Department of Physics, University of Ljubljana, Jadranska 19, 1000 Ljubljana, Slovenia

E-mail: `jernej.kamenik@ijs.si`

Abstract. Flavour physics plays a prominent role in our quest to understand and constrain possible phenomena beyond the standard model. It directly relates to two of the outstanding issues of contemporary particle physics: the so-called standard model and new physics flavour puzzles. At the same time it indirectly probes new physics scales up to 10^5 TeV through virtual effects. On the other hand, nonstandard flavour phenomena, if observed, could help shed light and constrain the nature of the electroweak symmetry breaking and the Higgs sector of the theory. In particular, such effects could help reduce fine-tuning in models addressing the electroweak hierarchy in light of null LHC new physics search results. Finally, exotic flavour-violating processes, such as those involving missing energy signatures may turn out to be the portals to possible hidden sectors beyond the standard model.

1. Why flavour matters in the LHC era?

In the past three decades, the standard model (SM) has withstood severe experimental scrutiny and emerged as a phenomenologically extremely successful theory. Nonetheless, theoretical arguments to consider it merely as an effective field theory remain relevant even after the recent discovery of the Higgs-like scalar at the LHC, completing the SM predicted particle content. In this context it is instructive to decompose the SM Lagrangian (including possible neutrino mass operators) in terms of the relevant gauge-kinetic terms and the effective potential

$$\mathcal{L}_{\nu\mathrm{SM}} = \mathcal{L}_{\mathrm{gauge}}(A_a, \psi_i) + D_\mu\phi^\dagger D^\mu\phi - V_{\mathrm{eff}}(\phi, A_a, \psi_i)\,, \tag{1}$$

where A_a, ψ_i and ϕ stand for the SM gauge fields, fermions and the Higgs field, respectively. Three of the most important SM theoretical puzzles concern the structure of V_{eff}

$$V_{\mathrm{eff}} = -\mu^2\phi^\dagger\phi + \lambda(\phi^\dagger\phi)^2 + Y^{ij}\psi^i_L\psi^j_R\phi + \frac{y^{ij}}{\Lambda}\psi^{iT}\psi^j_L\phi^T\phi + \ldots\,, \tag{2}$$

namely the electroweak (EW) hiearchy problem, i.e. why $|\mu^2| \ll \Lambda^2$ where Λ is the effective field theory cut-off scale; the SM flavour puzzle: why Y^{ij} are aligned (in the quark sector) and hierarchical (in both the quark and charged lepton sectors); and finally the mechanism of neutrino mass generation, i.e. the existence of the y^{ij}/Λ term. In absence of new light degrees of freedom, the outstanding task of high energy physics in general and flavour physics in particular is to understand and constrain the size of additional terms in the series (suppressed by increasing powers of $1/\Lambda$) with the hope of gaining new insights on these issues.

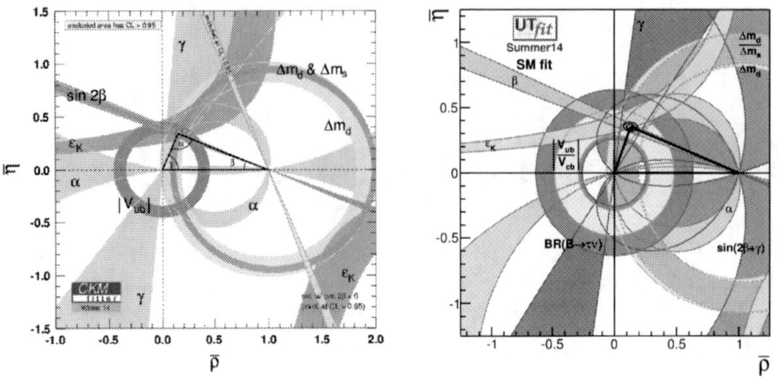

Figure 1. Result of the SM CKM fit projected onto the $\bar{\rho}-\bar{\eta}$ plane, as obtained by the UTFit (left) [1] and CKMfitter (right) [2] collaborations. Shown shaded are the 95% C.L. regions selected by the given observables.

2. Twofold role of flavour physics

In the SM quark sector (in absence of any additional terms in Eq. (2)), the relevant Yukawa matrices Y_u, Y_d are the only sources of the global quark flavour symmetry breaking which can be parametrized completely in terms of 10 physical parameters: 6 quark masses (given by the eigenvalues of the diagonal mass matrices $m_u \equiv v V_L^u Y_u V_R^{u\dagger}$ and $m_d \equiv v V_L^d Y_d V_R^{d\dagger}$, where $v \equiv \langle\phi\rangle$ and $V_{L,R}^{u,d}$ are unitary matrices), 3 CKM mixing angles and a single CP odd phase (all parametrized within a single physical unitary matrix $V \equiv V_L^u V_L^{d\dagger}$). The utility of flavour physics in constraining possible new effects in precision experiments is due to the fact, that (in principle) the above few parameters determine all the flavour phenomena in quark sector. This defines a twofold scope of flavour physics related to particle phenomena beyond SM: (1) it provides probes of new physics (NP) effects going well beyond the direct reach of high energy collider experiments; (2) it allows to test the sources of SM flavour symmetries and their violation.

Throughout the continuing improvement over the past few decades, the experimental measurements have generally exhibited excellent consistency with SM predictions. This is perhaps best exemplified by the two-dimensional projection of a recent compendulum of experimental constraints onto the SM quark flavour parameter space in the complex plane of $\bar{\rho}+i\bar{\eta} \equiv -(V_{ud}V_{ub}^*)/(V_{cd}V_{cb}^*)$ as performed by the UTFit [1] and the CKMFitter [2] collaborations in Fig. 1.

In order to interpret results of experimental measurements involving hadronic initial and final states, the theoretical predictions need to involve non-perturbative matching to an effective description in terms of QCD bound states. It has predominantly been due to the tremendous improvements in lattice QCD approaches to such calculations that propelled the field into the era of precision flavour constraints (for discussion on recent progress see [3]).

Given the multitude of complementary experimental results over-constraining the SM quark flavour sector, it has become possible to complete the above sketched programme even in presence of new sources of SM flavour symmetry breaking, i.e. flavour changing transitions among SM quarks mediated by new heavy degrees of freedom with masses $m_{NP} \gtrsim v$ and described by a Lagrangian $\mathcal{L}_{\rm BSM}$. At scales μ below the new particle thresholds but above the EW breaking scale ($v < \mu < m_{NP}$), any such effects can be described in complete generality in terms of local

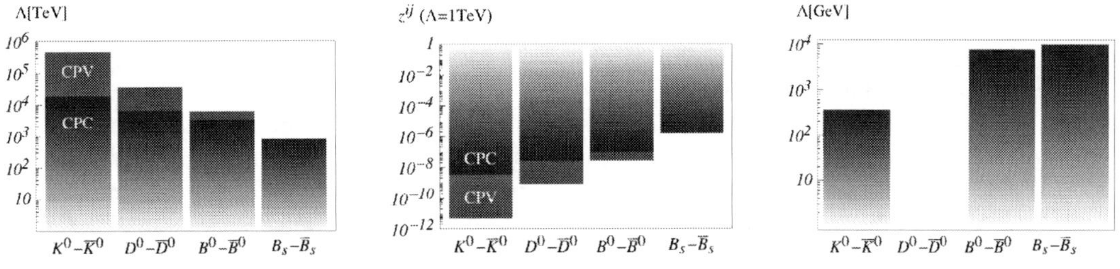

Figure 2. Current constraints of neutral meson oscillation measurements on new $\Delta F = 2$ dimension six operator contributions, given in terms of the effective operator scale (for generic flavour structures on the left and in the MFV limit on the right) or Wilson coefficients' size (in the centre). Bounds on the CP conserving and CP violating contributions are shown in blue and red, respectively (see text for details).

operators involving only SM fields [4] via the matching procedure

$$\mathcal{L}_{\rm BSM} \to \mathcal{L}_{\nu{\rm SM}} + \sum_{d>4} \frac{\mathcal{Q}_i^{(d)}}{\Lambda^{d-4}}, \tag{3}$$

where d is the canonical operator dimension. Below the EW breaking scale, these new contributions can lead to (a) shifts in the Wilson coefficients corresponding to $\mathcal{Q}_i$ present in $\mathcal{L}^{\rm eff}_{\rm weak}$ already within the SM; (b) the appearance of new effective local operators. In both cases, the resulting effects on the measured flavour observables can be computed systematically. Given the overall good agreement of SM predictions with current experimental measurements, such procedure typically results in severe bounds on the underlying NP flavour breaking sources in $\mathcal{L}_{\rm BSM}$.

Let us consider the canonical example of NP in $\Delta F = 2$ processes associated with oscillations of neutral mesons (for recent extended discussion see [5]). The leading ($d = 6$) NP operators are of the form $\mathcal{Q}^{(6)}_{AB} \sim z^{ij}[\bar{q}_i \Gamma^A q_j] \otimes [\bar{q}_i \Gamma^B q_j]$, where q_i denote the SM quark fields, while $\Gamma^{A,B}$ denote the Clifford algebra generators. Assuming z^{ij} to be generic $\mathcal{O}(1)$ complex numbers, $z \sim \exp(i\phi_{\rm NP})$, the reach of current constraints in terms the probed NP scales Λ are shown in Fig. 2 (left). It is important to stress that most of these constraints are currently limited by theory (i.e. lattice QCD inputs [6]) and parametric uncertainties. Consequently, significant future improvements will require a corroborative effort of mostly lattice QCD methods on the theory side, as well as improved experimental determinations of SM CKM parameters by flavour experiments, most notably LHCb and Belle II. Among the few $\Delta F = 2$ observables which remain largely free from theoretical uncertainties are those related to CP violation in D^0 and B_s oscillations. These are expected to remain effective experimental null-tests of the SM in the foreseeable future.

The current severe flavour bounds could be interpreted as a requirement on beyond SM (BSM) degrees of freedom to exhibit a large mass gap with respect to the EW scale (if the NP flavour and CP breaking sources are of order one and not aligned with $Y_{u,d}$). Conversely, TeV scale NP (c.f. Fig. 2 (centre)) can only be reconciled with current experimental results, provided it exhibits sufficient flavour symmetry or structure, such that $|z^{ij}| \ll 1$ (the extreme case being minimal flavour violation (MFV) [7], where one requires $Y_{u,d}$ to be the only sources of flavour breaking even BSM) . However, even in this most minimalistic scenario, the suggestive pattern of masses and mixing observed in both the quark and lepton (neutrino) sectors remains largely unexplained. It thus remains as one of the ultimate goals of flavour physics to determine whether the observed hierarchies and structures of flavour parameters are purely accidental, or rather determined dynamically or from symmetry considerations. A particularly interesting hint is provided by the fact that within the SM, flavour is only broken by Higgs interactions. On the other hand, beyond SM sources of flavour breaking in general may or may not be related to EW scale generation. At least in the minimal realisations of models addressing the electroweak hierarchy problem, a part of this intrinsic connection remains. In the minimal supersymmetric (SUSY) SM for example, new flavour sources appear due to supresymmetry breaking in the form of both squark and slepton soft masses as well as trilinear couplings. One of the peculiar features of this setup is that starting with a positive quadratic term in the Higgs potential at a very high scale, electroweak symmetry breaking is triggered radiatively by renormalization group running of the potential parameters. Incidentally, this running is dominated by the large top (and stop) Yukawa couplings [8]. Conversely, in models of a composite pseudo-Goldstone Higgs boson the SM flavour symmetry is broken via the mechanism of partial compositeness – kinetic mixing of SM chiral fermions with their massive vector-like partners [9]. However also in this case electroweak symmetry and SM flavour symmetries are connected since the breaking of the Goldstone shift symmetry generating a nontrivial potential for the SM Higgs doublet receives the dominant contribution from the top quark Yukawa [10].

In contrast to the quark sector, flavour mixing in the lepton sector, as exhibited by the entries of the PMNS matrix, is large. However, lepton flavour violating observables are generically tied to the breaking of chiral symmetry (for Dirac neutrinos) or lepton number breaking (for Majorana neutrinos) and thus suppressed by the smallness of observed neutrino masses. A case of observable lepton flavour violation can nonetheless be easily made. Assuming that, as suggested by the smallness of neutrino masses, lepton number is broken by Majorana terms at a very high scale, one can still decouple this mechanism from the breaking of the leptonic flavour symmetry, which can be broken at a much lower scale. The resulting neutrino spectrum is of pseudo-Dirac type and sizable rates for lepton flavour violating processes can result [11]. Interesting lepton flavour violating phenomenology is of course also expected in models where the breaking of the SM lepton flavour symmetries is not directly linked to neutrino masses, such as in SUSY and partial compositeness models discussed above.

3. Reclaiming flavourful NP at EW scale

In the SM, the global flavour symmetry in the quark sector $\mathcal{G}_F^q \equiv SU(3)_Q \times SU(3)_U \times SU(3)_D$ is broken by the two quark Yukawas, formally transforming under $\mathcal{G}_F^q$ as $Y_u \sim (3, \bar{3}, 1)$ and $Y_d \sim (3, 1, \bar{3})$. Since $Y_{u,d}$ alone preserve $SU(3)_{D,U}$, respectively, it is always possible to align any NP breaking of $SU(3)_{U,D}$ completely with the SM directions, leaving no observable effects. On the other hand, one can identify two unique sources of $SU(3)_Q$ breaking already in the SM given by $\mathcal{A}_u \equiv (Y_u Y_u^\dagger)_{\not{tr}}$ and $\mathcal{A}_d \equiv (Y_d Y_d^\dagger)_{\not{tr}}$, where $\not{tr}$ denotes the traceless part. Since $[\mathcal{A}_u, \mathcal{A}_d] \neq 0$, it is not possible to hide sizable new $SU(3)_Q$ breaking contributions by aligning them in flavour space with either $\mathcal{A}_u$ or $\mathcal{A}_d$.

What is more, even starting with a UV theory of NP coupling to SM matter fields in a flavour trivial way, the RG running between the high scale and the EW scale will necessarily induce some NP breaking of $\mathcal{G}_F^q$. The relevant contributions will be proportional to $\int d^4x\, T\{\mathcal{Q}_{\rm NP}\, \mathcal{H}_{\rm SM}\}$, where $\mathcal{Q}_{\rm NP}$ are the relevant NP operators and $T\{\}$ denotes the time-ordering operator. Fortunately, for flavour trivial $\mathcal{Q}_{\rm NP}$ the resulting low energy theory will be minimally flavour violating [7], easily satisfying most stringent flavour constraints even for NP degrees of freedom at the TeV scale [5, 12]. Considering again the explicit example of $\Delta F = 2$ transitions, in this limit only the $SU(3)_Q$ breaking contributions remain relevant generating $\mathcal{Q}_{LL}^{(6)} \sim [z_Q^{ij}(\bar{Q}_L^i \gamma_\mu Q_L^j)]^2$, where $\mathbf{z}_Q = \mathbf{1} + a_1 \mathcal{A}_{\mathbf{u}} + a_2 \mathcal{A}_{\mathbf{d}} + \ldots$, and $a_{i>2} \lesssim a_{1,2} \lesssim 1$. For $a_1 = 1$ and $a_{i\neq 1} = 0$ the resulting experimental constraints on the effective NP scale Λ are shown in Fig. 2 (right). Note that in this case, since $\mathrm{tr}[z_Q \cdot J] = 0$, where $J = [\mathcal{A}_u, \mathcal{A}_d]$ and $\det[J]$ is proportional to the SM Jarlskog invariant, no new CP violating effects are generated (c.f. Refs. [13, 14, 15] for a more detailed

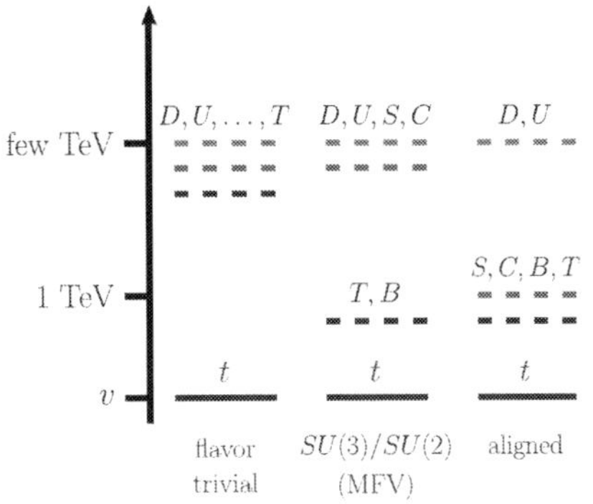

Figure 3. Schematic illustration of the current LHC direct search constraints on the masses of (SUSY) quark partners assuming different flavour structures: flavour triviality (left), respecting the approximate SM $SU(3)/SU(2)$ flavour breaking pattern (including MFV) (centre), and in flavour alignment models (right). See text for details.

discussion).

The absence so far of observable NP flavour effects when combined with stringent constraints on direct production of new (especially coloured) degrees of freedom at the LHC leads to important ramifications for the EW hierarchy problem. In the SM, the top Yukawa ($Y_t \equiv m_t/v$) imposes the largest fine-tuning in the Higgs potential, schematically parametrized as a UV sensitive shift in the μ^2 term in Eq. (2) ($\delta\mu^2 \sim Y_t^2\Lambda^2/16\pi^2$). Candidate NP theories hoping to address the SM naturalness problem introduce bosonic (SUSY theories) or fermionic (in composite pseudo-Goldstone Higgs models) top quark partners (T), whose contributions are related by symmetry to cancel exactly the Y_t induced quadratic sensitivity of the EW scale to Λ. It is instead replaced by a logarithmic one

$$\delta\mu^2 \sim \frac{Y_t^2}{16\pi^2}\Lambda^2 \to \frac{Y_t^2}{16\pi^2}m_T^2 \log\frac{\Lambda}{m_T} . \tag{4}$$

Natural EW NP (keeping $\delta\mu^2 \lesssim \mu^2$) with a high ultra-violet (UV) cut-off $\Lambda \gg v$ should thus contain relatively light top partners ($m_T \lesssim 1$ TeV). Such light new degrees of freedom seem to be at odds with the high NP scales implied by indirect flavour constraints unless the associated flavour structures are highly non-generic. While flavour triviality is able to resolve this tension, it necessarily also predicts degeneracy in the NP spectra (see Fig. 3, left). In the case of natural NP specifically, this implies the presence of $u-$ and $d-$quark partners (U, D) with $m_{U,D} \sim m_T$. In the recent years, the ATLAS and CMS experiments at the LHC have already mostly excluded the possibility of $m_{U,D} \lesssim 1$ TeV [16], thus putting the idea of flavour trivial natural NP under severe pressure.

Fortunately flavour triviality is actually not necessary for meeting the most stringent flavour requirements. This is because the dominant flavour symmetry breaking in the SM is driven by the large Y_t and is thus characterized by a $SU(3)/SU(2)$ symmetry breaking pattern [13]. It turns out that any NP respecting such a pattern can avoid the most stringent (CP violating) FCNC constraints in the K and D meson sectors, while simultaneously allowing for observable effects in B meson and top quark physics [13, 17]. Finally, a reduced approximate $SU(2)$ flavour symmetry is compatible with split spectra where third generation partners can be much lighter than their first two generation counterparts, thus improving the naturalness of such models in light of negative LHC NP search results (see also Fig. 3, centre).

An illustrative example of such models is the s.c. natural SUSY setup (c.f. [18]), where BSM flavour effects are mediated by 3rd generation squarks (and also sleptons). Together with the gaugino and higgsino sectors, these are also the only SUSY partners directly accessible at LHC energies. The key flavour observables constraining such scenarios have been identified to include meson mixing observables (ϵ_K, Δm_q, ϕ_q), rare (semi)leptonic B decays ($B \to (X)\ell^+\ell^-, \nu\bar{\nu}$), LFV as well as nuclear and leptonic EDMs [19]. In particular, the recently observed $B_s \to \mu^+\mu^-$ together with the first hints for $B_d \to \mu^+\mu^-$ [20] already represent a powerful probe of the MFV paradigm [12]. At the same time $B \to (X)\ell^+\ell^-$ sensitivity to SUSY Z-penguins is becoming competitive with direct LEP constraints on $Z \to b\bar{b}$ [21]. Finally, the best sensitivity to the slepton sector is possibly exhibited by leptonic dipole transitions. In particular, the recently improved experimental bound on the EDM of the electron [22] imposes non-trivial constraints on the Higgsino and sneutrino mass parameters in such models [19].

Another possibility is to align the directions (in flavour space) of SM and NP flavour breaking contributions [23]. This mechanism has several interesting features. In contrast to strict MFV-like structures [24], flavour alignment can easily be implemented within fundamental theories of flavour generation via (spontaneously broken) abelian horizontal symmetries [25], while offering comparable protection from dangerous right handed charged and neutral flavour changing currents (associated with the breaking of $SU(3)_{U,D}$). However as discussed in the beginning of this section, alignment is not possible in the case of $SU(3)_Q$ breaking and the most severe flavour constraints are then expected from measurements of CP violating FCNCs in the K and D sectors [14]. In the SM, light quark FCNCs are approximately CP conserving due to the smallness of both the masses of first two generation quarks as well as their mixings (CKM matrix elements) with the heavier third generation. It turns out that any new $SU(3)_Q$ breaking, which is sufficiently aligned with either $\mathcal{A}_u$ or $\mathcal{A}_d$ directions in the flavour (sub)space of the first two generations, inherits this CP protection, significantly relaxing the relevant flavour constraints [15]. Finally, since flavour alignment does not require non-abelian flavour symmetries, it allows for the possibility of split first two generation flavoured NP spectra.

In turn, NP close to the EW scale with significant (but sufficiently aligned) flavour breaking has several interesting implications for collider phenomenology. For example, in SUSY theories with not too heavy Majorana gluinos ($\tilde{g}$ with $m_{\tilde{g}} \lesssim$few TeV) the valence ($u-$ and $d-$) squark ($\tilde{u}$ and $\tilde{d}$) cross-sections at the LHC are significantly enhanced compared to non-valence contributions due to $t-$channel $\tilde{g}$ mediated $qq' \to \tilde{q}\tilde{q}'$ partonic processes (where $q^{(\prime)} = u, d$) [26, 27, 28]. Consequently, current severe LHC constraints on degenerate light generation squarks [16] are completely dominated by these first generation squark contributions. Conversely, in light of existing searches $s-$ and $c-$squarks ($\tilde{s}$ and $\tilde{c}$) could still be as light as $m_{\tilde{s},\tilde{c}} \gtrsim 500$ GeV (see also Fig. 3, right) [28]. The second intriguing possibility is that of large top and charm partner mixing. Such a scenario has been worked out in the context of the minimal SUSY SM [29] pointing out that large flavour breaking can significantly affect the sensitivity of existing experimental searches and also results in some reduction of the EW scale fine-tuning. (see also [30] for a similar study in the context of composite Higgs models).

4. Flavour portals to the dark sector

Despite the overwhelming success of the SM in describing the observed physical phenomena below the EW scale, the notion that the SM particles are the only dynamical degrees of freedom within the electroweak energy range is far from established. Not only is the existence of new light particles not excluded, since they would evade direct detection when sufficiently weakly interacting, but their presence could even be welcome. Indeed, many NP models are built upon some spontaneously broken symmetries, and do often have remnants at low-energy in the form of massless or very light Goldstone bosons. More crucially, there are now very strong indications that the universe is filled with dark matter, so there should be at least one new electrically neutral colourless particle, possibly lighter than the electroweak scale. Once opening that door, it is not such a drastic step to imagine a whole dark sector, i.e. a full-fledged set of darkly interacting dark particles only loosely connected to our own visible sector. Further, it should be stressed that adjoining a dark sector to the SM is always possible, does not need to be directly related to dark matter (so one would rather speak of a hidden sector), and is actually quite generic in SUSY models (c.f. for a recent review [31]).

Assuming that there are new particles, neutral under the SM gauge group, one can write down the gauge invariant operators coupling such particles to SM fields and thus construct the lowest-dimensional effective interactions parametrizing low-scale departures from the SM particle content [32]. In order to probe such BSM physics experimentally, one usually needs to resort to some further simplifying assumptions. In particular, the new particles may not be stable on cosmological scales and thus avoid dark matter, cosmic microwave background, cosmic ray and big bang nucleosynthesis constraints. On the other hand, if they are sufficiently long-lived, they may escape from particle experiments as missing energy. Being sufficiently weakly coupled to SM fields, they would also not significantly affect SM processes. Baring these restrictions in mind, the main impact then is to open new decay and production channels. Among the suitable decay probes, FCNC and other non-standard processes with missing energy, such as LFV decays of charged leptons and rare decays of quarkonia, kaons and B mesons are especially powerful due to their extreme suppression in the SM. In this way, non-trivial constraints on particular dark sector realizations in terms of hidden photons (A') mixing kinetically with the hypercharge, $\mathcal{L} \ni \epsilon B^{\mu\nu} A'_{\mu\nu}$, have been recently obtained by BaBar [33]. In the future, existing and updated constraints and measurements of rare kaon and B meson decay modes with missing energy, such as $K \to \pi E_{\rm miss}$ and $B \to K^{(*)} E_{\rm miss}$ might constitute unique windows into NP scenarios with light neutral long-lived particles [32]. Among the analogue flavour changing processes at high p_T, monotop production [34] currently seems to be most promising.

5. Conclusions

Success of SM in describing flavour-changing processes implies that large new sources of flavour symmetry breaking at TeV scale are mostly excluded. However, NP at TeV scale need not be flavour trivial. If (properly aligned) new sources of flavour breaking are present, precision flavour observables may still hide NP signals at the few 10% level in well motivated NP models (e.g. natural SUSY). At the same time, there are sectors of the theory that are just starting to be tested. The current measurements of $B_{s,d} \to \mu^+\mu^-$ for example are probing the Yukawa interactions at the loop level with a precision no better than 30%. New flavour dynamics can also significantly affect and guide NP searches at high p_T and have implications for EW fine-tuning. Finally, it is well worth investing in searches of exotic flavour-violating effects, such as missing energy modes, as they may turn out to be the portals to possible hidden BSM sectors.

References

[1] M. Ciuchini et al. [UTFit Group], *JHEP* **0107**, 013 (2001), updated results and plots available at: http://www.utfit.org

[2] J. Charles et al. [CKMfitter Group], *Eur. Phys. J. C* **41**, 1-131 (2005), updated results and plots available at: http:/ckmfitter.in2p3.fr

[3] C. Sachrajda, *PoS EPS* **-HEP2013**, 153 (2014).

[4] W. Buchmuller and D. Wyler, *Nucl. Phys. B* **268**, 621 (1986); B. Grzadkowski, M. Iskrzynski, M. Misiak and J. Rosiek, *JHEP* **1010**, 085 (2010).

[5] M. Bona *et al.* [UTfit Collaboration], *JHEP* **0803**, 049 (2008); G. Isidori, Y. Nir and G. Perez, *Ann. Rev. Nucl. Part. Sci.* **60**, 355 (2010); A. Lenz, U. Nierste, J. Charles, S. Descotes-Genon, H. Lacker, S. Monteil, V. Niess and S. T'Jampens, *Phys. Rev. D* **86** (2012) 033008.

[6] V. Bertone *et al.* [ETM Collaboration], *JHEP* **1303**, 089 (2013) [Erratum-ibid. **1307**, 143 (2013)]; N. Carrasco *et al.* [ETM Collaboration], *JHEP* **1403**, 016 (2014); N. Carrasco, M. Ciuchini, P. Dimopoulos, R. Frezzotti, V. Gimenez, V. Lubicz, G. C. Rossi and F. S. . L. Silvestrini *et al.*, arXiv:1403.7302 [hep-lat].

[7] R. S. Chivukula and H. Georgi, *Phys. Lett. B* **188**, 99 (1987); L. J. Hall and L. Randall, Phys. Rev. Lett. **65** (1990) 2939; A. J. Buras, *Acta Phys. Polon. B* **34**, 5615 (2003); G. D'Ambrosio, G. F. Giudice, G. Isidori and A. Strumia, *Nucl. Phys. B* **645**, 155 (2002).
[8] J. Ellis, D. V. Nanopoulos and K. Tamvakis, *Phys. Lett. B* **121**, 123 (1983); L. Alvarez-Gaumé, J. Polchinski and M. Wise, *Nucl. Phys. B* **221**, 495 (1983); L. E. Ibáñez and G. G. Ross, *Phys. Lett. B* **110**, 215 (1982).
[9] D. B. Kaplan, *Nucl. Phys. B* **365**, 259 (1991).
[10] R. Contino, Y. Nomura and A. Pomarol, *Nucl. Phys. B* **671**, 148 (2003); K. Agashe, R. Contino and A. Pomarol, *Nucl. Phys. B* **719**, 165 (2005).
[11] M. B. Gavela, T. Hambye, D. Hernandez and P. Hernandez, *JHEP* **0909** (2009) 038; D. Aristizabal Sierra, A. Degee and J. F. Kamenik, *JHEP* **1207**, 135 (2012).
[12] T. Hurth, G. Isidori, J. F. Kamenik and F. Mescia, *Nucl. Phys. B* **808**, 326 (2009).
[13] A. L. Kagan, G. Perez, T. Volansky and J. Zupan, *Phys. Rev. D* **80**, 076002 (2009).
[14] K. Blum, Y. Grossman, Y. Nir and G. Perez, *Phys. Rev. Lett.* **102**, 211802 (2009).
[15] O. Gedalia, J. F. Kamenik, Z. Ligeti and G. Perez, *Phys. Lett. B* **714**, 55 (2012).
[16] The ATLAS collaboration, ATLAS-CONF-2013-047.
[17] R. Barbieri, G. Isidori, J. Jones-Perez, P. Lodone and D. M. Straub, *Eur. Phys. J. C* **71**, 1725 (2011); R. Barbieri, P. Campli, G. Isidori, F. Sala and D. M. Straub, *Eur. Phys. J. C* **71**, 1812 (2011); R. Barbieri, D. Buttazzo, F. Sala and D. M. Straub, *JHEP* **1207**, 181 (2012); R. Barbieri, D. Buttazzo, F. Sala and D. M. Straub, *JHEP* **1210**, 040 (2012); R. Barbieri, D. Buttazzo, F. Sala, D. M. Straub and A. Tesi, *JHEP* **1305**, 069 (2013).
[18] S. Dimopoulos and G. F. Giudice, *Phys. Lett. B* **357** (1995) 573; A. G. Cohen, D. B. Kaplan and A. E. Nelson, *Phys. Lett. B* **388** (1996) 588; R. Auzzi, A. Giveon and S. B. Gudnason, *JHEP* **1202** (2012) 069; C. Csaki, L. Randall and J. Terning, *Phys. Rev. D* **86** (2012) 075009; N. Craig, M. McCullough and J. Thaler, *JHEP* **1206** (2012) 046; N. Craig, S. Dimopoulos and T. Gherghetta, *JHEP* **1204** (2012) 116.
[19] R. Barbieri, D. Buttazzo, F. Sala and D. M. Straub, *JHEP* **1405**, 105 (2014).
[20] CMS and LHCb Collaborations [CMS and LHCb Collaboration], CMS-PAS-BPH-13-007, LHCb-CONF-2013-012, CERN-LHCb-CONF-2013-012.
[21] D. Guadagnoli and G. Isidori, *Phys. Lett. B* **724** (2013) 63.
[22] J. Baron *et al.* [ACME Collaboration], *Science* **343** (2014) 6168, 269.
[23] Y. Nir and N. Seiberg, *Phys. Lett. B* **309**, 337 (1993).
[24] R. Alonso, M. B. Gavela, L. Merlo and S. Rigolin, *JHEP* **1107**, 012 (2011).
[25] C. D. Froggatt and H. B. Nielsen, Nucl. Phys. B **147**, 277 (1979).
[26] M. Heikinheimo, M. Kellerstein and V. Sanz, *JHEP* **1204**, 043 (2012).
[27] G. D. Kribs and A. Martin, *Phys. Rev. D* **85**, 115014 (2012).
[28] R. Mahbubani, M. Papucci, G. Perez, J. T. Ruderman and A. Weiler, *Phys. Rev. Lett.* **110**, no. 15, 151804 (2013).
[29] M. Blanke, G. F. Giudice, P. Paradisi, G. Perez and J. Zupan, *JHEP* **1306**, 022 (2013).
[30] C. Delaunay, T. Flacke, J. Gonzalez-Fraile, S. J. Lee, G. Panico and G. Perez, *JHEP* **1402** (2014) 055.
[31] J. Jaeckel and A. Ringwald, *Ann. Rev. Nucl. Part. Sci.* **60**, 405 (2010).
[32] J. F. Kamenik and C. Smith, *JHEP* **1203** (2012) 090.
[33] J. P. Lees *et al.* [BaBar Collaboration], arXiv:1406.2980 [hep-ex].
[34] J. Andrea, B. Fuks and F. Maltoni, *Phys. Rev. D* **84** (2011) 074025; J. F. Kamenik and J. Zupan, *Phys. Rev. D* **84** (2011) 111502; T. Aaltonen *et al.* [CDF Collaboration], *Phys. Rev. Lett.* **108** (2012) 201802; E. Alvarez, E. C. Leskow, J. Drobnak and J. F. Kamenik, *Phys. Rev. D* **89** (2014) 014016; J. L. Agram, J. Andrea, M. Buttignol, E. Conte and B. Fuks, *Phys. Rev. D* **89** (2014) 014028; CMS Collaboration [CMS Collaboration], CMS-PAS-B2G-12-022.

Status of the PIENU experiment

T. Numao[1], A.Aguilar-Arevalo[2], M. Aoki[3], M. Blecher[4], D.I. Britton[5], D.A. Bryman[6], D. vom Bruch[6], S. Chen[7], J. Comfort[8], S. Cuen-Rochin[6], L. Doria[1], P. Gumplinger[1], A. Hussein[9], Y. Igarashi[a], S. Ito[3], S. Kettell[b], L. Kurchaninov[1], L. Littenberg[b], C. Malbrunot[6], R. Mischke[1], D. Protopopescu[5], A. Sher[1], T. Sullivan[6], and D. Vavilov[1]

[1]TRIUMF, 4004 Wesbrook Mall, Vancouver, B.C. V6T 2A3, Canada
[2] Instituto de Ciencias Nucleares, Universidad Nacional Autónoma de Mexico, México
[3] Osaka University, Toyonaka, Osaka, 560-0043, Japan
[4] Virginia Tech., Blacksburg, VA 24061, USA
[5] University of Glasgow, Glasgow, UK
[6] University of British Columbia, Vancouver, B.C. V6T 1Z1, Canada
[7] Tsinghua University, Beijing, 100084, China
[8] Arizona State University, Tempe, AZ 85287, USA
[9] University of Northern British Columbia, Prince George, B.C. V2N 4Z9, Canada
[a] KEK, 1-1 Oho, Tsukuba-shi, Ibaraki, 305-0801, Japan
[b] Brookhaven National Laboratory, Upton, NY 11973-5000, USA

E-mail: toshio@triumf.ca

Abstract.
The branching ratio, $R_{e/\mu} = \Gamma(\pi \to e\nu + e\nu\gamma)/\Gamma(\pi \to \mu\nu + \mu\nu\gamma)$, provides a sensitive test of muon-electron universality in weak interactions. The status of the PIENU experiment at TRIUMF, which aims to improve the precision of the $R_{e/\mu}$ measurement by a factor of > 5, is presented.

1. Introduction

The branching ratio of pion decays, $R_{e/\mu} = \Gamma(\pi \to e\nu + e\nu\gamma)/\Gamma(\pi \to \mu\nu + \mu\nu\gamma)$, which has been precisely calculated by the Standard Model (SM) $R_{SM} = (1.2352 \pm 0.0001) \times 10^{-4}$ [1, 2], has provided one of the best tests of electron-muon universality in weak interactions. The current experimental values of the branching ratio are $R_{e/\mu} = (1.2265 \pm 0.0034(stat) \pm 0.0044(syst)) \times 10^{-4}$ (TRIUMF) [3] and $R_{e/\mu} = (1.235 \pm 0.005) \times 10^{-4}$ (PSI) [4], indicating that there is a room for improvement of almost two orders of magnitude. The goal of the PIENU experiment at TRIUMF is to improve the precision of the branching ratio measurement by a factor of > 5, confronting the SM prediction to better than 0.1 %, which allows access to new physics beyond the SM up to the mass scale of 1000 TeV. Examples of the new physics probed include the R-parity violating SUSY [5], heavy neutrino mixing [6], excited gauge bosons, leptoquarks [7], compositeness or charged Higgs bosons.

2. Experiment

In the PIENU experiment, the branching ratio $R_{e/\mu}$ is obtained from the ratio of positron yields from the $\pi^+ \to e^+\nu$ decay ($E_{e^+} = 70$ MeV) and the $\pi^+ \to \mu^+\nu$ decay followed by the $\mu^+ \to e^+\nu\overline{\nu}$ decay ($\pi^+ \to \mu^+ \to e^+$ decay, $E_{e^+} < 53$ MeV) using stopped π^+'s.

The π^+ beam was degraded by two thin plastic scintillators, and stopped in an 8-mm thick active target at a rate of 5×10^4 π^+/s [8]. A pion was tracked by wire chambers (WC1 and WC2) at the entrance of the detector system, and by two sets of single-sided 0.3-mm thick X and Y silicon strip counters located immediately upstream of the target.

The positron calorimeter located on the beam axis consisted of two thin plastic scintillators and a 48-cm (dia.) $\times$ 48-cm (length) single-crystal NaI(Tℓ) detector [9]. Two concentric layers of 97 pure CsI crystals [10] (9 radiation lengths radially) surrounded the NaI(Tℓ) crystal to capture the electromagnetic shower. Positrons were tracked by a set of X and Y Si-strip counters immediately downstream of the target counter, and three layers of wire chambers (WC3) in front of the NaI(Tℓ) crystal.

A coincidence of pion and positron signals with a time window of −300 ns to 540 ns with respect to the pion signal from the beam counters was the basis of the main trigger logic. This was prescaled by a factor of 16 to form an unbiased trigger. Selecting events in the early time window (2–40 ns) with respect to a pion stop or events with $E_{e^+} > 48$ MeV in the calorimeter provided enhanced triggers of $\pi^+ \to e^+\nu$ decays. The pulse shapes from the plastic counters, the CsI crystals, the Si-strip detectors, and the NaI(Tℓ) crystal were recorded, and time information from the wire chambers and logic signals were recorded with a time window of -4 to 4 μs.

The response function of the calorimeter to 70 MeV positrons for $\pi^+ \to e^+\nu$ was measured with a positron beam. The calorimeter was rotated to inject the beam at various entrance angles up to 48° with respect to the beam axis.

3. Analysis

Events originating from stopped pions are selected based on their energy losses in the beam counters. Any events with extra activity in the beam counters in the time region of −7 to 1 μs with respect to the pion stop are rejected. A fiducial cut, limiting one of position tracks at WC3 to be within 60 mm from the beam axis, is imposed to minimize the thickness variation of the material along the positron path and reduce the low energy tail of the $\pi^+ \to e^+\nu$ peak due to shower leakage from the NaI(Tℓ) crystal.

Events passing the above criteria are divided into two energy regions separated at $E_{cut} = 52$ MeV in the summed energies of the NaI and CsI. The positron time with respect to the pion for the time region from -300 to 540 ns is separately histogrammed for the fit. Figure 1 shows the time spectra of positrons for (a) the low energy region and (b) the high energy region.

In the low-energy time spectrum, the components are from $\pi^+ \to \mu^+ \to e^+$ decays and $\mu^+ \to e^+\nu\overline{\nu}$ (about 1 % of $\pi^+ \to \mu^+ \to e^+$) decays after decay-in-flight of a pion (πDIF), both starting at the pion stop, and $\mu^+ \to e^+\nu\overline{\nu}$ decays of "old" muons originating from the previous pion stops in the target area.

The main component in the high energy region is the $\pi^+ \to e^+\nu$ decay ($\lambda_\pi e^{-\lambda_\pi t}$). All backgrounds in the high-energy region originate from the same sources as in the low energy with additional effects that raise the observed energy. A scaled low-energy time spectrum comes from muon decays through the energy resolution of the detector, radiative muon decays in which the γ-ray raises the observed calorimeter energy, and a pile-up in the calorimeter with a flat time distribution (*e.g.* from neutrons from the production target). A combination of $\pi^+ \to \mu^+ \to e^+$ and "old" muon decays is another indispensable source of high-energy background. Radiative pion decays $\pi^+ \to \mu^+\nu\gamma$ (the branching ratio of 2×10^{-4}) followed by a $\mu^+ \to e^+\nu\overline{\nu}$ decay can also contribute to the high-energy spectrum when the γ-ray hits the

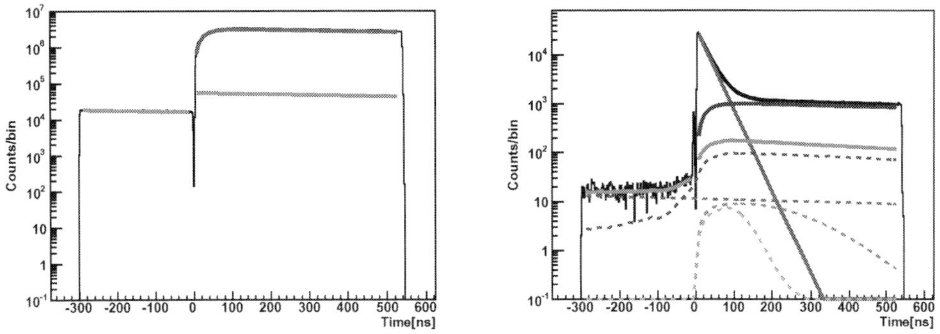

Figure 1. Time spectra of positrons for (a) low-energy and (b) high-energy regions. The full lines (red) indicate the signal components and other lines indicate various backgrounds.

calorimeter. The latter two contributions are estimated by Monte Carlo (MC) simulation using the observed pulse shapes.

A raw branching ratio is obtained by simultaneously fitting the low- and high-energy time spectra (the time region near the pion stop is excluded) to the expected time spectra.

4. Corrections

When a muon from $\pi^+ \to \mu^+\nu$ decays in flight (μDIF), Lorentz boosting may raise the positron energy above E_{cut}. Since this contribution has a time distribution of the pion decay, it is included in the raw branching ratio obtained by the time-spectrum fit. The amplitude of this component (about 0.2 %) is well estimated by the MC simulation.

A relative acceptance difference of low- and high-energy positron events due to small energy dependent effects in the energy-loss processes is estimated by a MC simulation. A series of systematic studies based on the measured uncertainties of the detector geometry indicate the acceptance correction can be estimated significantly better than that (0.11 %) of the previous experiment.

The largest correction to the raw branching ratio is for $\pi^+ \to e^+\nu$ events below the cut-off energy, which arise from the response function of the calorimeter and radiative pion decays. Because of the bump structure in the response function of the NaI detector [11] due to hadronic interactions, which cannot be reproduced by pure MC simulation at a satisfactory level, empirical measurements are required.

Special data taken with a 70-MeV/c positron beam are used for the study of the response function of the calorimeter. The effects of the geometry change in the measurement, the $\pi^+ \to e^+\nu$ angle distributions and radiative pion decays are corrected by MC. The fraction of the events below E_{cut} is obtained to be 3 % of the total events. Since the contribution from the low-energy tail of the positron beam cannot be ruled out completely, the tail correction obtained is treated as an upper bound.

In order to empirically estimate the lower bound, $\pi^+ \to e^+\nu$ events are enhanced by selecting the early decay-time region and using the target activity information. The contribution of πDIF to the background is further reduced by detection of a kink in the pion track when πDIF happens upstream of the target counter. Assuming that the background-suppressed spectrum in the low energy region contains a negligible $\pi^+ \to e^+\nu$ contribution, the area of the low-energy region can be scaled to the full background region ($< E_{cut}$) using the known background distribution. Due to the no-tail assumption, subtraction of this scaled sum from the total count in the suppressed spectrum below E_{cut} yields a lower bound on the tail. Since the target energy cut used in the suppression method tends to remove $\pi^+ \to e^+\nu$ events with Bhabha scattering which results in a larger energy deposit in the target, a correction to the effects estimated by the MC simulation is added to the tail correction. Combining all factors, the lower bound is also 3 %.

5. Status

Data taking of the PIENU experiment was completed in 2012, and the analysis of partial data taken in 2010 (representing 4×10^5 clean $\pi^+ \to e^+\nu$ events) is in progress.

Consistency of the measured branching ratio has been tested by changing various conditions in the analysis. When a variation beyond the statistical fluctuation is observed the source is sought until the cause is identified and corrected. Many cuts, which would potentially influence the background shape and amplitude in the present fitting function, are varied and the validity of the background functions and effectiveness of the cuts have been confirmed. This includes the energy threshold cuts and the pile-up rejection cuts for the incoming pion, which largely affect the amplitudes and/or shapes of the "old" muon background and other backgrounds in the high energy region. Also, E_{cut} has been varied to confirm the tail and μDIF corrections.

The next step is to fix all the cuts and analysis procedures and obtain all necessary correction factors before unblinding the branching ratio (shifted by a hidden value within 1 %). Following the completion of the 2010 analysis, a full data analysis will take place with further studies of backgrounds and corrections to a level of 0.01 % of the branching ratio.

6. Conclusions

The partial analysis of 2010 data by the PIENU experiment has confirmed the validity of the method and is expected to soon improve the measurement of $R_{e/\mu}$ by nearly a factor of two. The goal of the experiment is to improve the precision of the branching ratio measurement to a level of < 0.1 %.

References

[1] W.J. Marciano and A. Sirlin, Phys. Rev. Lett. **71**, 3629 (1993).
[2] V. Cirigliano and I. Rosell, JHEP **0710**, 005 (2007).
[3] D.I. Britton *et al.*, Phys. Rev. Lett. 68, 3000 (1992); and D.I. Britton *et al.*, Phys. Rev. D49, 28 (1994).
[4] G. Czapek *et al.*, Phys. Rev. Lett. 70, 17 (1993).
[5] M.J. Ramsey-Musolf, S. Su and S. Tulin, Phys. Rev. **D76**, 095017 (2007).
[6] H. Lacker and A. Menzel, JHEP 07, 006 (2010).
[7] S.Davidson, D. Bailey and B. Campbell, Z. Phys. C61, 613 (1994).
[8] A. Aguilar-Arevalo *et al.*, to be published.
[9] G. Blanpied *et al.*, Phys. Rev. Lett. **76** (1996) 1023.
[10] I-H. Chiang *et al.*, IEEE **NS-42** (1995) 394.
[11] A. Aguilar-Arevalo *et al.*, Nucl. Instrum. Method A621, 188 (2010).

Journal of Physics: Conference Series **556** (2014) 012003 doi:10.1088/1742-6596/556/1/012003

QCD tests with Kaons

Mauro Raggi[1]

INFN Laboratori Nazionali di Frascati, via E. Fermi 40, 00044 Frascati Italia

E-mail: mauro.raggi@lnf.infn.it

Abstract. Final results from an analysis of about 400 $K^{\pm} \to \pi^{\pm}\gamma\gamma$ rare decay candidates collected by the NA48/2 and NA62 experiments at CERN are presented. The results include a model-independent decay rate measurement and fits to Chiral Perturbation Theory (ChPT) description. The data support the ChPT prediction for a cusp in the di-photon invariant mass spectrum at the two pion threshold. The NA48/2 Collaboration at CERN has accumulated unprecedented statistics of rare kaon decays in the Ke4 modes $K^{\pm}_{e4} \to \pi^{+}\pi^{-}e\nu$ and $K^{00}_{e4} \to \pi^{0}\pi^{0}e\nu$ with one percent background contamination. The detailed study of the form factors brings new inputs to low energy QCD description and crucial tests of predictions from ChPT and lattice QCD calculations.

1. The NA48/2 and NA62-RK experiments

At an early stage the NA62 [1] experiment collected data, with modified beam conditions, using the NA48/2 detector, aiming to measure the ratio R_K of the rates of the leptonic kaon decays. The NA48/2 beam line has been designed to deliver simultaneous narrow momentum band K^+ and K^- beams derived from the 400 GeV/c protons extracted from the CERN SPS with central momenta of 60 GeV/c (for NA48/2) or 74 GeV/c (for NA62-RK). The beam kaons decayed in a fiducial volume contained in a 114 m long cylindrical vacuum tank. The momenta of charged decay products were measured in a magnetic spectrometer consisting of four drift chambers (DCHs), two upstream and two downstream of a dipole magnet. The dipole provided a horizontal transverse momentum kick of 120 MeV/c (for NA48/2) or 265 MeV/c (for NA62-RK) to charged particles. Each DCH was composed of eight planes of sense wires. A plastic scintillator hodoscope producing fast trigger signals and providing precise time measurements of charged particles was placed after the spectrometer. Further downstream was a liquid krypton electromagnetic calorimeter (LKr), an almost homogeneous ionization chamber with an active volume of 7 m^3 of liquid krypton, $27X_0$ deep, segmented transversally into 13248 projective 2x2 cm^2 cells and with no longitudinal segmentation. The LKr information is used for photon measurements and charged particle identification. An iron/scintillator hadronic calorimeter and muon detectors were located further downstream. A description of the detector can be found in [2].

[1] The NA48/2 Collaboration: Cambridge, CERN, Dubna, Chicago, Edinburgh, Ferrara, Firenze, Mainz, Northwestern, Perugia, Pisa, Saclay, Siegen, Torino, Wien. The NA62-RK collaboration: Birmingham, CERN. Dubna, Fairfax, Ferrara, Firenze, Frascati, Mainz, Merced, Moscow, Napoli, Perugia, Pisa, Protvino, Roma I, Roma II, Saclay, San Luis Potosi, Stanford, Sofia, Torino,TRIUMF

2. The $K^{\pm} \to \pi^{\pm}\gamma\gamma$ decay

Measurements of radiative non-leptonic kaon decays provide crucial tests for the ability of the Chiral Perturbation Theory (ChPT) to explain weak low energy processes. In the ChPT framework, the $K^{\pm}(p) \to \pi^{\pm}(p_3)\gamma(q_1)\gamma(q_2)$ differential decay rate can be parametrized as [3]:

$$\frac{\partial^2\Gamma}{\partial x \partial y} = \frac{m_K}{(8\pi)^3}\left[z^2(|A+B|^2+|C|^2)+\left(y^2-\frac{1}{4}\lambda(1,z,r_\pi^2)\right)^2\cdot\left(|B|^2+|D|^2\right)\right] \tag{1}$$

where $z = m_{\gamma\gamma}^2/m_K^2$ and $y = p\cdot(q_1-q_2)/m_K^2$. The dominant contribution at O(p^4) comes from the loop term $A(z,\hat{c})$, including the pion and kaon loop amplitudes, the B and D terms are zero and the Wess-Zumino-Witten term C accounts for ~10% of the total decay rate. Higher order unitarity corrections from $K^{\pm} \to 3\pi^{\pm}$ decays, including the main O(p^6) contribution as well as those beyond O(p^6), have been found to contribute significantly to the branching ratio (BR) (up to 30-40%) and to the shape of $M_{\gamma\gamma}$ spectrum [3]. The total decay rate is predicted to be $BR(K^{\pm} \to \pi^{\pm}\gamma\gamma) \sim 10^{-6}$. The only $K^{\pm} \to \pi^{\pm}\gamma\gamma$ experimental observation published so far is by the BNL E787 experiment [4] measuring the BR and $\hat{c}$:

$$BR(K^{\pm} \to \pi^{\pm}\gamma\gamma)_{E787} = (1.1 \pm 0.3_{stat} \pm 0.1_{syst})\cdot 10^{-6} \qquad with \qquad \hat{c} = 1.8 \pm 0.6 \tag{2}$$

based on 31 K^+ decay candidates in the kinematic region 100 MeV/c< p_π < 180 MeV/c (p_π is the π^+ momentum in the K^+ rest frame). The value of the BR quoted in (2) refers to the entire kinematical range and is obtained with a ChPT extrapolation using the value 1.8 for $\hat{c}$.

2.1. The data analysis

The measurements described here have been performed using minimum bias data sets collected during a 3-day special NA48/2 run in 2004 with 60 GeV/c $K^{\pm}$ beams, and a 3-month NA62 run in 2007 with 74 GeV/c $K^{\pm}$ beams. The effective kaon fluxes collected are similar, but the background conditions and resolution on kinematic variables differ significantly. Signal events are selected in the region z = $(m_{\gamma\gamma}/m_K)^2 > 0.2$ to reject the $K^{\pm} \to \pi^{\pm}\pi^0$ background peaking at z = 0.075. In total 149 (232) decays candidates are observed by NA48/2 (NA62), with backgrounds contaminations of 15.5 ± 0.7 (17.4 ± 1.1) dominated by $K^{\pm} \to \pi^{\pm}\pi^0(\pi^0)(\gamma)$ decays with merged photon clusters in the LKr calorimeter. The data spectra of the z kinematic variable, together with the signal and background expectations, are displayed in Figure 1: they support the ChPT prediction of a cusp at the two-pion threshold. The values of the $\hat{c}$ parameter in the framework of the ChPT $O(p^4)$ and $O(p^6)$ parameterizations according to [3] have been measured by performing likelihood fits to the data. The uncertainties are dominated

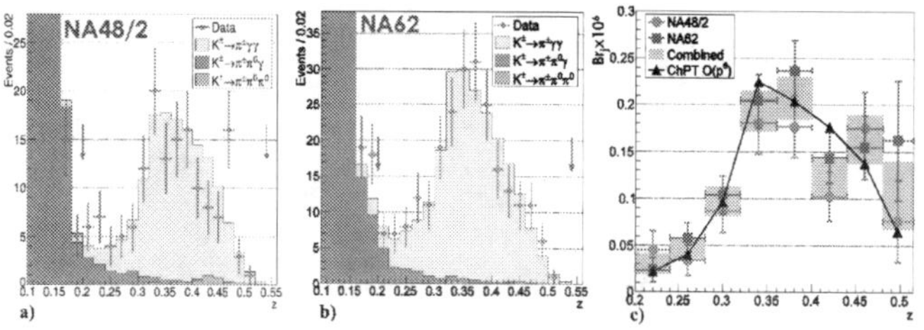

Figure 1. a),b) The $z = (m_{\gamma\gamma}/M_K)^2$ spectrum with data, MC best fit for signal, and backgrounds. c) model-independent BR B_j in z bins: NA48/2, NA62 and their combination.

by the statistical ones while the systematic ones are mainly due to the background estimates uncertainties. The O(p^6) parametrization involves a number of external inputs fixed as follows:

the polynomial contribution terms are η_1 = 2.06, η_2 = 0.24 and η_3 = -0.26 as suggested in [3], while the $K^\pm \to 3\pi^\pm$ amplitude parameters come from a fit to the experimental data [5]. Along with the separate 2004 and 2007 results, the combined NA48/2 and NA62 results are also presented in Table 1. The measured BR, in agreement with the E787 one, improves the

Table 1. Final results for the $\hat{c}$ and BR($\pi^+\gamma\gamma$) measurements.

	$\hat{c}$ $O(p^4)$ fit	$\hat{c}$ $O(p^6)$ fit	$BR_{MI}^{z>0.2}$(x10^6)	BR $O(p^6)$ (x10^6)
NA48/2[7]	1.37 ± 0.36	1.41 ± 0.40	$0.877 \pm 0.087_{stat} \pm 0.017_{syst}$	$0.910 \pm 0.072_{stat} \pm 0.022_{syst}$
NA62[8]	1.93 ± 0.27	2.10 ± 0.33	$1.088 \pm 0.093_{stat} \pm 0.027_{syst}$	$1.058 \pm 0.066_{stat} \pm 0.044_{syst}$
Combined[8]	1.72 ± 0.21	1.86 ± 0.25	$0.965 \pm 0.061_{stat} \pm 0.014_{syst}$	$1.003 \pm 0.051_{stat} \pm 0.024_{syst}$

precision of the measurement by a factor ~5. The obtained value of $\hat{c}$, for both the $O(p^4)$ and $O(p^6)$ fits, is in very good agreement with the previous measurement by E787 [4]. The same parameter has been measured by NA48/2 using the decay $K^\pm \to \pi^\pm e^+e^-\gamma$, with a compatible value: $\hat{c} = 0.90 \pm 0.45$ [6].

3. The K_{e4} decay formalism

Four-body final state decays are described by five kinematic variables historically called, for K_{e4} decays, Cabibbo-Maksymowicz variables [9]: two invariant masses $S_\pi = M^2_{\pi\pi}$ and $S_e = M^2_{e\nu}$ and three angles ϑ_π, ϑ_e and φ. The hadronic current is described by form factors which can be developed in a partial wave expansion as suggested in [10] Limiting the expansion to S- and P-waves and considering a unique phase δ_p for all P-wave form factors, two axial (F, G) and one vector (H) complex form factors contribute to the transition amplitude: $F = F_s\ e^{i\delta_s} +\ F_p\ e^{i\delta_p} cos\theta_\pi$, $G = G_p\ e^{i\delta_p}$, $H = H_p\ e^{i\delta_p}$. From the differential rate study in the 5-dimensional space, four real form factors (F_s, F_p, G_p and H_p) and a single phase difference ($\delta = \delta_s - \delta_p$) have been measured, together with their energy variation with S_π and S_e, by the NA48/2 experiment [11]. In the neutral pion mode (K^{00}_{e4}), the differential rate depends on a single hadronic form factor F_s whose variation with (S_π,S_e) is unknown and will be studied. No such study is available so far in the literature.

3.1. K^{00}_{e4} form factor measurement

The form factor (FF) study requires a sample free of large radiative effects which can pollute the original kaon decay amplitude. The event density in the plane (S_π,S_e) is proportional to F_s^2. The fit procedure minimizes a χ^2 expression in the two-dimensional space describing F_s by a polynomial expansion in $q^2 = (S_\pi/4m^2_{\pi+} - 1)$ and y^2=$(S_e/4m^2_{\pi+})$:

$$F(q^2, y^2)_{q^2 \geq 0} = N(1 + aq^2 + bq^4 + cy^2)^2 \qquad F(q^2, y^2)_{q^2<0} = N(1 + dC(q^2) + cy^2)^2 \tag{3}$$

The term $C(q^2) = \sqrt{|q^2/(1+q^2)|}$ parameterizes the cusp-like function. The results in Table 2 are in good agreement with those obtained in a high statistics measurement of the corresponding form factor of the $K^\pm_{e4}$ [11] mode see Figure 2b). Below $S_\pi = 4m^2_{\pi+}$, the observed deficit of events it is well described by a cusp-like function Figure 2a).

3.2. Measurement of the K^{00}_{e4} branching ratio

The K^{00}_{e4} rate is measured relative to the abundant $K^{00}_{3\pi}$ normalization channel. As the topologies of the two modes are similar the two samples are collected concurrently using the same trigger

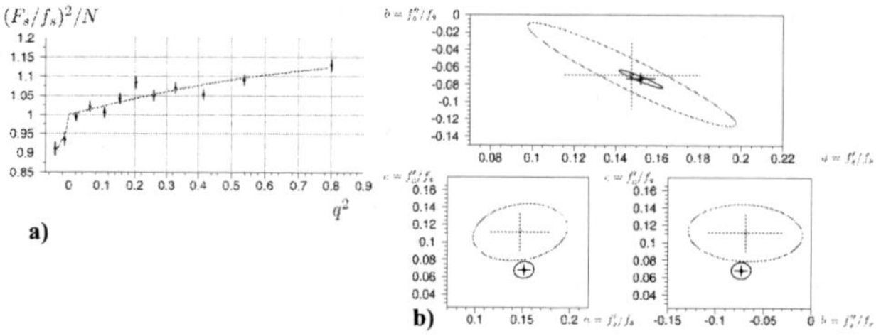

Figure 2. a) Ratio of the two q^2 distributions in equal population bins. b) FF description in 2-parameter planes obtained in K^{00}_{e4} [12] (red) and $K^{\pm}_{e4}$ [11] (black).

Table 2. Best description of the $F_s(K^{e4}_{00})$ form factor using Equation 3

Fit parameter	values
a	$0.149 \pm 0.033_{stat} \pm 0.014_{syst}$,
b	$-0.070 \pm 0.039_{stat} \pm 0.013_{syst}$,
c	$0.113 \pm 0.022_{stat} \pm 0.007_{syst}$,
d	$-0.256 \pm 0.049_{stat} \pm 0.016_{syst}$.
χ^2	101.4/107=0.95

logic and a common selection is employed as far as possible. This leads to partial cancellation of the systematic effects and avoids relying on the absolute kaon flux measurement. After the selection described in [12] a sample of 65210 candidate is identified with 1% background dominated by $K^{00}_{3\pi}$ with a misidentified pion as an electron. The BR measurement [12], using the value $BR(K^{00}_{3\pi}) = (1.761 \pm 0.022)\%$ as normalization, is obtained:

$$BR(K^{00}_{e4}) = (2.552 \pm 0.010_{stat} \pm 0.010_{syst} \pm 0.032_{ext}) \cdot 10^{-5} \quad (4)$$

This measurement improves the current world average precision by one order of magnitude. Both total rate and form factor description are used to obtain the absolute FF value at $S_\pi = 4m^2_{\pi+}$, $S_e = 0$ ($q^2 = 0$, $y^2 = 0$) [12]:

$$f_s = 6.079 \pm 0.012_{stat} \pm 0.027_{syst} \pm 0.046_{ext},$$

where the dominating external error comes from uncertainties on the normalization mode $K^{00}_{3\pi}$ branching ratio, on the mean kaon life time and on $|V_{us}|$. This value shows some tension with the corresponding form factor of the $K^{\pm}_{e4}$ mode $f^{\pm}_s = 5.705 \pm 0.003_{stat} \pm 0.017_{syst} \pm 0.031_{ext}$ [11].

References

[1] G. Anelli et al., CERN-SPSC-2005-013, CERN-SPSC-P-326, (2005).
[2] V. Fanti *et al.* [NA48 Collaboration], *Nucl. Instrum. Meth. A* **574**, 433 (2007).
[3] G. D'Ambrosio and J. Portoles, *Phys. Lett. B* **386**, 403 (1996)
[4] P. Kitching *et al.* [E787 Collaboration], *Phys. Rev. Lett.* **79**, 4079 (1997)
[5] J. Bijnens, P. Dhonte and F. Persson , *Nucl. Phys. B* **648**, 317 (2003)
[6] J. R. Batley *et al.* [NA48/2 Collaboration], *Phys. Lett. B* **659**, 493 (2008)
[7] J. R. Batley *et al.* [NA48/2 Collaboration], *Phys. Lett. B* **730**, 141 (2014)
[8] C. Lazzeroni *et al.* [NA62 Collaboration], *Phys. Lett. B* **732**, 65 (2014)
[9] N. Cabibbo and A. Maksymowicz, *Phys. Rev.* **137**, B438 (1965)
[10] A. Pais and S. B. Treiman, *Phys. Rev.* **168**, 1858 (1968).
[11] J. R. Batley *et al.* [NA48/2 Collaboration], *Phys. Lett. B* **715**, 105 (2012)
[12] J. R. Batley *et al.* [NA48/2 Collaboration], *J. High Energy Phys.* **08**, 159 (2014)

Progress and perspectives in the low-energy kaon-nucleon/nuclei interaction studies at the DAΦNE collider

M. Iliescu[1], M. Bazzi[1], G. Beer[2], C. Berucci[1,6], D. Bosnar[12], A.M. Bragadireanu[5], M. Cargnelli[6], C.Curceanu[1], A. d'Uffizi[1], L. Fabietti[10], C. Fiorini[3,4], F. Ghio[7], C. Guaraldo[1], R.S. Hayano[8], T. Ishiwatari[6], M. Iwasaki[9], J. Marton[6], S. Okada[9], D. Pietreanu[5], K. Piscicchia[1], M. Poli Lener[1], T. Ponta[5], R. Quaglia[3,4], A. Romero Vidal[11], P. Levi Sandri[1], E. Sbardella[1], F Schembari[3], A. Scordo[1], H. Shi[8], D.L. Sirghi[1,5], F. Sirghi[1,5], H. Tatsuno[9], I. Tucakovic[1], O. Vazquez Doce[10], E. Widmann[6], B. Wünschek[6] and J. Zmeskal[6]

1 INFN, Laboratori Nazionali di Frascati, Via Enrico Fermi 40, I-00044, Frascati (Roma), Italy
2 Department of Physics and Astronomy, University of Victoria, Victoria, BC, Canada
3 Politecnico di Milano,Dipartimento di Elettronica,Informazione e Bioingegneria, Milano,Italy
4 INFN Sezione di Milano, Milano, Italy
5 Horia Hulubei National Institute of Physics and Nuclear Engineering, Magurele, Romania
6 Stefan-Meyer-Institut für subatomare Physik, Vienna, Austria
7 INFN Sezione di Roma I and Istituto Superiore di Sanità, Roma, Italy
8 The University of Tokyo, Tokyo, Japan
9 RIKEN, The Institute of Physics and Chemical Research, Saitama, Japan
10 Technische Universität München, Physik Department, Garching, Germany
11 University of Santiago de Compostela, Santiago de Compostela, Spain
12 University of Zagreb, Zagreb, Croatia

(SIDDHARTA, SIDDHARTA-2 and AMADEUS collaborations)

E-mail: Mihai.Iliescu@lnf.infn.it

Abstract. Low-energy QCD is still lacking experimental results, fundamental for reaching a good understanding of the strangeness sector. The information provided by the low energy kaon-nucleon/nuclei interaction is accessible through the study of kaonic atoms and kaonic nuclear processes. The lightest atomic systems, namely the kaonic hydrogen and the kaonic deuterium, provide the isospin dependent kaon-nucleon scattering lengths by measuring the X-rays emitted during their de-excitation to the 1s level. The most precise kaonic hydrogen measurement to date, together with an exploratory measurement of kaonic deuterium and of upper-level transitions in kaonic helium 3 and kaonic helium 4 were carried out at the DAΦNE collider by the SIDDHARTA collaboration. Presently, a significantly upgraded setup developped by the SIDDHARTA-2 collaboration is ready to perform a precise measurement of kaonic deuterium and, afterwards, of heavier exotic atoms. In parallel, the kaon-nuclei interaction at momenta below 130 MeV/c is studied by the AMADEUS collaboration, using the KLOE detector and a dedicated setup inserted in the central region, near the interaction point. Preliminary results of the study of charged antikaons interacting with nuclei are shown, including an analysis of the controversial $\Lambda(1405)$.

1. Introduction

Reliable information on $\bar{K}N$ interaction at threshold was provided only in the 1990s by measuring the strong-interaction energy shift ϵ and width Γ of kaonic-hydrogen $1s$ state, at KEK-PS in Japan [1] and afterwards, at DAFNE in Italy [2]. From the two values, the K^-p scattering length was extracted, using a Deser-Trueman-like formula [3], lately improved to account for the isospin-breaking corrections [4]. The measurement allowed a consistent interpretation of the scattering data [5] and the calculation of chiral SU(3) meson-baryon effective Lagrangian. However, sub-threshold amplitudes were extrapolated with large uncertainties due to the experimental precision. Recently, the SIDDHARTA collaboration obtained more accurate values for both ϵ and Γ for kaonic hydrogen, thus providing stronger constraints on the theoretical models. The new data also motivated novel calculations regarding the structure of $\Lambda(1405)$ and theoretical investigations about the existence of kaonic nuclear clusters [6], both topics now under experimental study within the AMADEUS collaboration.

2. SIDDHARTA experiment

The SIDDHARTA experiment was performed at the upgraded DAΦNE e^+e^- collider in Frascati, Italy [7]. The beam energy is tuned to create ϕ mesons, which decay in K^+K^- pairs with a BR of 49.1%. The monochromatic, low-energy charged kaons are degraded, then stopped in a cryogenic gaseous target, producing kaonic atoms. The target is a critical item, the yields of kaonic atom X-rays decreasing sensitively with the gas density, due to Stark mixing. On the other hand, a low density does not permit efficient stopping and increases the kaons in-flight decay. For the SIDDHARTA optimized density, the yield was $\sim 1\%$ when filled with hydrogen and will presumably drop to $\sim 0.1\%$ for deuterium. The trigger was given by K^-K^+ coincident hits on fast scintillators. The kaonic atom X-rays were detected with 144 silicon drift X-ray detectors (SDDs), 1 cm^2 each, surrounding the target. The SDDs, developed by the collaboration, have an energy resolution of 180 eV FWHM at 6 keV, while their time response is below 800 ns FWHM. The trigger signal and the fast X-ray detectors confered a high background rejection, most of the background deriving from beam losses, uncorrelated with the Φ production.

2.1. SIDDHARTA results and ongoing upgrade

The data analysis used a global fit of both hydrogen and deuterium X-ray spectra to extract ϵ and Γ, thus improving the background subtraction. The fit lines are indicated in Fig. 1. The kaonic-hydrogen X-ray transitions were clearly observed, while kaonic deuterium ones were not. This agrees with the theoretical predictions, indicating one order of magnitude lower yield [8]. The repulsive-type shift observed is in agreement with the K^-p scattering data. Other characteristic X-rays, coming from the calibration foils and from the setup components, as well as Xrays from heavier kaonic-atoms produced in the target wall, were properly identified. The details of the analysis and the setup description are given in [9]. The fit results for the $1s$ strong-interaction shift and width in kaonic hydrogen are

$$\epsilon = -283 \pm 36(\text{stat}) \pm 6(\text{syst})\ \text{eV}$$

and

$$\Gamma = 541 \pm 89(\text{stat}) \pm 22(\text{syst})\ \text{eV},$$

respectively. The values are in agreement with the recent theoretical works [10], so yielding important constraints on the description of low-energy $\bar{K}N$ interactions [6]. Still, the full determination of the isospin-dependent $\bar{K}N$ scattering lengths requires a second measurement, namely the shift and the width of kaonic-deuterium $1s$ state. Therefore, the SIDDHARTA-2 experiment is being prepared to measure the -one order of magnitude lower yield- kaonic deuterium X-rays, by significantly improving the signal to background ratio [11]. The full results will grant an important progress of the theoretical interpretation.

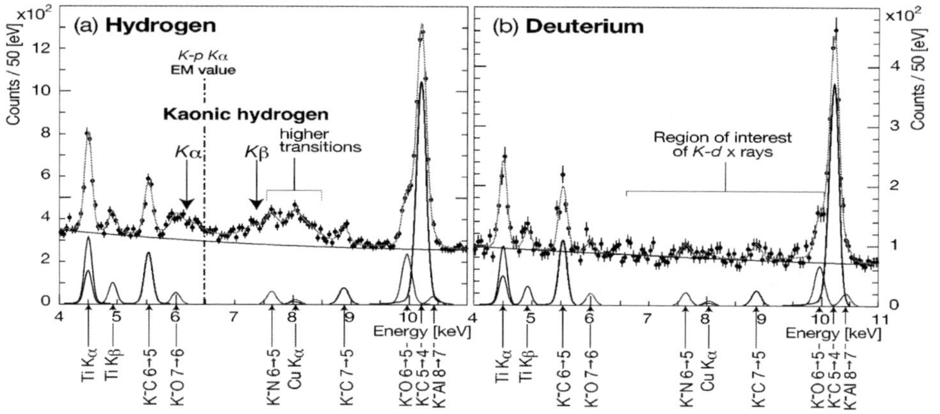

Figure 1. Global fit of the X-ray spectra from kaonic hydrogen (a) and kaonic deuterium (b). The dot-dashed vertical line indicates the EM energy of the kaonic-hydrogen Kα X-ray.

3. AMADEUS

A research line complementary to that of exotic atoms, the AMADEUS experiment at DAΦNE deals with the investigation of low-energy kaon-nucleon/nuclei hadronic interaction inside the nucleus. For that purpose, signals from K^- absorption at rest and in flight (with momenta below 127 MeV/c), are investigated using a sample of 1.4 fb^{-1} from the KLOE experiment data. In a first phase, components of the detector [12], like the gas filling the drift chamber (mainly ^{4}He) and the drift chamber wall (mainly ^{12}C), were used as active targets. The interpretation of the data will allow extracting qualitative and quantitative outputs for the search of kaonic nuclear clusters [13] and a new study of the debated $\Lambda(1405)$ resonance [6], to be added to the pp and heavy ion collisions data.

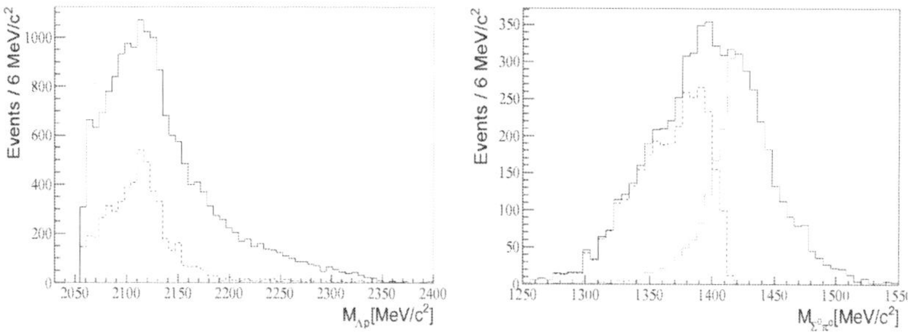

Figure 2. Left: Λp invariant mass for the inclusive Λp selection (continuous line) and for events with an extra π^- detected (dashed line). Right: $\Sigma^0\pi^0$ invariant mass (dashed and pointed lines represent the at-rest and in-flight components, respectively).

The invariant mass spectra for the Λp inclusive selection (expected decay channel of a K^-pp cluster) is shown in the left part of Fig. 2 for the ^{4}He events. The low invariant mass region is populated mainly by events coming from the single nucleon absorption (1NA) process, the proton in the final state coming from the participant nucleons trough the Σ/Λ nuclear conversion taking place inside the residual nucleus. The 1NA process is characterized by an additional negative pion in the final state. This subsample is represented by the dashed line in Fig. 2 (left), with arbitrary normalization, to be compared with the full spectrum. The ongoing analysis of Lambda-baryon channels will further provide the fraction of single/multi nucleon absorption, as well as the rates of nuclear Σ/Λ conversion for the given energy range.

For the study of the $\Lambda(1405)$, the $\Sigma^0\pi^0$ decay channel has been analyzed. The invariant mass distribution for events in ^{12}C, shown in the right part of Fig. 2, exhibits a significant fraction of events above the kinematical limit of the K^- absorption at rest (1416 MeV/c^2). A similar behaviour, explained by the small difference in the last nucleon's binding energies, was observed on ^{4}He. Consequently, the data set was split into in-flight and at rest candidate events. Further analysis indicates both distributions have a two-component structure; the lower momentum one correlated with high masses and vice versa. The ongoing work in extracting the resonant and non-resonant parts and the characterization of the line shapes of the resonance is important to correctly interpret the results and compare them to theoretical models.

Acknowledgments

We thank C.Capoccia, G.Corradi, B.Dulach and D.Tagnani from LNF-INFN and H.Schneider, L.Stohwasser, and D.Stückler from Stefan-Meyer-Institut for their fundamental contribution in designing and building the SIDDHARTA setup. We thank as well the DAΦNE staff for the fine working conditions and permanent support. Part of this work was supported by Hadron Physics I3 FP6 European Community program, Contract No.RII3 CT-2004-506078; the European Community-Research Infrastructure Integrating Activity Study of Strongly Interacting Matter (Hadron Physics 2, Grant Agreement No.227431 and HadronPhysics 3, Contract No.283286 under the Seventh Framework Programme of EU); Austrian Federal Ministry of Science and Research BMBWK 650962/0001 VI/2/2009; Romanian National Authority for Scientific Research,Contract No.2-CeX 06-11-11/2006; the Grant in-Aid for Specially Promoted Research (20002003), MEXT, Japan; the Austrian Science Fund (FWF):[P20651-N20]; and the DFG Excellence Cluster Universe of the Technische Universität München.

References

[1] M. Iwasaki *et al.* : Observation of the kaonic hydrogen K(alpha) x-ray. Phys. Rev. Lett. 78, 3067-3069 (1997). T.M. Ito *et al.* : Observation of kaonic hydrogen atom x rays. Phys. Rev. C 58 2366-2382 (1998).

[2] G. Beer *et al.* : Measurement of the kaonic hydrogen X-ray spectrum. DEAR Collaboration, Phys. Rev. Lett. 94 212302 (2005).

[3] S. Deser *et al.* : Energy level displacements in pi mesonic atoms. Phys. Rev. 96 774-776 (1954). T.L. Trueman : Energy level shifts in atomic states of strongly-interacting particles. Nucl. Phys. 26, 57-67 (1961). A. Deloff : S Matrix Approach to Hadronic Atom Level Shifts. Phys. Rev. C 13, 730-736 (1976).

[4] U.-G. Meißner, U. Raha, A. Rusetsky : Spectrum and decays of kaonic hydrogen. Eur. Phys. J. C 35, 349-357 (2004). A. Cieplý, J. Smejkal : Kaonic hydrogen versus the K-p low-energy data. Eur. Phys. J. A 34, 237-241 (2007).

[5] A.D. Martin: Kaon - Nucleon Parameters. Nucl. Phys. B 179, 33-48 (1981). D.N. Tovee *et al.* : Some properties of the charged sigma hyperons. Nucl. Phys. B 33, 493-504 (1971). R.J. Nowak *et al.* : Charged Sigma Hyperon Production by K- Meson Interactions at Rest. Nucl. Phys. B 139, 61-71 (1978).

[6] A. Cieplý *et al.* : Chirally motivated K^- nuclear potentials. Phys. Lett. B 702, 402-407 (2011). Y. Ikeda, T. Hyodo, W. Wise : Improved constraints on chiral SU(3) dynamics from kaonic hydrogen. Phys. Lett. B 706, 63-67 (2011). Y. Ikeda, T. Hyodo, W. Wise : Chiral SU(3) theory of antikaonnucleon interactions with improved threshold constraints. Nucl. Phys. A 881, 98-114 (2012).

[7] M. Boscolo *et al.* : Description and performances of luminosity and background detectors at the upgraded e+e-DAFNE collider. Nucl. Instrum. Methods Phys. Res. A 621, 157-170 (2010).

[8] T. Koike, T. Harada, Y. Akaishi : Cascade calculation of K-p and K-d atoms. Phys. Rev. C 53, 79-87 (1996).

[9] M. Bazzi *et al.* : A new measurement of kaonic hydrogen X-rays. SIDDHARTA Collaboration, Phys. Lett. B 704, 113-117 (2011). M. Bazzi *et al.* : Kaonic hydrogen X-ray measurement in SIDDHARTA. SIDDHARTA Collaboration, Nucl. Phys. A 881, 88-97 (2012).

[10] B. Borasoy, R. Nißler, W. Weise : Kaonic Hydrogen and K-p Scattering. Phys. Rev. Lett. 94, 213401 (2005). B. Borasoy, U.-G. Meißner, R. Nißler : K-p scattering length from scattering experiments. Phys. Rev. C 74, 055201 (2006).

[11] SIDDHARTA-2 Collaboration, Proposal of Laboratori Nazionali di Frascati of INFN (2010).

[12] M. Adinolfi et al.: The tracking detector of the KLOE experiment, Nucl. Instr. Meth. A 488, 51-73 (2002).

[13] Y. Akaishi and T. Yamazaki: Emission spectra and invariant masses of Λ and p in the stopped-K^-NN absorption process in ^{4}He and ^{6}Li, Nucl. Phys. A 792, 229 (2007).

Weak interactions of kaons from lattice QCD

Norman H Christ

Physics Department, Columbia University, New York, NY 10027, USA

E-mail: nhc@phys.columbia.edu

RBC and UKQCD Collaborations

Abstract. The methods of lattice QCD and available computer resources are now sufficient that predictions for many of the Standard Model properties of kaons can be made from first principles with accurately bounded uncertainties. We discuss two relatively new areas where lattice methods are having or will soon have a large impact: calculation of the two complex amplitudes A_0 and A_2 describing the decay of the kaon into two-pions with $I = 0$ and 2 and calculation of long-distance contributions to second-order electro-weak processes including the $K_L - K_S$ mass difference ΔM_K, the CP-violating parameter ϵ_K and certain rare kaon decays.

1. Introduction

Over the past thirty years lattice QCD has evolved from providing a first non-perturbative demonstration of confinement and chiral symmetry breaking in QCD to allowing much of low energy QCD to be accurately computed from first principles with controlled uncertainties. By replacing the space-time continuum by a regular, four-dimensional, hyper-cubic lattice and using powerful Monte Carlo methods to evaluate the Euclidean Feynman path integral which defines QCD, the previously intractable complexities of low-energy, strongly-coupled QCD are reduced to the problem of obtaining very large computational resources and developing increasingly powerful and efficient numerical strategies – two areas in which advancing technology and the growing importance of computational methods throughout science work to our advantage.

In the past two years, lattice QCD calculations have crossed an important threshold and can now determine directly the physics of the up, down and strange quarks using their physical masses, or in the case of most calculations, a single, isospin-symmetric average mass of ≈ 3 MeV, (expressed in the $\overline{\rm MS}$ renormalization scheme at 3 GeV) for the up and down quarks, giving three degenerate pions of mass 135 MeV. Earlier calculations used light quark masses as large as 30 or 40 MeV and relied on chiral perturbation theory to extrapolate to the physical value, introducing significant, uncontrolled errors.

This ability to work at a physical value for the light quark mass is especially advantageous for calculations performed with chiral fermions where the light quark mass can be the largest source of unphysical, chiral symmetry breaking. For example, the calculations of the RBC and UKQCD collaborations typically use the domain wall fermion formulation in which an extra fifth dimension, whose size varies between 12 and 32 lattice units, literally separates the right- and left-handed fermion chiralities. While the Wilson and staggered fermion formulations achieve physical chiral symmetry in the continuum limit, the breaking of chiral symmetry at short distances complicates the definition of the $\Delta S = 1$ weak Hamiltonian where the number of operators whose mixing must be controlled is dramatically reduced by chiral symmetry.

In addition, the explicit breaking of chiral symmetry by lattice artifacts for Wilson and staggered fermions is often the largest source of finite lattice spacing errors, errors which are absent when a chiral formulation is used. This is illustrated in Figure 1 where dimensionless ratios for a number of quantities are compared between two calculations, with inverse lattice spacings of $1/a = 1.73$ GeV and 2.28 GeV. These two calculations give results which agree on the $\approx 1\%$ level for lattice spacings which are not especially small by current standards.

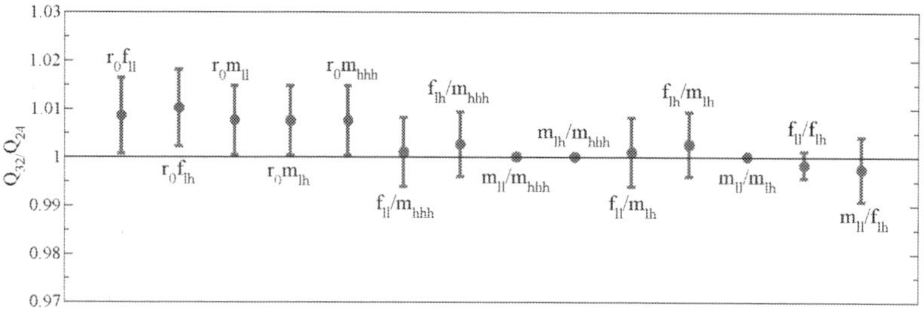

Figure 1. Comparison of a series of dimensionless ratios obtained from calculations at two inverse lattice spacings: $1/a = 1.73$ GeV and 2.28 GeV. The ratio of each ratio is plotted on the y axis. The subscripts indicate the masses of the quarks used where h indicates a near strange quark mass while l is a less massive quark. Further explanation can be found in Ref. [1].

This combined ability to work at physical quark masses and the empirically small finite lattice spacing errors found with chiral fermions implies that results accurate on the percent level can be obtained from a single lattice QCD calculation without chiral or even continuum extrapolations. For example, in a recent calculation on a $48^3 \times 96$ lattice with $1/a = 1.73$ GeV and light and strange quark masses chosen close to their physical values, we have obtained the result $f_\pi = 130.7(2)$ MeV directly from the computer with no extrapolations or corrections beyond the normalization factor for the axial current. The 0.2% error is statistical and a satisfactory agreement with the experimental value of $f_\pi = 130.4$ MeV is seen even for this simple, direct result. While added calculations with a smaller lattice spacing and heavier quark masses allow sub-percent corrections to be made that adjust for the small mismatch of the input and physical quark masses and $O(a^2)$ discretization errors [2], a result for f_π with percent accuracy can be obtained from a direct calculation of this quantity on a single lattice QCD ensemble.

Since the methods and available resources for lattice QCD are now sufficient to allow a basic quantity such as f_π to be so easily computed, it is natural to turn to more complex quantities which are more difficult to compute using lattice methods but also less well known and possibly of greater fundamental interest. In this paper we describe progress in two such directions: the calculation of the complex $\Delta I = 3/2$ and $1/2$, $K \to \pi\pi$ decay amplitudes A_2 and A_0 and the calculation of the long-distance contributions to $K^0 - \overline{K}^0$ mixing and rare kaon decays.

2. Computing A_0 and A_2 using lattice QCD

Three ingredients are needed for a first-principles lattice QCD calculation of amplitudes contributing to $K \to \pi\pi$ decay: the properly normalized, four-quark, $\Delta S = 1$ weak Hamiltonian $H_W^{\Delta S=1}$; the ability to exploit the energy quantization of two-pion, finite-volume states to create a final $\pi - \pi$ state with energy equal to M_K and an understanding of the finite volume effects which must be removed to obtain a physical, infinite-volume decay amplitude. Fortunately, reliable techniques are now available to address each of these issues.

The effective $\Delta S = 1$ weak Hamiltonian which should describe $K \to \pi\pi$ in the Standard Model is known from pioneering work in the 1970's and has been described in a comprehensive way [3] that can be accurately adapted to support lattice QCD calculations. The Rome-Southampton non-perturbative renormalization scheme [4] can be used to express the lattice-regularized versions of the seven independent four-quark operators which enter $H_W^{\Delta S=1}$ in terms of operators that have a well-defined continuum limit and can be related using continuum, QCD perturbation theory to the $\overline{\text{MS}}$ renormalization scheme used to determine the Wilson coefficients which appear in continuum expressions for $H_W^{\Delta S=1}$. These methods have been used successfully in recent lattice calculations [5] and are expected to be accurate at the 10-20% level.

This uncertainty is caused by the use of QCD perturbation theory at the scale of ≈ 2 GeV and the use of perturbation theory to remove the charm quark to obtain operators appropriate for a three-flavor theory. While this level of accuracy may be appropriate for present calculations, it can be improved to whatever extent is required by simply increasing the energy scale at which the lattice and continuum calculations are compared. A first step is to include the charm quark in the lattice calculation. By using a four-flavor theory we will avoid the problem of using perturbation theory at the charm quark scale and the uncertain validity of assuming that the charm quark mass is much larger than the energy scale relevant for K meson decay. Of course, if the charm quark is to be included in a lattice calculation, a lattice spacing must be used that is sufficiently small that $O\big((m_c a)^2\big)$ errors can be controlled. The second step in eliminating potential errors caused by the use of perturbation theory is to exploit "step-scaling" [6] and carry out the Rome-Southampton operator renormalization at a series of smaller lattice spacings and corresponding smaller lattice volumes until the scale of momentum employed is sufficiently large to guarantee sufficiently small perturbative errors. Note, this is much less demanding than performing the entire $K \to \pi\pi$ calculation at such small lattice spacings.

The next ingredient in this calculation is the creation of a physical two-pion final state. This can be done in a lattice calculation by following Lellouch and Lüscher [7], exploiting the finite-volume quantization of the two-pion energy and adjusting the volume so that the energy of a finite-volume, excited two-pion state matches M_K. In a Euclidean-space Green's function calculation the contribution of an excited two-pion state falls exponentially with increasing time relative to states with lower energy. This difficulty can be partially overcome if we introduce boundary conditions chosen to select the pion momentum that is present in the two-pion state of interest [8, 9]. For example, Figure 2 suggests the effects on the pion wave function of boundary conditions which are anti-periodic in one of the three spatial directions.

Of course, in a lattice calculation we can only impose boundary conditions on the underlying quarks and not directly on the pions. For the case of the $I = 2$, two-pion final state this is not difficult since we can use isospin symmetry to relate the amplitude of interest to the unique state in which both pions have charge $+1$ and then impose anti-boundary conditions on the down anti-quark, leaving the up quark to obey periodic boundary conditions. While this condition breaks isospin symmetry it cannot alter the $I = 2$ character of the finite-volume state in question because that state is uniquely determined by its charge.

The problem of imposing boundary conditions which will insure that the lowest energy, $I = 0$, $\pi - \pi$ state will have an energy equal to that of the K meson is much more challenging. Since the $I = 0$ state has the same electric charge as that with $I = 2$ we must impose boundary conditions which are consistent with isopsin symmetry to avoid mixing these two states. This can be done by imposing G-parity boundary conditions [8, 10] which have the unusual feature of mixing particle and anti-particle. This is illustrated in Figure 3.

In a calculation in which the effects of finite volume are being critically used to create a final two-pion state with the proper energy, we should be concerned that other finite-volume effects may introduce significant systematic errors. Fortunately in the paper [7] pointing out the utility of exploiting finite volume to create a physical two-pion state, Lellouch and Lüscher also provide

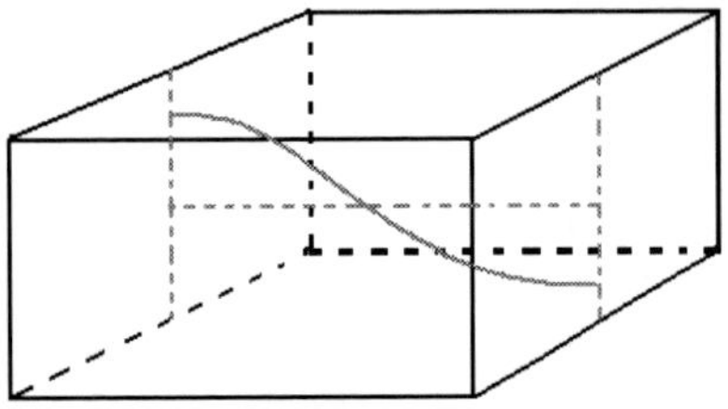

Figure 2. Diagram suggesting behavior of the pion wave function that will result if anti-periodic boundary conditions in one direction are obeyed by the pion.

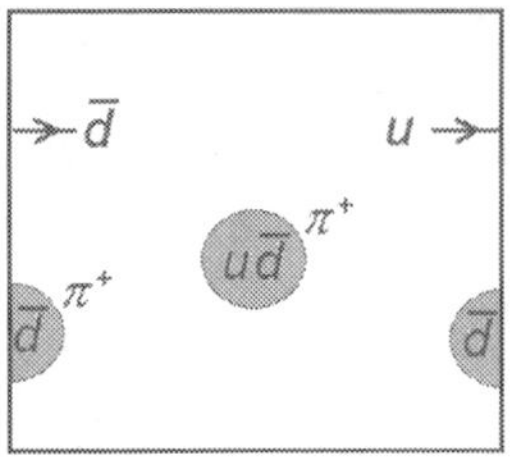

Figure 3. Diagram illustrating the unusual features of G-parity boundary conditions: a u quark passes through the boundary on the right and emerges on the left as a $\overline{d}$ while the $u-\overline{d}$, π^+ state in the center of the box is represented by a $\overline{d}\,\overline{d}$ pair when that pion state straddles the boundary.

a concrete formula for removing the leading finite-volume effects so that for volumes of linear extent L with $m_\pi L \geq 4$ one expects sub-percent residual finite-volume corrections.

While the easier calculation of the $\Delta I = 3/2$ amplitude was a pioneering effort in 2012 [11] it is now a well developed and routine part of the computational package run by the RBC and UKQCD Collaborations at increasingly small lattice spacing [12]. Our present preliminary result for A_2 at physical quark mass in the continuum limit is

$$\mathrm{Re}(A_2) = 1.606(61)_{\mathrm{stat}}(145)_{\mathrm{sys}} \times 10^{-8}\mathrm{GeV}, \quad \mathrm{Im}(A_2) = 7.35(24)_{\mathrm{stat}}(88)_{\mathrm{sys}} \times 10^{-13}\mathrm{GeV}. \quad (1)$$

The real part of A_2 agrees reasonably well with the experimental value of $1.436(4) \times 10^{-8}$ GeV while $\mathrm{Im}(A_2)$ cannot be directly measured in experiment and has not been previously computed.

Calculation of the $\Delta I = 1/2$ amplitude is much more difficult because the quantum numbers of the $I = 0$, $\pi - \pi$ final state are the same as those of the vacuum and the G-parity boundary conditions still allow the vacuum as a final state. Although this state can be subtracted, the noise left behind grows exponentially as $e^{M_K t}$ relative to the $\pi - \pi$ signal, when the final state propagates for the time t.

However, after much preparation we have begun a realistic calculation of A_0 using G-parity boundary conditions in all three spatial directions on a $32^3 \times 64$ lattice with $1/a = 1.37$ GeV and physical values for the light and strange quark masses. All seven operators entering $H_W^{\Delta S=1}$ are being evaluated. The two pions are absorbed on time slices separated by four time units and hydrogen atom wave functions with a radius of two lattice units are used for each of the pion states, an arrangement which is realized by using all-to-all propagators. These two features of the calculation have been shown in earlier studies [13, 14] to give improvements which reduce the noise coming from the vacuum subtraction by more than a factor of four. This calculation is now underway and we expect to have results with 20-30% errors within the coming year. If successful, this will give the first Standard Model prediction of the direct CP-violating parameter ϵ'. For the real parts of A_0 and A_2 we can already compare unphysical results for $\mathrm{Re}(A_0)$ with the physical calculation of $\mathrm{Re}(A_2)$ and recognize an emerging explanation for $\Delta I = 1/2$ rule [15]: the two amplitudes which add to give $\mathrm{Re}(A_0)$ cancel when combined to form $\mathrm{Re}(A_2)$.

3. Long distance contributions to second-order weak processes

Given the ability to work with physical quark masses and to control all systematic errors it is natural to ask if there are additional areas in kaon physics where lattice QCD might advance our understanding of Standard Model processes. One such new and promising direction is the use of lattice QCD to calculate what are often referred to as long distance, second-order weak phenomena. Such phenomena include the $K_L - K_S$ mass difference ΔM_K, the contribution of the up and charm quark loops to the indirect CP-violation parameter ϵ_K and certain rare kaon decays where the change of quantum numbers requires that the decay occur at second order in the weak interactions. Such second-order weak processes are of great interest because their small size implies increased sensitivity to other, non-Standard-Model phenomena.

Such second order weak phenomena usually involve internal loops containing W bosons and receive contributions from both short distances on the order of the inverse W or top quark mass and long distances on the order of the inverse charm quark mass $1/m_c$ or the QCD scale $1/\Lambda_{\rm QCD}$. Here the notions of short and long distances are best defined as the scales at which QCD perturbation theory is or is not applicable. While in the past, the charm quark scale has been included in the short-distance category, recent perturbative studies of ΔM_K [16] which is dominated by distances of order $1/m_c$ suggest that non-perturbative methods are required even at this scale to obtain reliable results.

The largest contribution to ϵ_K comes from a top-quark loop which implies that at long distances, this second-order weak process can be expressed as a local $\Delta S = 2$ operator multiplied by a Wilson coefficient which can be accurately computed in perturbation theory. However, up and charm quarks contribute at the few-percent level and such amplitudes involve two local W exchanges (each accurately represented by a $H_W^{\Delta S=1}$ or $H_W^{\Delta C=1}$ vertex), which may be separated by distances of the order of $1/\Lambda_{\rm QCD}$. For ΔM_K the top quark is suppressed by the relatively small size of the real parts of its CKM matrix element and the dominant contribution comes from the long-distance contributions of up and charm quarks.

While non-trivial, the computation of such long-distance mixing effects is possible using lattice QCD [17]. The challenge of such calculations arises both from the complexities of the second-order amplitudes involved and the fact that the calculation is performed in Euclidean space. The quantum mechanical strategy underlying the calculation is straight-forward. We compute the matrix element between K^0 and $\overline{K}^0$ as the product of two, four-quark, weak operators integrated over a volume of time extent T as is illustrated in Figure 4. In such a Euclidean time calculation there will be a number of physical process that contribute. The process of interest is simple propagation of a K-meson state with the exponential time dependence $e^{-(M_K+\Delta M_K)T}$ which when evaluated at second order in ΔH_W will give the term $\Delta M_K T$, linear in T, from which ΔM_K can be easily extracted.

Unfortunately this term of interest must be distinguished from exponentially larger terms in which the factors of H_W allow the energy-non-conserving decay of the kaon state to an intermediate vacuum, single pion state or $\pi - \pi$ state with energy below M_K. For example, the contribution of an intermediate, single-pion state will fall much less rapidly with increasing T as $e^{-M_\pi T}$. However, these larger terms can be computed independently and subtracted in a correlated fashion, yielding a linear term which can be accurately identified as shown in Figure 5. The data presented in this figure comes from a complete calculation [18], including all graphs, which obtains ΔM_K with a $\approx 15\%$ statistical error but which must be repeated at smaller lattice spacing if the $O\big((m_c a)^2\big)$ discretization errors associated with the charm quark mass are to be controlled. A similar calculation of the long-distance contributions to ϵ_K has now been started. This calculation is more difficult because there is reduced GIM suppression and a non-perturbative subtraction combined with a perturbative correction is needed to properly join the non-perturbative long-distance with the perturbative short-distance results.

For rare kaon decays such as $K_L \to \pi^0\ell^+\ell^-$ or $K^+ \to \pi^+\nu\overline{\nu}$ such non-perturbative, long-distance contributions are important at the few percent level and their calculation using lattice methods promises to extend the physics reach of experiments studying these decays. Calculation of such decays using lattice QCD [19] is now practical and exploratory calculations are now being carried out by RBC and UKQCD. These decays are more accessible to lattice QCD than the

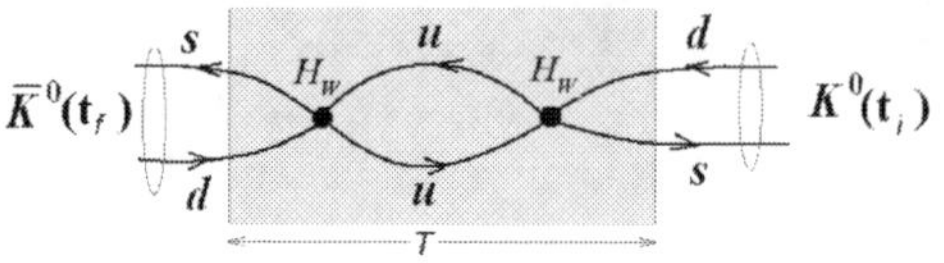

Figure 4. Diagram showing schematically a calculation of the long-distance parts of ΔM_K or ϵ_K using lattice QCD. The two first-order weak opeartors H_W are integrated over a region of time extent T lying between the operators which create the initial K^0 and destroy the final $\overline{K}^0$.

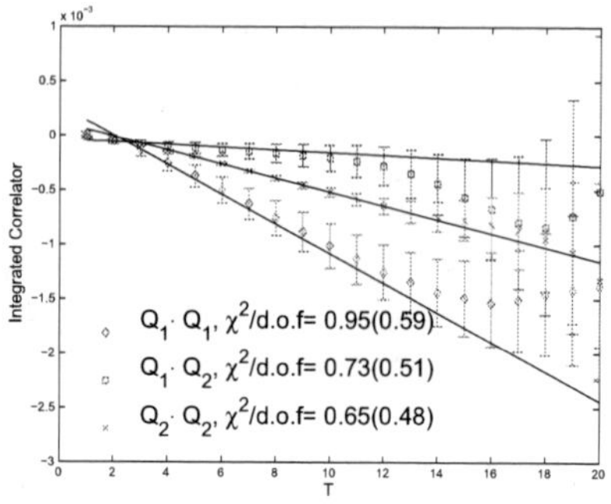

Figure 5. Plots of the linear T behavior in the calculation of ΔM_K for products of the operators O_1 and O_2 appearing in $H_W^{N_f=4}$.

$K \to \pi\pi$ calculations described in the previous section because of the absence of final state interactions, allowing the final state particles to be directly assigned the momenta carried by the physical decay products. However, the diagrams that must be evaluated are more complex and the presence of additional, Euclidean-time processes which are exponentially larger than the transition of interest (described above for the calculation of ΔM_K) must be overcome.

4. Conclusions

The ability to work directly with physical quark masses and to employ a fermion formulation which respects chiral symmetry makes possible the calculation of a variety of important quantities in kaon physics such as the complex, two-pion decay amplitudes A_0 and A_2, the $K_L - K_S$ mass difference and the long distance contributions to ϵ_K and rare kaon decays. Current calculations and future calculations at sufficiently small lattice spacing to allow accurate treatment of the charm quark, open the possibility to discover new phenomena beyond those predicted by the Standard Model at increasingly high energy and with increasing precision.

References

[1] Aoki Y *et al.* (RBC Collaboration, UKQCD Collaboration) 2011 *Phys.Rev.* **D83** 074508 (*Preprint* `1011.0892`)
[2] Blum T *et al.* (RBC Collaboration, UKQCD Collaboration) In preparation
[3] Buchalla G, Buras A J and Lautenbacher M E 1996 *Rev.Mod.Phys.* **68** 1125–1144 (*Preprint* `hep-ph/9512380`)
[4] Martinelli G, Pittori C, Sachrajda C T, Testa M and Vladikas A 1995 *Nucl.Phys.* **B445** 81–108 (*Preprint* `hep-lat/9411010`)
[5] Blum T, Boyle P, Christ N, Garron N, Goode E *et al.* 2011 *Phys.Rev.* **D84** 114503 (*Preprint* `1106.2714`)
[6] Arthur R and Boyle P (RBC Collaboration, UKQCD Collaboration) 2011 *Phys.Rev.* **D83** 114511 (*Preprint* `1006.0422`)
[7] Lellouch L and Luscher M 2001 *Commun. Math. Phys.* **219** 31–44 (*Preprint* `hep-lat/0003023`)
[8] Kim C and Christ N H 2003 *Nucl. Phys. Proc. Suppl.* **119** 365–367 (*Preprint* `hep-lat/0210003`)
[9] Kim C 2004 *Nucl. Phys. Proc. Suppl.* **129** 197–199 (*Preprint* `hep-lat/0311003`)
[10] Kelly C 2013 (*Preprint* `1310.0434`)
[11] Blum T, Boyle P, Christ N, Garron N, Goode E *et al.* 2012 *Phys.Rev.Lett.* **108** 141601 (*Preprint* `1111.1699`)
[12] Janowski T, Sachrajda C, Boyle P, Christ N, Mawhinney R *et al.* 2013 (*Preprint* `1311.3844`)
[13] Liu Q ISBN: 9781267290649
[14] Zhang D 2013 *PoS* **LATTICE2013** 3403
[15] Boyle P *et al.* (RBC Collaboration, UKQCD Collaboration) 2012 (*Preprint* `1212.1474`)
[16] Brod J and Gorbahn M 2012 *Phys.Rev.Lett.* **108** 121801 (*Preprint* `1108.2036`)
[17] Christ N, Izubuchi T, Sachrajda C, Soni A and Yu J 2012 *Phys. Rev.* **D88** 014508 (*Preprint* `1212.5931`)
[18] Bai Z, Christ N, Izubuchi T, Sachrajda C, Soni A *et al.* 2014 *Phys.Rev.Lett.* **113** 112003 (*Preprint* `1406.0916`)
[19] Isidori G, Martinelli G and Turchetti P 2006 *Phys.Lett.* **B633** 75–83 (*Preprint* `hep-lat/0506026`)

Measurement of the absolute branching ratio of the $K^+ \to \pi^+\pi^-\pi^+(\gamma)$ decay with the KLOE detector

Patrizia de Simone on behalf of the KLOE/KLOE-2 Collaboration
Laboratori Nazionali di Frascati dell' I.N.F.N.
E-mail: patrizia.desimone@lnf.infn.it

Abstract. The absolute branching ratio of the $K^+ \to \pi^+\pi^-\pi^+(\gamma)$ decay, inclusive of final-state radiation, has been measured using ~17 million tagged K^+ mesons collected with the KLOE detector at DAΦNE, the Frascati ϕ-factory. The result is:

$$BR(K^+ \to \pi^+\pi^-\pi^+(\gamma)) = 0.05565 \pm 0.00031_{stat} \pm 0.00025_{syst}$$

1. Introduction

The measurement of the branching ratio (BR) of the $K^+ \to \pi^+\pi^-\pi^+(\gamma)$ decay completes the KLOE program of precision measurements of the dominant kaon branching ratios, fully inclusive of radiation effects. The most recent BR($K^\pm \to \pi^\pm\pi^+\pi^-$) measurement, based on 2330 events from a sample of $\sim 10^5$ kaon decays, dates back to 1972 and gives no information on the radiation cut-off : BR($K^\pm \to \pi^\pm\pi^+\pi^-$) = $(5.56 \pm 0.20)\%$ [1].

The measurement of the absolute BR($K^+ \to \pi^+\pi^-\pi^+(\gamma)$) has been performed with the KLOE detector using data corresponding to an integrated luminosity $\int \mathcal{L}\, dt \simeq 174$ pb^{-1} collected at DAΦNE, the Frascati ϕ-factory[2]. DAΦNE is an e^+e^- collider operated at the energy of 1020 MeV, the mass of the ϕ-meson. The beams collide at the interaction point (IP) with a crossing angle $\theta_x \simeq 25$ mrad, producing ϕ-mesons with a small momentum of ~ 12.5 MeV/c in the horizontal plane. The ϕ-mesons decay in anti-collinear and monochromatic neutral (34%) and charged (49%) kaon pairs. The unique feature of a ϕ-factory is the tagging: detection of a $K^\pm$ (the tagging kaon) tags the presence of a $K^\mp$ (the tagged kaon) with known momentum and direction. The selection of the $K^\pm$ beams is done reconstructing the two-body decays $K^\pm \to \mu^\pm\nu(\gamma)$ ($K_{\mu 2}$ tags) and $K^\pm \to \pi^\pm\pi^0(\gamma)$ ($K_{\pi 2}$ tags) which provide two indipendent samples of pure kaons for the signal selection useful also for systematic uncertainties evaluation. These decays are easily identified as clear peaks in the distribution of $p^*_{m_\pi}$, the momentum of the charged secondary track in the kaon rest frame evaluated using the pion mass [3].

The KLOE detector consists of a large cylindrical drift chamber (DC) [4], surrounded by a lead scintillating fiber electromagnetic calorimeter (EMC) [5] both embedded in an axial 0.52 T magnetic field produced by a superconducting coil.

The DC tracking system has 25 cm internal radius, 4 m diameter and 3.3 m length, with a total of ~ 52000 wires, of which ~ 12000 are sense wires arranged in a stereo geometry. In order to minimize the multiple scattering and K_L regeneration, and to maximize the detection efficiency for low energy photons, the DC works with a helium-based gas mixture and its walls are made

of light materials, mostly carbon fiber composites. Spatial resolutions are $\sigma_{xy} \simeq 150\ \mu$m and $\sigma_z \simeq 2$ mm and the transverse momentum resolution is $\sigma(p^T)/p^T \leq 0.4\%$.
The calorimeter covers 98% of the solid angle and is composed by a barrel and two endcaps. For each impinging particle the calorimeter information consists of energy, position of impact point and arrival time, with corresponding accuracy of $\sigma_E/E = 5.7\%/\sqrt{E\ (\text{GeV})}$, $\sigma_z = 1.2$ cm$/\sqrt{E\ (\text{GeV})}$, $\sigma_\phi = 1.2$ cm, and $\sigma_t = 57$ ps$/\sqrt{E\ (\text{GeV})} \oplus 100$ ps.
The trigger [6] is based on energy deposits in the calorimeter and on hit multiplicity in the drift chamber. Only events triggered by the calorimeter have been used in the present analysis. The trigger system includes a second-level veto for cosmic-ray muons (cosmic-ray veto or CRV) based on energy deposits in the outermost layers of the calorimeter and followed by a third-level software trigger. A software filter (SF), based on the topology and multiplicity of energy clusters and drift chamber hits, is applied to filter out machine background. Both CRV and SF may be sources of events loss, and their effects on the BR measurement have been studied.

2. The analysis

To minimize the impact of the trigger efficiency on the signal side, we choose as normalization sample $K_{\mu 2}$ and $K_{\pi 2}$ tags which also provide the trigger of the event. The residual dependence of the signal sample on the tag selection, which we refer to as tag bias, has been evaluated for the BR measurement. Moreover, to avoid systematics due to kaon nuclear interaction, we use K^- as the tagging kaon ($K_{\mu 2}$ or $K_{\pi 2}$) and K^+ as the tagged kaon (signal), since the nuclear cross section for positive kaons with momenta $\simeq 100$ MeV is lower by a factor of $\sim 10^3$ with respect to that of negative kaons [7].
The track of the tagging kaon is backward extrapolated from its first hit in the DC to the IP. We use the momentum of the tagging kaon at the IP, $\mathbf{p}^{IP}_{K^-}$, and the momentum of the ϕ-meson measured run by run with Bhabha scattering events, $\mathbf{p}_\phi$, to evaluate the momentum of the tagged kaon at the IP, $\mathbf{p}^{IP}_{K^+} = \mathbf{p}_\phi - \mathbf{p}^{IP}_{K^-}$. Finally we extrapolate $\mathbf{p}^{IP}_{K^+}$ inside the DC (signal kaon path).
The kaon and the three charged pions from its decay have low momenta, less than ~ 200 MeV, and curl up in the KLOE magnetic field; this increases the probability to have poorly reconstructed tracks broken in more segments. We significantly improve the quality of the reconstruction requiring the K^+ decay to occur before it reaches the DC sensitive volume, i.e. inside a cylindrical fiducial volume centered at the IP and with a transverse radius $\rho_{xy} \leq 26$ cm. In this way only the pion tracks are reconstructed, and we extrapolate only two of them to search for a vertex along the signal kaon path. The decay vertices are selected cutting on the distance of closest approach between each extrapolated tracks and the signal kaon path, DCA< 3 cm, and on the distance of closest approach between the two selected tracks, DCA$_{12} < 3$ cm. To remove the background due to the two-body decays, each selected track must have a momentum in the kaon rest frame $\text{p}^*_{m_\pi} < 190$ MeV/c. After this selection, a residual background, mainly due to broken kaon tracks, is removed cutting almost collinear tracks, $|cos(\theta_{12})| < 0.90$, where θ_{12} is the opening angle between the two selected tracks.
To extract the number of $K^+ \to \pi^+\pi^-\pi^+(\gamma)$ we fit the missing mass spectrum $m^2_{miss} = E^2_{miss} - (\mathbf{p_{K^+}} - \mathbf{p_1} - \mathbf{p_2})^2$ where $\mathbf{p_1}$ and $\mathbf{p_2}$ are the momenta of the selected tracks, with MC-predicted shapes for the signal and the background. Fig 1 and Fig 2 show the results of the fits of the missing mass distributions compared to data for both $K^-_{\mu 2}$ and $K^-_{\pi 2}$ tagged samples.
The signal selection efficiency, ϵ_{sel}, is related to the track reconstruction efficiency of two charged secondaries from K^+ decays. We evaluate the selection efficiency from MC, and then we correct it to take into account data-MC differences in the track reconstruction. To this aim we select, both on data and MC, a control sample of $K^+ \to \pi^- X$ decays (for signal events X corresponds to $\pi^+\pi^+$) with a background contamination of $\simeq$10.7%, to measure the efficiency corrections as function of the total transverse momentum p^T_X, and of the total longitudinal momentum p^L_X of

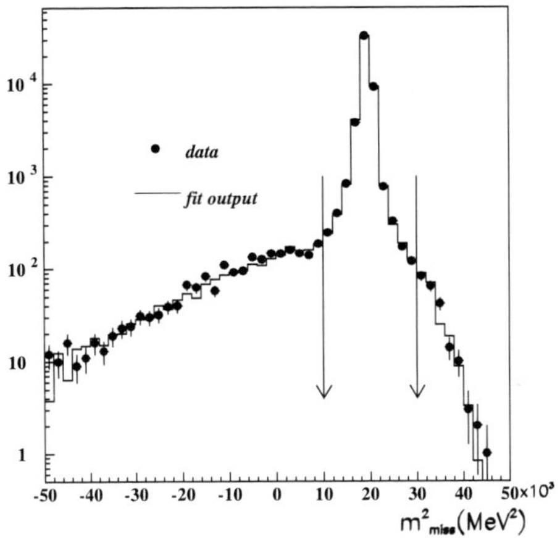

Figure 1. $K^-_{\mu 2}$ tagged sample: fit of the missing mass spectrum superimposed with data points, $\chi^2/\text{ndf} = 44.8/46$ ($P(\chi^2) = 0.52$). The arrows show the missing mass window for signal counting, where the signal over background ratio is $S/B \simeq 88$.

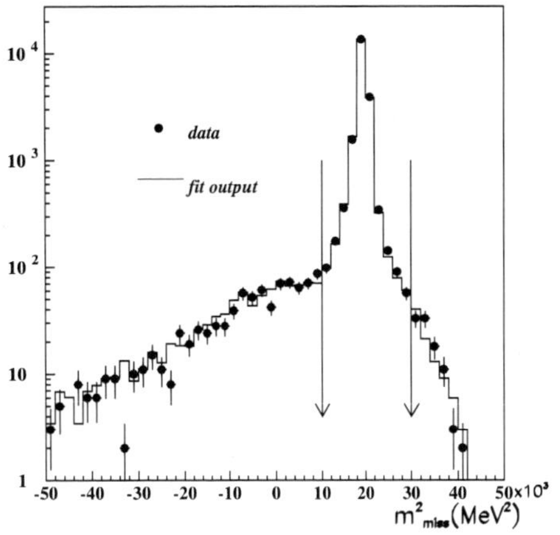

Figure 2. $K^-_{\pi 2}$ tagged sample: fit of the missing mass spectrum superimposed with data points, $\chi^2/\text{ndf} = 42.9/45$ ($P(\chi^2) = 0.56$). The arrows show the missing mass window for signal counting, where the signal over background ratio is $S/B \simeq 84$.

Table 1. Corrections to BR($K^+ \to \pi^+\pi^+\pi^-(\gamma)$) measurement.

corrections	$K^-_{\mu 2}$ tags	$K^-_{\pi 2}$ tags
cosmic ray veto C_{CRV}	1.00125 ± 0.00002	1.00049 ± 0.00001
software filter C_{SF}	1.0144 ± 0.0013	1.0003 ± 0.0005
tag bias C_{TB}	0.839 ± 0.001	0.802 ± 0.002

the $\pi^+\pi^+$ pair. The selection efficiency is finally obtained by folding the MC selection efficiency with the measured corrections: $\epsilon_{sel} = 0.0842 \pm 0.0003$ for the $K^-_{\mu 2}$ tagged sample, and $\epsilon_{sel} = 0.0866 \pm 0.0005$ for the $K^-_{\pi 2}$ tagged sample.
The corrections C_{CRV} and C_{SF} have been measured with data taken without the cosmic-ray veto and the software filter, respectively. The correction for the tag bias, C_{TB}, has been evaluated using MC. All correction values are reported in Table 1.
The sources of systematic uncertainties and the corresponding values are listed in Table 2.

3. The result

With a sample of $K^- \to \mu^-\bar{\nu}(\gamma)$ tagging events $N_{tag} = 12065087$ we found $N_{K\to 3\pi} = 48032 \pm 286$ signal events (the errors acconting for data and MC statistics), corresponding to:

$$BR(K^+ \to \pi^+\pi^-\pi^+(\gamma))|_{TagK_{\mu 2}} = 0.05552 \pm 0.00034_{stat} \pm 0.00034_{syst}.$$

With a sample of $K^- \to \pi^-\pi^0(\gamma)$ tagging events $N_{tag} = 5171239$ we found $N_{K\to 3\pi} = 20063 \pm 186$ signal events, corresponding to:

Table 2. Summary table of the fractional systematic uncertainties.

Source of systematic uncertainties	$K^-_{\mu 2}$ tags (%)	$K^-_{\pi 2}$ tags (%)
DCA, DCA$_{12}$, $\cos(\theta_{12})$ cuts	0.52	0.41
$p^*_{m_\pi}$ cut	0.08	0.11
m^2_{miss} cut	0.05	0.14
fiducial volume	0.11	0.10
selection efficiency estimate	0.16	0.16
tag bias	0.16	0.32
$K^\pm$ lifetime	0.12	0.12
Total fractional systematic uncertainty	0.60	0.59

$$BR(K^+ \to \pi^+\pi^-\pi^+(\gamma))|_{TagK_{\pi 2}} = 0.05587 \pm 0.00053_{stat} \pm 0.00033_{syst}.$$

Averaging these two results, accounting for correlations, we obtain:

$$BR(K^+ \to \pi^+\pi^-\pi^+(\gamma)) = 0.05565 \pm 0.00031_{stat} \pm 0.00025_{syst}.$$

This absolute branching ratio measurement is fully inclusive of final-state radiation and has a 0.72% accuracy [8], a factor $\simeq 5$ better with respect to the previous measurement [1].

References

[1] I.H. Chiang, *et al.*, Phys. Rev. D **6** (1972) 1254.
[2] A. Drago, *et al.* **LNF**-03/012 (2003).
[3] KLOE coll., F. Ambrosino, *et al.*, JHEP **02** (2008) 98.
[4] KLOE coll., M. Adinolfi *et al.*, Nucl. Inst. and Meth. A **488** (2002) 51.
[5] KLOE coll., M. Adinolfi *et al.*, Nucl. Inst. and Meth. A **482** (2002) 364.
[6] KLOE coll., M. Adinolfi *et al.*, Nucl. Inst. and Meth. A **492** (2002) 134.
[7] C.B. Dover and G.E. Walker, *The interaction of kaons with nucleons and nuclei*, Physics Report, **89** 1-177 (1982).
[8] KLOE/KLOE-2 coll. D. Babusci *et al.*, arXiv:1407.2028, accepted by Phys. Lett. B.

XI International Conference on Hyperons, Charm and Beauty Hadrons (BEACH 2014) IOP Publishing
Journal of Physics: Conference Series **556** (2014) 012007 doi:10.1088/1742-6596/556/1/012007

Spin observables in antihyperon-hyperon production with PANDA

K Schönning and E Thomé for the PANDA collaboration
Dept. of Physics and Astronomy, Uppsala University, Box 516, S-751 20 Uppsala, Sweden
E-mail: karin.schonning@physics.uu.se

Abstract. Spin observables provide a powerful tool in understanding the physics. Seven polarisation parameters of the Ω baryon can be extracted from the angular distributions of the hyperon decay and studied with the future PANDA experiment at FAIR. Simulation studies show that strange and single charmed hyperon channels have great prospects with PANDA.

1. Introduction

Hyperon production in $\overline{p}p \to \overline{Y}Y$ reactions gives important insights into strangeness and charm production. In this work, we consider single-, multi-strange and single-charmed hyperons. Their production from the light quarks in $\overline{p}p$ implies processes where they are replaced by heavier quarks. The relevant degrees of freedom of a given process are given by its energy scale, which for strangeness production is fixed by the mass of the strange quark, $m_s \approx 100$ MeV. This is close to the QCD cut-off, $\Lambda_{QCD} \approx 200$ MeV, where the strong coupling constant α_s grows so large that perturbative QCD breaks down. The production of strange hyperons can therefore probe QCD in the region below the perturbative regime, *i.e.* the confinement domain, about which very little is known so far. The scale of charm production is governed by $m_c \approx 1300$ MeV, more than ten times larger than m_s. The strong coupling constant in this region is $\alpha_s \approx 0.3$, just barely enough for a perturbative treatment to be valid. Comparing the production of strange and charmed hyperons could therefore give important insights into the differences in the production mechanism at these two energy scales. Theoretical models describing hyperon production in $\overline{p}p \to \overline{Y}Y$ reactions are often based on a constituent quark-gluon picture [1]. For strange hyperons, Kaon exchange models enter into play, where the production of single (multiple) strangeness hyperons are modeled by the single (multiple) exchange of a t-channel Kaon [2]. There have also been attempts to combine a quark-gluon approach with a Kaon exchange approach [3]. Spin variables are often very powerful in discriminating between different theoretical models. The hyperon spin variables can be related to the spin of individual quarks. The Λ hyperon can be modeled by a ud spin zero di-quark combined with an s-quark that carries the spin of the hyperon. A similar picture can be drawn for the Λ_c^+ hyperon, and by comparing spin observables of Λ and Λ_c^+, we can learn about the role of spin degrees of freedom in the creation of s- and c-quarks.

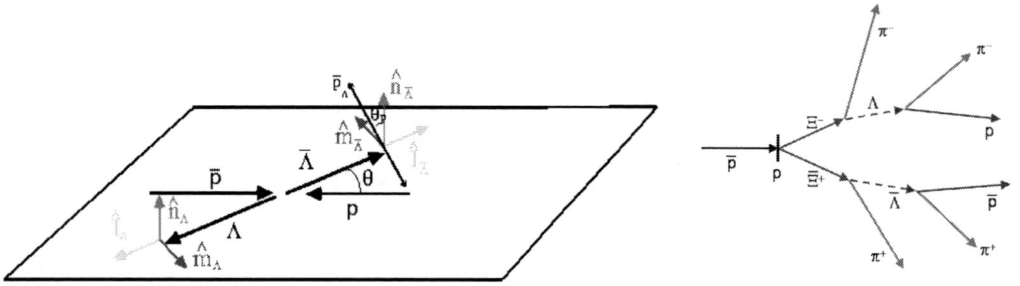

Figure 1. (a) The reference system of $\overline{p}p \to \overline{Y}Y$. (b) The $\overline{p}p \to \overline{\Xi}\Xi, \Xi \to \Lambda\pi, \Lambda \to p\pi$ decay.

2. Polarisation parameters for spin $\frac{1}{2}$ and spin $\frac{3}{2}$ hyperons

All physics information about a quantum mechanical ensemble is contained in the density matrix ρ. In an expansion in terms of hermitian matrices Q^L_M and polarisation parameters r^L_M [4], the density matrix of a particle of arbitrary spin j is given by

$$\rho = \frac{1}{2j+1}I + \sum_{L=1}^{2j} \frac{2j}{2j+1} \sum_{M=-L}^{L} Q^L_M r^L_M \tag{1}$$

where the first term denotes the unpolarised differential cross section I and the second the polarised part, containing the r^L_M parameters. L is the angular momentum and M its third component. In the case of spin $\frac{1}{2}$ particles, the Q^L_M are the Pauli matrices and the polarisation parameters correspond to the vector polarisations P_l, P_m and P_n. The indices are defined in the coordinate system shown in the left panel of figure 1. For parity conserving processes, *e.g.* $\overline{p}p \to \overline{\Lambda}\Lambda$ with unpolarised p and $\overline{p}$, symmetries of the spin density matrix imply that $P_l = P_m = 0$. This gives non-zero polarisation only perpendicular to the production plane.

The weak, parity violating decay of the hyperons means that the decay products are preferentially emitted along the direction of spin of the parent hadron. This makes the polarisation straightforward to measure. In the case of $\Lambda \to p\pi^-$, the angular distribution of the proton is related to the Λ polarisation by $I(\cos\theta_p) = \frac{1}{4\pi}(1 + \alpha_\Lambda P_n \cos\theta_p)$, where $\alpha_\Lambda = 0.64$ [5] is the asymmetry parameter. Some hyperons decay into other hyperons, *e.g.* the Ξ baryons, as illustrated in the right panel of figure 1. In the $\Xi^- \to \Lambda\pi^-, \Lambda \to p\pi^-$ process, additional asymmetry parameters β and γ of the Ξ^- hyperon are accessible *via* the angular distribution of the protons [6].

Not only the polarisation of individual hyperons $P_{y,Y}$ and antihyperons $P_{y,\overline{Y}}$ are of interest, but also the spin correlations $C_{i,j,\overline{Y}Y}$ $(i, j = x, y, z)$. These are all accessible with an unpolarised beam and an unpolarised target.

For spin $\frac{3}{2}$ hyperons, *e.g.* the Ω^-, the spin structure is more complicated. Here, we consider the polarisation parameters of individual spin $\frac{3}{2}$ hyperons only, and no correlations. The L number in equation 1 can be 1, 2 or 3 in this case. This gives three Q^1_M, five Q^2_M and seven Q^3_M matrices with fifteen corresponding r^L_M parameters. The corresponding spin density matrix was derived in Ref. [7]. Using symmetries imposed by strong interaction, eight polarisation parameters are identically zero.

The parameters r^2_0, r^2_1 and r^2_2 can been extracted from the angular distribution of the Λ in the $\Omega^- \to \Lambda K^-$ decay [7], assuming $\alpha_\Omega = 0$ according to previous measurements [5]. More information about the remaining four non-zero polarisation parameters, r^1_{-1}, r^3_{-1}, r^3_{-2} and r^3_{-3}, is obtained by studying the combined angular distribution of the Λ hyperons from the Ω^- decay and the protons from the subsequent Λ decay, $I(\theta_\Lambda, \phi_\Lambda, \theta_p, \phi_p)$ [8].

3. Previous measurements

The PS185 experiment at LEAR provided high-quality data of single-strangeness hyperons (Λ and Σ) up to an antiproton beam momentum of 2 GeV/c [9]. The spin structure of the $\bar{p}p \to \bar{\Lambda}\Lambda$ reaction was explored in detail but neither quark-gluon models nor hadron models fully succeeded in explaining the behaviour of the spin observables [10]. The existing data bank on multi-strange and charmed hyperons produced in $\bar{p}p$ annihilation is scarce. A handfull of bubble chamber $\bar{p}p \to \bar{\Xi}^+\Xi^-$ events exist [11] but neither the $\bar{p}p \to \bar{\Omega}^+\Omega^-$ nor the $\bar{p}p \to \bar{\Lambda}_c^-\Lambda_c^+$ reaction have ever been observed. New experimental data could serve as a guideline for the theoretical investigation which in turn could considerably widen our understanding of the underlying processes in this energy regime.

4. Prospects for PANDA

The future PANDA experiment at FAIR opens up new possibilies in hyperon physics. The antiproton beam from the HESR storage ring, operating in a momentum range between 1.5 GeV/c and 15 GeV/c, will interact with an internal hydrogen target. The PANDA experiment will profit from the HESR high luminosity and feature a near 4π acceptance with precise tracking and vertex reconstruction, sophisticated particle identification and calorimetry [12]. Simulation studies using a simplified Monte Carlo framework [13] show excellent prospects: high signal rate, low background rate and good detection efficiency over the full phase space for all single- and multi-strange and single-charmed hyperons [7],[14]. The estimated signal rates are given in table 1. They are calculated assuming the high luminosity operating mode of HESR, corresponding to *i.e.* $2{\cdot}10^{32}\text{cm}^{-2}\text{s}^{-1}$, as expected in the final version of FAIR. The only well known cross section is $\bar{p}p \to \bar{\Lambda}\Lambda$ at 1.64 GeV/c; for the other strange channels the cross sections are extrapolations from data collected at other energies. For Ω and Λ_c production in $\bar{p}p$ annihilations, no data exist. The quoted values for $\sigma(\bar{p}p \to \bar{\Omega}^+\Omega^-)$ and $\sigma(\bar{p}p \to \bar{\Lambda}_c^-\Lambda_c^+)$ represent a mean value of different theoretical calculations which can differ by even an order of magnitude [15],[16].

Table 1. Approximate foreseen production rates of various hyperon channels [7],[14]. The cross sections marked with a * have a very large uncertainty. Details are given in the text.

Beam mom. (GeV/c)	Reaction	σ (μb)	Eff (%)	Decay	Rate at $2{\cdot}10^{32}\text{cm}^{-2}\text{s}^{-1}$
1.64	$\bar{p}p \to \bar{\Lambda}\Lambda$	64	11	$\Lambda \to p\pi^-$	580 s^{-1}
4	$\bar{p}p \to \bar{\Lambda}\Sigma^0$	≈40	31	$\Sigma^0 \to \Lambda\gamma$	600 s^{-1}
4	$\bar{p}p \to \bar{\Xi}^+\Xi^-$	≈2	19	$\Xi^- \to \Lambda\pi^-$	30 s^{-1}
12	$\bar{p}p \to \bar{\Omega}^+\Omega^-$	≈0.002*	≈30	$\Omega \to \Lambda K^-$	≈80 h^{-1}
12	$\bar{p}p \to \bar{\Lambda}_c^-\Lambda_c^+$	≈0.1*	≈35	$\Lambda_c \to \Lambda\pi^+$	≈25 d^{-1}

Spin observables of Λ, Σ, Ξ, Ω and Λ_c hyperons have also been studied. Since no data on spin variables exist for these channels, a parametrisation based on sine functions was used. In this way, the functions are equal to 0 at $\cos\theta_Y = \pm 1$ which is required by the definition of the production plane (see figure 1). More specifically, the examples shown in figure 2 are obtained with parameterisations using $P_{y,\Xi} = \sin 2\theta_\Xi$, $C_{xz,\Xi} = \sin\theta_\Xi$ and $r^2_{0,\Omega} = \sin 2\theta_\Omega/\sqrt{3}$. These reconstructed spin observables show that their measurement can be satisfactorily performed with PANDA.

Journal of Physics: Conference Series **556** (2014) 012007 doi:10.1088/1742-6596/556/1/012007

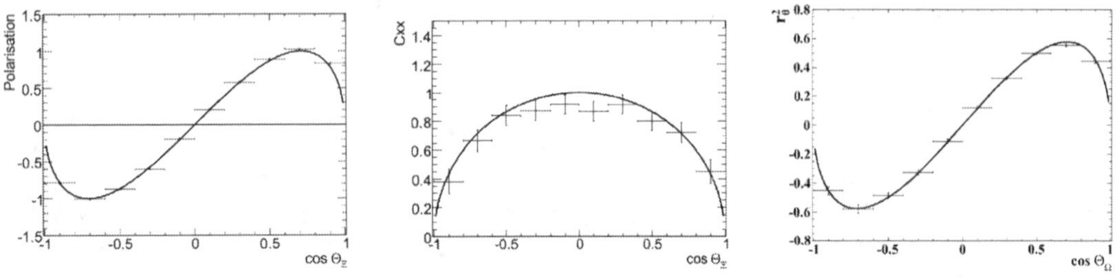

(a) (b) (c)

Figure 2. **(a)** The polarisation of the Ξ^- as a function of $\cos\theta_\Xi$. **(b)** The spin correlation in the x direction (m in Fig. 1) of the $\overline{\Xi}^+$ and the Ξ^-. **(c)** The polarisation parameter r_0^2 of the Ω as a function of $\cos\theta_\Omega$. The lines represent the input sine functions (see text) and the points and bars the reconstructed values for 10000 simulated events.

5. Summary and Outlook

The future PANDA experiment at FAIR will have a unique opportunity to give new insights into the strong interaction in the confinement domain. Spin observables in antihyperon-hyperon production provide a powerful tool for this purpose. In PANDA, the $\bar{p}p \to \overline{\Omega}^+\Omega^-$ and the $\bar{p}p \to \overline{\Lambda}_c^-\Lambda_c^+$ reaction will be studied for the first time. It will also be possible to measure the seven non-zero polarisation parameters of the Ω^- baryon that have recently been derived. The large foreseen hyperon data samples from PANDA could also open up the possibility of precise CP violation tests, hyperon Dalitz decays and other rare hyperon decays.

References

[1] Kohno M and Weise W, 1986 *Phys. Lett.* B **179** 15; Rubinstein H R and Snellman H 1985 *Phys. Lett.* B **165** 187; Furui S and Faessler A 1987 *Nucl. Phys.* A **486** 669; Burkardt M and Dillig M 1988 *Phys. Rev.* C **37** 1362; Alberg M A *et al.* 1988 *Z. Phys.* A **331** 207.
[2] Tabakin F and Eisenstein R A 1985 *Phys. Rev.* C **31** 1857; Kohno M and Weise W, 1986 *Phys. Lett.* B. **179** 15; La France P *et al.* 1988 *Phys. Lett.* B **214** 317; Timmermans R G E *et al.* 1992 *Phys. Rev.* D **45** 2288; Haidenbauer J *et al.* 1992 *Phys. Rev.* C **46** 2516.
[3] Ortega P G *et al.* 2011 *Phys. Lett.* B. **696** 352.
[4] Doncel M G *et al.* 1972 *Nucl. Phys* B **38** 477; Doncel M G *et al.* 1973 *Phys. Rev.* D **7** 815.
[5] Yao W M *et al.* [Particle Data Group] 2006 *J. Phys.* G **33** 1.
[6] Koch W 1968 *Analysis of scattering and decay* ed. Nikolic M (New York-London-Paris:Gordon and Breach).
[7] Thomé E 2012 *Multi-Strange and Charmed Antihyperon-Hyperon Physics for PANDA* Ph. D. Thesis, Uppsala University.
[8] Thomé E 2014 *A Method for Measurement of Polarisation Parameters in the* $\bar{p}p \to \overline{\Omega}^+\Omega^-$ *Reaction* (in preparation).
[9] Johansson T 2003 Antihyperon-hyperon production in antiproton-proton collisions *AIP Conf. Proc. 8th Int. Conf. on Low Energy Antiproton Physics* p 95.
[10] Klempt E *et al.* 2002 *Phys. Rep.* **368** 119.
[11] Musgrave B and Petmezas G 1965 *Nuovo Cim.* **35** 735; Baltay C *et al.* 1965 *Phys. Rev.* B **140** 1027.
[12] The PANDA collaboration, Technical Progress Report (2005).
[13] The PANDA collaboration, Physics Performance Report (2009) 43.
[14] Grape S 2009 *Studies of PWO Crystals and Simulations of the* $\bar{p}p \to \overline{\Lambda}\Lambda, \overline{\Lambda}\Sigma^0$ *Reactions for the PANDA experiment* Ph.D. Thesis, Uppsala University.
[15] Kaidalov A B and Volkovitsky P E 1994 *Z. Phys.* C **63** 51
[16] Titov A I and Kämpfer B 2008 *Phys. Rev.* C **78** 025201; Gornitschnig A T *et al.* 2009 *Eur. Phys. J.* A **42** 43; He J *et al.* 2011 *Phys. Rev.* D **84** 114010; Khodjamirian A *et al.* 2012 *Eur. Phys. J.* A **48** 31.

Quarkonia and Exotics at LHCb

Philippe Ghez, on behalf of the LHCb Collaboration
LAPP, 9 Chemin de Bellevue - BP 110, 74941 Annecy-le-Vieux Cedex - France
E-mail: ghez@lapp.in2p3.fr

Abstract. We present here the latest results from the LHCb collaboration on quarkonia and exotics. This summary includes three analyses in the charmonium sector, two of which concerning exotics, and one analysis in the bottomonium sector on χ_b production.

1. Introduction

Since charmonium and bottomonium production cross-sections lie in the [μb, pb] range [1], with a delivered integrated luminosity of several fb^{-1} for Run1, the LHC is a quarkonium factory. Its first years of running mark "The return of the Quarkonia" [2]. In this report, we address four of the many questions raised in the 2011 summary report of the Quarkonium Working Group [3]: $\psi(2S)$ polarisation, nature of the X(3872) state, existence and nature of the Z(4430) state and χ_b production. These four studies are all based on the detection of muon pairs.

2. Prompt $\psi(2S)$ polarisation

The measurement of the quarkonium polarisation is an important input to the many QCD-based models which have not yet been able to describe both quarkonium production and polarisation. At the LHC, the $\psi(2S)$ meson is produced in three ways: directly from hard scattering, from feed-down of higher mass quarkonia or via the decay of b hadrons. The first two modes are prompt. The feed-down mode is negligible. The $\psi(2S)$ is observed through its decay into a muon pair. The prompt $\psi(2S)$ polarisation is obtained from the angular distribution of the muons which is related to the polarisation parameters $(\lambda_\theta, \lambda_{\theta\phi}, \lambda_\phi)$ once a spin-quantization frame has been defined. In this analysis [4], two different choices of the spin-quantization frame have been used: the helicity frame (HX) and the Collins-Soper frame (CS).

The event sample is obtained from pp collisions at $\sqrt{s} = 7$ TeV with an integrated luminosity of 1 fb^{-1}. The event selection relies upon the use of a dedicated muon hardware trigger and a software trigger which refines the dimuon selection.The reconstruction of $\psi(2S)$ from the kinematic properties of the two muons is then performed and finally a cut is applied on the pseudo-decay time of the $\psi(2S)$, to reduce the non-prompt contamination from 20% to 3%. A mass fit in 5 $p_{\rm T}$ and 5 y (rapidity) bins of the $\psi(2S)$, labeled $p_{\rm T}{}^{\psi 2}$ and $y^{\psi 2}$ hereafter, is performed to extract the signal yield. After background subtraction, efficiency corrections obtained from simulated events are applied and a bin-by-bin likelihood is constructed.

The three polarisation parameters are obtained in each $(p_{\rm T}{}^{\psi 2}, y^{\psi 2})$ bin for the two choices of polarisation frames. Systematic and statistical errors are of the same magnitude (5% to 100% depending on the $(p_{\rm T}{}^{\psi 2}, y^{\psi 2})$ bin). Within errors and over the accessible kinematic range, $\lambda_{\theta\phi}$ and λ_ϕ are compatible with 0.

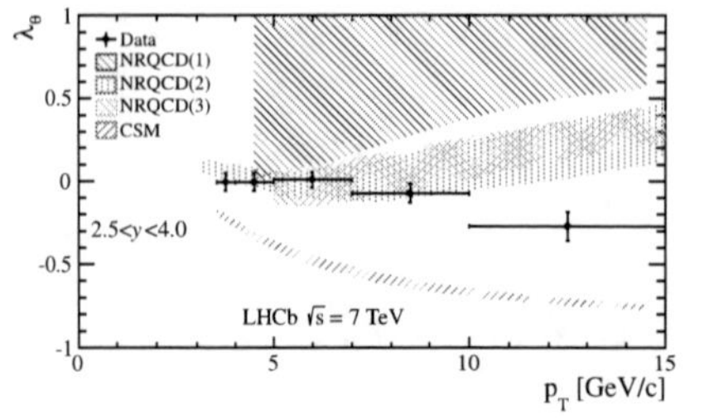

Figure 1. Polarisation parameter λ_θ in the HX frame as a function of ${p_T}^{\psi 2}$ for $y^{\psi 2}$ in the range [2.5,4.0]. The error bars represent the sum in quadrature of statistical and systematic uncertainties. The references for the models can be found in [4].

The λ_θ parameter in the HX frame does not show any large polarisation (longitudinal or transverse) effect which contradicts, at high ${p_T}^{\psi 2}$, several QCD-based models (Fig. 1). A frame-invariant polarisation parameter built from λ_θ and λ_ϕ is consistent with a negative polarisation with no strong ${p_T}^{\psi 2}$ and $y^{\psi 2}$ dependence. All results are consistent, within errors, between the HX and the CS choices of the polarisation frame.

3. Evidence for the decay X(3872)→ ψ(2S)γ

The X(3872) is the first exotic state discovered by Belle in 2003. Since its discovery, the nature of this state has been investigated with no clear conclusions. In this analysis [5], the decay X(3872)→$\psi(2S)\gamma$ is searched for. The X(3872)→$J/\psi\gamma$ decay is also investigated and the ratio of the two X(3872) branching fractions is measured. The value of this ratio is predicted by several models describing the X(3872) nature. The predictions range from ≈ 0 for a $D\overline{D}^*$ molecule to ≈ 15 for a pure charmonium state.

The data sample corresponds to the full statistics recorded by LHCb during Run1, i.e. to integrated luminosities of 1 fb^{-1} at $\sqrt{s} = 7$ TeV and 2 fb^{-1} at $\sqrt{s} = 8$ TeV. This also holds for the other analyses presented in the remainder of this document. The X(3872) particle is searched for in the $B^+ \to X(3872)K^+$ decay. The event selection is similar to that presented in the previous section except that the dimuon is requested to be non-prompt. The accompanying kaon is distinguished from a pion using the (RICH based) charged particle identification system and kinematic cuts are applied to ensure that the dimuon and the kaon are coming from the same displaced vertex. The dimuon invariant mass is constrained to be in the $\psi(2S)$ or J/ψ mass range. A photon is identified in the calorimeters as an electromagnetic cluster with no charged track pointing to it. Photons from $\pi^0 \to \gamma\gamma$ decays (main background to the isolated photons) are rejected if, when combined to other photons in the event, their invariant mass is within 25 MeV of the π^0 mass.

The signal yield of the $B^+ \to X(3872)K^+$ decays followed by X(3872)→ $\psi\gamma$ ($\psi \equiv \psi(2S)$ or J/ψ) is determined from a two-dimensional unbinned maximum likelihood fit in the ($\psi\gamma K^+$, $\psi\gamma$) invariant mass plane. In the $\psi(2S)$ channel, the X(3872) signal is observed with a significance of 4.4 standard deviations. The measured yields are corrected for efficiencies ($\epsilon_{J/\psi}, \epsilon_{\psi(2S)}$). Their ratio, determined from simulated events with similar topologies, is found to be 5.25 $\pm$ 0.04. The large value of this ratio is mainly due to the differences in the photon kinematics. The dominant systematic errors are coming from the X(3872) yield determination and the photon reconstruction ($\approx$ 6-7% each). The ratio of branching fractions is measured to be 2.46 $\pm$ 0.64 (stat) $\pm$ 0.29 (syst). This value agrees with expectations for both a pure charmonium and a molecular-charmonium mixture but does not support a pure $D\overline{D}^*$ molecular interpretation of the X(3872) state.

4. Observation of the resonant character of the Z(4430)$^-$ state

The Z(4430)$^-$ state has been the topic of many debates in the past years. Its existence and its nature are still controversial. In this analysis [6], we study the $B^0 \to \psi(2S)\pi^- K^+$ decay where the $\psi(2S)$ is observed through its decay to a pair of muons. In this decay channel , the dominant background to a $Z(4430)^- \to \psi(2S)\pi^-$ decay comes from the many $K^{*0} \to K^+\pi^-$ decays through their reflections in the mass peak. This study is a four-dimensional amplitude analysis where the amplitude depends on $(m^2_{K^+\pi_-}, m^2_{\psi(2S)\pi_-}, \cos\theta_{\psi(2S)}, \phi)$ where $\theta_{\psi(2S)}$ is the $\psi(2S)$ helicity angle and ϕ the angle between the K^{*0} and $\psi(2S)$ planes in the B^0 rest frame.

The B^0 selection uses similar techniques as that described in Section 3. The amplitude model includes all known K^{*0} resonances in various spin states (J=0,1,2,3) with nominal mass at or slightly above kinematic limits. The fits with all K^* components give a significance of 18.7 σ and $M_{Z^-} = 4475 \pm 7\,\text{MeV}, \Gamma_{Z^-} = 172 \pm 13\,\text{MeV}, f_{Z^-}$(amplitude fraction) $= (5.9 \pm 0.9)\%$, compatible with the latest Belle results. The significance goes down to 13.9 σ when all systematics are included. These systematics are obtained by varying the major components of the amplitude fits: K^* models, K^* S-wave representation, K^* masses and width, etc. Added in quadrature, the systematic errors are: (+15,-25) MeV for M_{Z^-}, (+37,-34) MeV for Γ_{Z^-} and (+1.5,-3.3) % for f_{Z^-}. The spin-parity of the Z(4430)$^-$ is found to be 1^+ with a very high significance.

An additional fit is performed where the Z(4430)$^-$ amplitude is represented as a combination of independent complex amplitudes (Re $A^{Z(4430)^-}$, Im $A^{Z(4430)^-}$) at six equidistant points in the $m^2_{\psi(2S)\pi^-}$ range. The resulting Argand diagram, shown in Fig. 2, exhibits a quasi circular pattern, characteristic of a resonance.

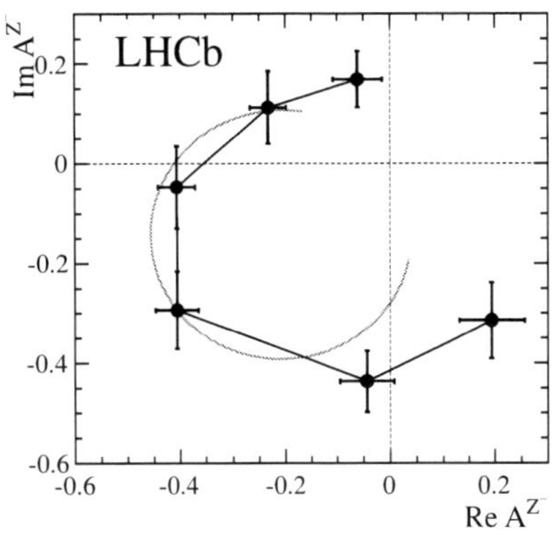

Figure 2. Fitted values of the Z^- amplitude in six $m^2_{\psi(2S)\pi^-}$ bins. The curve is the prediction from the Breit-Wigner formula with a resonance mass (width) of 4475 (172) MeV.

5. χ_b meson production

Feed-down contributions from P-wave quarkonium states to S-wave quarkonium states might be of crucial importance for the comparison between experimental results and theory. This analysis [7] studies the χ_b production in LHCb using the data sample described in Section 3 and measures the fraction of Υ(nS) originating from χ_b decays.

Prompt muon pairs are selected with each muon having a p_T larger than 1 GeV. Further cuts refine the quality and the purity of the muons. An identification purity ranging between 75% and 98% according to the muon p_T is achieved. Cuts improving the two-prong common vertex and the compatibility of this vertex with the primary vertex are applied. Three very clean peaks, $\Upsilon(1S)$, $\Upsilon(2S)$and $\Upsilon(3S)$, are observed in the $m_{\mu\mu}$ mass distribution. For each of these 3 states, the yield is obtained from an unbinned maximum likelihood fit.

Photons reconstructed using the electromagnetic calorimeter are combined to each muon pair having an invariant mass within 150 MeV of one of the Υ states to form a χ_b. A transverse energy cut of 600 MeV, information from the PS and SPD detectors and rejection of electromagnetic clusters matched to the extrapolation of a charge track are applied to improve the quality of photons. The reconstructed χ_b mass is corrected by replacing the reconstructed Υ mass by its known value. The χ_b yield is determined from an extended maximum likelihood fit. In the fit, the χ_{b0} contribution of the multiplet is neglected and the χ_{b1} and χ_{b2} contributions are set to be equal. This assumption is needed since the χ_b mass resolution does not allow to separate these two states. The assumption on the relative χ_{b1}/χ_{b2} contribution and the electromagnetic calorimeter energy scale are entering as the dominant systematic errors on the measurement of the χ_b (3P) mass. This mass is measured to be 10511.3 $\pm$ 1.7 (stat) $\pm$ 2.4 (syst) MeV.

The 'feed-down fractions' $R^{\chi_b(mP)}_{\Upsilon(nS)}$ of Υ originating from radiative χ_b decays are determined by computing the ratio of the Υ and χ_b yields in each ${p_T}^{\Upsilon}$ bin and for each decay: $\chi_b(mP) \rightarrow \Upsilon(nS)\gamma$ with m $\geq$ n. In this ratio, most of the systematic errors cancel. The remaining systematic errors are dominated by the uncertainties related to the photon reconstruction at low p_T and the various assumptions made in the χ_b mass fits: χ_{b1}/χ_{b2} ratio, p_T dependence on the mass fits and modelling of the mass resolution. The results of the feed-down fractions, for each Υ state as a function of their p_T are represented in the three plots of Figure 3. Contributions to the Υ(nS) production from radiative decays of χ_b(mP) states are far from negligible.

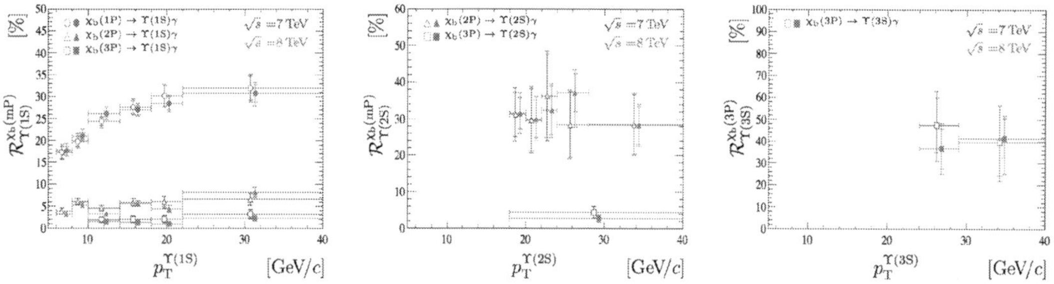

Figure 3. Feed-down fractions $R^{\chi_b(mP)}_{\Upsilon(nS)}$ as a function of ${p_T}^{\Upsilon}$. Open (solid) symbols correspond to data collected at $\sqrt{s}$=7(8) TeV respectively. Inner error bars represent statistical uncertainties while outer error bars represent the sum in quadrature of statistical and systematic uncertainties.

6. Conclusions

Four analyses in the sector of quarkonium production and exotic states, all based on the excellent reconstruction of muon pairs in LHCb, have been presented. These results bring improved knowledge and/or increased precision on the various topics considered. Given the sizes of the statistical errors, at the level of the systematic errors in most cases, new data expected soon with the start of Run2 will be most welcome.

References

[1] H. Wöhri presentation, LHCP 2013 Barcelona, https://indico.cern.ch/event/210555/
[2] P. Faccioli, The return of Quarkonia, Cern Courier, July 2013.
[3] Brambilla N. *et al* , Eur. Phys. J. C (2011)71:1534.
[4] Aaij R. *et al* , Eur. Phys. J. C (2014)74:2872
[5] Aaij R. *et al* , arXiv:1404.0275
[6] Aaij R. *et al* , Phys. Rev. Lett. 112, 222002 (2014)
[7] Aaij R. *et al* , arXiv:1407.7734

First Measurements of Form Factors of Pion, Kaon, Proton, and Hyperons for the Highest Timelike Momentum Transfers

Kamal K. Seth
Northwestern University, Evanston, IL, 60208, USA
E-mail: kseth@northwestern.edu

Abstract. Results of recent precision measurements of the electromagnetic form factors of the pion, kaon, proton, and the hyperons, Λ^0, Σ^0, Σ^+, Ξ^-, Ξ^0, Ω^-, at large timelike momentum transfers are presented. Evidence is found for diquark correlations in Λ^0, Σ^0 hyperons.

With the discovery of quarks hadrons were recognized as composite. It is generally agreed that electromagnetic form factors at large momentum transfer provide some of the best insight into the structure of the composite hadron. Four-momentum transfers defined as

$$Q^2(4\text{ mom}) = q^2(3\text{ mom})_{\text{space}} - (\text{energy})_{\text{time}}$$

can be *positive and spacelike*, or *negative and timelike*. I am going to talk about form factors for timelike momentum transfers as measured via the reactions $e^+e^- \to$ hadron − antihadron.

In 1960, in anticipation of electron-positron colliders, Cabibbo and Gatto wrote two classic papers [1] pointing out that these colliders would provide the unique opportunity to measure timelike form factors of any hadrons, mesons and baryons. Only 50 years later, we are now realizing the full promise of the vision of Cabibbo and Gatto in the measurements I am reporting here.

Before 2000 almost no experimental data with any precision existed for any hadron for timelike form factors for $|Q^2| > 5$ GeV^2. Recently, we made the first measurements of the form factors of pions, kaons, and protons with high precision for the large momentum transfers of $|Q^2| = 14.2$ and 17.4 GeV^2 [2].

The important results of these measurements are presented in Table 1 and Fig. 1 are:

(i) There is a remarkable agreement of the form factors with the dimensional counting rule prediction of QCD, that $|Q^2|F_{\pi,K}$ and $|Q^4|F_p$ are nearly constant, varying with $|Q^2|$ only weakly as $\alpha_S(|Q^2|)$ and $\alpha_S^2(|Q^2|)$. There is, however, an unexpected departure for protons at $|Q^2| = 14.2$ GeV^2.

(ii) The theoretical predictions for the magnitudes of $F_\pi(|Q^2|)$ and $F_p(|Q^2|)$ fail by factors, ≥ 2.

(iii) While pQCD predicts that $F_\pi/F_K = (f_\pi/f_K)2 = 0.67 \pm 0.01$, we find:

$$F_\pi/F_K = 1.21 \pm 0.03,\ \text{at } |Q^2| = 14.2\text{GeV}^2, \quad F_\pi/F_K = 1.09 \pm 0.04,\ \text{at } |Q^2| = 17.4\text{GeV}^2.$$

It has been suggested that this dramatic disagreement may be due to the kaon wave function being different from that of the pion due to $SU(3)$ breaking, the strange quark in the kaon having a 27 times larger mass than the up/down quarks in the pion. However, recent lattice calculations do not support this conjecture.

Table 1. Results for pion, kaon and proton timelike form factors.

Q^2 (GeV^2)	$Q^2F_\pi(\lvert Q^2\rvert)$ (GeV^2)	$Q^2F_K(\lvert Q^2\rvert)$ (GeV^2)	$Q^4F_p(\lvert Q^2\rvert)/\mu_p$ (GeV^4)
14.2	0.92 ± 0.04	0.76 ± 0.02	0.64(3)
17.4	0.84 ± 0.03	0.77 ± 0.03	0.82(5)

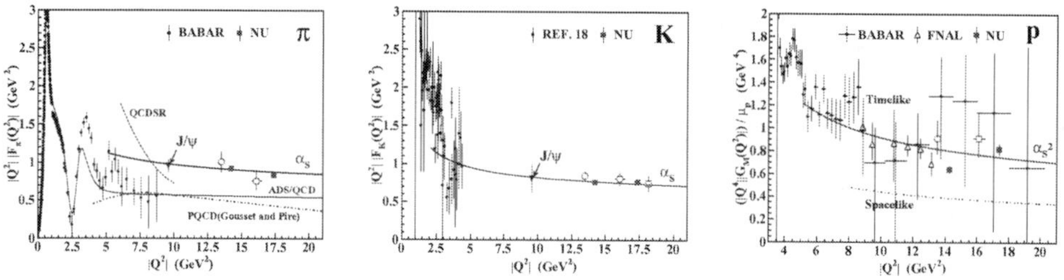

Figure 1. Results for pion, kaon and proton timelike form factors.

In the context of the quark-gluon structure of hadrons, it is particularly interesting to measure form factors of hyperons which may be expected to reveal the effects of $SU(3)$ breaking as successively one, two, and three of the up, down quarks in the nucleon are replaced by strange quarks in (Λ, Σ), Ξ, and Ω, respectively. The interest is further enhanced at large momentum transfers at which deeper insight is obtained into possible short-range correlations between the quarks. As Wilczek [4] has pointed out, "several of the most profound aspects of low-energy QCD dynamics are connected to diquark correlations," and the differences in quark-quark configurations between different hyperons make them an ideal laboratory to study diquark correlations.

The $e^+e^- \to$ hyperon − antihyperon cross sections were expected to be very small, and no experimental measurements were reported for 47 years after Cabibbo and Gatto's papers. In 2007, BaBar reported [3] form factor measurements for Λ and Σ^0 using the ISR method, but their statistically significant results were limited to $|Q^2| \lesssim 6$ GeV^2. Now we have measured form factors of all hyperons, Λ^0, Σ^0, Σ^+, Ξ^-, Ξ^0, Ω^- for the first time with good precision at the large momentum transfer of $|Q^2| = 14.2$ GeV^2.

The only existing theoretical study of hyperon form factors at large timelike momentum transfers is due to Körner and Kuroda [5]. These predictions were not constrained by any experimental measurements, and they turn out to be factors 10 to 80 larger than what we measure.

We use data taken with the CLEO-c detector at $\psi(3770)$, $\sqrt{s} = 3.77$ GeV, with the integrated luminosity $\mathcal{L} = 805$ pb^{-1}. Data taken at $\psi(3770)$ can only be used to determine hyperon form factors if it can be shown that the strong interaction yield of the hyperons at the $\psi(3770)$ resonance is negligible. We estimate it by using the pQCD prediction that the ratios of the branching fractions for the decay of any two vector resonances of charmonium to leptons (via virtual photons) and hadrons (via gluons) are identical. This relation leads to the estimates that the expected number of events for resonance decays of $\psi(3770)$ in our measurements are 1.3 p, 0.9 Λ^0, 0.2 Σ^+, Σ^0, Ξ^-, 0.05 Ξ^0, and 0.03 Ω^-. In other words, the contributions of strong decays are negligibly small in all decays, and the observed events arise from form factor decays.

We also use CLEO-c data taken at $\psi(2S)$, $\sqrt{s} = 3.686$ GeV, with luminosity $\mathcal{L} = 48$ pb^{-1}, which corresponds to $N(\psi(2S)) = 24.5 \times 10^6$, to measure the branching fractions for the decays $\psi(2S) \to B\overline{B}$. The large yield from resonance production of $B\overline{B}$ pairs from $\psi(2S)$ enables us to test the quality of our event selection criteria, and to determine contributions to systematic uncertainties.

We reconstruct the hyperons in their following major decay modes (with branching fractions [7] listed in parentheses): $\Lambda^0 \to p\pi^-$ (63.9%), $\Sigma^+ \to p\pi^0$ (51.6%), $\Sigma^0 \to \Lambda^0\gamma$ (100%), $\Xi^- \to \Lambda^0\pi^-$ (99.9%), $\Xi^0 \to \Lambda^0\pi^0$ (99.5%), $\Omega^- \to \Lambda^0 K^-$ (67.8%). We find that reconstructing back-to-back hyperons and anti-hyperons whose decay vertices are separated from the interaction point results in essentially background free spectra. We first identify single hyperons, and then construct hyperon-antihyperon pairs.

To determine the reconstruction efficiency of the above event selections, we generate Monte Carlo events using a GEANT-based detector simulation and the expected angular distribution.

The single hyperon mass spectra for $\psi(2S)$ decays are shown in Fig. 2 (left), for hyperons with $E(B)/E_{\text{beam}} > 0.95$. Clear peaks are seen for the reconstructed hyperons.

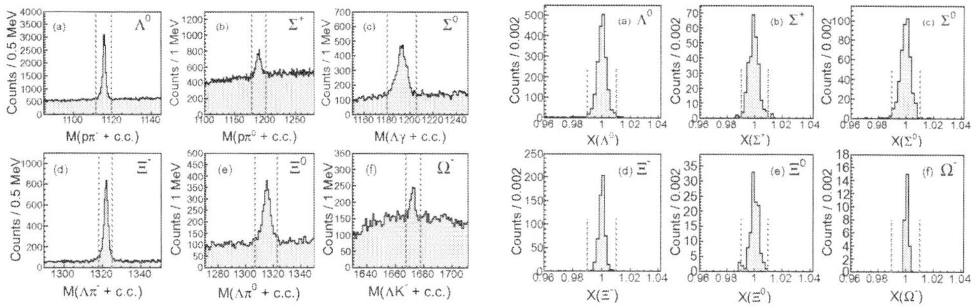

Figure 2. (left) Invariant masses of reconstructed final states in $\psi(2S)$ data. Single hyperons are accepted in the regions defined by vertical dashed lines. (right) Distributions of baryon-antibaryon events as function of, $X(B) \equiv (E(B) + E(\overline{B}))/\sqrt{s}$, in $\psi(2S)$ data. The vertical lines indicate the signal region $X = 0.99 - 1.01$.

The second step of analysis consists of constructing baryon antibaryon pairs. The distributions of these $B\overline{B}$ pairs is shown in Fig. 2 (right) for $\psi(2S)$ decays as a function of $X(B) \equiv [E(B) + E(\overline{B})]/\sqrt{s}$, which should peak at $X(B) = 1$. Clear peaks are seen for all decays with essentially no background. We define the signal region as $X(B) = 0.99 - 1.01$, with numbers of events in it as N_{data}. We calculate the radiative correction, $(1+\delta)$, using the method of Bonneau and Martin [8]. We obtain $(1 + \delta) = 0.77$ within 1% for all baryons for both $\psi(2S)$ and $\psi(3770)$. The Born cross sections are calculated as $\sigma_B = N_{\text{data}}/\epsilon_B\, \mathcal{L}(\psi(2S))\,(1 + \delta)$, and the branching fractions as $\mathcal{B}(\psi(2S) \to B\overline{B}) = N_{\text{data}}/\epsilon_B N(\psi(2S))$. The results are summarized in Table 2, including those for $\psi(2S) \to p\bar{p}$. The first uncertainties in σ_B and $\mathcal{B}$ are statistical, and the second uncertainties are estimates of systematic uncertainties. Our results for $\psi(2S)$ branching fractions are in agreement with the PDG averages [7] and previous small luminosity CLEO results [9], and have generally smaller errors.

We apply the same event selections to the $\psi(3770)$ decays as we do for $\psi(2S)$ decays. The $X(B)$ distributions for $\psi(3770)$ form factors decays are shown in Fig. 3 (left). Clear peaks are seen for each decay mode. The number of events, N_{ff}, in the region $X(B) = 0.99 - 1.01$, are used to calculate the cross sections as, $\sigma_0(e^+e^- \to B\overline{B}) = N_{\text{ff}}/(1 + \delta)\epsilon_B\, \mathcal{L}(3770)$, where ϵ_B are the MC-determined efficiencies, and $\mathcal{L}(3770) = 805$ pb^{-1} is the luminosity.

Table 2. Cross section and branching fraction results for $\psi(2S) \to B\overline{B}$.

$\mathcal{B}$	N_{data}	ϵ_B (%)	σ_B (pb)	$\mathcal{B} \times 10^4$
p	4475(78)	63.1	196(3)(12)	3.08(5)(18)
Λ^0	1901(44)	20.7	247(6)(15)	3.75(9)(23)
Σ^0	439(21)	7.96	148(7)(11)	2.25(11)(16)
Σ^+	281(17)	4.54	165(10)(11)	2.51(15)(16)
Ξ^-	548(23)	8.37	176(8)(13)	2.66(12)(20)
Ξ^0	112(11)	2.26	135(13)(10)	2.02(19)(15)
Ω^-	27(5)	2.32	31(6)(3)	0.47(9)(5)

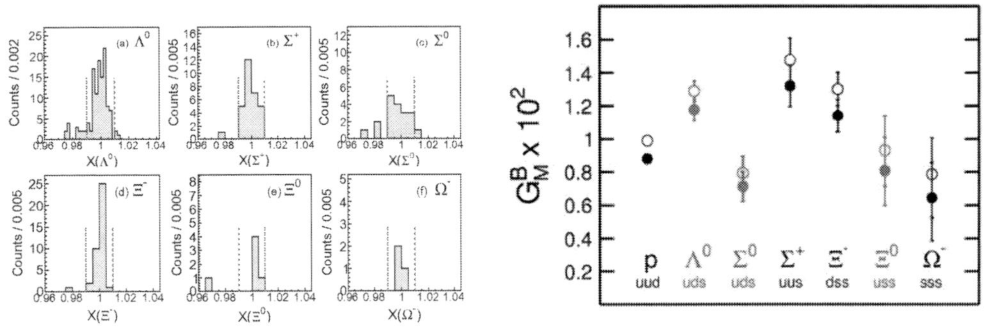

Figure 3. (left) Distributions of baryon-antibaryon scaled energy, $X(B) \equiv (E(B)+E(\overline{B}))/\sqrt{s}$, in $\sqrt{s} = 3770$ MeV data. The vertical lines indicate the signal region $X = 0.99 - 1.01$. (right) Magnetic form factors $|G_M^B| \times 10^2$ for protons and hyperons. The closed circles correspond to the assumption $G_M^B = G_E^B$, and the open circles to the assumption $G_E^B = 0$.

For the spin–1/2 baryons, the proton and the hyperons Λ, Σ, and Ξ, the well known relation between the cross sections and the magnetic form factor $G_M^B(s)$, and the electric form factor $G_E^B(s)$ is

$$\sigma_0^B = \left(\frac{4\pi\alpha^2\beta_B}{3s}\right)\left[|G_M^B(s)|^2 + (2m_B^2/s)|G_E^B(s)|^2\right] \tag{1}$$

where α is the fine structure constant, β_B is the velocity of the baryon in the center-of-mass system, and m_B is its mass. The statistics of the present measurements do not allow us to determine $|G_M^B|$ and $|G_E^B|$ separately. We therefore evaluate $|G_M^B(s)|$ under two commonly used extreme assumptions, $|G_E^B(s)|/|G_M^B(s)| = 0$, and 1. The results for $|G_E| = |G_M|$ are shown in Table 3. The efficiencies for the G_M and G_E components are determined assuming $1 + \cos^2\theta$ and $\sin^2\theta$ angular distributions, respectively. As shown in Fig. 3 (right), the values of $G_M^B(s)$ derived with the assumption $G_E^B = 0$ are found to all be nearly $\sim 12(2)\%$ larger than those for $G_E^B(s) = G_M^B(s)$.

The total systematic uncertainties from various sources are estimated to be 6.1% for Λ^0, 7.3% for Σ^0, 6.4% for Σ^+, 7.5% for Ξ^-, 7.3% for Ξ^0, and 10.2% for Ω^-.

Since no modern theoretical predictions for timelike form factors of hyperons at large momentum transfers exist, we can only discuss our experimental results qualitatively. Following are the main observations:

(a) The form factor cross sections in Table 3 are 150 to 500 times smaller than the resonance

Table 3. Results for proton and hyperon form factors at $|Q^2| = 14.2$ GeV^2, assuming $|G_E| = |G_M|$.

B	N_{ff}	ϵ_B, %	σ_0^B, pb	$\lvert G_M^B\rvert \times 10^2$	$\lvert G_M^B/\mu_B\rvert \times 10^2$
p	215(15)	71.3	0.46(3)(3)	0.88(3)(2)	0.31(1)(1)
Λ^0	105(10)	21.1	0.80(8)(5)	1.18(6)(4)	1.93(9)(6)
Σ^0	15(4)	8.36	0.29(7)(2)	0.71(9)(3)	0.91(11)(3)
Σ^+	29(5)	4.68	0.99(18)(6)	1.32(13)(4)	0.54(5)(2)
Ξ^-	38(6)	8.69	0.71(11)(5)	1.14(9)(4)	1.75(14)(7)
Ξ^0	$5^{+2.8}_{-2.3}$	2.30	$0.35^{+0.20}_{-0.16}(3)$	0.81(21)(3)	0.65(17)(2)
Ω^-	$3^{+2.3}_{-1.9}$	2.94	$0.16^{+0.13}_{-0.10}(2)$	$0.64^{+0.21}_{-0.25}(3)$	$0.32^{+0.11}_{-0.13}(2)$

cross sections in Table 2.

(b) As illustrated in Fig. 3 (right), the measured values of $|G_M^B|$ vary rather smoothly by approximately a factor two, except for $G_M(\Sigma^0)$.

(c) As shown in Table 3, there is no evidence for the proportionality of $G_M^B(s)$ to μ_B for hyperons.

The most significant result of the present measurements is that $G_M(\Lambda^0)$ is a factor 1.66(24) larger than $G_M(\Sigma^0)$, although Λ^0 and Σ^0 have the same *uds* quark content. We note that Σ^0 and Λ^0 differ in their isospin, with $I(\Sigma^0) = 1$, and $I(\Lambda^0) = 0$. Since only up and down quarks carry isospin, this implies that the pair of up/down quarks in Λ^0 and Σ^0 have different isospin configurations. This forces different spin configurations in Λ^0 and Σ^0. In Λ^0 the *ud* quarks have antiparallel spins coupled to $S = 0$, whereas in Σ^0 they couple to $S = 1$. The spatial overlap in the $S = 0$ configuration in Λ^0 is stronger than in the $S = 1$ configuration in Σ^0, and our measurement at large $|Q^2|$ is particularly sensitive to it.

Recently, Wilczek and colleagues [4, 10, 11] have emphasized the importance of diquark correlations in low-energy QCD dynamics, and have pointed out that for the non-strange quarks the favorable diquark configuration with attraction is the spin-isospin singlet, making what Wilczek calls a "good" diquark in Λ^0 as opposed to the repulsive spin-isospin triplet configuration in Σ^0. This results in a significantly larger cross section for the formation of Λ^0 than Σ^0, as anticipated by Selem and Wilczek [11]. We measure $\sigma(\Lambda^0)/\sigma(\Sigma^0) \approx 3$, and this results in the factor 1.66 larger form factor for Λ^0 than Σ^0. We believe that our observation of the large difference in the form factors of Λ^0 and Σ^0 is indeed due to significant "good" diquark correlation in Λ^0, and it constitutes an important example of significant diquark correlations in baryons.

References

[1] N. Cabibbo and R. Gatto, Phys. Rev. Lett. **313**, 4 (1960); Phys. Rev. **124**, 1577 (1961).
[2] Kamal K. Seth *et al.*, Phys. Rev. Lett. **110**, 022002 (2013).
[3] B. Aubert et al. [BaBar Collaboration], Phys. Rev. D **76**, 092006 (2007).
[4] F. Wilczek, "Diquarks as Inspiration and as Objects", in "From Fields to Strings", ed. M. Shifman et al., World Scientific (Singapore), 2005, vol. 1, p. 77 – 93 [arXiv:hep-ph/0409168].
[5] J. G. Körner and M. Kuroda, Phys. Rev. D **16**, 2165 (1977).
[6] B. Aubert *et al.* [BaBar Collaboration], Phys. Rev. D **76**, 092006 (2007).
[7] J. Beringer *et al.* [Particle Data Group], Phys. Rev. D **86**, 010001 (2012) and 2013 partial update for the 2014 edition.
[8] G. Bonneau and F. Martin, Nucl. Phys. B **27**, 381 (1971).
[9] T. K. Pedlar *et al.* [CLEO Collaboration], Phys. Rev. D **72**, 051108 (2005).
[10] R. Jaffe and F. Wilczek, Phys. Rev. Lett. **91**, 232003 (2003).
[11] A. Selem and F. Wilczek, Proc. Ringberg Workshop on "New Trends in HERA Physics", ed. G. Grindhammer et al., World Scientific (Singapore), 2006, p. 337 – 356; [arXiv:hep-ph/0602128].

Strangeness Production at LHC energies as measured with ALICE

D.D. Chinellato[1] **for the ALICE Collaboration**

[1] Universidade Estadual de Campinas (UNICAMP)

E-mail: `daviddc@ifi.unicamp.br`

Abstract. The main objective of the ALICE experiment is to study the properties and the evolution of the high-energy density system created in relativistic heavy-ion collisions at the LHC. For that purpose, ALICE has unique particle identification capabilities that enable the detection of weakly-decaying particles in the high-multiplicity environment created in lead-lead (Pb-Pb) collisions.

In this work, we discuss the systematic measurements of strange and multi-strange hadrons performed by ALICE for proton-proton (pp), proton-lead (p-Pb) and Pb-Pb collisions. Production rates measured for p-Pb and Pb-Pb are compared to predictions from statistical hadronization models and to measurements performed in pp collisions. These data are also used for a brief discussion on strangeness enhancement.

1. Introduction

The study of strange and multi-strange particle production is an important tool not only for the understanding of particle production mechanisms in proton-proton collisions but also for investigating the dynamics of the quark-gluon plasma (QGP) phase of nuclear collisions. The measurement of strangeness production rates relates to many signatures of this phase, such as strangeness enhancement and the thermalization of particle production rates. In these proceedings, we will briefly discuss the measurement techniques utilized to obtain the spectra of the weakly-decaying strange particles, K^0_S, Λ, $\bar{\Lambda}$, Ξ^-, $\bar{\Xi}^+$, Ω^- and $\bar{\Omega}^+$ throughout a variety of colliding systems, including pp, p-Pb and Pb-Pb. We will then discuss the implications of these measurements for the QGP properties.

2. Strange and multi-strange particle analysis

The strange and multi-strange hadrons K^0_S, Λ, $\bar{\Lambda}$, Ξ^-, $\bar{\Xi}^+$, Ω^- and $\bar{\Omega}^+$ are reconstructed via their characteristic weak decay topologies. In this process, tracks measured by the ALICE Time Projection Chamber (TPC) and the silicon Inner Tracking System (ITS) [1] are combined into candidates if they satisfy a set of quality and geometrical criteria (see [2, 3] for details). Particle identification based on the energy loss while traversing the TPC gas is also utilized to select particles of the expected decay daughter species in each case. Signal extraction is then performed in invariant mass space to isolate the desired particle yields, and efficiency corrections are calculated by making use of Monte Carlo simulations with full propagation of particles through the ALICE geometry with GEANT3 [4, 5].

The full analysis details as well as the resulting corrected transverse momentum spectra for weakly-decaying strange and multi-strange hadrons have been reported elsewhere [2, 3]. This discussion will focus on a few specific aspects related to the study of the QGP.

3. Strangeness and thermal models

It has been suggested that hadrons are produced in approximate chemical and thermal equilibrium in high-energy interactions. In nucleus-nucleus collisions, a grand-canonical statistical model is usually employed to describe the observed yields of each species. The successful description of strangeness production rates may require the addition of an empirical undersaturation factor γ_s, but recent studies have demonstrated that this factor is unnecessary (i.e. $\gamma_s \to 1$) at LHC energies [6]. The K^0_S yields measured in central Pb-Pb collisions are consistent with thermally equilibrated particle production rates, as can be seen in Fig. 1, while the strange and multi-strange baryons are slightly underpredicted by no more than 2σ by the models [7, 8, 9]. It has been suggested, based on lattice and Hadron Resonance Gas (HRG) calculations, that the transition temperature may exhibit flavour hierarchy which would explain these deviations from thermal expectations [10]. Testing for such a scenario is in principle possible using a net strangeness measurement; however, there exist a number of difficulties in theory, such as limitations in the HRG models, as well as from experiment, such as detector acceptance limitations.

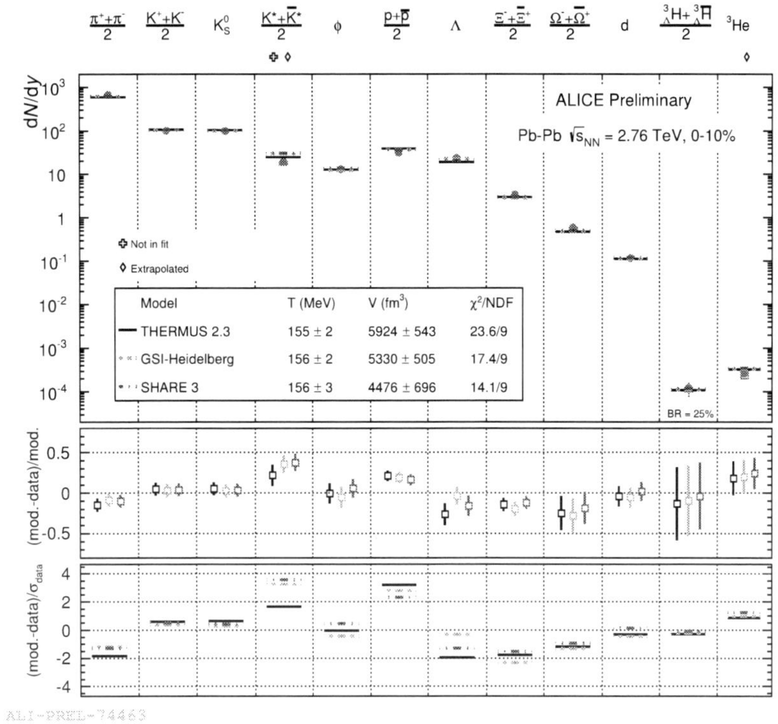

Figure 1. Grand canonical thermal fit to ALICE central Pb-Pb (0-10%) particle production rates using the GSI-Heidelberg model [6], THERMUS 2.3 [8], and the SHARE 3 model [9].

4. Strangeness enhancement

One of the proposed signatures of a deconfined partonic phase in the system evolution is an enhancement of strange quark production rates in a nuclear collision with respect to a hadron gas scenario [11]. Such an increase in strangeness production has been observed by STAR in gold-gold (Au-Au) [12] and also more recently by ALICE in Pb-Pb collisions. The strangeness enhancement factor is usually quantified by comparing the production rates of strange hadrons per participant nucleon as observed in nuclear collisions to the production rates per participant nucleon observed in pp at the same energies. These ratios are shown for Ξ and Ω in Fig. 2, where one observes an enhancement of strangeness production by up to a factor of approximately 3 for Ξ^- and 6 for $\Omega^- + \bar{\Omega}^+$. These factors are smaller than what was observed at lower energies, which is mostly attributed to the decrease of canonical suppression in the pp reference and the ensuing rise of Ξ and Ω production in this more elementary system.

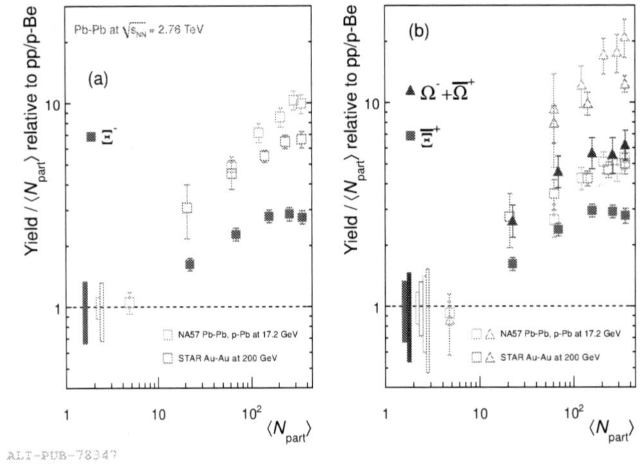

Figure 2. Ξ and Ω production rates per participant nucleon in nuclear collisions relative to the observed value in pp collisions. Left panel: enhancement factors for Ξ^-. Right panel: enhancement factors for $\bar{\Xi}^+$ and $\Omega^- + \bar{\Omega}^+$. In both panels, the enhancement factors are compared to values from nuclear collisions from NA57 [13, 14] and STAR [12]. Figure from [3].

However, recent measurements have shown that the number of participant nucleons, N_{part}, may not be an appropriate scaling variable to isolate the enhancement component that is due to strangeness content, since the production rates of charged particles do not scale linearly with N_{part} [15]. In order to factor out the overall increase of particle production and isolate the strangeness-related increase, one can compute the ratio of $(\Xi^- + \bar{\Xi}^+)$ and $(\Omega^- + \bar{\Omega}^+)$ production rates to $(\pi^- + \pi^+)$ yields. This has been done as a function of $dN_{ch}/d\eta$ not only for Pb-Pb but also for p-Pb collisions in ALICE, as shown in Fig. 3 and Fig. 4 for Ξ and Ω, respectively. The multi-strange baryon production relative to pions is shown to increase by up to a factor of about 2-3 going from pp collisions to Pb-Pb collisions, and seems to reach saturation for central Pb-Pb collisions. These saturation values compare favourably to predictions from statistical hadronization models using a chemical freezeout temperature of approximately 155 MeV [6, 8]. In this context, the p-Pb data show a monotonically increasing multi-strange particle production with respect to pions, with Ξ/π ratios that slightly exceed the saturation limit observed for Pb-Pb, while the Ω/π ratio is not higher than the one observed in peripheral Pb-Pb. This indicates that any thermal description of observed multi-strange yields will thus be worse in p-Pb than in Pb-Pb.

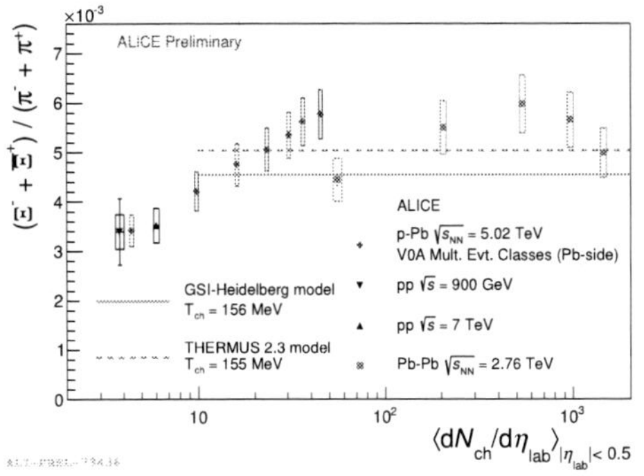

Figure 3. $(\Xi^- + \bar{\Xi}^+)/(\pi^- + \pi^+)$ ratio as a function of multiplicity for pp, p-Pb and Pb-Pb collisions shown together with predictions from statistical hadronization models.

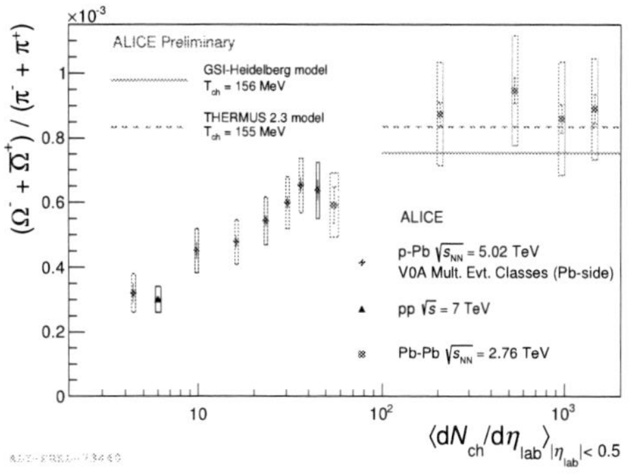

Figure 4. $(\Omega^- + \bar{\Omega}^+)/(\pi^- + \pi^+)$ ratio as a function of multiplicity for pp, p-Pb and Pb-Pb collisions shown together with predictions from statistical hadronization models.

5. Conclusions

ALICE has performed a comprehensive set of strange hadron measurements in which several quantities relevant to the onset of deconfinement, such as strangeness enhancement and statistical hadronization, have been studied in Pb-Pb. These measurements were also performed systematically in p-Pb collisions as a function of charged-particle multiplicity. Many qualitative observations traditionally related to the presence of a QGP phase, such as the monotonic increase of relative multi-strange baryon production according to multiplicity, have also been made in these smaller systems. These measurements represent an important milestone in advancing the understanding of hadrochemistry and nuclear collisions in general, and are expected to be further complemented by higher energy measurements starting in 2015 as well as by similar studies in pp collisions as a function of multiplicity.

6. Acknowledgements

We wish to thank Fundação de Amparo à Pesquisa do Estado de São Paulo, FAPESP, Brazil, for the support provided under grant number 2013/23144-8.

References

[1] Carminati F *et al.* (ALICE Collaboration) 2004 *Journal of Physics G: Nuclear and Particle Physics* **30** 1517
[2] Abelev B *et al.* (ALICE Collaboration) 2013 *Phys. Rev. Lett.* **111**(22) 222301
[3] Abelev B *et al.* (ALICE Collaboration) 2014 *Physics Letters B* **728** 216 – 227 ISSN 0370-2693
[4] Brun R *et al.* 1985 *CERN Data Handling Division DD/EE/841*
[5] Brun R *et al.* 1994 *CERN Program Library LongWrite- up, W5013*
[6] Andronic A, Braun-Munzinger P and Stachel J 2009 *Physics Letters B* **673** 142 – 145 ISSN 0370-2693
[7] Braun-Munzinger P, Redlich K and Stachel J 2003 *arXiv:nucl-th/0304013*
[8] Wheaton S, Cleymans J and Hauer M 2009 *Computer Physics Communications* **180** 84 – 106 ISSN 0010-4655
[9] Petran M, Letessier J, Petracek V and Rafelski J 2013 *Phys. Rev. C* **88**(3) 034907
[10] Bellwied R, Borsanyi S, Fodor Z, Katz S and Ratti C 2013 *Phys. Rev. Lett.* **111**(3) 202302
[11] Rafelski J and Müller B 1982 *Phys. Rev. Lett.* **48** 1066–1069
[12] Abelev B *et al.* (STAR Collaboration) 2009 *Phys. Rev. C* **79**(3) 034909
[13] Antinori F *et al.* (NA57 Collaboration) 2006 *J. Phys. G* **32** 427
[14] Antinori F *et al.* (NA57 Collaboration) 2010 *J. Phys. G* **37** 045105
[15] Aamodt K *et al.* (ALICE Collaboration) 2011 *Phys. Rev. Lett.* **106**(3) 032301

XI International Conference on Hyperons, Charm and Beauty Hadrons (BEACH 2014) IOP Publishing
Journal of Physics: Conference Series **556** (2014) 012011 doi:10.1088/1742-6596/556/1/012011

Production of quarkonium states at the ATLAS experiment

V. Kartvelishvili (on behalf of the ATLAS collaboration)

Department of Physics, Lancaster University, LA1 4YB, United Kingdom

E-mail: v.kartvelishvili@lancaster.ac.uk

Abstract. An overview is given of recent measurements of the production of various heavy quarkonium states in pp collisions at $\sqrt{s} = 7$ TeV, using the ATLAS detector at the LHC. These include the production cross sections of the C-even states χ_{c1} and χ_{c2} in their radiative decay modes into $J/\psi\gamma$, and of the radially excited vector state $\psi(2S)$ in its decay mode into $J/\psi\pi^+\pi^-$. Differential spectra in transverse momentum are measured separately for promptly produced charmonium states, originating from QCD sources, and for non-prompt states from the decays of B-hadrons. Results are compared to other existing measurements and various theoretical models.

1. Introduction

Almost 40 years after the discovery of the J/ψ meson, the investigation of hidden heavy flavour production processes in hadronic collisions still presents significant challenges to both theory and experiment. Large amounts of LHC data have now become available, thus opening a new energy frontier and a new chapter in such studies.

This note is an overview of recent results on quarkonium production from the ATLAS collaboration [1] at the LHC, based on the data collected at the c.m. energy of 7 TeV.

2. $\psi(2S)$ production

The production of the $\psi(2S)$ meson was studied in its decay mode into $J/\psi\pi^+\pi^-$ [2]. The J/ψ candidates, triggered by a low-p_T dimuon trigger, were combined offline with two oppositely charged tracks to form $\psi(2S)$ candidates. A kinematic fit constraining the two muon tracks to the J/ψ mass and all four tracks to a common vertex was used to improve resolution, both for the invariant mass of the candidates and the (pseudo-) proper decay time of the vertex. The $\psi(2S)$ candidates were divided into 30 analysis bins (3 in rapidity and 10 in transverse momentum). Candidates were weighted using acceptance and efficiency maps, derived using a combination of Monte Carlo and data-driven methods, and a 2D mass-lifetime unbinned maximum-likelihood fit was performed to extract the production cross section in each of the analysis bins. Fig. 1 shows mass and lifetime projections for a sample analysis bin, illustrating the mass resolution, and the separation of the $\psi(2S)$ signal into prompt and non-prompt components.

The measured fraction of non-prompt $\psi(2S)$ mesons (produced from long-lived sources such as b-hadron decays) is shown in Fig. 2. The fraction increases with increasing p_T, reaching about 70% at 100 GeV across the whole rapidity range studied here.

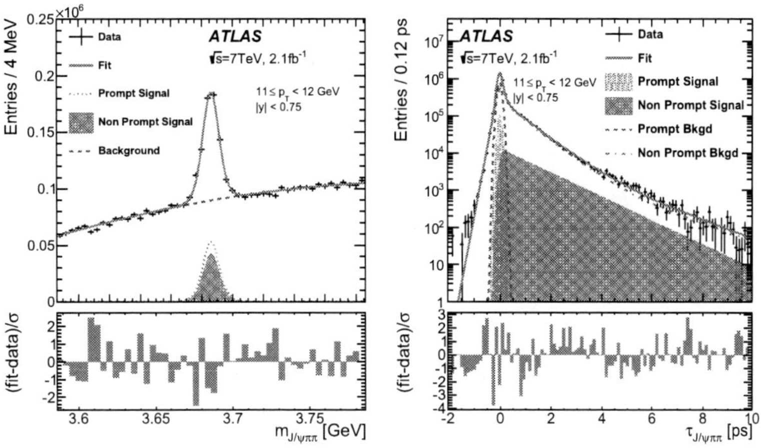

Figure 1. Mass and lifetime projections for the $\psi(2S)$ candidates from a sample analysis bin.

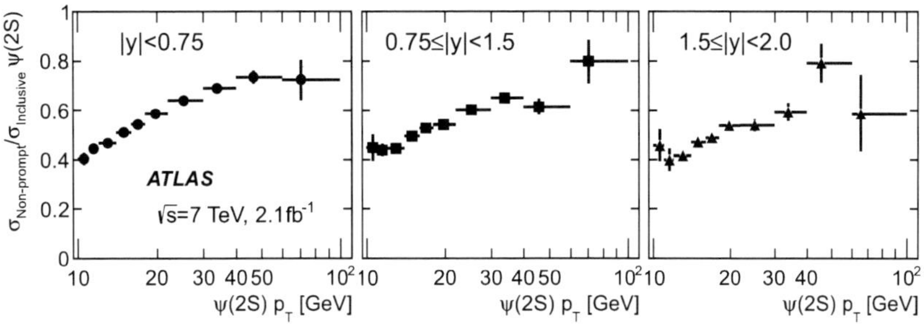

Figure 2. The measured fraction of non-prompt $\psi(2S)$ mesons.

The measured differential cross sections are shown in Fig. 3, separately for prompt and non-prompt $\psi(2S)$ mesons. Comparison with various theoretical models shows that NLO NRQCD and FONLL are doing a good job in describing data; more details can be found in [2].

3. χ_c production

The production the of C-even charmonium states χ_{c1} and χ_{c2} was studied in their radiative decay mode into $J/\psi\gamma$ [3]. As in the $\psi(2S)$ case, the J/ψ candidates were triggered by a low-p_T dimuon trigger and were combined offline with photon candidates. In order to achieve the best mass resolution, the analysis was confined to the central rapidity region, using the photons converted into e^+e^- pairs. The left plot in Fig. 4 shows the invariant mass distribution of all selected $\mu^+\mu^-\gamma$ candidates, where separate peaks from various χ_c states are clearly visible, while the right plot presents the lifetime distribution, used to separate the χ_c signal into prompt and non-prompt contributions.

The $\mu^+\mu^-\gamma$ candidates were weighted using acceptance and efficiency maps derived using a combination of Monte Carlo and data-driven methods, and a 2D mass-lifetime unbinned maximum-likelihood fit was performed to extract the production cross section in each of the five bins in transverse momentum, separately for prompt and non-prompt production.

The differential cross sections of promptly produced χ_{c1} and χ_{c2} states are shown in Fig. 5,

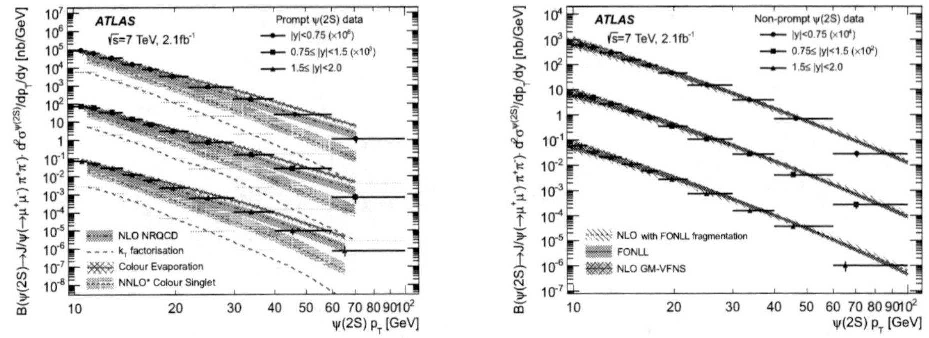

Figure 3. Measured differential cross-sections for prompt (left) and non-prompt (right) $\psi(2S)$ production, in comparison with theoretical predictions (see [2] for more details).

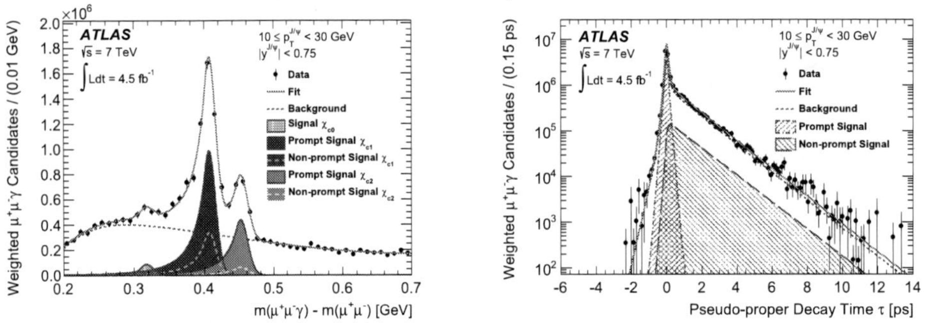

Figure 4. Mass (left) and lifetime (right) distributions of all selected $\mu^+\mu^-\gamma$ candidates.

while non-prompt production cross sections are presented in Fig. 6. Both are compared with

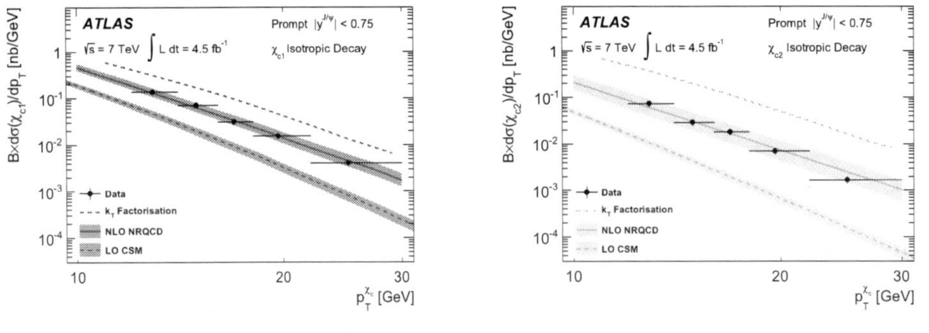

Figure 5. Differential cross sections for prompt production of χ_{c1} (left) and χ_{c2} (right).

various theoretical models (see [3] for details).

While trying to understand the dynamics of charmonium production, it is important to know what fraction of prompt J/ψ is produced from χ_c feed-down. These results, shown in Fig. 7, are in good agreement with predictions of NLO NRQCD.

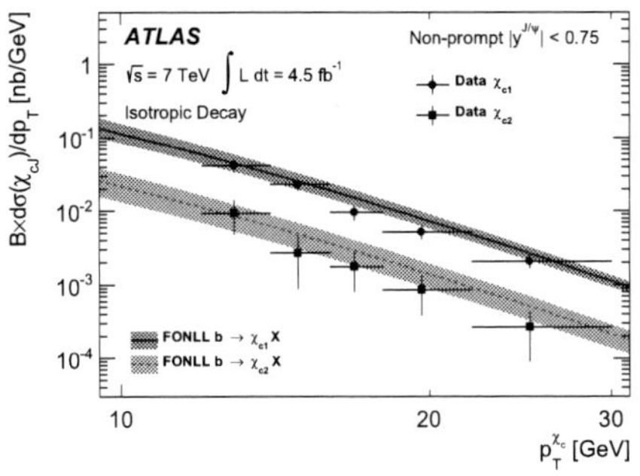

Figure 6. Differential cross sections for non-prompt production of χ_{c1} and χ_{c2}.

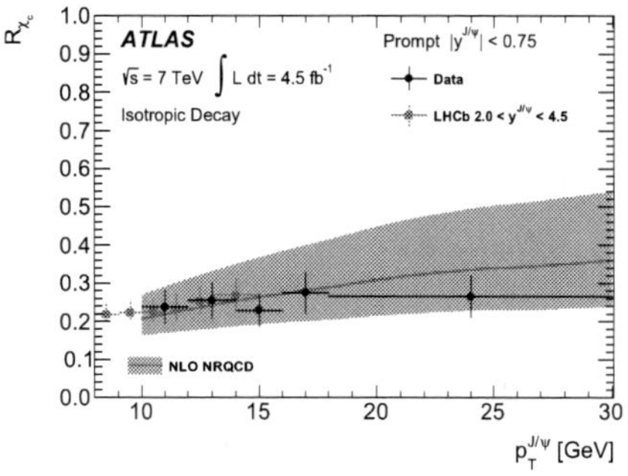

Figure 7. Fraction of J/ψ mesons produced from χ_c feed-down.

With the same data selection, ATLAS also measured the branching ratio of the decay $B^{\pm} \to \chi_{c1} K^{\pm}$, using the decay $B^{\pm} \to J/\psi K^{\pm}$ as a control mode. The mass spectrum of $B^{\pm} \to \chi_{c1} K^{\pm}$ candidates is shown in Fig. 8. The measured branching ratio, presented in Fig. 9,

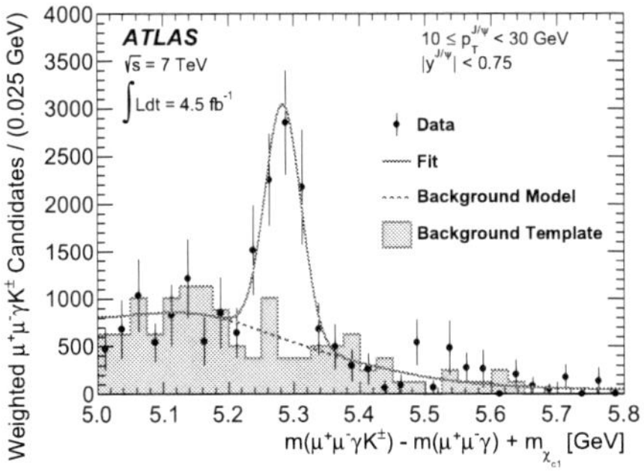

Figure 8. Invariant mass spectrum of $B^{\pm} \to \chi_{c1} K^{\pm}$ decay candidates.

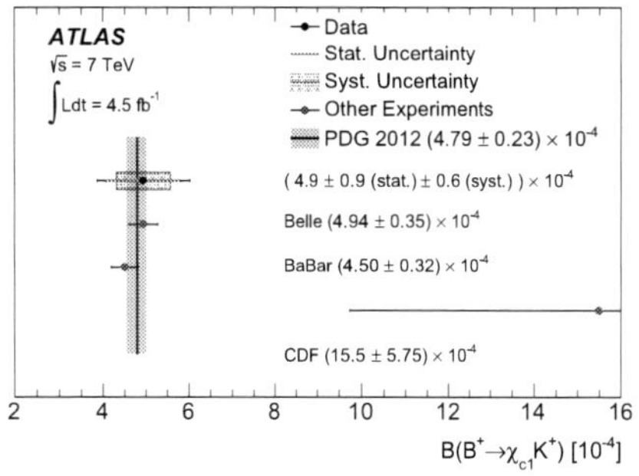

Figure 9. Branching ratio measurements for $B^{\pm} \to \chi_{c1} K^{\pm}$ decay.

is consistent with the best B-factory results and shows a significant improvement on the previous hadron collider measurement, with a great promise for Run II.

4. Summary

With the excellent performance of the LHC and ATLAS in Run I, new insights into the dynamics of quarkonium production in hadronic collisions have become possible. Ever wider variety and range of kinematic variables are being explored, thus providing new areas where the experimental measurements confront theoretical predictions.

More quarkonium-related results from ATLAS are covered in contributions from E. Bouhova-Thacker and M. Watson elsewhere in these proceedings.

References

[1] ATLAS collaboration, *JINST* **3** (2008) S08003.
[2] ATLAS collaboration, CERN-PH-EP-2014-095, *JHEP* (to be published), arXiv:1407.5532 [hep-ex].
[3] ATLAS collaboration, *JHEP* **07** (2014) 154, arXiv:1404.7035 [hep-ex].

Quarkonium production measurements and searches for exotic quarkonia at CMS

Ilse Krätschmer
On behalf of the CMS Collaboration
Institute of High Energy Physics Vienna, Austria
E-mail: ilse.kratschmer@cern.ch

Abstract. The CMS experiment at the LHC studies conventional and exotic quarkonia. CMS measured the quarkonium polarization of prompt Ψ(nS) and Υ(nS) states, using 2011 data taken in pp collisions at $\sqrt{s} = 7$ TeV. No strong polarization was observed in any of the states. Using the Ψ(nS) polarization results and other polarization hypotheses, the prompt Ψ(nS) differential production cross section was determined up to and even beyond 100 GeV. The production of X(3872) in its decay to $J/\psi\pi^+\pi^-$ was also measured, using data taken in pp collisions at $\sqrt{s} = 7$ TeV. The cross section times the branching fraction of the X(3872) relative to the one of ψ(2S) and the fraction of X(3872) originating from B decays were determined. With the help of these results, the prompt X(3872) differential cross section times branching fraction as a function of p_T was extracted. Furthermore, a search for the exotic quarkonium, X_b, decaying to Υ(1S) $\pi^+\pi^-$ in the mass range of 10 to 11 GeV was conducted, using data collected in pp collisions at $\sqrt{s} = 8$ TeV. No evidence of a X_b signal was observed. An upper limit on the relative inclusive production cross section times branching ratio of the X_b and Υ(2S) states at 95% CL was set.

1. Introduction
Quarkonia are an ideal system to study how the strong interaction binds quarks into hadrons since the formation of the initial quark-antiquark pair and the transformation into a bound state happens at very different timescales [1]. Quarkonium production and polarization are strongly intertwined. This paper starts with a short discussion of CMS quarkonium polarization measurements followed by recent results on the quarkonium production by CMS. CMS also conducted searches and studies of exotic quarkonia which are presented in Section 4.

2. Quarkonium polarization
Quarkonium polarization measurements are one of the most challenging ones in the description of quarkonium production. The polarization of $J^{PC} = 1^{--}$ quarkonium states can be studied through the angular decay distribution of the leptons (here muons). The most general observable distribution of a parity-conserving dilepton decay of vector particles can be written as

$$W(\cos\vartheta, \varphi|\vec{\lambda}) = \frac{3/(4\pi)}{(3+\lambda_\vartheta)}(1 + \lambda_\vartheta \cos^2\vartheta + \lambda_\varphi \sin^2\vartheta \cos 2\varphi + \lambda_{\vartheta\varphi} \sin 2\vartheta \cos\varphi)\,, \tag{1}$$

where $\vec{\lambda} = (\lambda_\vartheta,\ \lambda_\varphi,\ \lambda_{\vartheta\varphi})$ represents the frame-dependent polarization parameters and ϑ and φ are the polar and azimuthal angles of the positive muon with respect to the z-axis of the

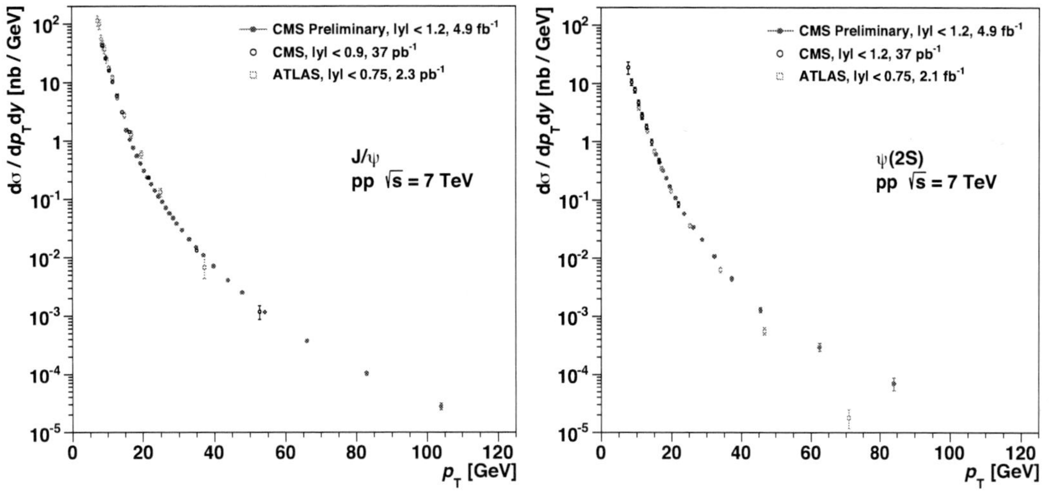

Figure 1. Differential production cross section for the prompt J/ψ (left) and ψ(2S) state (right), for $|y|$ <1.2, compared to previous CMS [10] and ATLAS measurements [11, 12]. In case of the ψ(2S), preliminary ATLAS results are used. The global uncertainties from the integrated luminosity and from the branching fractions are not shown.

chosen reference frame [2, 3]. CMS used three different reference frames: the center-of-mass helicity frame (HX) coinciding with the direction of the quarkonium momentum, the Collins-Soper frame [4] and the perpendicular helicity frame [5].

CMS determined the three frame-dependent polarization parameters and the frame-invariant quantity $\tilde{\lambda} = (\lambda_\vartheta + 3\lambda_\varphi)/(1 - \lambda_\varphi)$ in the three reference frames for the prompt ψ(nS) (n=1,2) and Υ(nS) (n=1,2,3) mesons. A dimuon data sample taken in pp collisions at $\sqrt{s}$ = 7 TeV corresponding to a total integrated luminosity of 4.9 fb^{-1} was used. Quarkonia are seen to be produced essentially unpolarized, with observed polarization levels much smaller than those predicted by most theoretical calculations [6, 7].

3. Quarkonium production

The experimental measurement of quarkonium production strongly depends on the polarization. The most recent CMS measurement of the prompt ψ(nS) production cross section [8] uses four different polarization scenarios in the acceptance calculation: no polarization, two extreme polarizations ($\lambda_\vartheta^{HX} = \pm 1$) and the polarization as measured by CMS averaged over all dimuon p_T and $|y|$ ranges. The same data sample as in the ψ(nS) polarization measurements was used. The efficiency and acceptance corrected yields of the prompt ψ(nS) mesons were extracted using an unbinned maximum likelihood fit to the dimuon invariant mass and decay length distribution [9].

Figure 1 shows the unpolarized double differential cross section of the prompt ψ(nS) mesons in comparison to CMS measurements using 2010 data [10] and results from ATLAS [11, 12]. The most recent CMS measurement considerably extends the reach in transverse momentum, p_T, with respect to the previous one. In case of the ψ(2S), the findings of CMS disagree with the preliminary result of ATLAS at high p_T. However, the final result of ATLAS [13] shows better agreement with the CMS measurements.

4. Exotic quarkonia

In the last decade, a plethora of exotic states decaying into heavy quarkonia have been observed. Amongst the first ones to be discovered was the X(3872) state [14].

CMS studied the production of X(3872) decaying to $J/\psi\pi^+\pi^-$ in pp collisions at $\sqrt{s}$ = 7 TeV [15]. The data used corresponds to a total integrated luminosity of 4.8 fb^{-1}. The inclusive cross section times branching fraction of X(3872) relative to the one of ψ(2S) was determined as a function of p_T. Additionally, the fraction of X(3872) originating from B decays was measured to be 6.56 ± 0.29 (stat.) ± 0.65 (syst.)%. Using these results and the prompt ψ(2S) cross section measurment [10], the differential production cross section of X(3872) was calculated. Figure 2 shows a comparison of the measured rate with Non Relativistic Quantum ChromoDynamics (NRQCD) calculations [16]. While the prediction is larger than the measurement, the dependence on p_T is reasonably well modelled.

CMS also looked for the bottomonium counterpart, X_b, of the X(3872) state [17]. The analysis was based on a data sample collected in pp collisions at $\sqrt{s}$ = 8 TeV, corresponding to an integrated luminosity of 20.7 fb^{-1}. The mass of the X_b has been predicted to lie between 10 and 11 GeV [18, 19, 20]. Therefore, a search for the X_b decaying to $\Upsilon(1S)\pi^+\pi^-$ was conducted in the mass range of 10 and 11 GeV, excluding the Υ(2S) and Υ(3S) regions. No excess signal was observed. Upper limits on the inclusive production cross section times the branching fraction of the X_b relative to the one of Υ(2S) were set at the 95% confidence level (CL). The upper limits range between 0.9 - 5.4%, as shown in Fig. 3. These are the first upper limits on the production of a possible X_b from a hadron collider.

5. Summary

CMS performed measurements of the polarizations of the prompt ψ(nS) and Υ(nS) states, using data collected in pp collisions at $\sqrt{s}$ = 7 TeV. No strong polarization was observed for any of the quarkonium states studied. Using the ψ(nS) polarization results and other polarization hypotheses, the double differential cross sections of the prompt ψ(nS) states were determined. These measurements considerably extend the reach in p_T up to and even beyond 100 GeV.

CMS also conducted studies with exotic quarkonia like the production of the X(3872) using data taken in pp collisions at $\sqrt{s}$ = 7 TeV. The fraction of X(3872) originating from B decays was measured to be (6.56 ± 0.29 (stat.) ± 0.65 (syst.))%. The inclusive cross section times

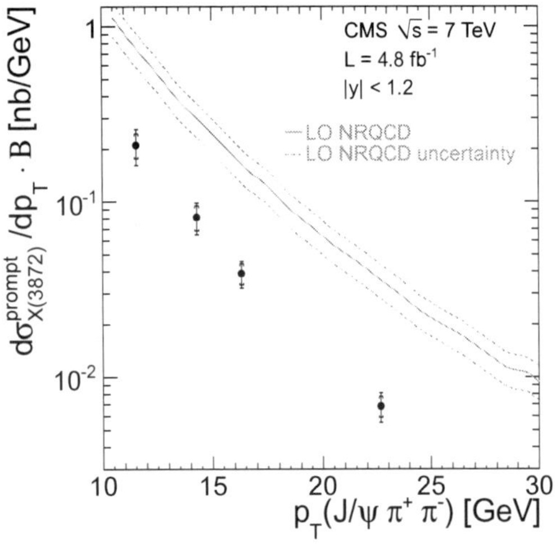

Figure 2. Measured differential cross section for the prompt X(3872) production times branching fraction, B, of the decay to $J/\psi\pi^+\pi^-$ as a function of p_T. The inner error bars represent the statistical uncertainty, while the outer error bars indicate the total uncertainty. The NRQCD calculations [16] are shown by the solid red line with the dotted lines representing the uncertainty.

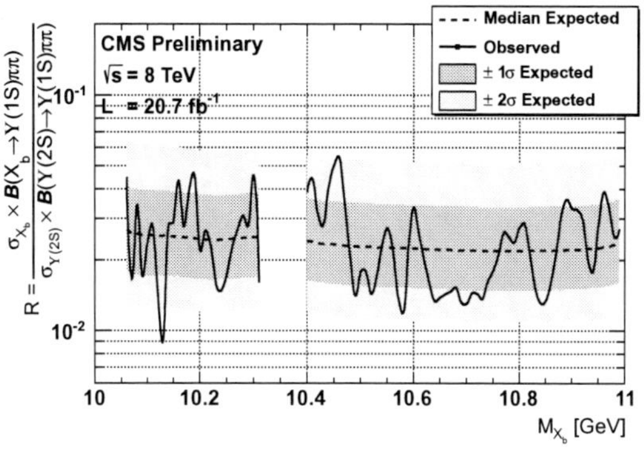

Figure 3. Exclusion limits on the production cross section times the branching fraction of the $X_b \to \Upsilon(1S)\pi^+\pi^-$ relative to the one of $\Upsilon(2S) \to \Upsilon(1S)\pi^+\pi^-$ as a function of the X_b mass, at 95% CL. The solid curve shows the observed limits, while the dashed curve represents the limits expected for a pure background hypothesis.

branching fraction of X(3872) relative to the one of $\psi(2S)$ and the cross section times branching fraction of the prompt X(3872) were determined as a function of p_T.

A search for the counter part of the X(3872) in the bottomonium spectrum, the X_b, was performed with $\sqrt{s} = 8$ TeV data collected in pp collisions. No evidence of a X_b signal was found, but upper limits on the relative inclusive production cross section times branching ratio of the X_b and $\Upsilon(2S)$ at 95% CL were set. These are the first upper limits for X_b with hadron collider data.

6. Acknowledgments
The author is supported by the Austrian Science Fund (FWF): P24167.

References
[1] Brambilla N et al 2011 *Eur. Phys. J.* C **71** 1534
[2] Faccioli P, Lourenço C and Seixas J 2010 *Phys. Rev.* D **81** 111502
[3] Faccioli P, Lourenço C, Seixas J and Wöhri H K 2010 *Eur. Phys. J.* C **69** 657
[4] Collins J C and Soper D E 1977 *Phys. Rev.* D **16** 2219
[5] Braaten E, Kang D, Lee J and Yu C 2009 *Phys. Rev.* D **79** 014025
[6] The CMS Collaboration 2013 *Phys. Lett.* B **727** 381
[7] The CMS Collaboration 2013 *Phys. Rev. Lett.***110** 081802
[8] The CMS Collaboration *CMS-PAS-BPH-14-001*
[9] CMS Collaboration 2011 *Eur. Phys. J.* C **71** 1575,
[10] The CMS Collaboration 2012 *J. High Energ. Phys.* **02** 011
[11] The ATLAS Collaboration 2011 *Nucl. Phys.* B **850** 387
[12] The ATLAS Collaboration *ATLAS-CONF-2013-094*
[13] The ATLAS Collaboration *arXiv:1407.5532* Sub. to *J. High Energ. Phys.*
[14] The Belle Collaboration 2003 *Phys. Rev. Lett.* **91** 262001
[15] The CMS Collaboration 2013 *J. High Energ. Phy.* **04** 154
[16] Artoisenet P and Braaten E 2010 *Phys. Rev.* D **81** 114018
[17] The CMS Collaboration 2013 *Phys. Lett.* B **727** 57
[18] Ebert D, Faustov R N and Galkin V O 2006 *Phys. Lett.* B **634** 214
[19] Ebert D, Faustov R N and Galkin V O 2009 *Mod. Phys. Lett.* A **24** 567
[20] Matheus R D, Narison S, Nielsen M and Richard J M 2007 *Phys. Rev.* D **75** 014005

Journal of Physics: Conference Series **556** (2014) 012013 doi:10.1088/1742-6596/556/1/012013

Latest results from the BESIII ex eriment

Chengping Shen

School of Physics and Nuclear Energy Engineering, Beihang University, Beijing, China

E-mail: shencp@buaa.edu.cn

Abstract.

In this article, we review the recent BESIII experimental studies. We discuss the results on the $X(3872)$, Z_c states [$Z_c(3900)^{\pm}$, $Z_c(3900)^0$ and $Z_c(4020)^0$], and the first observation of $e^+e^- \to \omega\chi_{c0}$ at $\sqrt{s} = 4.23$ and 4.26 GeV.

1. Introduction

Exotic hadrons including glueballs, hybrids, multi-quark states and hadron molecules have been searched for and many candidates were proposed. However, no solid conclusion was reached.

The BESIII [1] experiment at the BEPCII collider started data taking in 2009, and lots of data were accumulated at the peaks of the J/ψ, $\psi(2S)$, and $\psi(3770)$, as well as at above 4 GeV. These high energy data make the study of the exotic states (XYZ states) possible.

2. Observation of $e^+e^- \to \gamma X(3872)$

The $X(3872)$ was first observed by Belle in $B^{\pm} \to K^{\pm}\pi^+\pi^- J/\psi$ decays. Since its discovery, the $X(3872)$ has been interpreted as a candidate for a hadronic molecule or a tetraquark state. Until now, the $X(3872)$ was only observed in B meson decays and hadron collisions.

BESIII measured the process $e^+e^- \to \gamma X(3872) \to \gamma\pi^+\pi^- J/\psi$, $J/\psi \to \ell^+\ell^-$ ($\ell^+\ell^- = e^+e^-$ or $\mu^+\mu^-$) at e^+e^- center-of-mass (CM) energies from $\sqrt{s} = 4.009$ GeV to 4.420 GeV [2].

Figure 1(a) shows the $M(\pi^+\pi^- J/\psi)$ distribution (summed over all CM energy data sets) with the fitted results, which gives $M[X(3872)] = (3871.9 \pm 0.7_{\text{stat.}} \pm 0.2_{\text{sys.}})$ MeV/c^2. The statistical significance of $X(3872)$ is 6.3σ. Figure 1(b) shows the energy-dependent cross section together with the fitted results with a $Y(4260)$ resonance, a linear continuum, or a $E1$-transition phase space term. The $Y(4260)$ resonance describes the data better than the other two options, which strongly supports the existence of the radiative transition process $Y(4260) \to \gamma X(3872)$.

3. Observation of charged charmonium-like state $Z_c(3900)$

Motivated by the striking observations of charged charmonium-like and bottomonium-like states, Belle and BESIII investigate the existence of similar states as intermediate resonances in $Y(4260) \to \pi^+\pi^- J/\psi$ decays [3].

Unbinned maximum likelihood fits are applied to the distributions of $M_{\text{max}}(\pi^{\pm} J/\psi)$ from Belle and BESIII data. The signal shape is parameterized as an S-wave Breit-Wigner function convolved with a Gaussian with a mass resolution fixed at the MC simulated value. Figure 2 shows the fit results. The measured masses are $(3899.0 \pm 3.6 \pm 4.9)$ MeV/c^2 and $(3894.5 \pm 6.6 \pm 4.5)$ MeV/c^2 and the measured widths are $(46 \pm 10 \pm 20)$ MeV and $(63 \pm 24 \pm 26)$ MeV/c^2

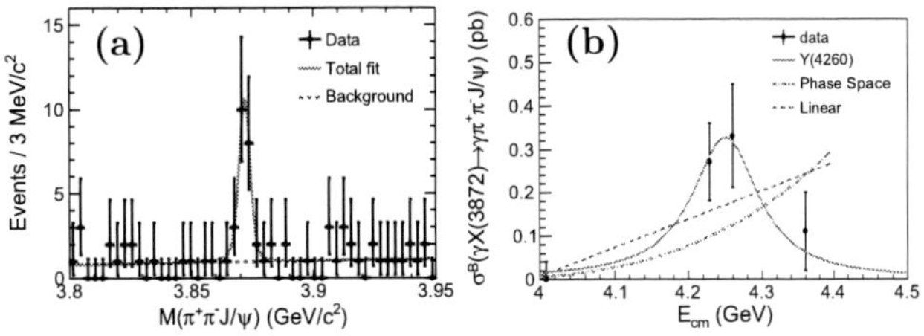

Figure 1. (a) Fit of the $M(\pi^+\pi^-J/\psi)$ distribution. The curve is the total fit result, and the dashed curve shows the background. (b) The fit to $\sigma^B[e^+e^- \to \gamma X(3872)] \times \mathcal{B}[X(3872) \to \pi^+\pi^-J/\psi]$.

from Belle and BESIII experiments, respectively. They are consistent with each other within the errors. The signal significance is greater than 5σ in both of the measurements. This state is close to the DD^* mass threshold. As the $Z(3900)^\pm$ state has a strong coupling to charmonium and is charged, it cannot be a conventional cc state. Since the observation of the $Z_c(3900)$, there have been a number of different interpretations, including tetraquark state, hadronic molecule, hadron-charmonium state and so on.

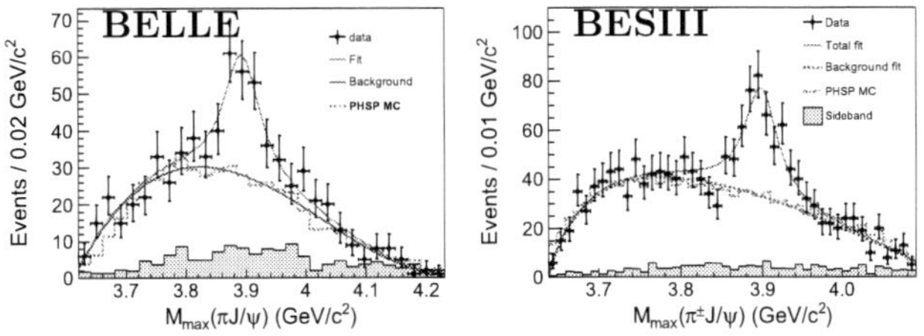

Figure 2. Unbinned maximum likelihood fits to the distributions of the $M_{\max}(\pi J/\psi)$ from Belle and BESIII experimental data. The solid curves are the best fits, the dashed histograms represent the results of phase space distribution and the shaded histograms are J/ψ sidebands.

4. Observation of $e^+e^- \to \pi^0\pi^0J/\psi$, $\pi^0\pi^0h_c$ and $Z_c(3900)^0$, $Z_c(4020)^0$

After charged charmonium-like states $Z_c(3900)^\pm$ [3] and $Z_c(4020)^\pm$ [4] were observed, an important question is whether isospin=0 neutral Z_c partners exist.

Figures 3(a), 3(b) and 3(c) show $M(\pi^0J/\psi)$ invariant mass distributions including both combinations of π^0J/ψ for events from $\sqrt{s} = 4.26$, 4.36 and 4.23 GeV data samples, respectively. There is a significant peak around 3.9 GeV/c^2 that corresponds to $Z_c(3900)^0$. An unbinned maximum simultaneous likelihood fit is performed to the distributions of $M(\pi^0J/\psi)$ to extract $Z_c(3900)^0$ parameters. Here the Breit-Wigner functions for the signal shapes are constrained to be the same for all energy points, while the Argus functions as backgrounds have independent parameters. The fitted results are shown in Figs. 3(a)-(c). The statistical significance of $Z_c(3900)^0$ is 10.4σ and the measured mass and width are 3894.8 ± 2.3 MeV/c^2 and 29.6 ± 8.2 MeV/c^2, respectively, where the errors are statistical only.

The selections of $\pi^0\pi^0h_c$ follow closely the analysis of $e^+e^- \to \pi^+\pi^-h_c$ [4] with the selection of $\pi^+\pi^-$ replaced with the selection of a pair of π^0s. The h_c is reconstructed via $h_c \to \gamma\eta_c$ with $\eta_c \to X_i$, where X_i denotes 16 hadronic final states. After the event selections, in the η_c signal region, a clear peak at the h_c mass is observed. The measured Born cross sections together with other information including the ratios of the Born cross sections for the neutral and charged $e^+e^- \to \pi\pi h_c$ modes are listed in Table 1. The combined ratio $\mathcal{R}_{\pi\pi h_c}$ is determined to be (0.63 ± 0.09), which is within 2σ of the expectation of isospin symmetry, i.e., 0.5.

Table 1. Energies ($\sqrt{s}$), luminosities ($\mathcal{L}$), numbers of events ($n^{\rm obs}_{h_c}$), Born cross sections $\sigma^{\rm B}(e^+e^- \to \pi^0\pi^0 h_c)$ and ratios $\mathcal{R}_{\pi\pi h_c} = \frac{\sigma(e^+e^- \to \pi^0\pi^0 h_c)}{\sigma(e^+e^- \to \pi^+\pi^- h_c)}$.

$\sqrt{s}$ (GeV)	$\mathcal{L}$ (pb^{-1})	$n^{\rm obs}_{h_c}$	$\sigma^{\rm B}(e^+e^- \to \pi^0\pi^0 h_c)$ (pb)	$\mathcal{R}_{\pi\pi h_c}$
4.230	1090.0	82.5 ± 15.6	$25.6 \pm 4.8 \pm 2.6 \pm 4.0$	$0.54 \pm 0.11 \pm 0.06$
4.260	826.8	62.8 ± 13.3	$24.4 \pm 5.2 \pm 3.2 \pm 3.8$	$0.63 \pm 0.14 \pm 0.10$
4.360	544.5	64.3 ± 11.5	$36.2 \pm 6.5 \pm 4.1 \pm 5.7$	$0.73 \pm 0.14 \pm 0.10$

The h_c signal events are selected by requiring 3.51 GeV/c^2 $< M^{\rm recoil}_{\pi^0\pi^0} <$3.55 GeV/c^2. Figure 3(d) shows the $M^{\rm recoil}_{\pi^0}|_{\rm max}$ distribution where there is an obvious peak near 4.02 GeV/c^2, which corresponds to the expected position of a $Z_c(4020)^0$ signal.

An unbinned maximum likelihood fit is applied to the $M^{\rm recoil}_{\pi^0}|_{\rm max}$ distribution summed over all 16 η_c decay modes. The data at $\sqrt{s} = 4.23$, 4.26, and 4.36 GeV are fitted simultaneously with the same signal function with common mass and width. Because of the limited statistics of the $Z_c(4020)^0$ signal, its width is fixed to that of its charged partner [4]. The solid curve in Fig. 3(d) shows the fit results, which yields a $Z_c(4020)^0$ mass of $4023.6 \pm 2.2_{\rm stat.} \pm 3.9_{\rm sys.}$. The statistical significance is greater than 5.0σ.

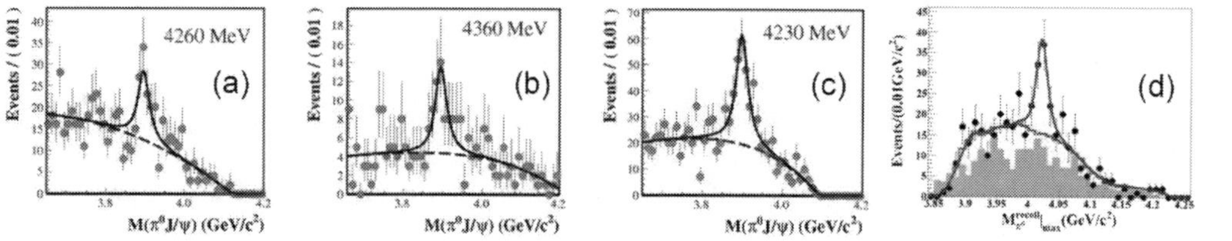

Figure 3. Distributions of $\pi^0 J/\psi$ invariant mass including both combinations of $\pi^0 J/\psi$ from $\sqrt{s} =$ (a) 4.26, (b) 4.36 and (c) 4.23 GeV data samples, respectively, for $e^+e^- \to \pi^0\pi^0 J/\psi$. Sum of the simultaneous fit to the $M^{\rm recoil}_{\pi^0}|_{\rm max}$ distribution at $\sqrt{s} = 4.23$, 4.26 and 4.36 GeV is shown in (d) for $e^+e^- \to \pi^0\pi^0 h_c$. The solid line shows the simultaneous fit result, and the dashed line is the fitted background shape.

5. Study of $e^+e^- \to \omega\chi_{c0}$ at $\sqrt{s}$ from 4.21 to 4.42 GeV

Since the $Y(4260)$ was observed by BaBar, CLEO and Belle experiments [5], many theoretical models were proposed to interpret the $Y(4260)$, *e.g.*, as a quark-gluon charmonium hybrid, a tetraquark state, a hadroncharmonium, or a hadronic molecule. Very recently, the charged charmonium-like states $Z_c(3900)$ [3] was observed in the data around the $Y(4260)$, which may hint a complicated structure of the $Y(4260)$. Searching for new decay modes and measuring the line shape are very important for understanding the nature of the $Y(4260)$.

In Ref. [6], authors predicted a sizeable coupling between the $Y(4260)$ and the $\omega\chi_{c0}$ channel and suggested the $Y(4260)$ resonance to be a conventional cc state renormalized by the $\omega\chi_{c0}$ continuum which plays a role in reducing the decay rates into open charm channels.

Based on data samples collected with the BESIII detector at 9 CM energies from 4.21 to 4.42 GeV, we search for the production of $e^+e^- \to \omega\chi_{c0}$, where ω is reconstructed with $\pi^+\pi^-\pi^0$ decay mode, χ_{c0} is reconstructed with $\pi^+\pi^-$ and K^+K^- decay modes.

After all the event selections, an unbinned maximum likelihood fit is performed to the $M(\pi^+\pi^-)$ and $M(K^+K^-)$ distributions at $\sqrt{s} = 4.23$ and 4.26 GeV on the $\pi^+\pi^-$ and K^+K^- modes simultaneously. For data at $\sqrt{s} = 4.23$ GeV, the total signal yield of the two modes is 125.3 ± 13.5 with a statistical significance 11.9σ. For data at 4.26 GeV, the total signal yield is 45.5 ± 10.2 with a statistical significance of 5.5σ. The Born cross sections are measured to be $(55.4 \pm 6.0 \pm 5.9)$ pb and $(23.7 \pm 5.3 \pm 3.5)$ pb, respectively. For other energy points, no significant signals are found and the upper limits on the cross section at the 90% C.L. are determined. Figure 4 shows the Born cross sections of $e^+e^- \to \omega\chi_{c0}$ in the full energy region.

We fit the cross section by using a $Y(4260)$ resonance, the fit gives $\chi^2/ndf = 40.2/8$, so the $Y(4260)$ resonance cannot describe the data well. It indicates that the $\omega\chi_{c0}$ signals are very unlikely from the $Y(4260)$ decays. By assuming the $\omega\chi_{c0}$ signals come from a resonance, the fitting results are $\Gamma_{ee}\mathcal{B}(\omega\chi_{c0}) = (2.7 \pm 0.7)$ eV, $M = (4229 \pm 11)$ MeV/c^2, and $\Gamma_t = (40 \pm 14)$ MeV/c^2, which are consistent with the $Y(4220)$ state observed in the cross section of $e^+e^- \to \pi^+\pi^-h_c$ [7]. The data may suggest that the $Y(4260)$ observed in $e^+e^- \to \pi^+\pi^-J/\psi$ [5] have fine structures as observed in $e^+e^- \to \pi^+\pi^-h_c$ [7], and the lower mass structure at about 4230 MeV/c^2 has a sizable coupling to the $\omega\chi_{c0}$ channel as predicted in Ref. [6].

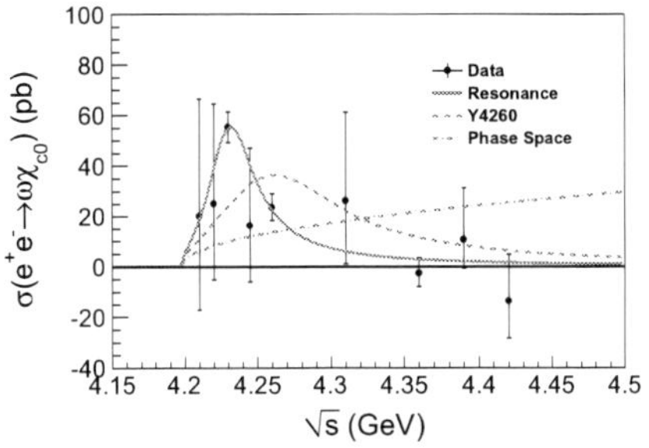

Figure 4. Fit to $\sigma(e^+e^- \to \omega\chi_{c0})$ with a resonance (solid curve), the $Y(4260)$ (dashed curve), or a phase space term (dot-dashed curve). Dots with error bars are the Born cross sections.

6. Conclusion

Some progresses in XYZ studies at BESIII are reviewed. For some XYZ states, the nature are still unknown. A systematic study of their quantum numbers, decay modes, excited states, comparison with their beauty-counterparts is crucial to achieve a satisfactory explanation. With more upcoming data, an exciting future is expected.

References

[1] Ablikim M *et al.* [BESIII Collaboration] 2010 *Nucl. Instrum. Methods Phys. Res., Sect.* A **614** 345
[2] Ablikim M *et al.* [BESIII Collaboration] 2014 *Phys. Rev. Lett.* **112** 092001
[3] Ablikim M *et al.* [BESIII Collaboration] 2013 *Phys. Rev. Lett.* **110** 252001 Liu Z Q *et al.* [Belle Collaboration] 2013 *Phys. Rev. Lett.* **110** 252002
[4] Ablikim M *et al.* [BESIII Collaboration] 2013 *Phys. Rev. Lett.* **111** 242001
[5] Aubert B *et al.* [BaBar Collaboration] 2005 *Phys. Rev. Lett.* **95** 142001 He Q *et al.* [CLE Collaboration] 2006 *Phys. Rev.* D **74** 091104 Yuan C Z *et al.* [Belle Collaboration] 2007 *Phys. Rev. Lett.* **99** 182004
[6] Dai L Y, M. Shi, Tang G Y and Zheng H Q, arXiv:1206.6911
[7] Yuan C Z 2014 *Chin. Phys.* C **38** 043001

Recent bottomonium stu ies at Belle

Xionghong He
Peking University
E-mail: hexionghong@pku.edu.cn

Abstract. We report the recent studies of bottomonium spectroscopy and transitions using the data sample collected with the Belle detector. Unambiguous $\pi^+\pi^-\pi^0\chi_{b1}$ ($J = 1,\ 2$), $\omega\chi_{b1}$ signals are observed near the $\Upsilon(5S)$ resonance mass region. In the $\pi^+\pi^-\pi^0$ invariant mass distribution, some $(\pi^+\pi^-\pi^0)_{\text{non}-\omega}$ events are observed besides clear ω signal events. A X(3872)-like state in bottom sector is searched for in the final state $\omega\Upsilon(1S)$. The Belle collaboration also searches for the η transitions with the $\Upsilon(4S)$ and $\Upsilon(5S)$ data samples. The $\Upsilon(4S) \to \eta h_b(1P)$ process is observed and evidences of η transitions from $\Upsilon(5S)$ to lower bottomonium are found. The $\eta_b(1S)$ resonance parameters are updated based on the decay mode $h_b(1P) \to \gamma\eta_b(1S)$ in the same work.

1. Study of $e^+e^- \to \pi^0\pi^+\pi^-\gamma\Upsilon(1S)$ at $\Upsilon(5S)$

1.1. Observation of $e^+e^- \to \pi^0\pi^+\pi^-\chi_{bJ}$

Heavy quarkonium systems are basically non-relativistic and hadronic transition between them have largely been successfully described by the QCD multipole expansion model [1]. The Belle collaboration has observed the anomalously large width of the $\Upsilon(5S) \to \pi^+\pi^-\Upsilon(mS)$ (m = 1, 2, 3) and $\pi^+\pi^-h_b(nP)$ ($n = 1,\ 2$) transitions [2], which is interpreted as the rescattering of the B mesons [3] or the existence of a tetraquark state with a mass close to that of the $\Upsilon(5S)$ [4]. A detailed study [5] revealed the presence of two charged bottomonium-like states, $Z_b(10610)^\pm$ and $Z_b(10650)^\pm$. It is natural to investigate the similarity to $\pi^0\pi^+\pi^-$ hadronic transitions between the $\Upsilon(5S)$ and χ_{bJ} ($J = 0,\ 1,\ 2$).

The study [6] is based on a 118 fb^{-1} data sample at $\sqrt{s} \sim 10.867$ GeV collected with the Belle detector [7] at the KEKB asymmetric-energy e^+e^- collider [8]. Fig. 1 shows the scatter plot of $M(\pi^0\pi^+\pi^-)$ versus $M(\gamma\Upsilon(1S))$ after event selection criteria. A unbinned two-dimensional fit is performed to extract signal yields. There is a clear ω signal and some $(\pi^+\pi^-\pi^0)_{\text{non}-\omega}$ events, as shown in Fig. 2.

Fig. 3 shows the $\gamma\Upsilon(1S)$ mass projections within and outside ω signal region. The signal statistical significance and the calculated Born cross sections and branching fractions are summarized in Table 1.

The measured branching fractions of $\Upsilon(5S) \to \pi^+\pi^-\pi^0\chi_{b1}$ and $\pi^+\pi^-\pi^0\chi_{b2}$ are at the same order as the $\Upsilon(5S) \to \pi^+\pi^-\Upsilon(mS)$ ($m = 1,\ 2,\ 3$) [2]. The ratio of the branching fractions of $\Upsilon(5S)$ decays or the cross sections of e^+e^- to $\omega\chi_{b2}$ and $\omega\chi_{b1}$ is $0.38 \pm 0.16(\text{stat.}) \pm 0.09(\text{syst.})$. It is significantly lower than the expectation from the heavy quark symmetry [9].

Journal of Physics: Conference Series **556** (2014) 012014 doi:10.1088/1742-6596/556/1/012014

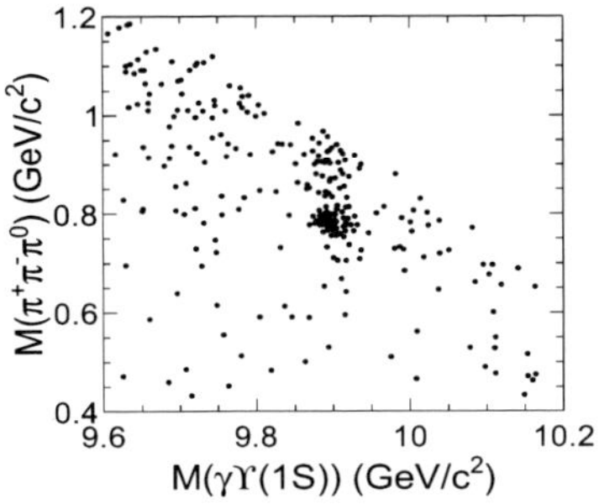

Figure 1. The scatter plot of $M(\pi^+\pi^-\pi^0)$ versus $M(\gamma\Upsilon(1S))$ for selected $e^+e^- \rightarrow \pi^+\pi^-\pi^0\gamma\Upsilon(1S)$ candidate events.

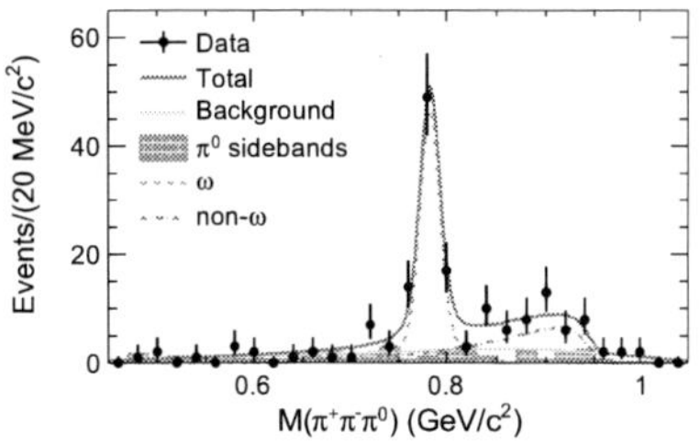

Figure 2. The $M(\pi^+\pi^-\pi^0)$ invariant mass distribution, where the dashed and dash-dotted curves represent the ω and $(\pi^+\pi^-\pi^0)_{\text{non}-\omega}$ events; the hatched histogram is the normalized π^0 sideband events.

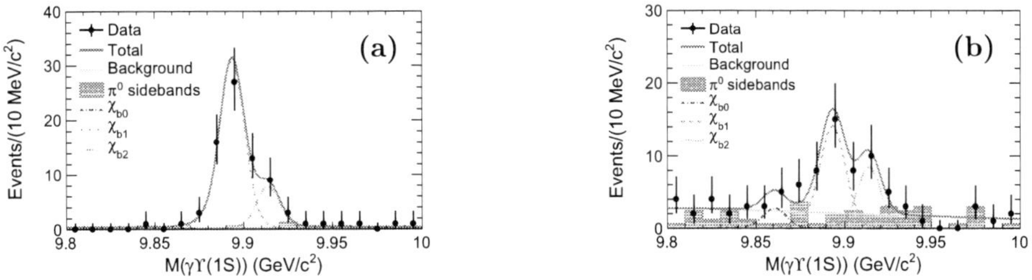

Figure 3. Projections of $M(\gamma\Upsilon(1S))$ (a) in the ω signal region and (b) outside of ω signal region, where the dash-dotted, dashed and dotted curves represent the χ_{b0}, χ_{b1} and χ_{b2} signals, respectively.

Table 1. The signal significance ( ), Born cross section (σ_B) and branching fraction ($\mathcal{B}$). The upper limits are given at 90% confidence level (C.L.) if the significance is lower than 3σ.

Mode	(σ)	σ_B (pb)	$\mathcal{B}$ (10^{-3})
$(\pi^+\pi^-\pi^0)\chi_{b0}$	1.0	< 3.1	< 6.3
$(\pi^+\pi^-\pi^0)\chi_{b1}$	12	$0.90 \pm 0.11 \pm 0.13$	$1.85 \pm 0.23 \pm 0.23$
$(\pi^+\pi^-\pi^0)\chi_{b2}$	5.9	$0.57 \pm 0.13 \pm 0.08$	$1.17 \pm 0.27 \pm 0.14$
$\omega\chi_{b0}$	0.5	< 1.9	< 3.9
$\omega\chi_{b1}$	12	$0.76 \pm 0.11 \pm 0.11$	$1.57 \pm 0.22 \pm 0.21$
$\omega\chi_{b2}$	3.5	$0.29 \pm 0.11 \pm 0.08$	$0.60 \pm 0.23 \pm 0.15$
$(\pi^+\pi^-\pi^0)_{\text{non}-\omega}\chi_{b0}$	0.4	< 2.3	< 4.8
$(\pi^+\pi^-\pi^0)_{\text{non}-\omega}\chi_{b1}$	4.9	$0.25 \pm 0.07 \pm 0.06$	$0.52 \pm 0.15 \pm 0.11$
$(\pi^+\pi^-\pi^0)_{\text{non}-\omega}\chi_{b2}$	3.1	$0.30 \pm 0.11 \pm 0.14$	$0.61 \pm 0.22 \pm 0.28$

1.2. Search for X_b decaying to $\omega\Upsilon(1S)$

The CMS Collaboration reported a null search of the $X(3872)$ counterpart in bottomonium sector (called X_b) in the $\pi^+\pi^-\Upsilon(1S)$ final state [10]. Unlike the $X(3872)$, the $X_b \rightarrow \omega\Upsilon(1S)$ is a isospin-conserving process.

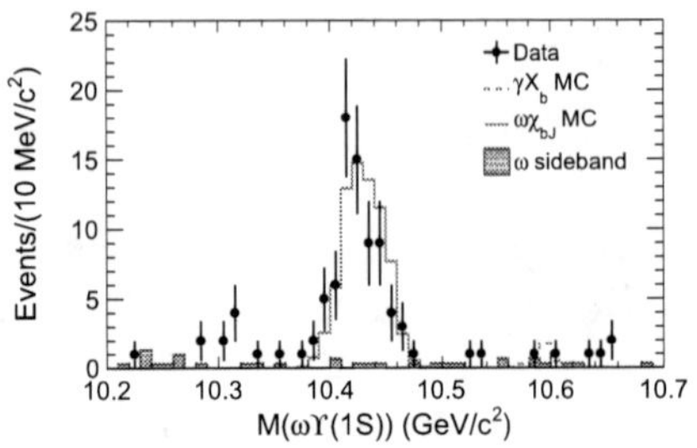

Figure 4. The $\omega\Upsilon(1S)$ invariant mass distribution from selected $\pi^0\pi^+\pi^-\gamma\Upsilon(1S)$ events.

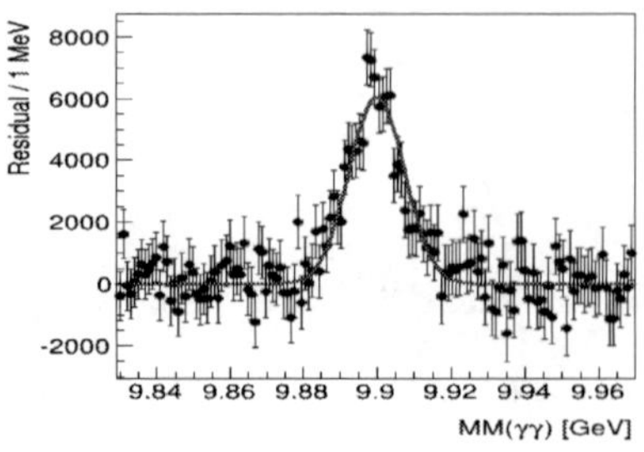

Figure 5. The η recoil mass distribution with the background subtracted in the $\Upsilon(4S)$ dataset. The blue curve represents the best fit.

The $\omega\Upsilon(1S)$ invariant mass distribution is shown in Fig.4, where the solid histogram represents the reflection from the contribution of $e^+e^- \to \omega\chi_{bJ}$ ($J = 0,\ 1,\ 2$), the shaded histogram is the normalized ω sideband events and the dashed histogram is from the MC signal sample with X_b mass fixed at 10.6 GeV/c^2 and yield fixed at the upper limit at 90% C.L. No significant signal is observed for X_b with mass between 10.55 and 10.65 GeV/c^2. The 90% C.L. upper limits on the product branching fraction $\mathcal{B}(\Upsilon(5S) \to \gamma X_b)\mathcal{B}(X_b \to \omega\Upsilon(1S))$ varies from 2.6×10^{-5} to 3.8×10^{-5}.

2. Study on the η transitions

The processes $\Upsilon(nS) \to \eta\Upsilon(mS)$ are spin flipping, and the branch fractions are suppressed with respect to the $\pi^+\pi^-$ transitions in QCD mutipole expansion approach. For $\Upsilon(4S)$ case, the ratio $\mathcal{B}(\Upsilon(4S) \to \eta\Upsilon(1S))/\mathcal{B}(\Upsilon(4S) \to \pi^+\pi^-\Upsilon(1S)) = 2.4$ [11], which shows the strong discrepancy with the theoretical prediction [12]. However, for $\Upsilon \to h_b$ transitions, the similar ratio should be not less than 1.0.

Recently, Belle collaboration searched for the η transitions from $\Upsilon(4S)$ and $\Upsilon(5S)$ to the $h_b(nP)$ [13]. The results presented here are preliminary. Fig. 5 shows η recoil mass distribution with the combinatorial background subtracted in the $\Upsilon(4S)$ dataset. A clear $h_b(1P)$ peak is observed, with a statistical significance greater than 5σ. The corresponding branching fraction is $\mathcal{B}(\Upsilon(4S) \to \eta h_b(1P)) = (1.83 \pm 0.16 \pm 0.17)\times 10^{-3}$. The ratio $\mathcal{B}(\Upsilon(4S) \to \eta h_b(1P))/\mathcal{B}(\Upsilon(4S) \to \pi^+\pi^- h_b(1P))$ is greater than 2.0 at 90% C.L., which is in reasonable agreement with the QCD multipole expansion model expectation.

For $\Upsilon(5S)$ dataset case, the η recoil mass, after the subtraction of combinatorial background, shows significant signals corresponding to the $\Upsilon(5S) \to \eta\Upsilon(2S)$ and $\Upsilon(5S) \to \eta\Upsilon(1D)$, as shown in Fig. 6. The statistical significances are 3.1σ and 4.4σ for the two processes, respectively. The measured branching fractions are

$$\mathcal{B}(\Upsilon(5S) \to \eta\Upsilon(2S)) = (2.1 \pm 0.7 \pm 0.3) \times 10^{-3},$$
$$\mathcal{B}(\Upsilon(5S) \to \eta\Upsilon(1D)) = (2.8 \pm 0.7 \pm 0.4) \times 10^{-3},$$
$$\mathcal{B}(\Upsilon(5S) \to \eta h_b(1P)) < 3.3 \times 10^{-3},$$
$$\mathcal{B}(\Upsilon(5S) \to \eta h_b(2P)) < 3.7 \times 10^{-3},$$

where the upper limits are obtained at 90% C.L. Since the $\Upsilon(5S) \to \pi^+\pi^- h_b(nP)$ proceed mainly through the charged Z_bs, the interpretation of $\mathcal{B}(\Upsilon(5S) \to \eta h_b(nP))/\mathcal{B}(\Upsilon(4S) \to \pi^+\pi^- h_b(nP))$ of $\Upsilon(5S)$ results is not straightforward.

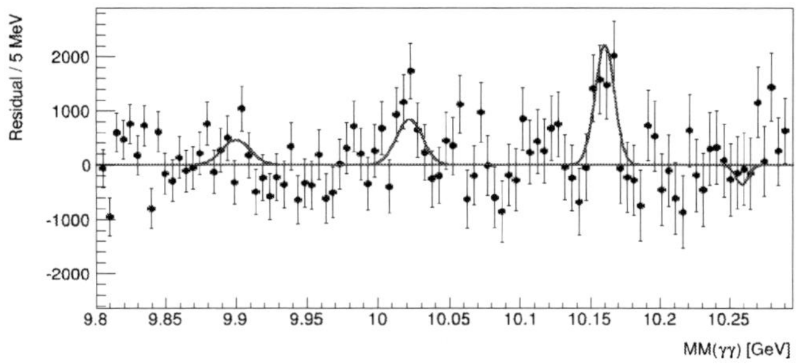

Figure 6. The η recoil mass distribution with the combinatorial background subtracted in the $\Upsilon(5S)$ dataset. The blue curve represent the best fit.

The $\eta_b(1S)$ resonance parameters are measured using the process $\Upsilon(4S) \to \eta h_b(1P) \to \eta\gamma\eta_b(1S)$. A variable $\Delta MM = MM(\gamma\eta) - MM(\eta)$ is defined, where $MM(\gamma\eta)$ and $MM(\eta)$ are the missing masses of $\gamma\eta$ and η. The ΔMM distribution is shown in Fig. 7. The measured $M(\eta_b(1S))$ is $9405.3 \pm 1.3 \pm 3.0$ MeV/c^2 and corresponding hyperfine splitting $\Delta M_{HF}(1S) = M(\Upsilon(1S)) - M(\eta_b(1S))$ is $55.0 \pm 1.3 \pm 3.2$ MeV/c^2 which is in agreement with the lattice and potential non-relativistic QCD expactations [14]. The measured width is $\Gamma(\eta_b(1S)) = 11^{+8}_{-6} \pm 3$ MeV/c^2.

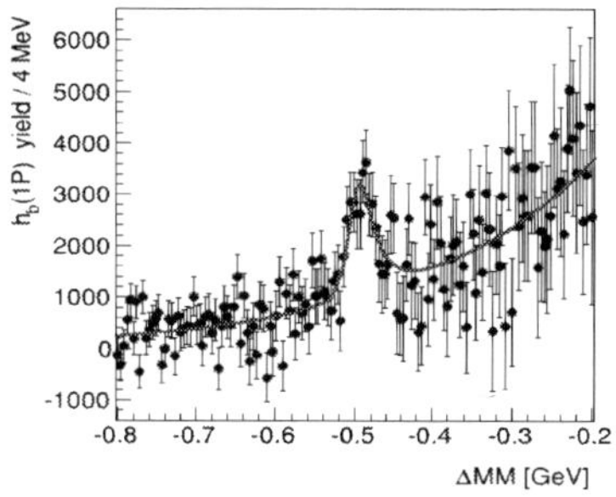

Figure 7. The ΔMM distribution. The blue curve represents the best fit.

References

[1] Y. P. Kuang, *Front. Phys. China* **1**, 19 (2006).
[2] K.-F. Chen *et al.* (Belle Collaboration), *Phys. Rev. Lett.* **100**, 112001 (2008)
I. Adachi *et al.* (Belle Collaboration), *Phys. Rev. Lett.* **108**, 032001 (2012).
[3] Yu. A. Simonov, *JETP Lett.* **87**, 147 (2008)
C. Meng and K. T. Chao, *Phys. Rev.* D **78**, 034022 (2008).
[4] A. Ali, C. Hambrock, and M. J. Aslam, *Phys. Rev. Lett.***104**, 162001, (2010). [Erratum-ibid. **107** 049903 (2011)].
[5] A. Bondar *et al.* (Belle Collaboration), *Phys. Rev. Lett.* **108**, 122001 (2012).
[6] X. H. He *et al.* (Belle Collaboration), *Phys. Rev. Lett.***113**, 142001, (2014b).
[7] A. Abashian *et al.* (Belle Collaboration), *Nucl. Instrum. Methods Phys. Res., Sect.* A **479**, 117 (2002) also see detector section in J. Brodzicka *et al.*, *Prog. Theor. Exp. Phys.* (2012) 04D001.
[8] S. Kurokawa and E. Kikutani, *Nucl. Instrum. Methods Phys. Res., Sect.* A **499**, 1 (2003), and other papers included in this volume
T. Abe *et al.*, *Prog. Theor. Exp. Phys.* (2013) 03A001 and following articles up to 03A011.
[9] P. L. Cho and M. B. Wise, *Phys. Lett.* B **346**, 129 (1995).
[10] S. Chatrchyan *et al.* (CMS Collaboration), *Phys. Lett.* B **727**, 57 (2013).
[11] B. Aubert *et al.* (Belle Collaboration), *Phys. Rev. Lett.* **108**, 122001 (2012).
[12] Yu. A. Simonov and A. I. Veselov, *Phys. Lett.* B **673**, 211 (2009).
[13] Umberto T. 2014. *Spin effects in bottomonium,* Talk presented at DIS2014, Vancouver.
[14] B. A. Kniehl *et al. Phys. Rev. Lett.* **92**, 242001 (2004) [Erratum-ibid. **104**, 199901 (2010)].

Journal of Physics: Conference Series **556** (2014) 012015 doi:10.1088/1742-6596/556/1/012015

Long-distance structure of the $X(3872)$

F.-K. Guo[1], C. Hidalgo-Duque[2], J. Nieves[2], A. Ozpineci[3] and M. Pavón Valderrama[4]

[1] Helmholtz-Institut für Strahlen- und Kernphysik and Bethe Center for Theoretical Physics, Universität Bonn, D-53115 Bonn, Germany

[2] Instituto de Física Corpuscular (IFIC), Centro Mixto CSIC-Universidad de Valencia, Institutos de Investigación de Paterna, Aptd. 22085, E-46071 Valencia, Spain

[3] Middle East Technical University - Department of Physics TR-06531 Ankara, Turkey

[4] Institut de Physique Nucléaire, Université Paris-Sud, IN2P3/CNRS, F-91406 Orsay Cedex, France

E-mail: jmnieves@ific.uv.es

Abstract. We investigate heavy quark symmetries for heavy meson hadronic molecules, and explore the consequences of assuming the $X(3872)$ and $Z_b(10610)$ as an isoscalar $D\bar{D}^*$ and an isovector $B\bar{B}^*$ hadronic molecules, respectively. The symmetry allows to predict new hadronic molecules, in particular we find an isoscalar 1^{++} $B\bar{B}^*$ bound state with a mass about 10580 MeV and the isovector charmonium partners of the $Z_b(10610)$ and the $Z_b(10650)$ states. Next, we study the $X(3872) \to D^0\bar{D}^0\pi^0$ three body decay. This decay mode is more sensitive to the long-distance structure of the $X(3872)$ resonance than its $J/\psi\pi\pi$ and $J/\psi 3\pi$ decays, which are mainly controlled by the short distance part of the $X(3872)$ molecular wave function. We discuss the $D^0\bar{D}^0$ final state interactions, which in some situations become quite important. Indeed in these cases, a precise measurement of this partial decay width could provide precise information on the interaction strength between the $D^{(*)}\bar{D}^{(*)}$ charm mesons.

1. Heavy quark symmetries and hidden charm meson molecules

Heavy hadron molecules are a type of exotic hadron theorized more than thirty years ago [1, 2]. Their main component is a pair of heavy hadrons instead of a quark–antiquark pair. The experimental advances in heavy quarkonium spectroscopy have identified several molecular candidates among the recently observed XYZ states. The most promising ones are the $X(3872)$ [3] and the twin $Z_b(10610)$ and $Z_b(10650)$ states, to be called Z_b and Z_b', respectively [4, 5]. Among many different interpretations of the $X(3872)$, the one assuming it to be[1] a $(D\bar{D}^* - D^*\bar{D})/\sqrt{2}$, hadronic molecule with quantum numbers $J^{PC} = 1^{++}$ [6] is the most promising. Heavy quark symmetries deduced from QCD provide new insights into the hadron spectrum. Thus, heavy quark spin symmetry (HQSS) implies that molecular states may appear in HQSS multiplets. From heavy flavor symmetry (HFS), we know that the interaction among heavy hadrons is roughly independent on whether they contain a charm or a bottom quark. Combining both HQSS and HFS, various partners of the $X(3872)$ and the isovector $Z_b'^s$ states can be predicted [7, 8, 9, 10, 11, 12, 13, 14]. Some of these predictions from [14]

[1] From now on, when we refer to $D^0\bar{D}^{*0}$, D^+D^{*-}, or in general $D\bar{D}^*$ we are actually referring to the combination of these states with their charge conjugate ones in order to form a state with well-defined C-parity.

are collected in table 1. Moreover, owing to heavy antiquark-diquark symmetry, the doubly heavy baryons have approximately the same light-quark structure as the heavy antimesons. As a consequence and thanks to this approximate symmetry the existence of a heavy meson-antimeson molecules implies the possibility of a partner composed of a heavy meson and a doubly-heavy baryon (triply-heavy pentaquarks) [19]. Indeed, HQSS heavily constrains also the low-energy interactions among heavy hadrons [8, 10, 12, 13, 20]. As long as the hadrons are not too tightly bound, they will not probe the specific details of the interaction binding them at short distances. Moreover, each of the constituent heavy hadrons will be unable to see the internal structure of the other heavy hadron. This separation of scales can be used to formulate an effective field theory (EFT) description of hadronic molecules [10, 12] compatible with the approximate nature of HQSS. At leading order (LO) the EFT is particularly simple and it only involves energy-independent contact range interactions, since pion exchanges and coupled-channel effects can be considered subleading [12]. Thus, at very low energies, the interaction between a heavy and anti-heavy meson ($D^{(*)}\bar{D}^{(*)}$ or $B^{(*)}\bar{B}^{(*)}$) can be accurately described just in terms of a contact-range potential. The LO Lagrangian respecting HQSS contains four independent terms in the SU(3) flavor limit [13]. These are determined by the isoscalar C_{0A} and C_{0B} and the isovector C_{1A} and C_{1B} low energy constants (LEC's). The contact interaction potential is used as kernel of a two body elastic Lippmann-Swinger equation (LSE). The LSE shows an ill-defined ultraviolet (UV) behaviour, and it requires a regularization and renormalization procedure (we employ a standard Gaussian regulator as in [13, 14]). Bound states ($D^{(*)}\bar{D}^{(*)}$ or $B^{(*)}\bar{B}^{(*)}$ molecules) correspond to poles of the T-matrix below threshold on the real axis in the first Riemann sheet of the complex energy.

If we assume that the $X(3872)$ and the isovector $Z_b(10610)$ resonances are $\left(D\bar{D}^* - D^*\bar{D}\right)/\sqrt{2}$ and $\left(B\bar{B}^* + B^*\bar{B}\right)/\sqrt{2}$ bound states, respectively, and use the isospin breaking information of the decays of the $X(3872)$ into $J/\psi\pi\pi$ and $J/\psi\pi\pi\pi$, we can determine three linear combinations among the four LECs C_{0A}, C_{0B}, C_{1A} and C_{1B} with the help of HQSS and HFS [13, 14]. Note the complex isospin dynamics of the $X(3872)$ state implied by the experimental ratio $\mathcal{B}_X = \Gamma\left[X \to J/\Psi\,\rho\right]/\Gamma\left[X \to J/\Psi\,\omega\right] = 1.3 \pm 0.5$ [21]. The isospin properties of the $X(3872)$ molecule are mainly determined by its mass, which is only few hundreds of keV below the $D^0\bar{D}^{*0}$ threshold, making relevant the around 8 MeV difference between the threshold of the neutral and of the charged (D^+D^{*-}) channels. Further details are discussed in [13, 22].

2. $X(3872) \to D^0\bar{D}^0\pi^0$ decay

As it is discussed in [23], in the hadronic molecular picture, the decay channels of the $X(3872)$ with a charmonium in the final state ($J/\psi\pi\pi$, $J/\psi 3\pi$, $J/\psi\gamma$ and $\psi'\gamma$) are mainly sensitive to the short distance part of the $X(3872)$ wave-function. This is because the heavy quarks of the $D\bar{D}^*$ meson pair have to recombine to get the charmonium in the final state. The transition from the charm–anti-charm meson pair into the J/ψ plus pions (or a photon), occurs at a distance much smaller than both the size of the $X(3872)$ as a hadronic molecule and the range of forces between the D and $\bar{D}^*$ mesons. However, in the case of the $X(3872) \to D^0\bar{D}^0\pi^0$ decay, one of the constituent hadrons (D^0) is in the final state and the rest of the final particles are products of the decay of the other constituent hadron ($\bar{D}^{*0}$) of the $X(3872)$ molecule. Thus, in this decay the relative distance between the $D\bar{D}^*$ mesons can be as large as allowed by the size of the $X(3872)$ resonance, since the final state is produced by the decay of the $\bar{D}^*$ meson instead of a rescattering transition. Actually, it can be proved that within some approximations, the $d\Gamma/d|\vec{p}_{D^0}|$ distribution is related to the $X(3872)$ wave-function $\Psi(\vec{p}_{D^0})$ [23].

To estimate the $X(3872) \to D^0\bar{D}^0\pi^0$ decay width, we have evaluated the diagrams depicted in figure 1. The tree level contribution corresponds to the mechanism depicted in the first diagram

Table 1. Heavy meson–heavy meson combinations having the same contact term as the $X(3872)$ and $Z_b(10610)$, and the predictions of the masses, which are understood to correspond to bound states except if we write "V" in parenthesis for denoting a virtual state. †: increasing the strength of the potential to account for the various uncertainties, the virtual pole evolves into a bound state. Masses are given (MeV units) for two UV regulators. For further details see [14].

$I(J^{PC})$	States	Thresholds	M ($\Lambda = 0.5$ GeV)	M ($\Lambda = 1$ GeV)	Measurements
$0(1^{++})$	$\frac{1}{\sqrt{2}}(D\bar{D}^* - D^*\bar{D})$	3875.87	3871.68 (input)	3871.68 (input)	3871.68 ± 0.17 [15]
$0(2^{++})$	$D^*\bar{D}^*$	4017.3	4012^{+4}_{-5}	4012^{+5}_{-12}	?
$0(1^{++})$	$\frac{1}{\sqrt{2}}(B\bar{B}^* - B^*\bar{B})$	10604.4	10580^{+9}_{-8}	10539^{+25}_{-27}	?
$0(2^{++})$	$B^*\bar{B}^*$	10650.2	10626^{+8}_{-9}	10584^{+25}_{-27}	?
$0(2^{+})$	D^*B^*	7333.7	7322^{+6}_{-7}	7308^{+16}_{-20}	?
$1(1^{+-})$	$\frac{1}{\sqrt{2}}(B\bar{B}^* + B^*\bar{B})$	10604.4	10602.4 ± 2.0 (input)	10602.4 ± 2.0 (input)	10607.2 ± 2.0 [4] 10597 ± 9 [16]
$1(1^{+-})$	$B^*\bar{B}^*$	10650.2	10648.1 ± 2.1	$10648.1^{+2.1}_{-2.5}$	10652.2 ± 1.5 [4] 10649 ± 12 [16]
$1(1^{+-})$	$\frac{1}{\sqrt{2}}(D\bar{D}^* + D^*\bar{D})$	3875.87	3871^{+4}_{-12} (V)	3837^{+17}_{-35} (V)	$3899.0 \pm 3.6 \pm 4.9$ [17] $3894.5 \pm 6.6 \pm 4.5$ [18]
$1(1^{+-})$	$D^*\bar{D}^*$	4017.3	4013^{+4}_{-11} (V)	3983^{+17}_{-32} (V)	?
$1(1^{+})$	D^*B^*	7333.7	$7333.6^{\dagger}_{-4.2}$ (V)	7328^{+5}_{-14} (V)	?

of this figure. The amplitude is fully determined by the $D^0\bar{D}^{*0}\pi$ vertex, the $X(3872)$ mass and its coupling constant to the neutral $D^0\bar{D}^{*0}$ channel, which is determined by the residue of the $T-$matrix element at the $X(3872)$ pole. We find [23] $\Gamma(X(3872) \to D^0\bar{D}^0\pi^0)_{\rm tree} = 44.0^{+2.4}_{-7.2}\left(42.0^{+3.6}_{-7.3}\right)$ keV, where the values outside and inside the parentheses are obtained with UV cutoffs of $\Lambda = 0.5$ and 1 GeV, respectively, and the uncertainty reflects the uncertainty in the inputs ($M_{X(3872)}$ and the ratio of decay amplitudes for the $X(3872) \to J/\psi\rho$ and $X(3872) \to J/\psi\omega$ decays), and it is represented by the grey bands in figure 2.

The last two diagrams in figure 1 account for the $D\bar{D} \to D\bar{D}$ final state interaction (FSI) effects, which are considered by means of the appropriated linear combinations of the isoscalar and isovector $T-$matrices. Those are obtained by solving a LSE in coupled channels with a LO contact potential determined by the LEC's C_{0A}, C_{0B}, C_{1A} and C_{1B} introduced above [13, 23]. As mentioned, the inputs (masses of the $X(3872)$ and $Z_b(10610)$ resonances and the ratio of $X(3872) \to J/\psi\pi\pi$ and $X(3872) \to J/\psi\pi\pi\pi$ branching fractions) determine only three of the four couplings, that describe the heavy meson-antimeson S-wave interaction at LO in the heavy quark expansion. The value of the contact term parameter C_{0A} is undetermined, and thus the $D\bar{D}$ FSI effects on this decay are not fully determined. As can be seen in figure 2, these effects might be quite large, because for a certain range of C_{0A} values, a near-threshold isoscalar pole could be dynamically generated in the $D\bar{D}$ system [13, 12]. If the partial decay width is measured in future experiments, a significant deviation from the predicted value at tree level will indicate a FSI effect, which could eventually be used to extract the value of C_{0A}.

Acknowledgments
C. H.-D. thanks the support of the JAE-CSIC Program. This work is supported in part by the DFG and the NSFC through funds provided to the Sino-German CRC 110 "Symmetries and the Emergence of Structure in QCD", by the NSFC (Grant No. 11165005), by the Spanish Ministerio de Economía y Competitividad and European FEDER funds under the contract FIS2011-28853-C02-02 and the Spanish Consolider-Ingenio 2010 Programme CPAN (CSD2007-00042), by Generalitat Valenciana under contract PROMETEOII/2014/0068 and by the EU HadronPhysics3 project, grant agreement no. 283286.

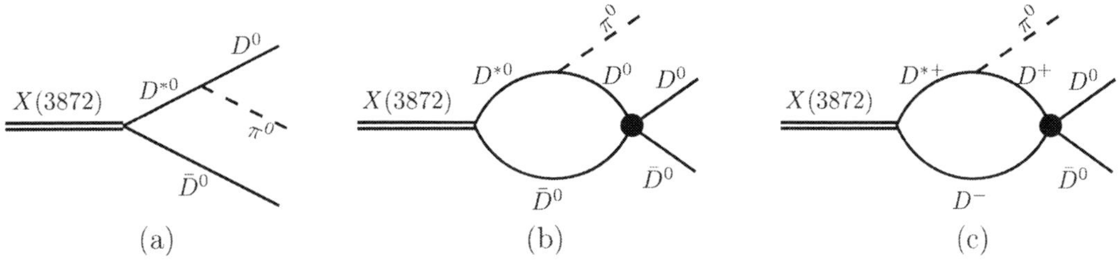

Figure 1. Feynman diagrams for the decay $X(3872) \to D^0\bar{D}^0\pi^0$. The charge conjugate channel is not shown but included in the calculations.

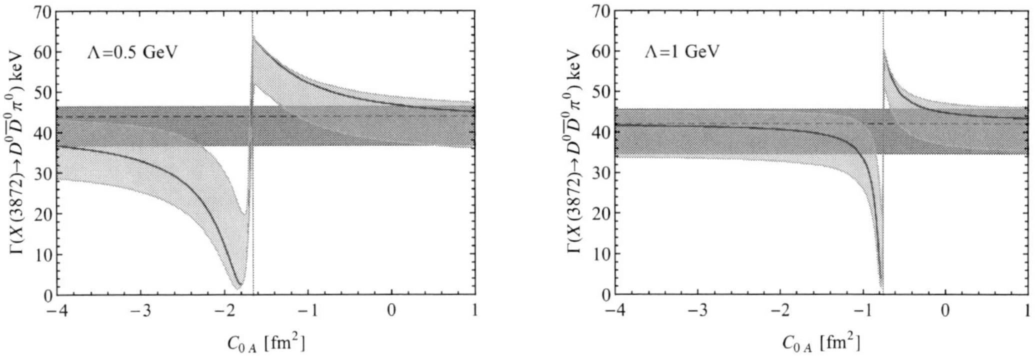

Figure 2. $X(3872) \to D^0\bar{D}^0\pi^0$ partial decay width as a function of C_{0A}. The UV cutoff is set to $\Lambda = 0.5$ GeV (1 GeV) in the left (right) panel. The blue error bands contain $D\bar{D}$ FSI effects, while the grey bands stand for the tree level predictions (see Ref [23] for details).

References

[1] Voloshin M B and Okun L B 1976 *JETP Lett.* **23** 333
[2] De Rujula A, Georgi H and Glashow S L 1977 *Phys. Rev. Lett.* **38** 317
[3] Choi S K, et al. 2003 *Phys. Rev. Lett.* **91** 262001
[4] Bondar A, et al. 2012 *Phys. Rev. Lett.* **108** 122001
[5] Adachi I, et al. 2012 Evidence for a $Z_b^0(10610)$ in Dalitz analysis of $\Upsilon(5S) \to \Upsilon(nS)\pi^0\pi^0$ *Preprint* arXiV:1207.4345
[6] Aaij R, et al. 2013 *Phys. Rev. Lett.* **110** 222001
[7] Guo F K, Hanhart C and Meißner U.G. 2009 *Phys. Rev. Lett.* **102** 242004
[8] Bondar A E, Garmash A, Milstein A I, Mizuk R and Voloshin M B (2011) *Phys. Rev.* D **84** 054010
[9] Voloshin M B 2011 *Phys. Rev.* D **84** 031502
[10] Mehen T and Powell J W 2011 *Phys. Rev.* D **84** 114013
[11] Nieves J and Valderrama M P 2011 *Phys. Rev.* D **84** 056015
[12] Nieves J and Valderrama M P 2012 *Phys. Rev.* D **86** 056004
[13] Hidalgo-Duque C, Nieves J and Valderrama M P 2013 *Phys. Rev.* D **87** 076006
[14] Guo F K, Hidalgo-Duque C, Nieves J and Valderrama M P 2013 *Phys. Rev.* D **88** 054007
[15] Beringer J, et al. 2012 *Phys. Rev.* D **86** 010001
[16] Adachi I. et al. 2012 Study of Three-Body Υ (10860) Decays *Preprint* arXiV:1209.6450
[17] Ablikim M, et al. 2013 *Phys. Rev. Lett.* **110** 252001
[18] Liu Z, et al. 2013 *Phys. Rev. Lett.* **110** 252002
[19] Guo F K, Hidalgo-Duque C, Nieves J and Valderrama M P 2013 *Phys. Rev.* D **88** 054014
[20] AlFiky M T, Gabbiani F and Petrov A A 2006 *Phys. Lett.* B **640** 238
[21] Choi S K, et al. 2011 *Phys. Rev.* D **84** 052004
[22] Gamermann D, Nieves J, Oset E and Ruiz Arriola E 2010 *Phys. Rev.* D **81** 014029
[23] Guo F K, Hidalgo-Duque C, Nieves J, Ozpineci A and Valderrama M P 2014 *Eur. Phys. J.* C **74** 2885

XI International Conference on Hyperons, Charm and Beauty Hadrons (BEACH 2014) IOP Publishing
Journal of Physics: Conference Series **556** (2014) 012016 doi:10.1088/1742-6596/556/1/012016

Properties of doubly and triply heavy baryons

K. Azizi[1], T. M. Aliev[2] and M. Savci[3]

[1] Department of Physics, Doğuş University, Acıbadem-Kadıköy, 34722 Istanbul, Turkey
[2,3]Department of Physics, Middle East Technical University, 06531 Ankara, Turkey

E-mail: [1]`kazizi@dogus.edu.tr`, [2] `taliev@metu.edu.tr`, [3]`savci@metu.edu.tr`

Abstract. We calculate the mass and residue of the doubly/triply heavy spin–1/2 and spin–3/2 baryons containing two/three heavy b or c quarks in the framework of QCD sum rules. We use the most general interpolating currents in symmetric and anti-symmetric forms with respect to the exchange of heavy quarks, to calculate the two-point correlation functions describing the baryons under consideration. A comparison of the obtained results with the existing experimental data as well as predictions of other theoretical approaches is also made.

1. Introduction

The quark model predicts heavy baryons containing single, doubly or triply heavy charm or bottom quarks having either spin–1/2 or spin–3/2. So far, all heavy baryons with single heavy quark have been discovered in the experiments except the Ω_b^* baryon with spin–3/2. In the case of doubly heavy baryons only the doubly charmed Ξ_{cc} baryon has been discovered by SELEX Collaboration [1, 2]. The experimental discovery of the doubly and triply heavy baryons and study of their properties constitute one of the main directions of the physics program at LHC. Hence, theoretical studies on the spectroscopy and decay properties of these baryons can help us in this respect.

2. QCD sum rules for the masses and residues of the doubly and triply heavy baryons

In this section we apply the QCD sum rule approach [3] to calculate the spectroscopic properties, namely masses and residues (characterizing the overlap between the baryonic states and the vacuum) of the doubly and triply heavy baryons. For this aim, we start with the following correlation functions:

$$\Pi(q) = i \int d^4x e^{iqx} \langle 0|\mathcal{T}\Big\{\eta(x)\bar{\eta}(0)\Big\}|0\rangle, \tag{1}$$

and

$$\Pi_{\mu\nu}(q) = i \int d^4x e^{iqx} \langle 0|\mathcal{T}\{\eta_\mu(x)\bar{\eta}_\nu(0)\}|0\rangle \ , \tag{2}$$

where $\eta(x)$ and $\eta_\mu(x)$ are interpolating currents of the spin–1/2 and spin–3/2 baryons, respectively. The general expressions of the interpolating currents for the spin–1/2 doubly heavy baryons in their symmetric and anti-symmetric forms can be written as

$$\begin{aligned}
\eta^S &= \frac{1}{\sqrt{2}}\epsilon_{abc}\Big\{(Q^{aT}Cq^b)\gamma_5 Q'^c + (Q'^{aT}Cq^b)\gamma_5 Q^c + \beta(Q^{aT}C\gamma_5 q^b)Q'^c + \beta(Q'^{aT}C\gamma_5 q^b)Q^c\Big\}, \\
\eta^A &= \frac{1}{\sqrt{6}}\epsilon_{abc}\Big\{2(Q^{aT}CQ'^b)\gamma_5 q^c + (Q^{aT}Cq^b)\gamma_5 Q'^c - (Q'^{aT}Cq^b)\gamma_5 Q^c + 2\beta(Q^{aT}C\gamma_5 Q'^b)q^c \\
&+ \beta(Q^{aT}C\gamma_5 q^b)Q'^c - \beta(Q'^{aT}C\gamma_5 q^b)Q^c\Big\}, \qquad (3)
\end{aligned}$$

where β is an arbitrary auxiliary parameter. The case, $\beta = -1$ corresponds to the Ioffe current. Here C stands for the charge conjugation operator, T denotes the transposition, a, b, and c are color indices; and $Q^{(')}$ and q correspond to the heavy and light quarks fields, respectively. The interpolating current with the light quark u or d corresponds to the $_{QQq}$, and with s to the $_{QQq}$ baryons, respectively. Here, we would like to note that in the symmetric part, both heavy quarks may be identical or different, but in the anti-symmetric part two heavy quarks must be different.

The interpolating current for doubly heavy baryons with spin–3/2 is written as

$$\eta_\mu = \frac{1}{\sqrt{3}}\epsilon^{abc}\Big\{(q^{aT}C\gamma_\mu Q^b)Q'^c + (q^{aT}C\gamma_\mu Q'^b)Q^c + (Q^{aT}C\gamma_\mu Q'^b)q^c\Big\}, \qquad (4)$$

where the quark content of the doubly heavy spin–3/2 baryons is shown in Table 1.

Table 1. The quark content of the spin–3/2 doubly heavy baryons.

baryon	Light quark q	Heavy quark Q	Heavy quark Q'
$^*_{QQ}$	u or d	b or c	b or c
$^*_{QQ}$	s	b or c	b or c
$^*_{QQ'}$	u or d	b	c
$^*_{QQ'}$	s	b	c

The interpolating current for the triply heavy spin–1/2 baryons can be written as

$$\eta_{QQQ'} = 2\epsilon_{abc}\Big\{\Big(Q^{aT}CQ'^b\Big)\gamma_5 Q^c + \beta\Big(Q^{aT}C\gamma_5 Q'^b\Big)Q^c\Big\}, \qquad (5)$$

where the heavy Q and Q' quarks contents of the triply heavy baryons predicted by the quark model is given in Table 2. From the current given in Eq. (5) one can formally obtain the interpolating current of the proton (neutron) by replacing $Q \to u$ and $Q' \to d$ ($Q \to d$ and $Q' \to u$). Finally, the interpolating field for the triply heavy spin–3/2 baryons can written as

Table 2. The quark contents of the triply heavy spin–1/2 baryons.

Baryon	Q	Q'
$_{bbc}$	b	c
$_{ccb}$	c	b

$$\eta_\mu = \frac{1}{\sqrt{3}}\epsilon^{abc}\Big\{2(Q^{aT}C\gamma_\mu Q'^b)Q^c + (Q^{aT}C\gamma_\mu Q^b)Q'^c\Big\} , \tag{6}$$

where the quark contents for all members of the triply heavy spin–3/2 baryons are given in Table 3.

Table 3. The quark contents of the triply heavy spin–3/2 baryons.

Baryon	Q	Q'
$^{*}_{bbc}$	b	c
$^{*}_{ccb}$	c	b
$^{*}_{bbb}$	b	b
$^{*}_{ccc}$	c	c

According to the general philosophy of the method, the correlation functions in Eqs. (1) and (2) can be calculated in two different ways. On the hadronic side, they are calculated in terms of the hadronic parameters, while on the QCD side they are evaluated in terms of QCD degrees of freedom. Matching these two representations, then, gives us QCD sum rules for the masses and residues of the baryons under consideration. To suppress the contributions of the higher states and continuum we apply Borel transformation, as well as continuum subtraction to both sides of the obtained sum rules. These procedures introduce two more auxiliary parameters, the Borel mass parameter M^2 and continuum threshold s_0. We determine their working regions such that the physical quantities depend weakly on these parameters. For details of calculations see [4, 5, 6, 7].

3. Numerical results and discussion

The numerical results for the masses and residues of the doubly and triply heavy baryons obtained from QCD sum rules are depicted in this section. Note that because of the restrictions in number of pages of this article, we only show the results of sum rules for the masses and residues of the doubly heavy baryons in tables 4 and 5 as examples. In these tables, we also present predictions of other non-perturbative approaches as well as existing experimental data.

When we look at tables 4 and 5, we see that our prediction on the mass of the $_{cc}$ baryon is in a good consistency with the only existing experimental data provided by SELEX Collaboration. Our results on the masses of doubly heavy baryons are mainly in good consistency with the predictions of other non-perturbative approaches like different quark models within the errors. In the case of residues, our predictions are comparable with the results of [9] within the errors, but considerably differ from those of [8]. These differences can be attributed to the point that [8] uses different interpolating current and obtains different spectral densities compared to our results (for more discussion see [4]). Our numerical calculations on the masses and residues of the doubly heavy spin–3/2 [5], triply heavy spin–1/2 [6] as well as triply heavy spin–3/2 [7] also show that our predictions on the masses are mainly in good consistency with the predictions of other non-perturbative approaches, but in the case of residue, somehow we see some differences between our results and existing predictions in the literature. Considering the recent progress at different hadron colliders especially LHC, we hope that we will have more experimental data on these baryons, whose comparison with the predictions of theoretical works can give us valuable information not only about the internal structures of these baryons but also about the perturbative and non-perturbative aspects of QCD.

Table 4. The mass of the doubly heavy spin–1/2 baryons (in units of GeV). The numbers inside the parentheses show the uncertainties.

Baryon	Present work	[8]	[9]	[10]	[11]	[12]	Exp [13]
$_{bb}$	9.96(0.90)	9.78(0.07)	10.17(0.14)	9.94(0.91)	10.202	–	–
$_{bb}$	9.97(0.90)	9.85(0.07)	10.32(0.14)	9.99(0.91)	10.359	–	–
$_{bc}$	6.72(0.20)	6.75(0.05)	–	6.86	6.933	7.053	–
$_{bc}$	6.75(0.30)	7.02(0.08)	–	6.864	7.088	7.148	–
$_{cc}$	3.72(0.20)	4.26(0.19)	3.57(0.14)	3.52(0.06)	3.620	3.676	3.5189(0.0009)
$_{cc}$	3.73(0.20)	4.25(0.20)	3.71(0.14)	3.53(0.06)	3.778	3.787	–
$'_{bc}$	6.79(0.20)	6.95(0.08)	–	–	6.963	7.062	–
$'_{bc}$	6.80(0.30)	7.02(0.08)	–	–	7.116	7.151	–

Table 5. The residues of the doubly heavy spin–1/2 baryons (in units of GeV^3).

Baryon	Present work	[8]	[9]
$_{bb}$	0.44(0.08)	0.067 ÷ 0.057	0.252(0.064)
$_{bb}$	0.45(0.08)	–	0.311(0.077)
$_{bc}$	0.28(0.05)	0.046 ÷ 0.021	–
$_{bc}$	0.29(0.05)	–	–
$_{cc}$	0.16(0.03)	0.042 ÷ 0.026	0.115(0.027)
$_{cc}$	0.18(0.04)	–	0.138(0.030)
$'_{bc}$	0.30(0.05)	–	–
$'_{bc}$	0.31(0.06)	–	–

References

[1] M. Mattson et al., SELEX Collaboration, Phys. Rev. Lett. 89, 112001 (2002).
[2] A. cherashvili et al., SELEX Collaboration, Phys. Lett. B 628, 18 (2005).
[3] M. A. Shifman, A. I. Vainshtein, V. I. Zakharov, Nucl. Phys. B 147, 448 (1979).
[4] T. M. Aliev, K. Azizi, M. Savci, Nucl.Phys. A 895, 59 (2012).
[5] T. M. Aliev, K. Azizi, M. Savci, J. Phys. G 40, 065003 (2013).
[6] T. M. Aliev, K. Azizi, M. Savci, JHEP 1304, 042 (2013).
[7] T. M. Aliev, K. Azizi, M. Savci, J. Phys. G 41, 065003 (2014).
[8] J. R. Zhang, M. Q. Huang, Phys. Rev. D 78, 094007 (2008).
[9] Z. G. Wang, Eur. Phys. J. A 45, 267 (2010).
[10] E. Bagan, M. Chabab, S. Narison, Phys. Lett. B 306 (1993) 350 R. M. Albu uer ue and S. Narison, Phys. Lett. B 694, 217 (2010).
[11] D. Ebert, R. N. Faustov, V. . Galkin and A. P. Martynenko, Phys. Rev. D 66, 014008 (2002).
[12] D. B. Lichtenberg, R. Roncaglia, and E. Predazzi, Phys. Rev. D 53, 6678 (1996).
[13] J. Beringer et al., (Particle Data Group), Phys. Rev. D 86, 010001 (2012).

Associated production of heavy flavors and W, Z bosons at CMS

Vieri Candelise

Università degli Studi di Trieste e INFN Trieste

Email: vieri.candelise@ts.infn.it

Abstract. The mechanism of production of heavy-flavoured jets, originated by the hadronization of b or c quarks, in association with vector bosons, W or Z, in the Standard Model is of primary importance. The study of events with one or two well-identified and isolated leptons accompanied by b-jets or displaced secondary vertices is therefore crucial to refine the theoretical calculations in perturbative QCD, as well as validate associated predictions from simulation. The understanding of these processes is furthermore required by Higgs and Beyond the Standard Model searches with similar final states. Using the LHC proton-proton collision data collected at a center of mass energy of 7 TeV by the CMS detector, measurements of the W+b, W+c, Z+b and Z+B hadrons cross sections are presented, comparing experimental data with several theoretical predictions in quantum chromodynamics.

1. Introduction

The associated production of vector bosons V (V=W,Z) and heavy flavoured quarks (HF) in proton-proton collisions at LHC is of prime interest in the physics program of the CMS experiment [1]. This process is important for several physics aspects. First of all in perturbative QCD (pQCD) studies, where direct measurement on the quark content of the proton and parton density functions (PDF) can be made through the measurement of the production cross sections. The prediction in pQCD for the strange quark content of the proton can be probed by the W+c production, and the different interpretation of the b quark density inside the proton can be tested in the Z+b measurements. Higher order pQCD effects can also be studied. From the Standard Model point of view, the W, Z + heavy flavours production allows to understand the background component in the associated production of the Higgs boson together with a vector boson. Furthermore, many Beyond the Standard Model searches (such as supersymmetric bottom partners, new Higgs bosons, 4th generation of heavy quarks) have a similar topology with respect to the V+HF production, and thus the control on their backgrounds might be highly improved by means of these measurements.

2. W+b and W+c differential cross section at 7 TeV

2.1. Associated production of W boson and c quarks

The W+c production in proton-proton collisions allows to directly probe the strange quark content of the proton. The cross section measurement is made with an integrated luminosity of 5 fb^{-1} with a center of mass energy of 7 TeV [2]. The selection is made by requiring an isolated lepton (electron, muon) with high momentum (p_T) within the pseudorapidity (η) of $|\eta|<2.1$ coming from a W boson

Journal of Physics: Conference Series **556** (2014) 012017 doi:10.1088/1742-6596/556/1/012017

decay. The c-jets are reconstructed exploiting the decays of charmed mesons and reconstructing the displaced secondary vertex of the charm quark. The transverse momentum of the c-jet is then required to be greater than 25 GeV in the pseudorapidity range of $|\eta|<2.5$. Selected events are then combined with the integrated luminosity and the efficiency for the leptons and jets identification in order to give the cross section, after the background subtraction. Results are compared to the theoretical predictions of the next-to-leading order event generator MCFM [3]. Several PDF sets have been used including MSTW08 [4], CT10 [5], NNPDF23 [6]. The main systematic uncertainty for this measurement comes from the jet energy scale. Results for the combination of electron and muon final states are presented in Fig.1, showing good agreement between measured data and the theoretical predictions. The measured cross section shows no discrepancy between data and the theoretical predictions of the different strange quark and anti-quark parton distribution functions, which are probed at an energy significantly higher than those of the previous experiments.

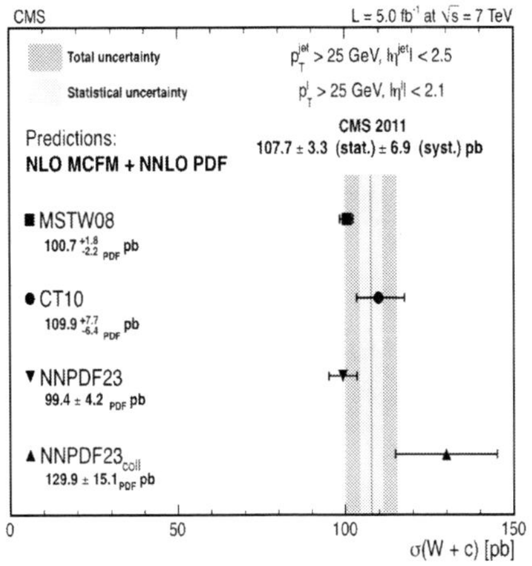

Figure 1. Comparison of the theoretical predictions for the W+c production cross section computed with MCFM and several sets of PDFs with the average of the experimental measurements, combining the electron and muon final states.

2.2. Associated production of W boson and b quarks

The production of W bosons in association with b quarks is studied using proton proton collisions at 7 TeV with the CMS experiment at the LHC [7] in a data sample corresponding to an integrated luminosity of 5.0 fb^{-1}. The W+b events are selected in the $W \rightarrow \mu\nu$ decay mode with a muon of $p_T > 25$ GeV and $|\eta|<2.1$, and exactly two b-tagged jets with $p_T > 25$ GeV and $|\eta|<2.1$. The measured particle-level cross section times the $W \rightarrow \mu\nu$ branching ratio is $0.53\pm 0.05(stat) \pm 0.10$ (sys) ± 0.01 (lumi) pb, in agreement with the Standard Model prediction. The cross section is unfolded to the level of final-state particles, and in the fiducial regime, requiring a muon with $p_T(\mu)>25$ GeV and $|\eta(\mu)|<2.1$ and exactly two particle jets, reconstructed using the anti-k_T jet algorithm with $\Delta R=0.5$ and with $p_T(bjet)>25$ GeV and $|\eta(bjet)|<2.1$, and each containing at least one b hadron with $p_T(B)>5$ GeV.

A correction factor to bring the hadron level to the level of final partons (before hadronization) is applied, computed with the MADGRAPH event generator [8]. The cross section is measured from the reconstructed objects unfolded to the level of final-state particles, corresponding to events with a muon and two particle jets matching the kinematic criteria defined above, requiring the particle jets to be matched to b hadrons. The corrected cross section is compared to the NLO cross section calculated with the MCFM event generator. The NLO cross section computed with MCFM is 0.52 ± 0.03 pb. This cross section is evaluated using the MSTW2008 NNLO PDF set, and setting the renormalization and factorization scales to $\mu(F) = \mu(R) = m_W + 2m_b$.

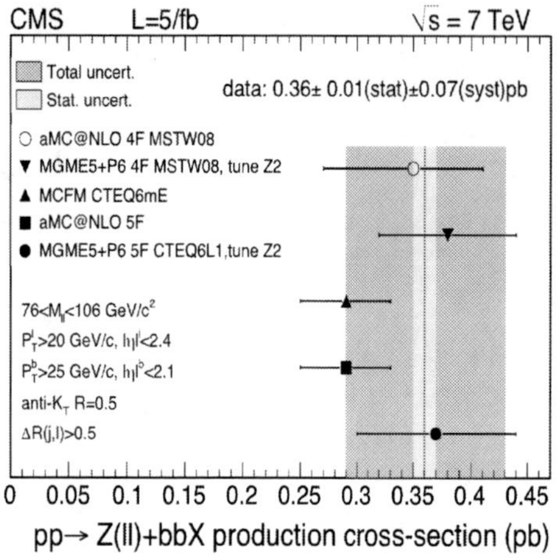

Figure 2. The inclusive Z+bb production cross section measurement compared with the theoretical expectations of the 4- and 5-flavour schemes tree level MADGRAPH calculations, the 5FS aMC@NLO, MCFM. Measurements are presented together with the relative statistical and statistical plus systematic uncertainty.

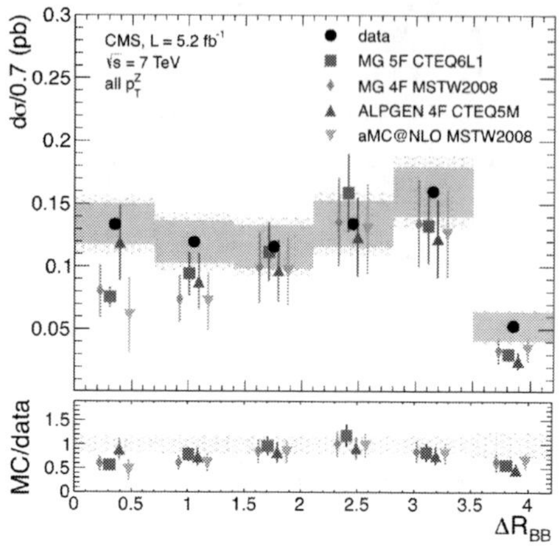

Figure 3. Differential Z+B hadrons production cross section as a function of the angular separation of the two B hadrons in the η-φ space, ΔR, compared with the 4- and 5-flavour schemes MADGRAPH event generator at tree level, aMC@NLO, ALPGEN 4FS.

3. Z+b and Z+B hadrons cross section at 7 TeV

3.1. Associated production of a Z boson and b jets

The production cross section for the process involving a Z boson decaying into two leptons (electrons and muons) in association with one or more b jets is measured in proton-proton collisions at a centre-of-mass energy of 7 TeV [9]. The data were recorded in 2011 data-taking with the CMS detector and correspond to an integrated luminosity of 5 fb^{-1}. The Z boson is selected from isolated, high momentum dilepton pairs in the invariant mass range of 76-106 GeV within the pseudorapidity range of 2.4. The b-jets are identified by means of a discriminator based on secondary vertex measurements and are selected if they have at least 25 GeV and pseudorapidity less than 2.1. The Z+b-jets cross sections are measured for a Z boson produced with exactly one or at least two b jets. The measured cross sections are compared to several theoretical predictions, and the data show better agreement with the 5-flavour scheme, where b quarks are assumed massless. The measured data are compared with the predictions from the tree-level calculation by MADGRAPH event generator rescaled to the next-to-next to leading order precision, with MSTW08 parton density function set using the PYTHIA parton shower simulation. The comparison with aMC@NLO [10] (both in the 4 and 5-flavour schemes) and MCFM event generators is also performed, using respectively MSTW08 and CTEQ6mE parton density function sets. Results are presented in Fig.2.

3.2. Associated production of Z boson and B hadrons

The inclusive and differential cross sections as function of the angular separation between B hadrons produced in association with a Z boson are measured in proton-proton collisions at a center of mass energy of 7 TeV using an integrated luminosity of 5 fb^{-1} [11]. The B hadrons are identified by means of their displaced secondary vertex only, which allows to study B-hadron pair production even at small angular separation. The differential cross section as function of the angular separation and as a function of the Z boson transverse momentum are compared to several calculations made at tree-level and NLO precisions. MADGRAPH in the 4- and 5- flavour schemes, ALPGEN and aMC@NLO in the 4-flavour scheme have been tested. The Z boson candidates are selected requiring two isolated charged leptons with high transverse momentum, making an invariant mass in the Z boson mass range, 81 to 111 GeV, and of two B hadrons with a transverse momentum greater than 15 GeV inside the pseudorapidity range of 2.0. The measurements of the differential cross section are presented in Fig.3, showing the comparison between measured data and the above mentioned theoretical predictions. The cross section result shows a discrepancy in the collinear region ($\Delta R<1$), which is dominated by diagrams q $\rightarrow$ ZbbX, involving the contribution of the gluon splitting process initiated by a quark q. In this region only ALPGEN is able to describe data, while the other predictions underestimate the cross section.

Conclusions

Standard Model W, Z boson measurements in association with heavy flavours in LHC proton-proton collision at a center of mass energy of 8 TeV with the CMS experiment have been presented. Results on the W+c process cross section, W+b jets differential cross sections, Z+b jets and Z+B hadrons have been measured and compared with different theoretical expectation from several Monte Carlo generators and PDF tunings, up to the NNLO precision in perturbative QCD when available.

References

[1] CMS Collaboration, "The CMS experiment at the CERN LHC", JINST 3 (2008) S08004, doi:10.1088/1748-0221/3/08/S08004.
[2] CMS Collaboration, "Measurement of associated W + charm production in pp collisions at sqrt(s) = 7 TeV". arXiv:1310.1138 [hep-ex] (2013).
[3] Campbell, John M. *et al.* "MCFM for the Tevatron and the LHC" (2010) Nucl.Phys.Proc.Suppl. arXiv:1007.3492 [hep-ph] FERMILAB-CONF-10-244-T
[4] Gao J., Guzzi M., Huston J., Lai H.-L., Li Z. *et al.*, "The CT10 NNLO Global Analysis of QCD" (2013) 1302.6246 arXiv hep-ph SMU-HEP-12-23.
[5] Martin A. D., Stirling W. J., Thorne R. S. and Watt G., Eur. Phys. J. C, 63 (2009) 189.
[6] CMS Collaboration, "Measurement of the production cross section for a W boson and two b jets in pp collisions at $\sqrt{s}$ = 7 TeV" arXiv:1312.6608 [hep-ex].
[7] J. Alwall *et al.*, "MadGraph 5 : Going Beyond", JHEP 1106 (2011) 128, 390 doi:10.1007/JHEP06(2011)128, arXiv:1106.0522.
[8] CMS Collaboration Collaboration, "Measurement of the cross section and angular correlations for associated production of a Z boson with b hadrons in pp collisions at $\sqrt{7}$ TeV", arXiv:0904.0391 (2013).
[9] Valentin Hirschi, Rikkert Frederix, Stefano Frixione, Maria Vittoria Garzelli, Fabio Maltoni, *et al.* "Automation of one-loop QCD corrections". JHEP, 1105: 044, 2011.
[10] Ball R. D. *et al.*, "Unbiased determination of polarized parton distributions and their uncertainties", ArXiv:1303.7236 (2013).
[11] CMS Collaboration, "Measurement of the cross section and angular correlations for associated production of a Z boson with b hadrons in pp collisions at $\sqrt{s}$=7 TeV", JHEP 12 (2013) 039.

Measurements of vector bosons with charm and beauty at ATLAS

Miriam Watson, on behalf of the ATLAS Collaboration
University of Birmingham, Edgbaston, Birmingham, B15 2TT, UK
E-mail: Miriam.Watson@cern.ch

Abstract. Several recent results from the ATLAS experiment at the Large Hadron Collider are presented, using proton-proton collisions at $\sqrt{s} = 7$ TeV. Measurements of W bosons in association with b-jets and single c-quark jets are presented, in addition to a new measurement of the production of b-jets in association with Z bosons. These measurements allow QCD predictions to be tested and also probe the parton density of the proton. The associated production of quarkonia and vector bosons is observed in the $J/\psi + W$ channel, and compared to colour singlet and colour octet production models.

1. Introduction

The production of heavy quarks in conjunction with vector bosons at hadron colliders allows the validity of quantum chromodynamics (QCD) predictions and calculations to be tested. Understanding such processes is vital in the measurement of Higgs boson decays and in searches for physics beyond the Standard Model. Vector bosons produced with bottom or charm quarks give a unique sensitivity to the heavy quark density of the proton, while the associated production of vector bosons with quarkonium states (e.g. J/ψ) can provide insight into the quarkonium production mechanism in a new environment.

All the analyses presented here are based on the data recorded in the ATLAS detector [1] at the LHC in 2011 at $\sqrt{s} = 7$ TeV, with an integrated luminosity of 4.6 fb^{-1}. Samples of W or Z bosons are selected in their muon or electron final states, which give clean signatures for triggering and lepton identification. Events including Z bosons are identified using pairs of oppositely charged leptons, close to the expected Z mass, while W identification uses the missing transverse energy as a signature. The leptons are required to be isolated, to reduce multi-jet backgrounds, using either tracking or calorimeter information around the lepton.

2. W boson production in association with b-jets

The cross-section for W boson production in association with b-quark jets has been measured for one-jet and two-jet final states [2]. This channel is an important background for the Higgs production process $WH(\to b\bar{b})$. The b-jet contribution is extracted from a template fit to weights from a b-tagging algorithm, and corrected for backgrounds and detector effects. Figure 1 shows the fiducial W+b-jets cross-sections, which are compared with next-to-leading order predictions from MCFM [3] and POWHEG [4], and leading-order ALPGEN [5] predictions. The one-jet data lie slightly above the predictions, but are consistent within 1.5σ, while the two-jet measurements agree well with the predictions. Differential cross-sections as a function of the leading b-jet p_{T} show that the predictions underestimate the one-jet data slightly at high p_{T}.

3. *W* boson production in association with charm

The production of a W boson in association with a single charm quark has also been studied [6]. To leading order, this process is dominated by the gs and $g\bar{s}$ production channels, at the level of 90%, and is therefore an excellent probe of the s-quark distribution function in the proton. Previous analyses have shown a preference for s-quark suppression relative to d-quarks in the sea; however, a recent analysis of ATLAS W and Z production and HERA DIS data [7] favoured a flavour-symmetric sea at LHC scales and $x \sim 0.01$. The W+charm data allow these properties to be tested further.

Charm quarks are tagged either using charmed hadrons in the decay modes $D^+ \to K^-\pi^+\pi^+$ and $D^{*+} \to D^0\pi^+$ (and charge conjugates), or requiring a particle jet containing a soft semileptonic decay to a muon. For the signal process, the charges of the W boson and the charm quark are expected to be opposite in sign. Since most background processes are charge-symmetric, for example W+$b\bar{b}$ or W+$c\bar{c}$, signal distributions can be extracted by subtracting same-sign from opposite-sign distributions.

The W+charm cross-sections are measured within fiducial phase space regions, and are compared with predictions from the next-to-leading order aMC@NLO [8] in combination with HERWIG++ [9] and a number of different PDF sets. These NLO and NNLO PDF sets give rise to variations of approximately 25% in the predicted cross-sections. The uncertainties on the data cover a wide range of predictions, but tend to disfavour those which have a suppressed s-quark component (MSTW2008, NNPDF2.3 and HERAPDF1.5), as shown in Figure 2 for the W^++$\bar{c}$-jet cross-section. The ratio of the s-to-d sea-quark distributions is evaluated in the context of HERAPDF1.5 as $0.96^{+0.26}_{-0.30}$ at $Q^2 = 1.9$ GeV^2. This is consistent with an SU(3)-symmetric light-quark sea. The ratio of cross-sections $(W^+ + \bar{c})/(W^- + c)$ is consistent within uncertainties with PDFs including a small s:$\bar{s}$ asymmetry and no asymmetry.

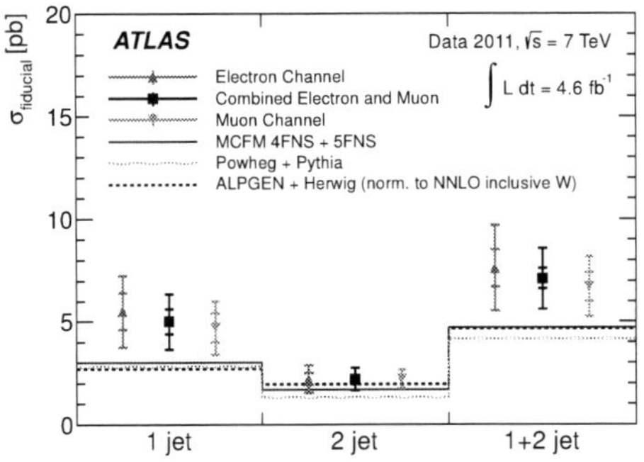

Figure 1. W+b-jet cross-sections in the 1-jet, 2-jet, and 1+2-jet fiducial regions [2].

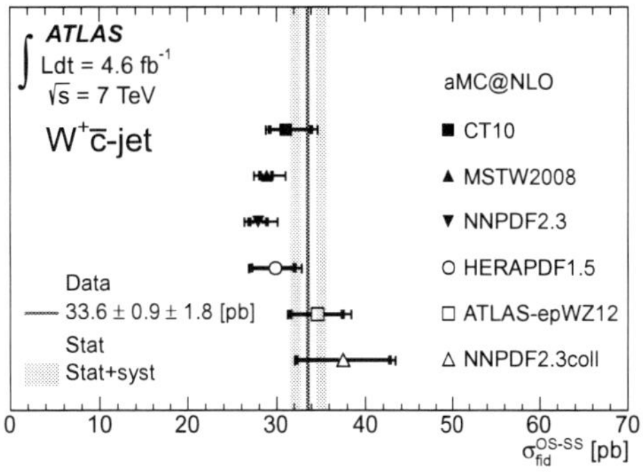

Figure 2. Measured W^++$\bar{c}$-jet fiducial cross-section compared to various PDF predictions based on aMC@NLO [6].

4. *Z* boson production in association with *b*-jets

A recent analysis of the production of Z bosons in association with b-jets [10] allows several aspects of QCD predictions to be tested. For example, the four- and five-flavour number schemes (4FNS, 5FNS) can be compared, where the 5FNS includes b-quark density in the proton, as can predictions involving massive or massless b-quarks, or calculations to different orders. The b-jets are identified using template fits to appropriate b-tagging variables in either the one-b-tag or two-b-tag cases, then yields and differential distributions are corrected for reconstruction efficiencies and detector resolution effects.

Cross-sections for $Z+\geq 1b$-jet and $Z+\geq 2b$-jets are compared with various predictions, as shown in Figure 3 for the 1-b-jet case. Firstly, the data are compared with MCFM [3], an NLO prediction in the 5FNS, evaluated for several different PDF sets. All the MCFM predictions agree with the data within uncertainties. Two different predictions from aMC@NLO [8] plus HERWIG++ [9] are also considered. One is a $Z+b$ calculation in the 5FNS and the other is $Z+b\bar{b}$ in the 4FNS. In the 1-b-jet case, both calculations are next-to-leading order and the dominant difference is the number of flavours; the data prefer the 5FNS. For 2-b-jets, the $Z+b$ 5FNS calculation is now a leading-order approximation and does not describe the data so well (in contrast to the NLO 4FNS). Leading-order predictions from ALPGEN [5] and SHERPA [11] are also shown. These underestimate the data, but do not include the theoretical uncertainties. Twelve differential cross-section distributions are measured, allowing more stringent tests of the theoretical predictions. For example, the 2-b-jet cross-section as a function of the opening angle between the b jets (Figure 4) shows some discrepancies between data and predictions at small angles.

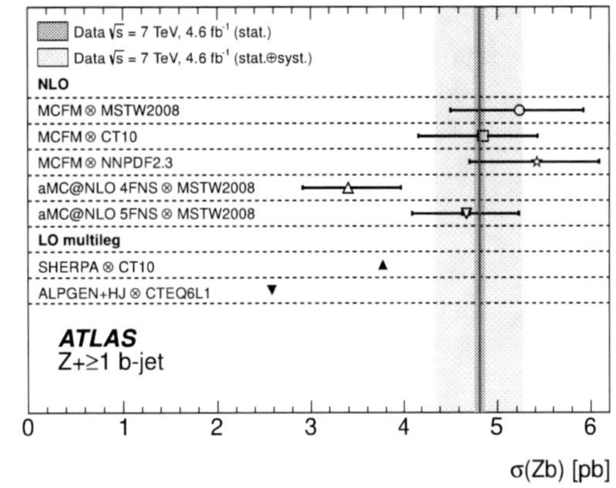

Figure 3. Cross-section for $Z+\geq 1b$-jet and comparison with theoretical predictions [10].

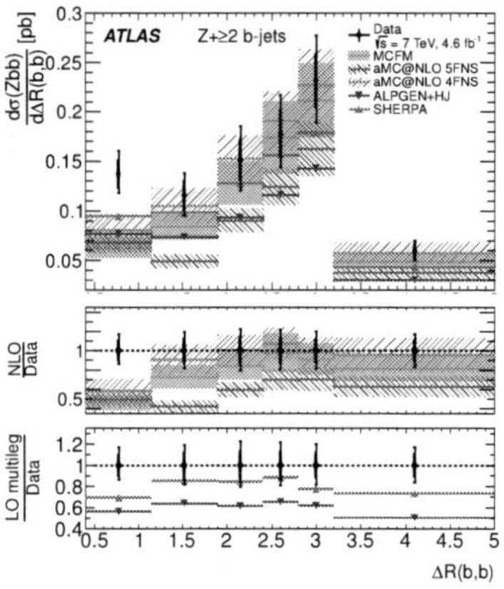

Figure 4. The cross-section for $Z+\geq 2b$-jets as a function of the opening angle ΔR(b, b) [10].

5. W boson production with prompt J/ψ mesons

An alternative measurement, which looks for quarkonium formation instead of b- or c-jets, is the associated production of W bosons with prompt J/ψ [12], where the J/ψ and W boson are both identified in their muon decay modes. The measurement probes the quarkonium production mechanism and the possible contributions from colour singlet and colour octet processes. This signature will also be sensitive to multiple parton interactions, in which the W and J/ψ are produced by two pairs of partons from the same proton-proton collision, and the current measurement includes an estimate of the double parton scattering (DPS) contribution.

As with many J/ψ analyses, the invariant mass and pseudo-proper time are fitted simultaneously to extract the prompt J/ψ component. Each prompt candidate then receives a weight in the W transverse mass distribution, which is fitted with W and multi-jet templates, to estimate a multi-jet contamination of 0.1 ± 4.6 events. The associated prompt J/ψ plus W signal is ~ 27 events, which is incompatible with the background-only hypothesis at 5.1σ, and is the first observation of this process.

The cross-section of W+J/ψ is measured relative to the inclusive W cross-section in the same dataset. The contribution from DPS is estimated under the assumption that the W and J/ψ contributions are independent and uncorrelated, and are scaled by the effective cross-section obtained from W+ 2-jet events at ATLAS [13]: its magnitude is $\sim 40\%$ of the total. The azimuthal angle between the W and the J/ψ

will be uniform for independent DPS processes but will peak at $\sim \pi$ for single scattering (SPS) events. The data appear to show both components, as expected, and the differential cross-section as a function of p_{T} (Figure 5) indicates that SPS dominates at low p_{T} of the J/ψ. The cross-section ratio is corrected for J/ψ acceptance and DPS, then the resulting SPS cross-section is compared with predictions from two models: a LO colour singlet model [14] and NLO colour octet model [15]. The theoretical predictions lie below the data, but both are within 2σ of the large experimental uncertainties, as shown in Figure 6.

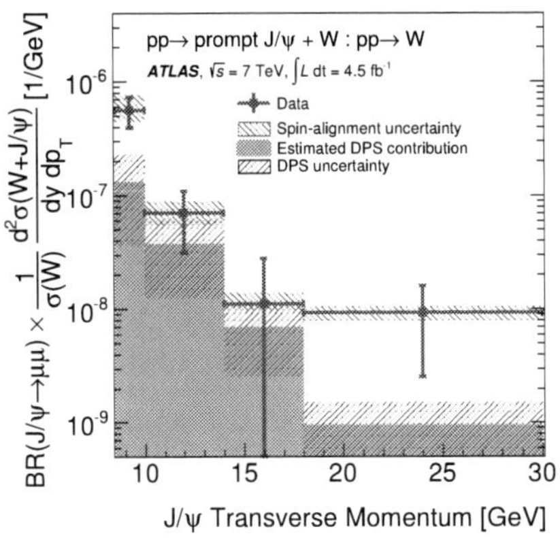

Figure 5. The W+prompt J/ψ to inclusive W production cross-section ratio as a function of J/ψ p_{T} [12].

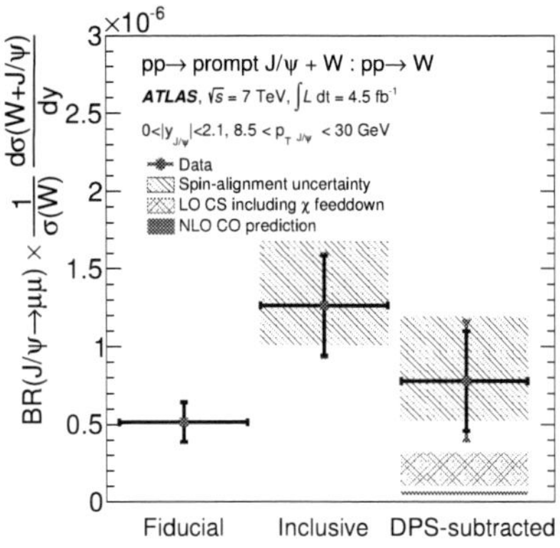

Figure 6. The W+prompt J/ψ to inclusive W production cross-section ratio [12].

6. Summary

Recent measurements of heavy flavour and vector boson production in ATLAS have been presented. These allow more stringent tests of QCD predictions in a new regime: measurements with *b*- or *c*-jets give access to the proton structure function and heavy flavour densities, while measurements of quarkonium with vector bosons open a new field in understanding both quarkonium production and multiple parton interactions.

Acknowledgments

The author is supported by a Royal Society Research Fellowship.

References

[1] ATLAS Collaboration 2008 JINST **3** S08003
[2] ATLAS Collaboration 2013 *J. High Energy Phys.* JHEP06(2013)084
[3] Campbell J M and Ellis R 2010, *Nucl. Phys. Proc. Suppl.* **205-206** 10
[4] Alioli S, Nason P, Oleari C and Re E 2010 *J. High Energy Phys.* JHEP06(2010)043
[5] Mangano M L, Moretti M, Piccinini F, Pittau R and Polosa A D 2003 *J. High Energy Phys.* JHEP07(2003)001
[6] ATLAS Collaboration 2014 *J. High Energy Phys.* JHEP05(2014)068
[7] ATLAS Collaboration 2014 *Phys. Rev. Lett.* **109** 012001
[8] Alwall J et al. 2014 *J. High Energy Phys.* JHEP07(2014)079
[9] Bahr M et al. 2008 *Eur. Phys. J.* C **58** 639
[10] ATLAS Collaboration 2014, submitted to *J. High Energy Phys.*; arXiv:1407.3643
[11] Gleisberg T et al. 2009 *J. High Energy Phys.* JHEP02(2009)007
[12] ATLAS Collaboration 2014 *J. High Energy Phys.* JHEP04(2014)172
[13] ATLAS Collaboration 2013, *New J. Phys.* **15** 033038
[14] Lansberg J P and Lorcé C 2013 *Phys. Lett.* B **726** 218-222
[15] Gang L, Mao S, Ren-You Z and Wen-Gan M 2011 *Phys. Rev.* D **83** 014001

Mass and Lifetime measurements at LHCb

Lucio Anderlini
on behalf of the LHCb Collaboration.
Università di Firenze, INFN Sezione di Firenze, via G. Sansone 1, Sesto Fiorentino, Italy
E-mail: Lucio.Anderlini@fi.infn.it

Abstract. The data collected at LHCb during the first run of the LHC are analysed to perform precision measurement of mass and lifetime of hadrons containing heavy quarks. I report here about recent mass measurements of the Ξ_c^0, Λ_b^0, Ξ_b^-, and Ω_b^- baryons, and of charged and neutral D mesons. Recent precision measurements of the lifetime of Λ_b^0, Ξ_b^0 and Ω_b^- baryons are also reported. I conclude this report with an overview on the experimental knowledge of the $B^0_{s,H}$, $B^0_{s,L}$, and B^0_s flavour specific effective lifetimes. LHCb has performed the world best measurements of most of these quantities.

The large amount of hadrons containing heavy quarks produced at the LHC allows to perform measurements of the properties of b- and c- hadrons with unprecedented precision. The LHCb detector, described in Ref. [1], offers an excellent resolution on the momentum of charged tracks and on the position of the secondary vertices, and therefore on mass and lifetime of the reconstructed particles.

The recent mass and lifetime measurements summarized in this paper have been obtained analyzing the dataset of pp collisions collected with the LHCb detector in 2011 and 2012, at 7 and 8 TeV of energy in the center of mass frame, and corresponding to an integrated luminosity of 1 and 2 fb^{-1}, respectively.

1. Mass measurements

The LHCb Collaboration has recently reported the mass measurements of the charmed mesons D^0, D^+ and D_s^+ performed analysing the dataset collected in 2011 [2]. The following decays have been studied: $D^0 \to K^+K^-\pi^+\pi^-$, $D^0 \to K^+K^-K^-\pi^+$, $D^+ \to K^+K^-\pi^+$, and $D_s^+ \to K^+K^-\pi^+$; for which 4608 ± 89, 849 ± 36, $68\,787 \pm 321$, and 248694 ± 540 candidates were observed, respectively. The mass measurements are obtained with extended unbinned maximum likelihood fits to the distributions of the invariant mass of the daughters. The D^0 mass measurements obtained with the 3- and 4-body decays are combined to give $M(D^0) = 1864.75 \pm 0.15\,(\text{stat}) \pm 0.11\,(\text{syst})$ MeV/c^2, where the systematic uncertainty, correlated between the two channels, is dominated by the limited knowledge of the momentum scale. Such a contribution partially cancels when evaluating mass differences, so that it is preferable to quote the results on D^+ and D_s^+ masses as

$$M(D^+) - M(D^0) = 4.76 \pm 0.12\,(\text{stat}) \pm 0.07\,(\text{syst})\ \text{MeV}/c^2, \text{ and}$$
$$M(D_s^+) - M(D^+) = 98.68 \pm 0.03\,(\text{stat}) \pm 0.04\,(\text{syst})\ \text{MeV}/c^2.$$

To measure the mass of the Ξ_c^+ baryon LHCb has studied the decays $\Xi_b^0 \to \Xi_c^+\pi^-$ with $\Xi_c^+ \to pK^-\pi^+$, using the decay $\Lambda_c^+ \to pK^-\pi^+$ as a control sample [3]. The mass difference measured using the combined 2011 and 2012 datasets is

$$M(\Xi_c^+) - M(\Lambda_c^+) = 181.51 \pm 0.14\,(\text{stat}) \pm 0.10\,(\text{syst})\ \text{MeV}/c^2.$$

In the same analysis, the mass of the Ξ_b^0 baryon relative to the Λ_b^0 mass was measured to be

$$M(\Xi_b^0) - M(\Lambda_b^0) = 172.44 \pm 0.39\ (\text{stat}) \pm 0.17\ (\text{syst})\ \text{MeV}/c^2.$$

Other interesting LHCb results on b-baryons include the analysis of the 2011 dataset to measure the masses of the Λ_b^0, Ξ_b^-, and Ω_b^- baryons [4]. The mass measurements are obtained with maximum likelihood fits on the invariant mass distributions of 6870 ± 110 $\Lambda_b^0 \to J/\psi\Lambda^0$, 111 ± 12 $\Xi_b^- \to J/\psi\Xi^-$, and 19 ± 5 $\Omega_b^- \to J/\psi\Omega^-$ candidates. The statistical significance of the decay $\Omega_b^- \to J/\psi\Omega^-$ is 6 standard deviations. The measured values are

$$\begin{aligned} M(\Lambda_b^0) &= 5619.53 \pm 0.13\,(\text{stat}) \pm 0.45\,(\text{syst})\ \text{MeV}/c^2, \\ M(\Xi_b^-) &= 5795.8 \pm 0.9\,(\text{stat}) \pm 0.4\,(\text{syst})\ \text{MeV}/c^2, \\ M(\Omega_b^-) &= 6046.0 \pm 2.2\,(\text{stat}) \pm 0.5\,(\text{syst})\ \text{MeV}/c^2. \end{aligned}$$

As for the measurements discussed above, the systematic uncertainty is dominated by the limited knowledge of the momentum scale.

2. Measurements of baryon lifetimes

Combining the optical theorem, the operator product expansion, and some results of the Heavy Quark Effective Theory (HQET) leads to a theoretical prediction for the decay width, and hence lifetime, for each heavy hadron, dominated by the "free-quark lifetime" with QCD corrections suppressed by the squared mass of the heavy quark [5]. Testing the theoretical estimates of these corrections is important to validate this approach, named *Heavy Quark Expansion* (HQE).

Historically, HQE could hardly accommodate the LEP experimental results of the Λ_b^0 lifetime. Indeed, while the ratio $\tau(\Lambda_b^0)/\tau(\bar{B}^0)$ was predicted to be larger than ~ 0.9 [6], the world average in 2003 was 0.798 ± 0.052 [7]. Since then, many updates to the experimental average and to the theoretical prediction were made, the latter including modern matrix element calculations with lattice QCD, but inconsistency remained important until the first measurements by LHCb.

The first LHCb analysis, measuring the Λ_b^0 lifetime using the $\Lambda_b^0 \to J/\psi\Lambda^0$ decay channel was limited by the systematic uncertainty on the proper time dependence of the efficiency function, due to the trigger requirements on the impact parameter of the daughter tracks [8]. Relative lifetime measurements are much less affected by this uncertainty. If B_1 and B_2 are two generic b hadrons decaying to the similar final states f_1 and f_2; τ_1 and τ_2 their lifetimes, and $\epsilon_1(t)$ and $\epsilon_2(t)$ the reconstruction efficiencies of the two final states dependent on the decay time t, one can write the number of reconstructed events for f_1 and f_2 as $N_1(t) \propto \epsilon_1(t)\exp(-t/\tau_1)$ and $N_2(t) \propto \epsilon_2(t)\exp(-t/\tau_2)$, where the experimental resolution on the decay time is neglected. If the final states are similar, the ratio $\epsilon_1(t)/\epsilon_2(t)$ is nearly unity, and small corrections can be safely obtained from simulation. A fit to

$$R(t) = \frac{N_1(t)}{N_2(t)} = \frac{\epsilon_1(t)}{\epsilon_2(t)} \exp\left[\left(\frac{1}{\tau_2} - \frac{1}{\tau_1}\right)t\right]$$

is used to extract τ_1 known τ_2 (or vice-versa).

Applying this technique to $\Lambda_b^0 \to J/\psi pK^-$ decays in the 2011 dataset [9] before, and then updating the result including the data collected in 2012 [10], LHCb measured the $\tau(\Lambda_b^0)$ relative to the lifetime $\tau(\bar{B}^0)$ of the $\bar{B}^0$ meson as reconstructed in the $J/\psi\pi^+K^-$ final state. The results,

$$\frac{\tau(\Lambda_b^0)}{\tau(\bar{B}^0)} = 0.974 \pm 0.006\ (\text{stat}) \pm 0.004(\text{syst}), \quad \text{and} \quad \tau(\Lambda_b^0) = 1.479 \pm 0.009\ (\text{stat}) \pm 0.010\ (\text{syst})\ \text{ps},$$

Journal of Physics: Conference Series **556** (2014) 012019 doi:10.1088/1742-6596/556/1/012019

are consistent with a modern HQE prediction $\tau(\Lambda_b^0)^{\rm HQE} = 1.41 \pm 0.08$ ps [5].

The same method has been applied to the 2011 and 2012 datasets to measure the Ξ_b^- lifetime using the decay $\Lambda_b^0 \to \Lambda_c^+\pi^-$ as normalization [4]. The result, $\tau(\Xi_b^-) = 1.55^{+0.10}_{-0.09}$ (stat) $\pm$ 0.03 (syst) ps, is consistent with the HQE prediction $\tau(\Xi_b^-)^{\rm HQE} = 1.56 \pm 0.10$ ps [5].

In some cases, there are no suitable normalization channels to perform a relative lifetime measurement and the determination of the efficiency as a function of the decay time requires other strategies. The optimization of the simulation reliability has allowed to obtain lifetime precision measurements for the Ξ_b^0 and Ω_b^- with a systematic uncertainty smaller than the statistical error. The decays $\Xi_b^- \to J/\psi\Xi^-$ and $\Omega_b^- \to J/\psi\Omega^-$ have been studied in the 2011 and 2012 datasets. 313 ± 20 Ξ_b^- candidates, and 58 ± 8 Ω_b^- candidates were used to determine the lifetimes of the two heavy baryons:

$$\tau(\Xi_b^-) = 1.55^{+0.10}_{-0.09}\ (\text{stat}) \pm 0.03\ (\text{syst})\ \text{ps},$$
$$\tau(\Omega_b^-) = 1.54^{+0.26}_{-0.21}\ (\text{stat}) \pm 0.05\ (\text{syst})\ \text{ps}.$$

The HQE prediction for the Ξ_b^- lifetime [5], $\tau(\Xi_b^-)^{\rm HQE} = 1.56 \pm 0.10$ ps is found in good agreement with the experimental result.

3. Measurements of the lifetimes of the B_s^0-$\bar{B}_s^0$ system

Experimental techniques similar to those applied to measure the baryon lifetimes can be used to determine the lifetimes of B_s^0 states. However, the eigenstates of the hamiltonian and time-evolution operators are not the flavour eigenstates B_s^0 $(\bar{b}s)$ and $\bar{B}_s^0$ $(b\bar{s})$, but rather their combinations $|B_{s,L}^0\rangle = p|B_s^0\rangle + q|\bar{B}_s^0\rangle$, and $|B_{s,H}^0\rangle = p|B_s^0\rangle - q|\bar{B}_s^0\rangle$. The $B_{s,L}^0$ and $B_{s,H}^0$ eigenstates have decay width Γ_L and Γ_H, respectively. Their average and difference are indicated as $\Gamma_s = (\Gamma_L + \Gamma_H)/2$ and $\Delta\Gamma_s = \Gamma_L - \Gamma_H$. Defining A_f and $\bar{A}_f$ as the decay amplitudes to the generic final state f: $B_s^0 \to f$ and $\bar{B}_s^0 \to f$, respectively; and $\lambda_f = \frac{q}{p}\frac{\bar{A}_f}{A_f}$, one neglects production asymmetries to writes the decay width of a B_s^0 mixture to f as

$$\Gamma[f,t] = \Gamma(B_s^0(t) \to f) + \Gamma(\bar{B}_s^0(t) \to f) \propto e^{-\Gamma_L t}|\langle f|B_{s,L}^0\rangle|^2 + e^{-\Gamma_H t}|\langle f|B_{s,H}^0\rangle|^2, \tag{1}$$

and approximated at the second order in $\Delta\Gamma_s t$ as [11]

$$\Gamma[f,t] = e^{-\Gamma_s t}\left[1 + \frac{1}{2}\left(\frac{\Delta\Gamma_s}{2}t\right)^2 + \frac{-2\mathcal{R}e(\lambda_f)}{1+|\lambda_f|^2}\left(\frac{\Delta\Gamma_s}{2}t\right)\right]. \tag{2}$$

Neglecting CP violation, measured to be small [12], when f is an even (odd) CP-eigenstate, $\langle f|B_{s,H}^0\rangle$ $(\langle f|B_{s,L}^0\rangle)$ is null for CP conservation, and time dependence in Equation 1 simplifies into a single exponential whose slope depends on Γ_L (Γ_H). For "flavour specific" B_s^0 decays, where $\bar{B}_s^0 \to \bar{f}$ and $B_s^0 \to f$, the decay is the sum of two exponentials that when fitted with a single exponential allows to evaluate an *effective lifetime* approximated by [13]

$$\tau_{\rm fs} = \frac{1}{\Gamma_s}\frac{1 + \left(\frac{\Delta\Gamma_s}{2\Gamma_s}\right)^2}{1 - \left(\frac{\Delta\Gamma_s}{2\Gamma_s}\right)^2}.$$

Experimentally, LHCb reported recently an analysis of the 2011 dataset to measure the (effective) lifetimes of the B_d^0 and B_s^0 mesons in charmless 2-body decays [14]. The time-dependent acceptance function was obtained with a data-driven technique named "swimming" already in use at LEP and at TeVatron experiments [15, 16]. It consists of an event-by-event determination of the *turning points* of the efficiency at which the single candidate efficiency changes from *not-selected* to *selected* by gradually increasing the distance between the production and decay vertices. The B_s^0 decay channels studied are $B_s^0 \to K^+K^-$, yielding the measurement of τ_L possibly polluted by penguin-diagram contributions $\tau_{B_s^0 \to K^+K^-} = 1.407 \pm 0.016$ (stat) $\pm$ 0.007 (syst) ps, and $B_s^0 \to \pi^+K^-$, yielding the measurement of the flavour specific B_s^0 lifetime $\tau_{B_s^0 \to \pi^+K^-} = 1.60 \pm 0.06$ (stat) $\pm$ 0.01 (syst) ps. The lifetime of the B_d^0 meson is also measured using the decay channel $B_d^0 \to K^+\pi^-$ to be $\tau_{B_d^0 \to K^+\pi^-} = 1.524 \pm 0.011$ (stat) $\pm$ 0.004 (syst) ps.

Some sensitivity to physics beyond the Standard Model is offered by the comparison of the results for $\tau_{B_s^0 \to K^+K^-}$ with a previous measurement by LHCb determining the lifetime of the $B_{s,L}^0$ eigenstate through a decay channel less polluted by penguin contributions: $B_s^0 \to D_s^- D_s^+$ [17]. The analysis, performed on the combined 2011 and 2012 datasets, used $B^- \to D^- D_s^+$ as normalization channel and obtained the result $\tau_{B_s^0 \to D_s^- D_s^+} = 1.379 \pm 0.026$ (stat) $\pm$ 0.017 (syst) fs, in good agreement with $\tau_{B_s^0 \to K^+K^-}$.

In the same paper, LHCb reported the measurement of the flavour specific lifetime $\tau_{\bar{B}_s^0 \to D^- D_s^+}$ relative to $\tau_{B_d^0}$ as measured in the decay channel $B_d^0 \to D^- D_s^+$. The result, $\tau_{\bar{B}_s^0 \to D^- D_s^+} =$ 1.52$\pm$0.15 (stat) $\pm$0.01(syst) ps, is consistent with a more recent and more precise determination of the flavour specific B_s^0 lifetime obtained analysing the decays $B_s^0 \to D_s^- \pi^+$, followed by $D_s^- \to K^+K^-\pi^+$ in the combined 2011 and 2012 LHCb datasets [18]. The B_s^0 lifetime was measured relative to three normalization channels: $B^- \to D^0\pi^-$ with $D^0 \to K^-\pi^+$, $B^- \to D^0\pi^-$ with $D^0 \to K^-\pi^+\pi^-\pi^+$, and $B^0 \to D^+\pi^-$ with $D^+ \to K^-\pi^+\pi^+$. The results obtained with the three normalization channels are perfectly consistent, but have fully correlated uncertainties and therefore their combination would not improve the precision of the measurement. The best result was obtained using the $B^0 \to D^+\pi^-$ channel as normalization: $\tau_{\bar{B}_s^0 \to D_s^+\pi^-} = 1.535 \pm 0.015$ (stat) ± 0.012 (syst) $\pm 0.007(\tau_{B^-})$ ps, with the third uncertainty being on the world average of the lifetime measurements of the B^- meson.

4. Conclusion

The excellent performance of the LHC and of the LHCb detector has allowed many world leading results on b- and c-hadron masses and lifetimes. Consistency with theoretical predictions is good for both these important quantities, and notably for lifetimes where the recent results restore confidence in the HQE formalism.

The future run of LHC will allow to further challenge new physics with lifetime measurements of both mesons and baryons, and to include new heavier probes such as Ξ_b, Ω_b^-, and B_c^+ states.

References

[1] Alves A Augusto J *et al.* (LHCb Collaboration) 2008 *JINST* **3** S08005
[2] Aaij R *et al.* (LHCb collaboration) 2013 *JHEP* **1306** 065 (*Preprint* 1304.6865)
[3] Aaij R *et al.* (LHCb collaboration) 2014 *Phys.Rev.Lett.* **113** 032001 (*Preprint* 1405.7223)
[4] Aaij R *et al.* (LHCb collaboration) 2013 *Phys.Rev.Lett.* **110** 182001 (*Preprint* 1302.1072)
[5] Lenz A 2014 (*Preprint* 1405.3601)
[6] Uraltsev N 2000 (*Preprint* hep-ph/0010328)
[7] Battaglia M, Buras A, Gambino P, Stocchi A, Abbaneo D *et al.* 2003 (*Preprint* hep-ph/0304132)
[8] Aaij R *et al.* (LHCb collaboration) 2014 *JHEP* **1404** 114 (*Preprint* 1402.2554)
[9] Aaij R *et al.* (LHCb collaboration) 2013 *Phys.Rev.Lett.* **111** 102003 (*Preprint* 1307.2476)
[10] Aaij R *et al.* (LHCb Collaboration) 2014 *Phys.Lett.* **B734** 122 (*Preprint* 1402.6242)
[11] Stone S 2014 (*Preprint* 1406.6497)
[12] Amhis Y *et al.* (Heavy Flavor Averaging Group) 2012 (*Preprint* 1207.1158)
[13] Hartkorn K and Moser H 1999 *Eur.Phys.J.* **C8** 381–383
[14] Aaij R *et al.* (LHCb collaboration) 2014 (*Preprint* 1406.7204)
[15] Adam W *et al.* (DELPHI Collaboration) 1995 *Z.Phys.* **C68** 363–374
[16] Aaltonen T *et al.* (CDF Collaboration) 2011 *Phys.Rev.* **D83** 032008 (*Preprint* 1004.4855)
[17] Aaij R *et al.* (LHCb Collaboration) 2014 *Phys.Rev.Lett.* **112** 111802 (*Preprint* 1312.1217)
[18] Aaij R *et al.* (LHCb Collaboration) 2014 (*Preprint* 1407.5873)

Journal of Physics: Conference Series **556** (2014) 012020 doi:10.1088/1742-6596/556/1/012020

Open heavy-flavour and quarkonium measurements with ALICE at the LHC

Nicole Bastid for the ALICE Collaboration

Laboratoire de Physique Corpusculaire, Clermont Université, Université Blaise Pascal, CNRS-IN2P3, Clermont-Ferrand, France

E-mail: nicole.bastid@clermont.in2p3.fr

Abstract. The ALICE Collaboration at the LHC measured open heavy flavours and quarkonia over a wide rapidity and transverse momentum range, and in several decay channels in pp, p–Pb and Pb–Pb collisions. An overview of recent results in p–Pb collisions at $\sqrt{s_{\rm NN}} = 5.02$ TeV and Pb–Pb collisions at $\sqrt{s_{\rm NN}} = 2.76$ TeV is presented. A particular emphasis is placed on the nuclear modification factors.

1. Introduction

The ALICE experiment [1] at the LHC aims at investigating the properties of the hot and dense state of strongly-interacting matter, the Quark Gluon Plasma (QGP), produced in high-energy heavy-ion collisions. Heavy quarks (charm and beauty) that are created in a short time scale with respect to that of the QGP formation and subsequently interact with the medium, are regarded as efficient probes of the QGP properties. Open heavy-flavour hadrons are expected to be sensitive to the energy density of the system through the mechanism of in-medium energy loss of heavy quarks. According to QCD, quarks should lose less energy than gluons, and due to the dead cone effect [2], heavy-quark energy loss should be further reduced with respect to that of light quarks. Quarkonia could provide an estimate of the medium temperature through their dissociation in the QGP, due to colour screening [3]. In this scenario, the different binding energy of the various states should lead to a sequential suppression pattern where the strongly bound states (J/ψ, Υ(1S)) melt at higher temperatures than the more loosely bound states (ψ(2S), $\chi_{\rm c}$, Υ(2S), Υ(3S)). On the other hand, due to the abundant production of charm quarks in heavy-ion collisions at LHC energies, charmonium suppression could be counteracted by regeneration from recombination of c and $\bar{\rm c}$ quarks in the QGP and/or at hadronization [4]. The medium properties are usually investigated by means of the nuclear modification factor, $R_{\rm AA}$, defined as the ratio of the particle yield measured in Pb–Pb to that observed in pp collisions scaled with the number of binary nucleon-nucleon collisions. Further insights into the properties of the medium created in heavy-ion collisions can be gained from the study of the azimuthal anisotropy of open heavy flavours and quarkonia in semi-central collisions (not discussed here, see [5]). The heavy-ion physics program requires also the study of pp and p–Pb collisions. Besides providing the essential reference for the nuclear modification factor, pp collisions allow a sensitive test of models describing the production of heavy flavours in elementary hadronic collisions at LHC energies. The study of p–Pb collisions is crucial to access cold nuclear matter (CNM) effects in the initial and final state, assuming that an extended QGP is not formed in these collisions.

2. Open heavy-flavour and quarkonium measurements with ALICE

The ALICE experiment, described in detail in [1], consists of a central barrel ($|\eta| < 0.9$), a muon spectrometer ($-4 < \eta < -2.5$) and a set of smaller detectors located in the forward and backward pseudo-rapidity regions for global collision characterization and triggering purposes. At mid-rapidity, tracking is performed by combining information from the Time Projection Chamber (TPC) and the Inner Tracking System (ITS). Charged particle identification is provided by their specific energy loss in the TPC and their time of flight measured with the TOF detector. The electron identification can include, in addition, information from the Transition Radiation Detector (TRD) and the Electromagnetic Calorimeter (EMCAL). Muons with $p > 4$ GeV/c, p being the momentum, are identified at forward rapidity with the muon spectrometer.

Open heavy flavours are measured in the charm hadronic decay channels (including $\mathrm{D}^0 \to \mathrm{K}^-\pi^+$, $\mathrm{D}^+ \to \mathrm{K}^-\pi^+\pi^+$, $\mathrm{D}^{*+} \to \mathrm{D}^0\pi^+$, $\mathrm{D}_s^+ \to \phi\pi^+$ with $\phi \to \mathrm{K}^+\mathrm{K}^-$ and charge conjugates in $|y| < 0.5$) and with electrons from semi-leptonic decays of charm and/or beauty at mid-rapidity, and in the semi-muonic decay of charm and beauty hadrons at forward rapidity. Electrons from beauty decays can be isolated at mid-rapidity by exploiting the precise measurement of the track impact parameter and their displacement from the interaction vertex. J/ψ are reconstructed down to $p_\mathrm{T} = 0$, p_T being the transverse momentum, via unlike-sign dielectrons and unlike-sign dimuons at mid-rapidity ($|y| < 0.9$) and forward rapidity ($2.5 < y < 4$), respectively. In addition, ψ(2S) and Υ(1S) are also measured down to $p_\mathrm{T} = 0$ at forward rapidity via unlike-sign muon pairs. Details on the reconstruction of D mesons, J/ψ, ψ(2S) and Υ(1S), and on the identification of electrons and muons from heavy-flavour hadron decays can be found in [6, 7, 8, 9, 10, 11].

3. Results

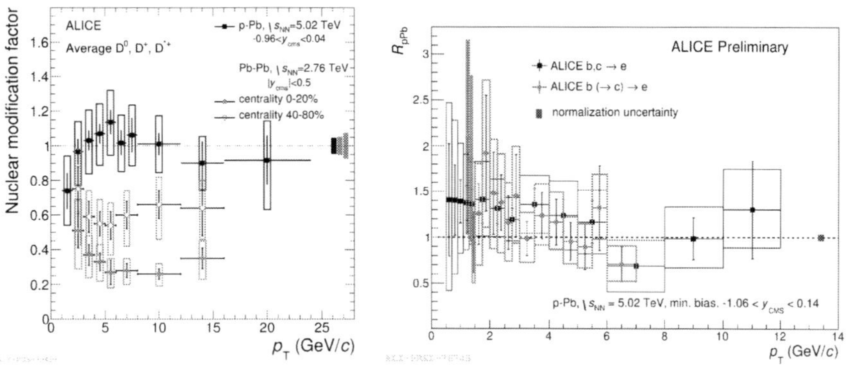

Figure 1. Left: p_T-differential R_pPb of D mesons in p–Pb collisions at $\sqrt{s_\mathrm{NN}} = 5.02$ TeV and in Pb–Pb collisions at $\sqrt{s_\mathrm{NN}} = 2.76$ TeV for the 0–20% and 40–80% centrality classes [6]. Right: p_T-differential R_pPb of electrons from heavy-flavour (c, b) and beauty-hadron decays. Statistical (bars), systematic (empty boxes) and normalization (filled boxes) uncertainties are displayed.

Figure 1 (left) presents the average of the nuclear modification factors of prompt D^0, D^+ and D^{*+} as a function of p_T, measured at mid-rapidity in minimum bias p–Pb collisions at $\sqrt{s_\mathrm{NN}} = 5.02$ TeV (R_pPb), and central (0–20%) and peripheral (40–80%) Pb–Pb collisions at $\sqrt{s_\mathrm{NN}} = 2.76$ TeV. In the region $p_\mathrm{T} > 2$ GeV/c, R_pPb is consistent with unity within uncertainties (see also [6]). A similar behaviour is also reported for electrons from heavy-flavour (c, b) and beauty-hadron decays (Fig. 1, right). This indicates that cold nuclear matter effects are small and that the significant suppression observed in central collisions for D mesons (Fig. 1, left) and electrons from heavy-flavour (b,c) and beauty-hadron decays [12] is a final state effect due to the dense

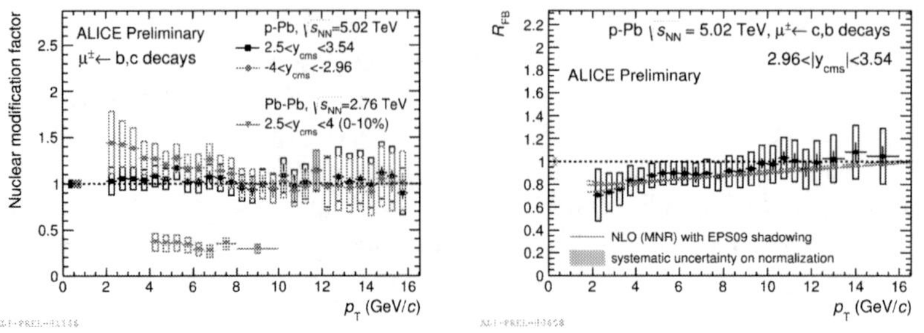

Figure 2. Left: $p_{\rm T}$-differential $R_{\rm pPb}$ of muons from heavy-flavour decays measured at forward and backward rapidity, compared to the $R_{\rm AA}$ in central Pb–Pb collisions. Right: $p_{\rm T}$-differential $R_{\rm FB}$ compared to model calculations including shadowing. Statistical (bars), systematic (empty boxes) and normalization (filled boxes) uncertainties are displayed.

medium. Cold nuclear matter effects have been also investigated at forward (p-beam direction, $2.5 < y_{\rm cms} < 3.5$) and backward (Pb-going direction, $-4 < y_{\rm cms} < -2.96$) rapidity with muons from heavy-flavour decays in the range $2 < p_{\rm T} < 16$ GeV/c, as shown in Fig. 2. $R_{\rm pPb}$ is compatible with unity over the whole $p_{\rm T}$ range at forward rapidity. This evidences that the strong suppression measured in central Pb–Pb collisions results from the dense medium [10]. At backward rapidity, the $R_{\rm pPb}$ is slightly larger than unity in the range $2 < p_{\rm T} < 4$ GeV/c and close to unity at higher $p_{\rm T}$. Cold nuclear matter effects can be also quantified using the forward-to-backward ratio, $R_{\rm FB}$, defined as the ratio of the $p_{\rm T}$-differential production cross sections in a rapidity range symmetric with respect to $y_{\rm cms} = 0$. This common y interval is limited to $2.96 < |y_{\rm cms}| < 3.53$. $R_{\rm FB}$ is slightly smaller than unity in $2 < p_{\rm T} < 4$ GeV/c and close to unity at higher $p_{\rm T}$ (Fig. 2, right). Next-to-leading order perturbative calculations including the EPS09 parameterization of nuclear parton distribution functions (PDFs) [13] describe reasonably well the $R_{\rm FB}$ ratio. A similar agreement between the measured $R_{\rm pPb}$ of D mesons and electrons from heavy-flavour decays and these predictions was also reported in [6, 14].

The ALICE Collaboration studied inclusive J/ψ and ψ(2S) production in p–Pb collisions over a wide rapidity range. The $p_{\rm T}$-integrated $R_{\rm pPb}$ of inclusive J/ψ shows a suppression of about a factor 1.4 at forward rapidity ($2.03 < y_{\rm cms} < 3.53$) and mid-rapidity ($-1.37 < y_{\rm cms} < 0.43$), while the measurement at backward rapidity ($-4.46 < y_{\rm cms} < -2.96$) is consistent with unity (Fig. 3, left). Charmonium production was also investigated as a function of $p_{\rm T}$, as shown in Fig. 3 (middle) for the results at forward rapidity. The $p_{\rm T}$-differential $R_{\rm pPb}$ of inclusive J/ψ increases significantly with increasing $p_{\rm T}$ and becomes close to unity for $p_{\rm T} > 5$ GeV/c. Furthermore, the ψ(2S) $R_{\rm pPb}$ indicates a stronger suppression than the J/ψ, with a slight $p_{\rm T}$ dependence. Theoretical predictions valid for both resonances [8], based on shadowing [15] or coherent energy loss [16] describe the J/ψ $R_{\rm pPb}$ behaviour but overestimate the ψ(2S) $R_{\rm pPb}$. This indicates the presence of sizeable final state CNM effects on the ψ(2S) production. Calculations based on the Color Glass Condensate [17] do not reproduce the inclusive J/ψ $R_{\rm pPb}$ at forward rapidity (Fig. 3, left). The inclusive J/ψ suppression measured at forward rapidity in central Pb–Pb collisions increases with increasing $p_{\rm T}$ and is smaller than that measured in Au–Au collisions at $\sqrt{s_{\rm NN}} = 0.2$ TeV (Fig. 3, right). These trends are consistent with a significant fraction of J/ψ from regeneration at low $p_{\rm T}$. In the high-$p_{\rm T}$ region, one can conclude that the measured suppression is due to the hot medium ($R_{\rm pPb}$ shows that cold nuclear matter effects are small in this $p_{\rm T}$ region).

Bottomonia have also been measured in the dimuon channel. These studies [9, 18], not

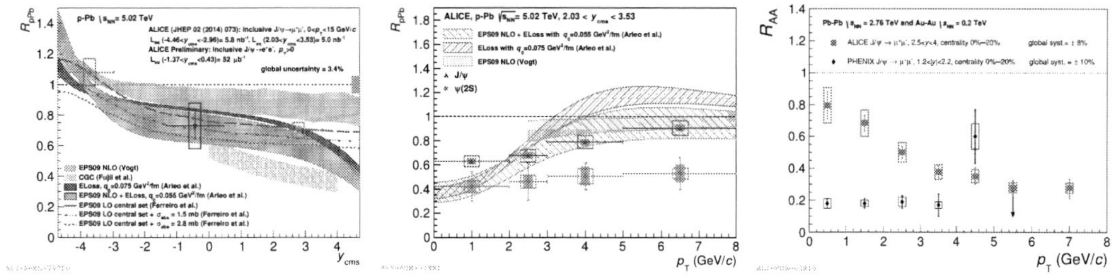

Figure 3. Left: p_{T}-integrated R_{pPb} of inclusive J/ψ as function of y_{cms}. Middle: p_{T}-differential R_{pPb} of inclusive J/ψ and ψ(2S) at forward rapidity. Right: p_{T}-differential R_{AA} of inclusive J/ψ in central Pb–Pb collisions at $\sqrt{s_{\mathrm{NN}}} = 2.76$ TeV and Au–Au collisions at $\sqrt{s_{\mathrm{NN}}} = 200$ GeV. Statistical (bars), systematic (boxes) and normalization (filled boxes) uncertainties are displayed. The results in p–Pb collisions are compared to model predictions.

reported here, demonstrated in particular that the strong suppression of the Υ(1S) yield observed in Pb–Pb collisions (R_{AA} of about 0.3 in the 0–90% centrality class) cannot be attributed to CNM effects only and is underestimated by transport models.

4. Conclusion

Measurements of open heavy flavours and quarkonia have been successfully carried out with the ALICE experiment at the LHC. Open heavy-flavour R_{pPb} results provide evidence that CNM effects are small and confirm that the strong suppression seen in Pb–Pb collisions is a final state effect due to in-medium parton energy loss. The inclusive J/ψ and Υ(1S) R_{pPb} behaviour at forward rapidity is described by models including initial state CMN effects such as nuclear modification of the PDFs and/or energy loss. Final state effects are evidenced on the ψ(2S) production in p–Pb collisions. The study of the inclusive J/ψ R_{AA} in Pb–Pb collisions suggests a non-negligible contribution from J/ψ produced via combination of charm quarks in the QGP and/or at hadronization. High expectations are placed in the upcoming LHC run 2 which should allow to perform more precise and differential measurements.

References

[1] Aamodt K *et al.* (ALICE Collaboration) 2008 *JINST* **3** S08002.
[2] Dokshitzer Y L and Kharzeev D E 2001 *Phys. Lett. B* **519** 199.
[3] Matsui T and Statz H 1986 *Phys. Lett B* **178** 416.
[4] Braun-Munzinger P and Stachel J 2000 *Phys. Lett. B* **490** 196.
[5] Abelev B *et al.* (ALICE Collaboration) 2013 *Phys. Rev. Lett.* **111** 102301; *Phys. Rev. Lett.* **111** 162301.
[6] Abelev B *et al.* (ALICE Collaboration) 2012 *JHEP* **9** 112; 2014 arXiv:1405.3452.
[7] Abelev B*et al.* (ALICE Collaboration) 2014 *Phys. Lett. B* **743** 314; 2014 JHEP **02** 073.
[8] Abelev B *et al.* (ALICE Collaboration) 2014 arXiv:1405.3796.
[9] Abelev B *et al.* (ALICE Collaboration) 2014 arXiv:1405.4493.
[10] Abelev B *et al.* (ALICE Collaboration) 2012 *Phys. Rev. Lett.* **109** 112301.
[11] Abelev B *et al.* (ALICE Collaboration) 2014 arXiv:1405.4144.
[12] Festanti A for the ALICE Collaboration, arXiv:1407.6541.
[13] Mangano M L *et al.* 1992 *Nucl. Phys. B* **373** 295; Eskola K *et al.* 2009 *JHEP* **0904** 065.
[14] Li S for the ALICE Collaboration 2014 arXiv:1408.1915.
[15] Vogt R 2013 *Int. J. Mod. Phys. E* **22** 1330007.
[16] Arleo F and Peigné S 2013 *JHEP* **1303** 122; Ferreiro E G *et al.* 2013 *Eur. Phys. J. C* **73** 2427 and 2013 *Phys. Rev. C* **88** 047901.
[17] Fujii H and Watanabe K 2013 *Nucl. Phys. A* **915** 1; 2013 *Nucl. Phys. A* **920** 78.
[18] Hadjidakis C for the ALICE Collaboration 2014 arXiv:1405.1177.

Journal of Physics: Conference Series **556** (2014) 012021 doi:10.1088/1742-6596/556/1/012021

Particle Production in Nuclear Collisions: Hadronization and QCD

J. Steinheimer[1], T. Lang[1], E. Grossi[4], H. van Hees[2], F. Becattini[4], R. Stock[1,3], M. Bleicher[1,2]

[1] Frankfurt Institute for Advanced Studies (FIAS), Frankfurt, Germany
[2] Institut für Theoretische Physik, Goethe-Universität, Frankfurt, Germany
[3] Institut für Kernphysik, Goethe-Universität, Frankfurt, Germany
[4] Università di Firenze and INFN Sezione di Firenze, Firenze, Italy

E-mail: `steinheimer@fias.uni-frankfurt.de`

Abstract. We discuss recent progress made in the description of light and heavy quark observables, regarding the properties of the quark gluon plasma which is produced in relativistic nuclear collisions. In particular we highlight the effects of the hadronic phase on the QCD hadronization temperature and the modifications of charm quark properties in the dense plasma and hadronic phase.

1. Introduction

The production and the properties of strange and charmed hadrons was always considered a useful probe for the hot and dense systems produced in relativistic nuclear collisions [1, 2]. In this paper we will present recent progress made in the description of light and heavy quark observables, and how they can be used to shed light on the properties of QCD.

2. Strangeness production and the hadronization curve

The determination of the hadronization temperature of QCD, via an experimental observable, is one of the goals of high energy nuclear physics. It has been conjectured [3] that experimentally measured hadron multiplicities represent such an observable: they depend on the temperature at, or close to QGP hadronization. Statistical approaches are able to reproduce the measured hadronic yields, both in elementary [4] and in relativistic nucleus-nucleus collisions [5]. This has led to the formulation of the Statistical Hadronization Model (SHM) [6], which assumes that hadrons are emitted from the fireball in chemical equilibrium. One hopes to take advantage of this phenomenon to obtain the position of the parton-hadron coexistence line of QCD matter in the plane of temperature and baryo chemical potential (T, μ_B) [7]. A recent outcome from LHC heavy-ion data has been the relatively low p/π ratio measured by the ALICE experiment in central Pb+Pb collisions [8] at $\sqrt{s_{NN}}$ = 2.76 TeV, compared to the expectation from the SHM [9]. A similar result was obtained earlier by the SPS experiment NA49 [10]. This has been interpreted [11, 12, 13, 14] as an evidence of post-hadronization baryon-antibaryon annihilation. An alternative explanation has been put forward in ref. [15]. Recently, the ALICE experiment at the LHC has provided [16] a set of measurements of hadronic multiplicities also as a function of centrality. The improved accuracy and the increased total multiplicity with respect to lower

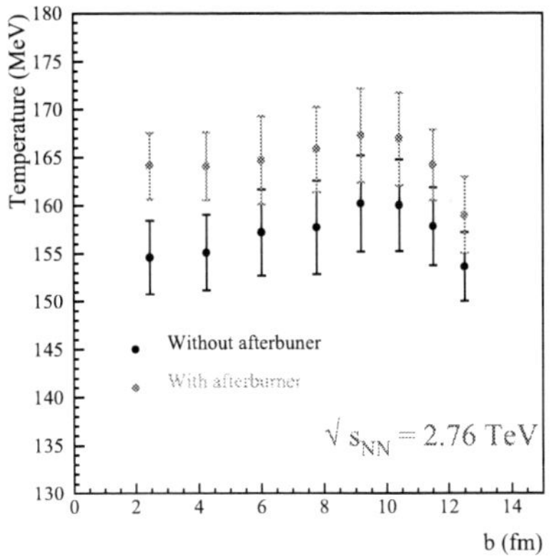

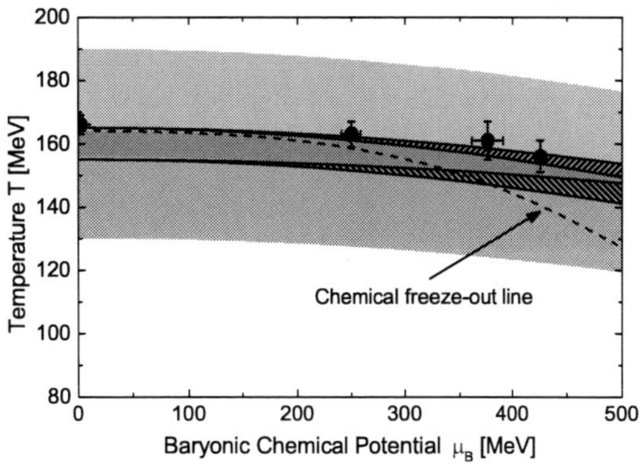

Figure 1. Left: Temperature as a function of the impact parameter b (central values corresponding to centralities measured by ALICE). Black dots: chemical freeze-out temperature. Red dots: LCEP (see text) temperature obtained by including UrQMD modification factors. Right: Predictions from lattice-QCD calculations on the curvature of the crossover line (in T and μ): the upper solid line is the critical temperature defined through strange quark susceptibility, the lower one defined through the chiral condensate) [23, 24] and the colored areas represent the widths of the crossover transitions. The dashed line is the chemical freeze-out line [25], while the reconstructed original chemical equilibrium points in this work are shown as closed circles.

beam energy, allows us to identify a dependence of the chemical freez eout temperature T_{chem} on centrality. We have analyzed the multiplicities measured by the ALICE experiment [16] to determine the chemical freeze-out parameters with fits to the usual SHM (described in ref. [17]) predictions and to the same formula corrected for the *modification factors*, namely the ratios between the particle yields with and without hadronic rescattering. The modification factors have been estimated with a hybrid version of the Ultra relativistic Quantum Molecular Dynamics (UrQMD) code [18, 19] taking into account hadronic rescatterings after hadron generation according to local thermodynamical equilibrium (Cooper-Frye formula).

The results of the fit including corrections for hadronic rescattering are shown in the left frame of figure 1. The theoretical yields are calculated multiplying the output from SHM with the modification factors [20]. Therefore, the fitted temperature pertains to the source at its latest chemical equilibrium point (LCEP, see ref. [21]), before hadronic collisions set in. The fit quality, in terms of $\chi^2/d.o.f.$, improves after the implementation of hadronic phase corrections. The fitted temperature rises by several MeV, as shown in fig. 1 (middle), in agreement with our previous findings [21]. Furthermore, the LCEP temperature is less centrality dependent than the plain chemical freeze-out temperature, which supports the idea of a universal hadronization temperature [17, 22]. Figure 1 (right) finally shows the resulting new chemical freeze out curve, for central collisions at several beam energies, when the effects of the hadronic final state is taken into account properly, compared with the curvature of the lattice QCD pseudo critical line. One can observe a correspondence of the two lines over a wide range in T and μ_B.

3. Charm Quark transport

Charm quarks are an ideal probe for the QGP. They are produced, in hard processes, early in the collision and therefore probe the medium during its entire evolution. When the system dilutes they hadronize, and their decay products can be detected. The two most interesting observables

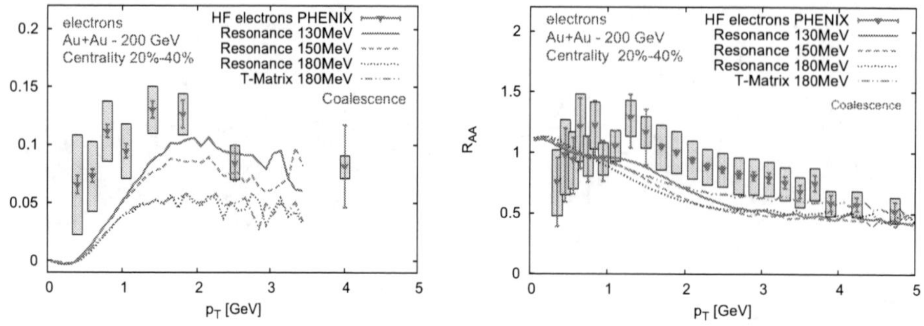

Figure 2. Elliptic flow v_2 (left) and nuclear modification factor R_{AA} (right) of electrons from heavy quark decays in Au+Au collisions at $\sqrt{s_{NN}} = 200$ GeV using a coalescence mechanism. We use a rapidity cut of $|y| < 0.35$. For a decoupling temperature of 130 MeV we obtain a reasonable agreement to data [29].

are the nuclear modification factor, R_{AA}, and the elliptic flow, v_2. Experimentally, the nuclear modification factor shows a large suppression of the open heavy-flavor meson spectra at high transverse momenta (p_T) compared to the findings in pp collisions. The measured large elliptic flow, v_2, of open heavy-flavor mesons indicates that charm quarks take part in the collective expansion of the bulk medium. We explore the modification of heavy-flavor p_T spectra, using a hybrid model, consisting of the Ultra-relativistic Quantum Molecular Dynamics (UrQMD) model [18] and a full (3+1)-dimensional ideal hydrodynamical model [26, 27] to simulate the medium expansion. A more detailed description of the hybrid model including parameter tests and results can be found in [19]. The heavy-quark propagation in the medium is modeled by a relativistic Langevin approach [28] which describes the diffusion of a "heavy particle" in a medium consisting of "light particles". One approximates the collision term of the corresponding Boltzmann equation, which can be mapped into an equivalent stochastic (relativistic) Langevin equation:

$$\mathrm{d}x_j = \frac{p_j}{E}\mathrm{d}t \quad \text{and} \quad \mathrm{d}p_j = -\Gamma p_j \mathrm{d}t + \sqrt{\mathrm{d}t}\, C_{jk}\rho_k. \tag{1}$$

Here $\mathrm{d}t$ is the time step in the Langevin calculation, $\mathrm{d}x_j$ and $\mathrm{d}p_j$ are the coordinate and momentum changes in each time-step, $E = \sqrt{m^2 + \boldsymbol{p}^2}$, and Γ is the drag or friction coefficient. The covariance matrix, C_{jk}, of the fluctuating force is related to the diffusion coefficients. Both Γ and C_{jk} depend on $(t, \boldsymbol{x}, \boldsymbol{p})$ and are defined in the (local) rest-frame of the fluid. The ρ_k are Gaussian-normal distributed random variables. For the production of open charm D mesons we us a quark coalescence approach. To implement this coalescence we perform the Langevin calculation until a decoupling temperature is reached. Subsequently we coalesce a light quark with a heavy quark. As the light quarks constitute the medium propagated by hydrodynamics, the average velocities of the light quarks are approximated by the local flow-velocities of the medium. The mass of the light quarks is assumed to be 370 MeV so that the D-meson mass becomes 1.87 GeV when the masses of the light quarks and the charm quarks (1.5 GeV) are added.

The decay to electrons is calculated using PYTHIA to compare to experimental measurements from the PHENIX collaboration. Fig. 2 (left) shows our results for v_2. The high elliptic flow is due to the momentum kick of the light quarks in the recombination process, which provides additional flow from the medium. For a decoupling temperature of 130 MeV we obtain a reasonable agreement with the experimental data. In Fig. 2 (right) the nuclear modification factor for non-photonic single electrons is depicted. Also here we obtain a good agreement with the data, especially at moderate $p_T \sim 2\,\text{GeV}$. The coalescence mechanism pushes the heavy quarks to higher p_T. As seen before, we obtain the best agreement to data for rather low decoupling temperatures. In conclusion we observe that the coalescence mechanism is required to describe experimental data with our Langevin model. Only with the coalescence model one is able to describe both R_{AA} and v_2 consistently in the present model. Within our study we find the best agreement to experimental data using a low decoupling temperature of 130 MeV. This emphasizes the need to study effects of hadronic rescattering on the D-meson properties, as the effective degrees of freedom at such low temperatures should be hadrons.

4. Acknowledgements

This work was supported by the Hessian LOEWE initiative through the Helmholtz International Center for FAIR (HIC for FAIR), the Deutsche Forschungsgemeinschaft (DFG), the Instituto Nazionale di Fisica Nucleare (INFN) and the Bundesministerium für Bildung und Forschung (BMBF) grant No. 06FY7083. The computational resources were provided by the Frankfurt LOEWE Center for Scientific Computing (LOEWE-CSC).

References

[1] P. Koch, B. Müller and J. Rafelski, Phys. Rept. **142**, 167 (1986).
[2] T. Matsui and H. Satz, Phys. Lett. B **178**, 416 (1986).
[3] R. Stock, Phys. Lett. B **456**, 277 (1999).
[4] F. Becattini, J. Phys. G **23**, 1933 (1997) and references therein.
[5] J. Cleymans, H. Satz, Z. Phys. C **57** 135 (1993); P. Braun-Munzinger, J. Stachel, J. P. Wessels and N. Xu, Phys. Lett. B **365** 1 (1996); F. Becattini, M. Gazdzicki and J. Sollfrank, Eur. Phys. J. C **5** 143 (1998); P. Braun-Munzinger, D. Magestro, K. Redlich and J. Stachel, Phys. Lett. B **518**, 41 (2001);
[6] F. Becattini, arXiv:0901.3643 [hep-ph] and references therein
[7] A. Andronic *et al.*, Nucl. Phys. A **837**, 65 (2010).
[8] L. Milano, ALICE Coll., Nucl. Phys. A **904-905** (2013) 531c; B. Abelev *et al.* [ALICE Collaboration], Phys. Rev. C **88**, 044910 (2013).
[9] A. Andronic, P. Braun-Munzinger, K. Redlich and J. Stachel, Nucl. Phys. A904-905 **2013**, 535c (2013).
[10] T. Anticic *et al.* [NA49 Collaboration], Phys. Rev. C **83**, 014901 (2011).
[11] F. Becattini *et al.* , Phys. Rev. C **85**, 044921 (2012).
[12] J. Steinheimer, J. Aichelin and M. Bleicher, Phys. Rev. Lett. **110**, 042501 (2013).
[13] R. Stock, *et al.*, PoS CPOD **2013**, 011 (2013).
[14] Y. Pan and S. Pratt, arXiv:1210.1577 [nucl-th].
[15] M. Petran, J. Letessier, V. Petracek and J. Rafelski, Phys. Rev. C **88**, 034907 (2013).
[16] B. B. Abelev *et al.* [ALICE Collaboration], Phys. Lett. B **728**, 216 (2014); B. B. Abelev *et al.* [ALICE Collaboration], Phys. Rev. Lett. **111**, 222301 (2013);
[17] F. Becattini, arXiv:0901.3643 [hep-ph] and references therein.
[18] S. A. Bass *et al.*, Prog. Part. Nucl. Phys. **41**, 255 (1998); M. Bleicher *et al.*, J. Phys. **G25**, 1859 (1999);
[19] H. Petersen, J. Steinheimer, G. Burau, M. Bleicher and H. Stöcker, Phys. Rev. C **78**, 044901 (2008)
[20] F. Becattini, E. Grossi, M. Bleicher, J. Steinheimer and R. Stock, arXiv:1405.0710 [nucl-th].
[21] F. Becattini *et al.*, Phys. Rev. Lett. **111**, 082302 (2013).
[22] H. Satz, Int. J. Mod. Phys. E **21**, 1230006 (2012) and references therein.
[23] G. Endrodi, Z. Fodor, S. D. Katz and K. K. Szabo, JHEP **1104**, 001 (2011)
[24] O. Kaczmarek *et al.*, Phys. Rev. D **83**, 014504 (2011)
[25] J. Cleymans, H. Oeschler, K. Redlich and S. Wheaton, Phys. Rev. C **73**, 034905 (2006)
[26] D. H. Rischke, S. Bernard and J. A. Maruhn, Nucl. Phys. A **595**, 346 (1995)
[27] D. H. Rischke, Y. Pursun and J. A. Maruhn, Nucl. Phys. A **595**, 383 (1995)
[28] R. Rapp and H. van Hees, arXiv:0903.1096 [hep-ph].
[29] A. Adare *et al.* [PHENIX Collaboration], Phys. Rev. C **84**, 044905 (2011)

XI International Conference on Hyperons, Charm and Beauty Hadrons (BEACH 2014) IOP Publishing
Journal of Physics: Conference Series **556** (2014) 012022 doi:10.1088/1742-6596/556/1/012022

Heavy Flavour Production at HERA

Ringailė Plačakytė
(on behalf of the H1 and ZEUS Collaborations)
Deutsches Elektronen Synchrotron, Notkestrasse 85, 22603 Hamburg, Germany
E-mail: ringaile@mail.desy.de

Abstract. Measurements of open charm and beauty production at HERA provide important input for stringent tests of quantum chromodynamics and are used to constrain the parton distribution functions of the proton. The recent results on the heavy flavour production obtained by the H1 and ZEUS experiments at HERA are reviewed in this contribution.

1. Heavy Flavour Production at HERA

The production of beauty and charm quarks in *ep* collisions at HERA provides important tests of the theory of perturbative Quantum Chromodynamics (pQCD). At leading order, the dominant process for heavy-quark production at HERA is boson-gluon fusion (BGF) where a virtual photon emitted by the incoming electron (positron) interacts with a gluon from the proton forming a heavy quark-antiquark pair. Besides the masses of the heavy quarks, additional scales involved in the heavy flavour production are the virtuality, Q^2, of the exchanged photon in case of deep inelastic scattering (DIS) and the transverse momenta, p_T, of the outgoing quarks. Depending on the details how the scales are treated, different approaches are available in pQCD for the treatment of the heavy flavours in DIS:

- The zero-mass variable-flavour-number-scheme (ZM-VFNS): in this scheme only three light flavours are participating in the parton evolution and the charm quark is treated as a massless parton above the threshold at $Q^2 \sim$ charm quark mass, m_C^2.
- Fixed Flavour Number Scheme (FFNS): heavy quarks are treated as massive at all scales but they are not considered as partons in the proton. The numbers of active flavours are fixed to three (four) for charm (beauty). Heavy quarks are assumed to be produced only in the hard scattering process. In the FFNS calculations [1, 2] the pole mass [3] or the running mass definition [4] is used (later adopted from the recent ABM variant of the FFN scheme).
- General Mass Variable Flavour Number Scheme (GM-VFNS): an interpolation scheme where a low Q^2 region with FFNS and high Q^2 region with ZM-VFN schemes are combined. GM-VFNS profits from the advantages of those two schemes, however, at the interpolation region some level of arbitrariness is introduced for the treatment of heavy quarks.

At HERA different techniques have been used to measure open charm production cross sections in DIS: the full reconstruction of D or D* mesons, the long lifetime of heavy flavoured hadrons and semi-leptonic decays of heavy hadrons. The best signal-to-background ratio is observed using the full reconstruction of D mesons albeit having the small branching ratios. The method using semi-leptonic decays profits from larger branching ratios but has worse signal-to-background ratio. Finally, inclusive analyses using lifetime information have no limitation on branching ratios but yield the worst signal-to-background ratio. In the following, the recent HERA heavy flavour measurements are reviewed.

2. Combination of Charm Production Measurements and QCD Analysis

Nine different H1 and ZEUS charm data sets measured with different charm tagging techniques are combined into one consistent set taking into account correlated systematic uncertainties and the normalisation [5]. The effect of including the combined HERA charm data in the QCD analysis of the parton distribution functions (PDFs) of the proton was also studied in [5]. In this study, the charm and the inclusive data are used with different implementations of the GM-VFN schemes, which can have different impacts on the charm contribution to the sea quark distribution and thus affect the flavour composition of the quark PDFs. Therefore the accuracy of the charm data influences the uncertainties on the $W^{\pm}$ and Z production cross section predictions at LHC. Figure 1 (left) summarises the study by showing the charm quark mass M_C (which, due to freedom introduced by an interpolation approach in GM-VFN schemes, can be treated as a parameter in the fit) for all schemes used in the fits. It is interesting to observe that firstly, different schemes have different optimal charm quark mass parameter M_C^{opt}, and secondly, the χ^2 minimum values are comparable for all schemes despite different optimal values of M_C. The PDFs with M_C^{opt} were then propagated to the calculation of $W^{\pm}$ and Z boson production cross section predictions for the LHC. As an example, the W^+ production cross section as a function of M_C for the different schemes is shown in figure 1 (right). Good agreement between these predictions is observed at M_C^{opt} which results in a reduction of the uncertainties due to the heavy flavour treatment to below 1.5%. In addition, the running mass of the charm quark $m_C(m_C)$ is determined using the FFN scheme where, unlike in GM-VFNS, the charm quark mass corresponds directly to a physical mass (figure 2, left).

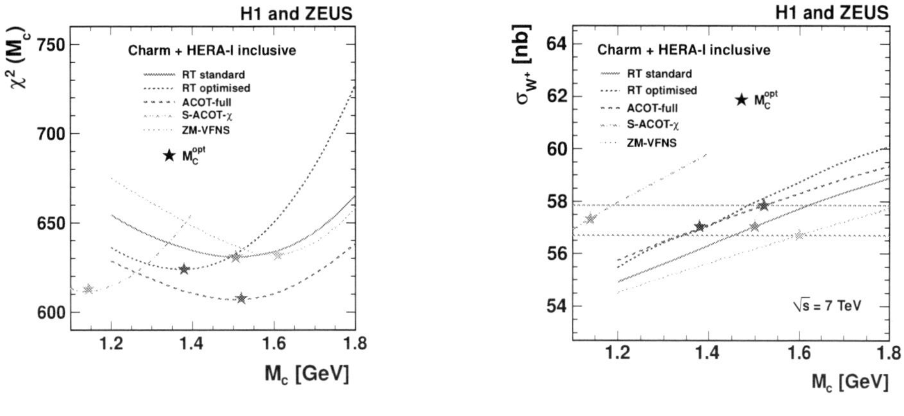

Figure 1. Left: The values of $\chi^2(M_C)$ for the PDF fit to the combined HERA inclusive DIS and charm measurements. Right: W^+ production cross section σ_{W^+} at the LHC for $\sqrt{s} = 7$ TeV obtained from the resulting PDF fits as a function of M_C. The lines represent different heavy flavour schemes and the stars indicate the values of M_C^{opt} for each scheme.

3. Measurement of the Charm Mass Running

Similarly to running of α_s, the scale dependence of the masses of heavy quarks can be calculated perturbatively. For example, the running of the $\overline{MS}$ beauty quark mass has been established

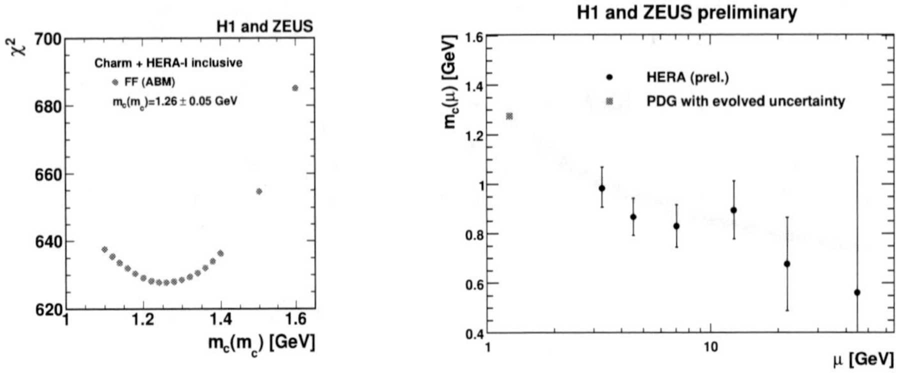

Figure 2. Left: The values of χ^2 for the PDF fit to the combined HERA DIS and charm data as a function of the running charm quark mass $m_C(m_C)$. Right: The running of the charm mass $m_C(\mu)$ obtained in the $\overline{MS}$ mass scheme as a function of the scale μ. The PDG world average [6] and the band of its expected running [7] are also shown.

from measurements at the LEP collider [8] but no similar measurement has been performed for charm quark so far. The combined HERA charm data [5] have been used to show the running of the charm quark mass for the first time. The basic idea is the following: the value of the charm quark mass $m_C(m_C)$ is extracted from the χ^2 distribution of data to theory comparison using PDFs obtained from inclusive DIS HERA data and the same setup as in [5] but varying the charm mass. The extraction is performed from charm data subdivided into several kinematic intervals according to Q^2 and assuming the running of α_s and m_C as predicted by QCD. The measured $m_C(m_C)$ are reinterpreted in terms of a value of $m_C(\mu)$ at the typical charm measurement scale $\mu = \sqrt{Q^2 + 4m_C^2}$. In this way, the effect of the initial assumption of QCD running on the interpretation of the measurement is minimised. The comparison of $m_C(\mu)$ at different scales with the expected running behavior is illustrated in figure 2 (right). More details about the extraction of the charm mass running at HERA can be found in [9].

4. Latest Heavy Flavour Measurements at HERA

After the combined analysis of the charm data [5] was performed, new measurements of charm production in DIS have been reported at HERA, the most recent of which are discussed below. The charm production has been measured with the ZEUS detector using an integrated luminosity of 354 pb^{-1} by reconstructing $D^{\pm}$ mesons in the $D^{\pm} \rightarrow K^{\pm}\pi^{\pm}\pi^{\pm}$ decay channel [10]. These new precise data have the potential to constrain further the parton densities in the proton.

In another recent ZEUS measurement, the cross sections for the photoproduction of D^* mesons have been measured at three different *ep* centre-of-mass energies, $\sqrt{s}$ = 318, 251 and 225 GeV [11], see figure 3 (left). The dependence on $\sqrt{s}$ in this study is presented as normalised to the high-statistics measurement at $\sqrt{s}$ = 318 GeV leading to the cancellation of a number of systematic effects. Predictions from the next-to-leading order QCD describe the $\sqrt{s}$ dependence of the data well demonstrating consistency of the gluon distribution probed here with that extracted in PDF fits to inclusive DIS data. These results enhance the confidence in the QCD predictions of charm production rates and QCD processes for a future TeV-scale *ep* collider.

The new, most precise D^* data measured with integrated luminosity of 363 pb^{-1} with the ZEUS detector have been used to extract the charm contribution to the proton structure functions [12]. These data together with the measurements from H1 [13, 14] were used in the D^* combination at the visible cross section level [15]. The cross sections were combined

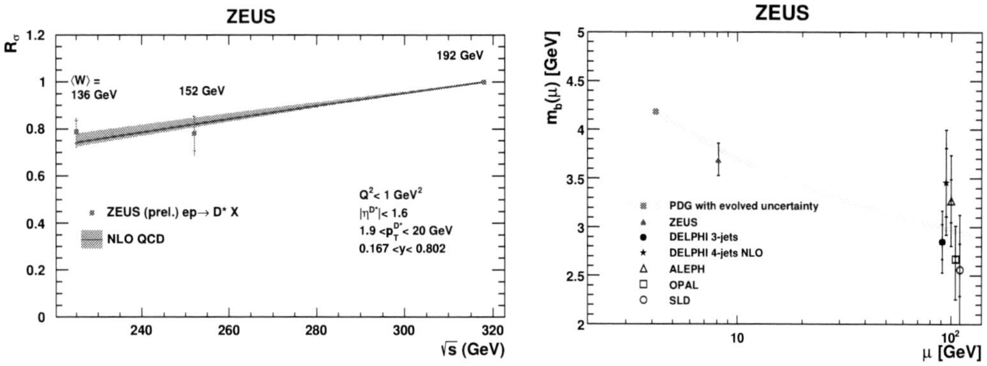

Figure 3. Left: Normalised D* visible photoproduction cross sections as a function of the *ep* centre-of-mass energy. The data points are shown with statistical uncertainties (inner error bars) and statistical and systematic uncertainties added in quadrature (outer error bars). The predictions from NLO QCD are shown as a line with the band representing its uncertainty. Right: The running of the beauty mass $m_b(\mu)$ obtained in the $\overline{MS}$ mass scheme as a function of the scale μ and compared with the PDG world average [6].

taking into account all relevant correlations and thus significantly reducing the uncertainties. The combined cross sections can be compared directly to differential NLO predictions without the need for extrapolation, however NNLO calculations are desired to match the data precision. Finally, the inclusive jet cross sections in beauty and charm events were used to extract the heavy-quark contribution to the proton structure function F_2 in the ZEUS collaboration [16]. In this analysis the long lifetimes of the weakly decaying b and c hadrons and their large masses were exploited. The running beauty-quark mass in the $\overline{MS}$ scheme was determined (figure 3, right) from an NLO QCD fit in the fixed-flavour-number scheme following a procedure similar to that used for extraction of the charm-quark mass [5]. This represents the first measurement of the b-quark mass using HERA or any other hadron collider data.

References

[1] E. Laenen et al., Phys. Lett. **B291**, (1992) 325; E. Laenen et al., Nucl. Phys. **B392**, (1993) 162; E. Laenen et al., Nucl. Phys. **B392**, (1993) 229; S. Riemersma, J. Smith and W. L. van Neerven, Phys. Lett. **B347**, (1995) 143 [hep-ph/9411431].
[2] B. W. Harris and J. Smith, Phys. Rev. **D57**, (1998) 2806 [hep-ph/9706334].
[3] K. Nakamura *et al.*, [Particle Data Group], J. Phys. G37, (2010) 075021.
[4] S. Alekhin and S. Moch, Phys. Lett. **B699**, (2011) 345 [arXiv:1011.5790].
[5] [H1 and ZEUS Collaborations], H. Abramowicz *et al.*, Eur.Phys.J. **C73** (2013) 2311 [arXiv:1211.1182].
[6] [Particle Data Group], J. Beringer *et al.*, Phys. Rev. **D86** (2012) 1.
[7] B. Schmidt, M. Steinhauser, CRunDec: C++ package for running and decoupling of the strong coupling and quark masses [arXiv:1201.6149]; K.G. Chetyrkin, J.H. Kuhn, M. Steinhauser, RunDec: Mathematica package for running and decoupling of the strong coupling and quark masses, Comp. Phys. Comm. **133** (2000) 43.
[8] [DELPHI Collaborations], J. Abdallah *et al.*, Eur.Phys.J. **C55** (2008) 525.
[9] [H1 and ZEUS Collaborations], H1-prelim-14-071, ZEUS-prel-14-006.
[10] [ZEUS Collaboration], I. Abt *et al.*, JHEP 1305, **023** 2013.
[11] [ZEUS Collaboration], H. Abramowicz *et al.*, DESY-14-082, [arXiv:1405.5068].
[12] [ZEUS Collaboration], H. Abramowicz *et al.*, JHEP 05 **097** 2013.
[13] [H1 Collaboration], F.D. Aaron *et al.*, Phys. Lett. **B686** (2010) 91.
[14] [H1 Collaboration], F.D. Aaron *et al.*, Eur. Phys. J. **C71** (2011) 1509.
[15] [H1 and ZEUS Collaborations], H1-prelim-13-171, ZEUS-prel-13-002.
[16] [ZEUS Collaboration], DESY-14-083, to be published in JHEP [arXiv:1405.6915].

B hadron production and decay at ATLAS

E Bouhova-Thacker, on behalf of the ATLAS Collaboration
Physics Department, Lancaster University, Lancaster LA1 4YB, UK

E-mail: e.bouhova-thacker@lancaster.ac.uk

Abstract. Selected recent B-physics results published by the ATLAS experiment at the LHC are presented. The Λ_b^0 parity-violating decay asymmetry parameter, α_b, is measured to be $0.30 \pm 0.16(\text{stat}) \pm 0.06(\text{syst})$. A new excited state of the $B_c^\pm$ meson is observed with a mass $6842 \pm 4(\text{stat}) \pm 5(\text{syst})$ MeV, consistent with expectations for the second S-wave state of the $B_c^\pm$ meson, $B_c^\pm(2S)$.

1. Introduction

Results by the ATLAS Collaboration on the measurement of the parity-violating asymmetry parameter α_b and the helicity amplitudes for the decay $\Lambda_b^0 \to J/\psi\Lambda^0$ [1] and on the observation of an excited $B_c^\pm$ meson state [2] are reported.

2. Parity violation and helicity amplitudes in $\Lambda_b^0 \to J/\psi(\mu^+\mu^-)\Lambda^0$ decays

Parity violation in hadron weak decays is not maximal and depends on the hadron's constituents. For heavy baryons, theoretical predictions are possible using perturbative QCD and the Factorisation Approximation [3] or Heavy Quark Effective Theory (HQET) [4]. The decay asymmetry parameter, α, of a two-body spin-$\frac{1}{2}$ particle decay can be extracted from the angular distribution $\frac{dN}{d\cos\theta} = \frac{1}{2}(1+P\alpha\cos\theta)$, where θ is the angle between the polarisation vector and the direction of the decay product in the particle's rest frame. Due to the symmetries of the initial state and the ATLAS detector [5], $P(p_T,\eta) = -P(p_T,-\eta)$, the average polarisation is zero, and a full angular analysis is required to extract α. The $\Lambda_b^0 \to J/\psi\Lambda^0$ decay can be described by four helicity amplitudes $A(\lambda_{\Lambda^0}, \lambda_{J/\psi})$: $a_+ = |a_+|e^{i\rho_+} \equiv A(1/2,0)$, $a_- = |a_-|e^{i\rho_-} \equiv A(-1/2,0)$, $b_+ = |b_+|e^{i\omega_+} \equiv A(-1/2,-1)$, $b_- = |b_-|e^{i\omega_-} \equiv A(1/2,1)$, where λ_{Λ^0} and $\lambda_{J/\psi}$ are the helicities of Λ^0 and J/ψ. The full angular probability density function (PDF) of the decay angles $\Omega = (\theta,\theta_1,\phi_1,\theta_2,\phi_2)$, where θ_1 (θ_2) and ϕ_1 (ϕ_2) are the polar and azimuthal angles of the proton (μ^+) in the Λ^0 (J/ψ) rest frame with respect to the Λ^0 (J/ψ) direction in the Λ_b^0 rest frame, is given as

$$w(\Omega, A, P) = \frac{1}{(4\pi)^3}\sum_{i=0}^{19} f_{1i}(A) f_{2i}(P, \alpha_{\Lambda^0}) F_i(\Omega).$$

The F_i are orthogonal functions of the decay angles, multiplied by functions of the helicity amplitudes and the polarisation. As $\langle P \rangle = 0$ only six terms are non-zero. The five independent parameters are chosen as: $\alpha_b = |a_+|^2 - |a_-|^2 + |b_+|^2 - |b_-|^2$, $k_+ = \frac{|a_+|}{\sqrt{|a_+|^2+|b_+|^2}}$, $k_- = \frac{|a_-|}{\sqrt{|a_-|^2+|b_-|^2}}$, $\Delta_+ = \rho_+ - \omega_+$ and $\Delta_- = \rho_- - \omega_-$. The analysis extracts α_b and the helicity amplitudes from the measured averages $\langle F_i \rangle = \frac{1}{N^{data}}\sum_{n=1}^{N^{data}} F_i(\Omega_n)$ with a least squares fit.

The analysis uses 4.6 fb^{-1} of 7 TeV data collected with single-muon and $J/\psi \to \mu\mu$ triggers. A mass constrained cascade fit is performed with a J/ψ and Λ^0 mass constraints to the world average values [6] and requiring that the Λ^0 momentum points to the di-muon vertex. Cuts are applied on the fit quality, $\chi^2/N_{dof} < 3$, the transverse momentum of Λ^0, $p_T^{\Lambda^0} > 3.5$ GeV, the transverse decay length of Λ^0, measured from the Λ_b^0 vertex, $L_{xy}^{\Lambda^0} > 10$ mm, and the Λ_b^0 proper decay time, $\tau > 0.35$ ps. A B_d hypothesis is also considered and candidates are kept if the Λ_b^0 hypothesis has the larger cumulative χ^2 probability, $\mathcal{P}_{\Lambda_b^0} > \mathcal{P}_{B_d}$. Figure 1 shows the invariant mass of the reconstructed Λ_b^0 and $\bar{\Lambda}_b^0$ candidates. CP invariance is assumed.

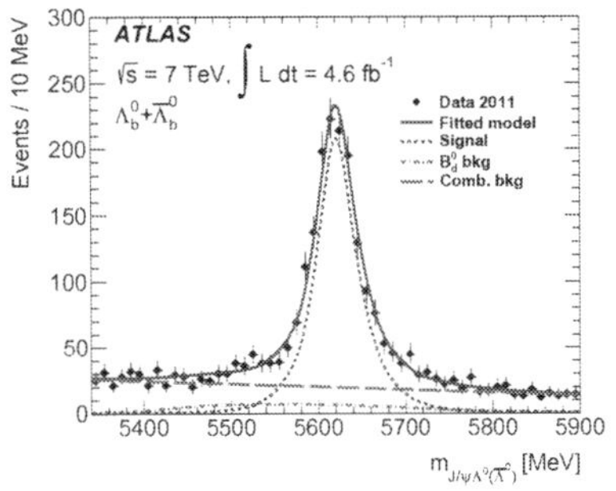

Figure 1. The reconstructed mass of Λ_b^0 and $\bar{\Lambda}_b^0$ candidates.

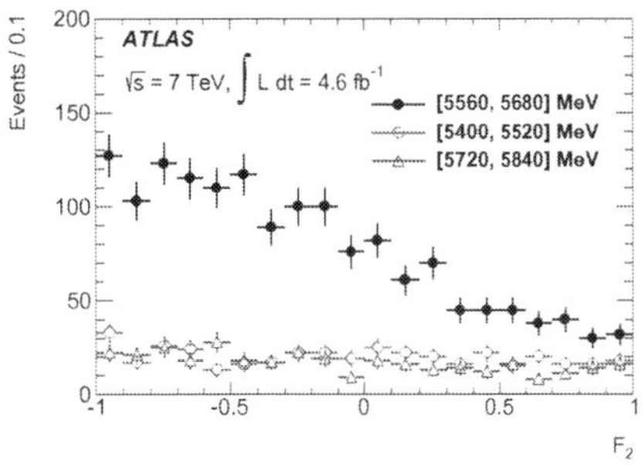

Figure 2. F_2 distribution for the signal and sideband regions.

An extended binned maximum likelihood fit is performed to extract the signal. Monte Carlo templates are used to describe the Λ_b^0 and B_d mass shapes and the combinatorial background is parametrised by a first-order polynomial. The distribution of F_2 in the signal and background regions is shown in figure 2. The two sideband distributions are similar and the average is used to subtract from the signal. The parameters are extracted by minimising $\chi^2 = \sum_i \sum_j (\langle F_i\rangle^{exp} - \langle F_i\rangle) V_{ij}^{-1} (\langle F_j\rangle^{exp} - \langle F_j\rangle)$, where V is the covariance matrix of the moments, $\langle F_i\rangle$. The expected values are given by $\langle F_i\rangle^{exp} = \sum_j f_{1j}(A) f_{2j}(\alpha_{\Lambda^0}) C_{ij}$, where $C_{ij} = \frac{1}{(4\pi)^3} \iint F_i(\Omega') T(\Omega', \Omega) F_j(\Omega) d\Omega' d\Omega \approx \frac{\epsilon_T}{NMC} \sum_{n=1}^{NMC} F_i(\Omega_n') F_j(\Omega_n)$. $T(\Omega', \Omega)$ accounts for acceptance and detector effects, while ϵ_T is determined using the identity $\langle F_0\rangle^{exp} = 1$.

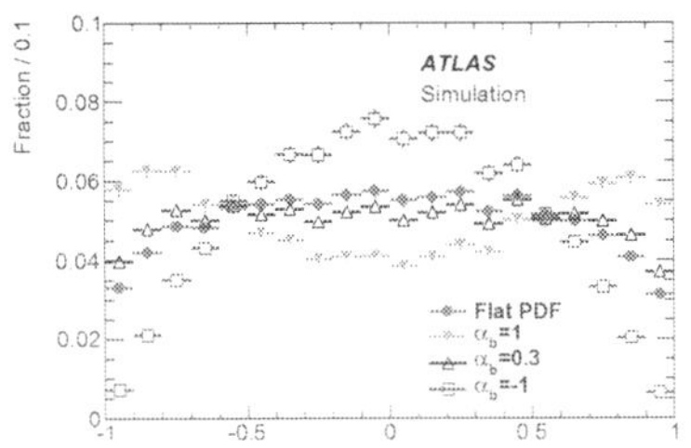

Figure 3. $\cos\theta_2$ distribution, after detector effects, for different values of α_b, $k_+ = 0.21$, $k_1 = 0.134$, $\Delta_\pm = 0$.

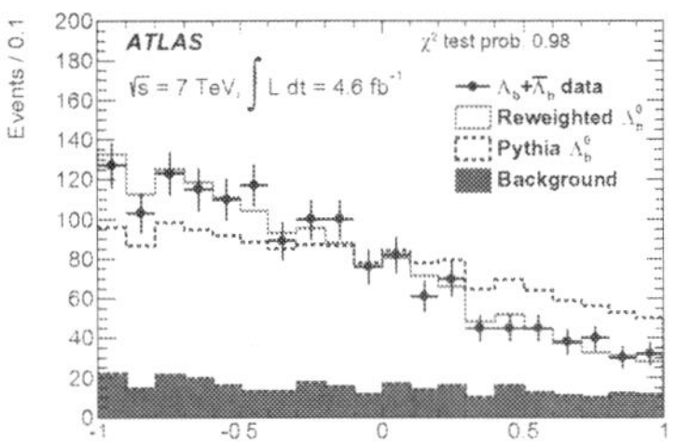

Figure 4. F_2 distribution from the sum of the weighted MC events and the background, compared to data.

Figure 3 shows one of the angular distributions after acceptance, efficiency and resolution effects are taken into account. To illustrate the ATLAS sensitivity the events are weighted with different values for α_b. The Monte Carlo (MC) events are generated with uniform distributions. After extracting the parameters, the MC signal distributions are re-weighted using the measured parameters, added to the background distributions and compared to data. Figure 4 shows good agreement between data and re-weighted MC. The main sources of systematic uncertainty are from the background shape modelling and the limited MC statistics.

The measured values of $\langle F_i \rangle$, after sideband subtraction and B_d correction, yield the following values for the parameters and helicity amplitudes:

$$\alpha_b = 0.30 \pm 0.16(\text{stat}) \pm 0.06(\text{syst}) \qquad |a_+| = 0.17^{+0.12}_{-0.17}(\text{stat}) \pm 0.09(\text{syst})$$

$$k_+ = 0.21^{+0.14}_{-0.21}(\text{stat}) \pm 0.13(\text{syst}) \qquad |a_-| = 0.59^{+0.06}_{-0.07}(\text{stat}) \pm 0.03(\text{syst})$$

$$k_- = 0.13^{+0.20}_{-0.13}(\text{stat}) \pm 0.15(\text{syst}) \qquad |b_+| = 0.79^{+0.04}_{-0.05}(\text{stat}) \pm 0.02(\text{syst})$$

$$|b_-| = 0.08^{+0.13}_{-0.08}(\text{stat}) \pm 0.06(\text{syst}).$$

The value of α_b is consistent with a recent LHCb measurement, α_b=0.05±0.17(stat)±0.07(syst) [7], and lies between the pQCD [3], (α_b in the range -0.17 to -0.14) and HQET [4] (α_b=0.78) predictions. The large values of $|a_-|$ and $|b_+|$ suggest a negative helicity for Λ^0.

3. Observation of an excited $B_c^\pm$ meson state

The spectrum and properties of the B_c family are predicted by non-relativistic potential models, perturbative QCD and lattice calculations [8]. The second S-wave state is expected to have a mass in the range $m_{B_c(2S)} \sim 6835-6917$ MeV. The pseudo-scalar and vector spin states cannot be separated by this analysis. The ATLAS experiment observe a peak structure in the mass difference distribution

$$Q = m(B_c^\pm \pi\pi) - m(B_c^\pm) - 2m(\pi^\pm).$$

The ground state is observed in the $B_c^\pm \to J/\psi(\mu^+\mu^-)\pi^\pm$ decay mode.

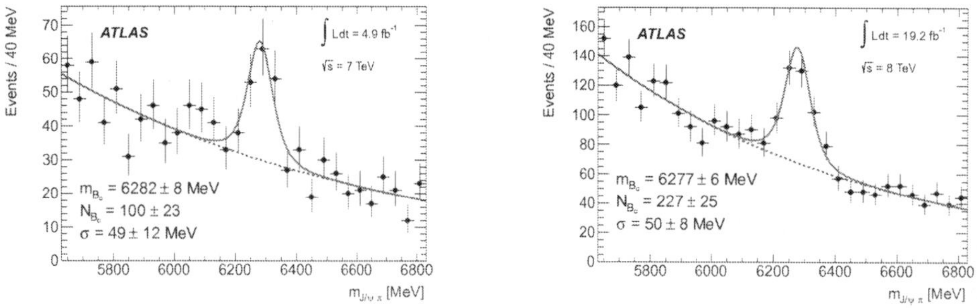

Figure 5. Invariant mass distribution of the reconstructed $B_c^\pm \to J/\psi(\mu^+\mu^-)\pi^\pm$ candidates in 7 (left) and 8 (right) TeV data.

The analysis uses 4.9 fb^{-1} of 7 TeV and 19.2 fb^{-1} of 8 TeV data collected with $J/\psi \to \mu\mu$ triggers. The analysis is optimised separately for the two centre-of-mass energies due to different pile-up conditions. $B_c^\pm$ candidates are reconstructed using a mass constrained fit. The combinatorial background is reduced using an impact parameter cut, $d_0^{\pi^\pm}/\sigma(d_0^{\pi^\pm}) > 5\,(4.5)$ for the 7 (8) TeV data. The fitted $B_c^\pm$ mass, shown in figure 5, is consistent with the world average [6]. Both distributions are fitted separately using an extended unbinned maximum likelihood fit, with a Gaussian (exponential) function modelling the signal (background) shape. The stability of the $B_c^\pm$ yield is checked through its normalisation to $B^\pm \to J/\psi(\mu^+\mu^-)K^\pm$ decays, reconstructed with similar requirements.

$B_c^\pm$ candidates in $\pm 3\sigma$ mass window of the fitted mass value of each dataset are combined with two tracks from the corresponding primary vertex in a cascade fit with a J/ψ mass constraint and requiring that the $B_c^\pm$ momentum points to the $B_c^\pm(2S)$ vertex. An unbinned maximum likelihood fit gives the following values for the mass of the observed structure: 288.2 ± 5.1 MeV and 288.4 ± 4.8 MeV for the 7 and 8 TeV datasets. The corresponding values for the widths of the structure are 18.2 ± 3.8 MeV and 17.6 ± 4.0 MeV. All uncertainties are statistical only. The Q distributions for the 7 and 8 TeV data are shown in figure 6.

The fit includes a third-order polynomial to model the background and a Gaussian function for the structure. The relative $B_c^\pm(2S)/B_c^\pm$ yield ratio is verified to be statistically consistent between the 7 TeV and 8 TeV data. The main contributions to the systematic uncertainty come from the pion momentum scale, the uncertainty on the mass of the ground state and the signal and background models. The total averaged systematic uncertainty is 4.1 MeV.

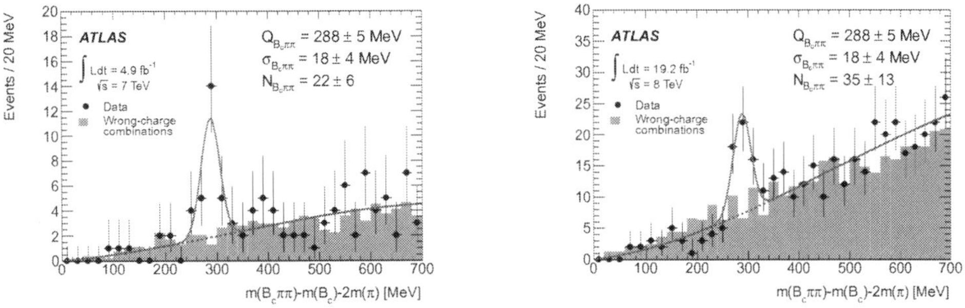

Figure 6. The $m(B_c^\pm\pi\pi) - m(B_c^\pm) - 2m(\pi^\pm)$ distribution in 7 (left) and 8 (right) TeV data.

The significance of the observed structure is evaluated with pseudo-experiments. The mean mass value of the signal contribution is not constrained within the theoretically motivated range to account for a "look elsewhere effect". The significance of the observation is 3.7σ in the 7 TeV data and 4.5σ in the 8 TeV data. The total significance for the combined 7 and 8 TeV dataset is 5.2σ. The error weighted mean mass of the observed structure is $Q = 288.3 \pm 3.5(\text{stat}) \pm 4.1(\text{syst})$ MeV, corresponding to a mass of $6842 \pm 4(\text{stat}) \pm 5(\text{syst})$ MeV. In conclusion, a new state is observed in the ATLAS detector with $4.9\,\text{fb}^{-1}$ of 7 TeV and $19.2\,\text{fb}^{-1}$ of 8 TeV data, consistent with an excited $B_c^\pm(2S)$ state with a predicted mass $m_{B_c(2S)} \sim 6835$–6917 MeV.

References

[1] ATLAS Collaboration 2014 *Phys. Rev.* D **89** 092009
[2] ATLAS Collaboration 2014 arXiv:1407.1032 [hep-ex], *submitted to Phys. Rev. Lett.*
[3] Cheng H-Y 1997 *Phys. Rev.* D **56** 2799
Fayyazuddin and Riazuddin 1998 *Phys. Rev.* D **58** 014016
Mohanta R, Khanna M P, Ishida M, Ishida S and Oda M 1999 *Prog. Theor. Phys.* **101** 959
Chou C-H, Shih H-H, Lee S-C and Li H-N 2002 *Phys. Rev.* D **65** 074030
Wei Z-T, Ke H-W and Li X-Q 2009 *Phys. Rev.* D **80** 094016
Gutsche T, Ivanov M A, Korner J G, Lyubovitskij V E and Santorelli P 2013 *Phys. Rev.* D **88** 114018
[4] Leitner O, Ajaltouni Z and Conte E 2005 *Nucl. Phys.* A **755** 435
Ajaltouni Z, Conte E and Leitner O 2005 *Phys. Lett.* B **614** 165
[5] ATLAS Collaboration 2008 *JINST* **3** S08003
[6] Beringer J *et al* (Particle Data Group) 2012 *Phys. Rev.* D **86** 010001
[7] Aaij R *et al* (LHCb Collaboration) 2013 *Phys. Lett.* B **724** 27
[8] Godfrey S 2004 *Phys. Rev.* D **70** 054017
Eichten E and Quigg C 1994 *Phys. Rev.* D **49** 5845
Kiselev V V, Likhoded A K and Tkabladze A V 1995 *Phys. Rev.* D **51** 3613
Penin A A, Pineda A, Smirnov V A and Steinhauser M 2004 *Phys. Lett.* B **593** 124
Dowdall R J, Davis C T H, Hammant T C and Horgan R R 2012 *Phys. Rev.* D **86** 094510

Measurement of open-charm hadrons with ALICE

M A S Figueredo, for the ALICE Collaboration
University of Liverpool, Oxford Street, Liverpool L69 7ZE, UK
E-mail: marcel.figueredo@cern.ch

Abstract. Heavy-flavour hadrons constitute a very good probe of the hot and dense medium created in high-energy Pb–Pb collisions. ALICE results on open-charm hadron production in Pb–Pb and p–Pb collisions are presented. Additionally, the perspectives for heavy-flavour measurements after the ALICE upgrade are discussed.

1. Introduction

ALICE [1] is the LHC experiment dedicated to the study of the nuclear matter at high temperatures and densities as those reached in the most central heavy ion collisions, where a transition of the QCD matter from a nuclear state to a QGP (Quark Gluon Plasma) is expected.

Heavy quarks, i.e. charm and beauty, are produced in hard scattering processes in the early stages of high-energy nucleus-nucleus collisions. The produced quarks propagate through the medium and interact with its constituents, losing energy via elastic collisions and gluon radiation. Therefore, they represent an effective probe to study the QGP. The measurement of D^0, D^+, D^{*+}, D_s^+ and Λ_c^+ production in p–Pb and Pb–Pb collisions can be used to investigate the energy-loss mechanisms of the heavy quarks in the QGP and also their hadronisation. For the latter, in particular, the production of Λ_c^+ baryons and D_s^+ mesons relative to non-strange D mesons at low and intermediate p_T is expected to be sensitive to hadronisation via quark recombination [2].

One experimental observable used to study such effects in Pb–Pb collisions and to assess Cold Nuclear Matter (CNM) effects in p–Pb collisions is the nuclear modification factor, which is defined as $R_{AA}(p_T) = \frac{1}{\langle N_{coll} \rangle} \times \frac{dN_{AA}/dp_T}{dN_{pp}/dp_T}$ or $R_{pPb}(p_T) = \frac{1}{A} \times \frac{d\sigma_{pPb}/dp_T}{d\sigma_{pp}/dp_T}$, where $\langle N_{coll} \rangle$ represents the average number of binary collisions and A is the mass number of the Pb nucleus. A deviation of the nuclear modification factor from unity could be due to energy loss in the QGP (Pb–Pb) or CNM effects in p–Pb collisions. The energy loss in Pb–Pb collisions is expected to follow a mass hierarchy $\Delta E_g > \Delta E_{u,d,s} > \Delta E_c > \Delta E_b$ [3, 4, 5] (where ΔE_g represents the energy loss by gluons and $\Delta_{u,d,s,c,b}$ the energy loss by the respective quarks), which could lead to $R_{AA}(B) > R_{AA}(D) > R_{AA}(\pi)$, with some caveats due to the fact that R_{AA} is sensitive also to the different production and fragmentation kinematics of light partons, c and b quarks.

In Pb-Pb collisions, a collective expansion of the medium is established due to multiple interactions among its constituents. In non-central collisions, these interactions can convert the initial spatial anisotropy of the overlap region of the colliding nuclei into a final-state particle momentum anisotropy [6]. The elliptic flow, v_2, is the second order Fourier coefficient of the particle azimuthal distribution relative to the reaction plane [6]. At low p_T, a positive D-meson v_2 can reflect the participation of charm quarks in the collective motion of the medium, and, at high p_T, v_2 can be used to study the path-length dependence of in-medium energy loss.

2. Analysis and Results

The open-charm hadron decays $D^0 \to K^-\pi^+$ (BR $3.88 \pm 0.05\%$), $D^+ \to K^-\pi^+\pi^+$ (BR $9.13 \pm 0.19\%$), $D^{*+} \to D^0\pi^+$ (BR $67.7 \pm 0.5\%$), $D_s^+ \to \pi^+\phi \to K^+K^-\pi^+$ (BR $2.28 \pm 0.12\%$), as well as their charge conjugates, were reconstructed in the central rapidity region ($|\eta| < 0.9$) in pp collisions at $\sqrt{s}$=7 and 2.76 TeV, p-Pb collisions at $\sqrt{s_{\rm NN}} = 5.02$ TeV and Pb-Pb collisions at $\sqrt{s_{\rm NN}} = 2.76$ TeV.

The analysis is based on the selection of fully reconstructed decay topologies displaced from the interaction vertex. A combination of topological cuts and particle identification (PID) of the decay products was applied to reduce the combinatorial background. The PID of pions, kaons and protons was based on information on the specific ionisation dE/dx in the Time Projection Chamber (TPC) gas and a time-of-flight measurement with the TOF detector [7]. The topological cuts were based on the displaced vertex analysis, exploiting the high resolution on the track position in the vicinity of the interaction vertex provided by the Inner Tracking System (ITS). The pp reference cross section for the nuclear modification factors $R_{\rm AA}$ and $R_{\rm pPb}$ was obtained by scaling the D-meson cross sections measured at $\sqrt{s} = 7$ TeV to the $\sqrt{s_{\rm NN}}$ energies of Pb-Pb and p-Pb collisions based on FONLL pQCD calculations, as explained in [8].

The average $R_{\rm AA}$ of D^0, D^+ and D^{*+} as a function of $p_{\rm T}$ for Pb-Pb collisions in the centrality class 0-7.5% is shown in Fig. 1, along with the $R_{\rm AA}$ of charged particles and pions in the 0-10% centrality class. The D-meson yield is strongly suppressed at high $p_{\rm T}$ with respect to binary-scaled pp collisions and the $R_{\rm AA}$(D) is compatible with $R_{\rm AA}(\pi)$ within the current uncertainties. The available statistics are not sufficient to conclude on the mass hierarchy, that is, $R_{\rm AA}$(D) $> R_{\rm AA}(\pi)$ at moderate $p_{\rm T}$. In the high $p_{\rm T}$ region, as shown in the Fig. 2, $R_{\rm AA}$ of D mesons measured with ALICE is lower than that of J/ψ from B decays measured with CMS [9]. These data are described by model calculations that include mass-dependent energy loss.

D_s^+ $R_{\rm AA}$ was also measured [10], showing a suppression similar to the non-strange D-meson $R_{\rm AA}$ in the interval $8 < p_{\rm T} < 12$ GeV/c. At lower $p_{\rm T}$, a conclusion cannot be drawn with the current uncertainties.

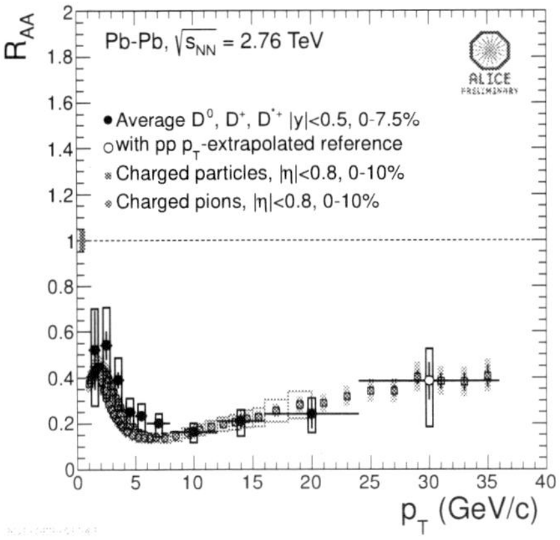

Figure 1: Average $R_{\rm AA}$ of D^0, D^+ and D^{*+} mesons as a function of $p_{\rm T}$ compared with that of charged particles and pions.

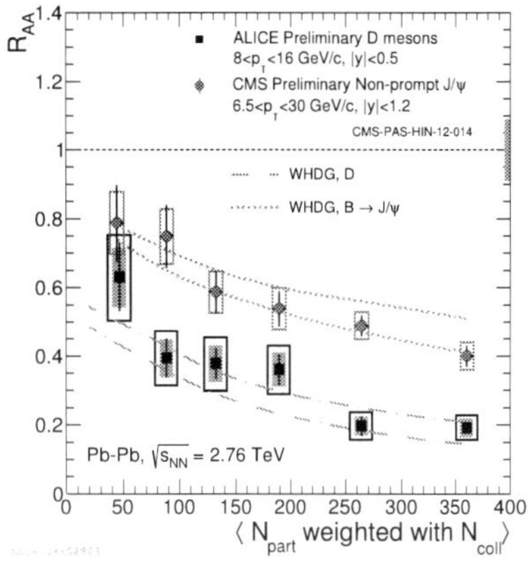

Figure 2: $R_{\rm AA}$ of D mesons compared with $R_{\rm AA}$ of non-prompt J/ψ measured with CMS [9] at high $p_{\rm T}$, as a function of the centrality of the collision expressed in terms of number of participant nucleons, compared with model calculations [3].

The D-meson v_2 was measured using the event plane method, as explained in [11]. Fig. 3 shows the elliptic flow of charmed hadrons measured in the centrality class 30-50% compared with the charged particle v_2. The v_2 of D mesons is positive in the region $2 < p_{\rm T} < 6$ GeV/c (with 5σ significance), suggesting that charm quarks participate in the collective motion. On the other hand, given the current statistics, it is not possible to come to a conclusion on the path-length dependence of the energy loss at high $p_{\rm T}$. As shown in Fig. 4, some of the in-medium energy-loss models describe the measured v_2 reasonably well within uncertainties. However, a simultaneous description of both v_2 and R_{AA} still represents a challenge for the models (for a discussion see e.g.[11]).

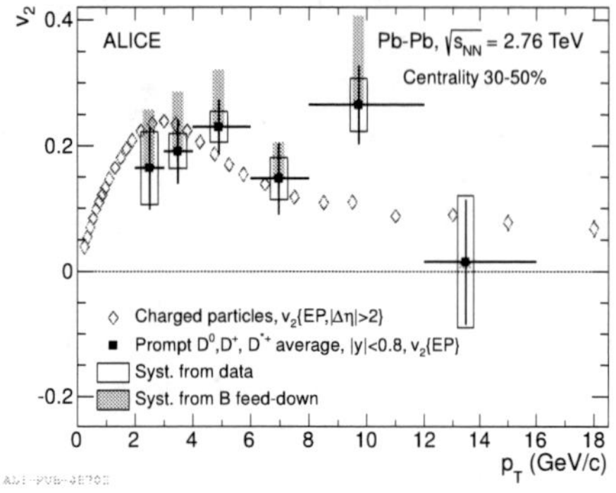

Figure 3: Average v_2 of D^0, D^+ and D^{*+} mesons as a function of p_T compared with v_2 of charged particles [11].

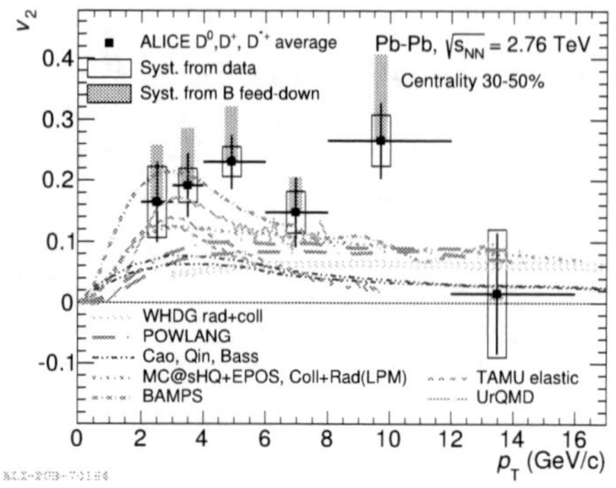

Figure 4: Average v_2 of D^0, D^+ and D^{*+} mesons as a function of p_T compared with model calculations [11].

Fig. 5 shows the measurement of the average R_{pPb} of D^0, D^+ and D^{*+} as a function of p_T for p-Pb collisions at $\sqrt{s_{NN}}$ =5.02 TeV. The R_{pPb} is compatible with unity for $p_T > 2$ GeV/c, which suggests that CNM effects are small and that the suppression observed in Pb–Pb collisions is dominated by the influence of the hot and dense medium. Models including CNM effects describe the data within uncertainties (for a more detailed discussion see e.g.[12]).

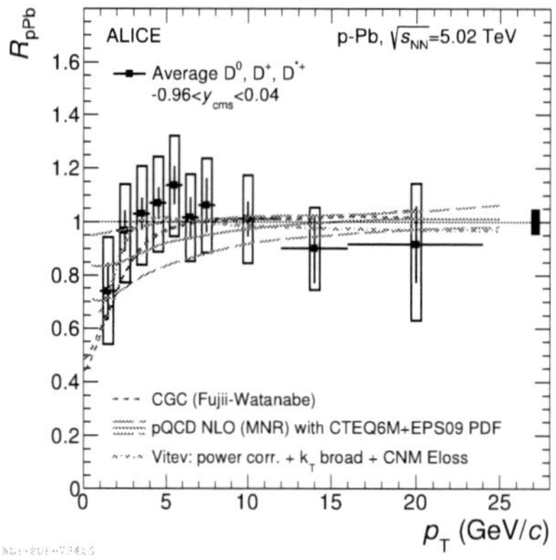

Figure 5: Average R_{pPb} of D^0, D^+ and D^{*+} as a function of p_T [12], compared with model calculations [13, 14, 15, 16, 17].

3. ALICE upgrade studies

An upgrade of the ALICE experiment is planned for the next long shutdown of the LHC (2018-2019). The heavy-flavour measurements will benefit in particular from the upgrade of the Inner Tracking System (ITS) and the Time Projection Chamber (TPC). As one of the main objectives, the resolution of the track impact-parameter measurement will be improved by a factor of 3 in the plane transverse to the beam direction. This will largely enhance the rejection of the combinatorial background in the heavy-flavour reconstruction. The upgrade of the readout capabilities of the TPC and several other detectors will allow to record minimum bias Pb–Pb collisions – which are used for open charm measurements at low momentum – at hundred times the rate compared to the current detector.

The expected performance for the open charm and beauty measurements with the upgraded ALICE detector was studied with simulations of Pb-Pb collisions at $\sqrt{s_{NN}}$ = 5.5 TeV. These simulations included the upgraded Inner Tracking System, a seven-layer silicon tracker with billions of pixel cells [18]. The studies shown here consider the expected integrated luminosity after the ALICE upgrade ($\mathcal{L}_{int} = 10$ nb^{-1}). As shown in Fig. 6 and Fig. 7, the $R_{\rm AA}$ of charmed mesons will be measured with percent-level precision and down to low $p_{\rm T}$, enabling a precise comparison between strange and non-strange D mesons, as well as open beauty and charm. As discussed in [18], the $R_{\rm AA}$ of Λ_c^+ will be measured for the first time, as well as the v_2 of Λ_c^+ and D_s^+.

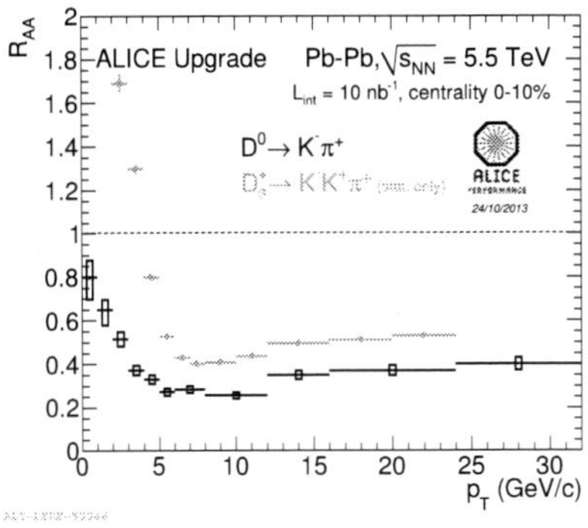

Figure 6: Perspective for $R_{\rm AA}$ of D^0 in comparison with $R_{\rm AA}$ of D_s^+ with the upgraded ALICE detector [18].

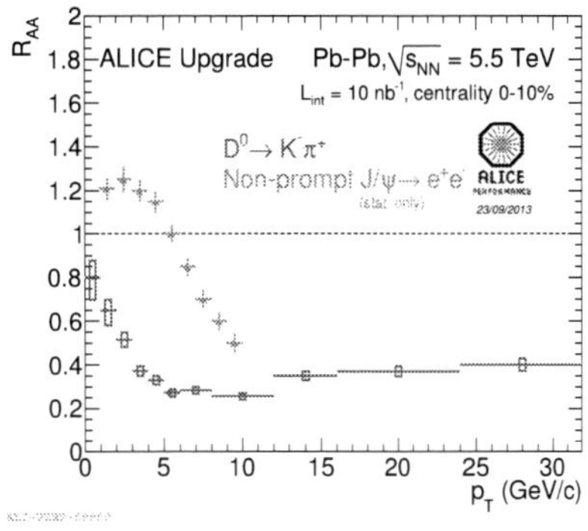

Figure 7: Perspective for $R_{\rm AA}$ of D^0 in comparison with $R_{\rm AA}$ of non-prompt J/ψ with the upgraded ALICE detector [18].

4. Conclusions

Open charm production cross sections in pp collisions are well described by pQCD calculations and in p-Pb by models including CNM effects collisions. The strong suppression observed at high $p_{\rm T}$ in Pb–Pb collisions is due to final state effects, since CNM effects are small as indicated by $R_{\rm pPb}$ being close to unity. The different $R_{\rm AA}$ of D mesons and J/ψ from B-meson decays is described by model calculations including mass-dependent energy loss. Nevertheless, more statistics are needed to draw a conclusion on the comparison between the $R_{\rm AA}$ of D mesons and pions. The positive elliptic flow suggests collective motion of charm quarks in the medium. More statistics are needed to put more stringent constraints to models. As a perspective for the ALICE upgrade, high precision measurements of $R_{\rm AA}$ and v_2 of several heavy-flavour hadron species will become available for the first time.

References

[1] ALICE Collaboration 2008 *JINST* **3** S08002
[2] Kuznetsova I and Rafelski J 2007 *Eur. Phys. J. C* **51** 113–133
[3] Wicks S, Horowitz W, Djordjevic M and Gyulassy M 2007 *Nuc. Phys. A* **784** 426 – 442
[4] Djordjevic M and Heinz U 2008 *Phys.Rev.* **C77** 024905
[5] Dokshitzer et al 2001 *Phys. Lett. B* **519** 199–206
[6] Snellings R 2011 *New J. Phys.* **13** 055008
[7] ALICE Collaboration 2006 *Phys. G* **32** 1295
[8] Averbeck R, Bastid N, del Valle Z C, Crochet P, Dainese A *et al.* 2011 *arXiv:1107.3243*
[9] 2012 Tech. Rep. CMS-PAS-HIN-12-014 CERN Geneva
[10] Innocenti G M 2013 *Nuclear Physics A* **904905** 433c – 436c
[11] Abelev B *et al.* (ALICE Collaboration) 2014 *Phys. Rev. C* **90**(3) 034904
[12] Abelev B *et al.* (ALICE Collaboration) 2014 *arXiv:1405.3452*
[13] Mangano M L, Nason P and Ridolfi G 1992 *Nucl. Phys. B* **373** 295 – 345
[14] Eskola K, Paukkunen H and Salgado C 2009 *J. High Energy Phys.* **2009** 065
[15] Herzog C P and Son D T 2003 *J. High Energy Phys.* **2003** 046
[16] Fujii H and Watanabe K 2013 *Nucl. Phys. A* **920** 78 – 93
[17] Sharma R, Vitev I and Zhang B W 2009 *Phys. Rev. C* **80**(5) 054902
[18] ALICE Collaboration 2013 Tech. Rep. CERN-LHCC-2013-024. ALICE-TDR-017 CERN Geneva

Angular analysis and CP violation studies in B decays at CMS

Kai-Feng Chen

Department of Physics, National Taiwan University, Taipei, Taiwan

E-mail: Kai-Feng.Chen@cern.ch

Abstract. Angular distributions of the decay $B^0 \to K^{*0}\mu^+\mu^-$ are a clean probe of physics beyond the standard model. In particular the forward-backward asymmetry and other variables which can be determined as a function of q^2, provide fruitful information from a single decay channel. An angular and proper decay time analysis is applied to the $B_s \to J/\psi\phi$ decays. The B_s signal candidates are reconstructed and used to extract the mixing phase ϕ_s. The theoretical prediction of the ϕ_s angle is particularly robust, thus any deviation from the prediction can be a smoking gun signal of new physics.

1. Introduction

The flavor-changing neutral-current decays such as $B^0 \to K^{*0}\mu^+\mu^-$ are of particular interest for new physics searches, since new physics phenomena beyond the standard model can be observed indirectly through their influence in the loop diagram. The decay $B^0 \to K^{*0}\mu^+\mu^-$ can provide many observables, such as the branching fraction, forward-backward asymmetry of the muons (A_{FB}), as well as the longitudinal polarization fraction of the K^{*0} (F_L), and in particular these observables can be measured as a function of $q^2 = M^2(\mu^+\mu^-)$. Theoretical calculations for these observables are relatively robust, hence any deviation between experimental measurements and the predictions can be interpreted as a hint of the new phenomena (See Ref. [1, 2, 3, 4, 5] for example).

The weak phase ϕ_s arises from the interference between direct B_s meson decays into a CP eigenstate $B_s \to J/\psi\phi$ and decays through mixing to the exactly same state. The weak phase ϕ_s is related to the CKM matrix elements, where $\phi_s \approx -2\beta_s$, $\beta_s = \arg(-V_{ts}V_{tb}^*/V_{cs}V_{cb}^*)$. Within the standard model, the ϕ_s is predicted to have a small value of $-0.0363^{+0.0016}_{-0.0015}$ rad [6]. Since the prediction is rather precise, any deviation of the measured value would be a hint of new physics in the B_s mixing. In particular, many new physics scenarios may enhance the value of ϕ_s [7, 8].

The analyses for $B^0 \to K^{*0}\mu^+\mu^-$ and for $B_s \to J/\psi\phi$ are performed using the data collected by the CMS detector. The CMS detector is a general-purpose detector. A detailed description can be found in Ref. [9]. The central feature of the detector is a superconducting solenoid that provides an axial magnetic field of 3.8 T. Charged particle trajectories are measured within the field volume by pixel and silicon strip tracker. The calorimeter enclosing the tracker includes a lead tungstate crystal electromagnetic calorimeters (ECAL), which is composed of a barrel part and two endcaps, a lead and silicon preshower detector in front of the ECAL endcaps, and a brass/scintillator hadron calorimeter (HCAL) that together provide an energy measurement for electrons, photons, and hadronic jets. Muons are identified and measured in gas-ionization detectors embedded in the steel return yoke outside the solenoid. The detector is nearly hermetic, allowing accurate energy balance measurements in the plane transverse to the beam direction. The direction of particles measured inside the CMS detector is described using the azimuthal angle (ϕ) and the pseudorapidity (η), which is defined as $\eta \equiv -\ln[\tan\theta/2]$, where θ is the polar angle relative to the counterclockwise proton beam direction, as measured from the nominal interaction vertex.

2. Angular analysis of $B^0 \to K^{*0}\mu^+\mu^-$

In this section, the measurements of A_{FB}, F_L, and the differential branching fraction $d\mathcal{B}/dq^2$ from $B^0 \to K^{*0}\mu^+\mu^-$ decays, using data collected from pp collisions at $\sqrt{s} = 7\,\text{TeV}$ and corresponding to an integrated luminosity of $5\,\text{fb}^{-1}$, are presented. The details of the analysis can be found in Ref. [10].

The signal $B^0 \to K^{*0}\mu^+\mu^-$ decays are recorded with a trigger, which requires two identified muons of opposite charge to form a vertex that is displaced from the beamspot. The trigger configurations were updated during the data taking period with tightening requirements to preserve an acceptable trigger rate. The signal decay mode used in this analysis require two reconstructed muon candidates and two charged hadrons. The muon candidates are required to match the triggered muon in the event and to pass several quality requirements. A muon track must be matched with at least one muon segment, χ^2 per degrees of freedom of track-fit must be less than 1.8, and there must be at least 11 (2) hits in the tracker (pixel detector).

The charged hadron tracks are required to fail the muon selection criteria. The invariant mass of the two charged hadrons, with either $K^+\pi^-$ or $K^-\pi^+$ hypothesis, must be within a ± 80 MeV mass window around the nominal K^{*0} mass. If the invariant mass with a hypothesis of two charged kaons is below 1.035 GeV, the pair of hadron tracks is rejected for possible ϕ meson contributions. The B meson candidates are obtained by requiring the four charged tracks from a common vertex with a constrained fit. The invariant mass of the B^0 candidate must be within ± 280 MeV from the world-average mass of B^0 meson. The candidate is identified as a B^0 or $\overline{B}^0$ if the $K^+\pi^-$ or $K^-\pi^+$ invariant mass is closer to the nominal K^{*0} mass, respectively. In cases where both K-π hypotheses are within ± 50 MeV from the K^{*0} mass, the event is rejected from the analysis. With the full selection criteria, there are 8% of the candidates with incorrect b-flavor assignment.

Figure 1 shows the angular observables which define the decay: θ_K is the helicity angle of the K^{*0} candidate, θ_l is the helicity angle for the dimuon system, and ϕ is the angle between the K^{*0} and dimuon planes. Extraction of the physics parameters, including A_{FB} and F_L, do not require the angle ϕ and therefore it is integrated out to reduce the complexity. The angular distribution of $B^0 \to K^{*0}\mu^+\mu^-$ can be expressed as:

$$\frac{1}{\Gamma}\frac{d^3\Gamma}{d\cos\theta_K d\cos\theta_L dq^2} \propto \left\{ \left[\tfrac{2}{3}F_S + \tfrac{4}{3}A_S\cos\theta_K\right](1-\cos^2\theta_L) + (1-F_S)\left[2F_L\cos^2\theta_K(1-\cos^2\theta_L) + \tfrac{1}{2}(1-F_L)(1-\cos^2\theta_K)(1+\cos^2\theta_L) + \tfrac{4}{3}A_{FB}(1-\cos^2\theta_K)\cos\theta_L\right]\right\}, \quad (1)$$

where the F_S and A_S are the S-wave faction and the interference amplitude between the S- and P-wave decays, respectively. The result of the analysis is extracted from unbinned extended maximum-likelihood fits in bins of q^2 to the invariant mass of the B^0 meson candidate, and to the two angles θ_K and θ_l, using the angular distribution model introduced above. After excluding the q^2 regions corresponding to J/ψ and ψ' resonances, there are a total 415 ± 30 signal events distributed in 6 q^2 bins.

The measured values of F_L, A_{FB}, and differential branching fractions in bins of q^2 are summarized in Fig. 2, along with the standard model predictions [11]. The results are found to be

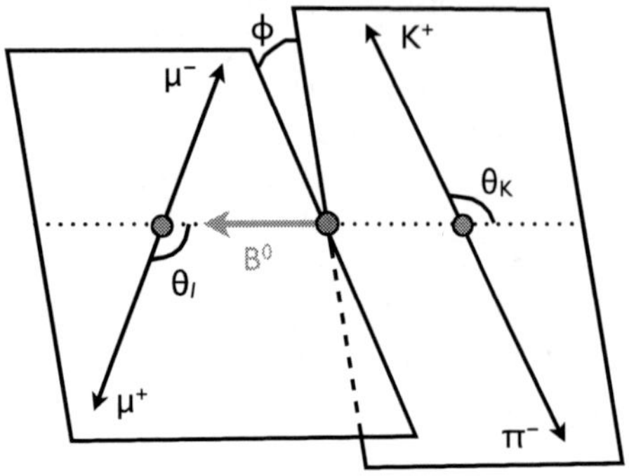

Figure 1. The definition of the angular observables for the $B^0 \to K^{*0}\mu^+\mu^-$ decay.

consistent with the predictions, as well as with measurements from other experiments [12, 13, 14]. The CMS results for this decay channel are competitive with other experiments, in particular for the higher q^2 bins. The analysis with full 2012 data sets will provide further information in this decay. These results can be used to constrain the new physics beyond the standard model in the near future.

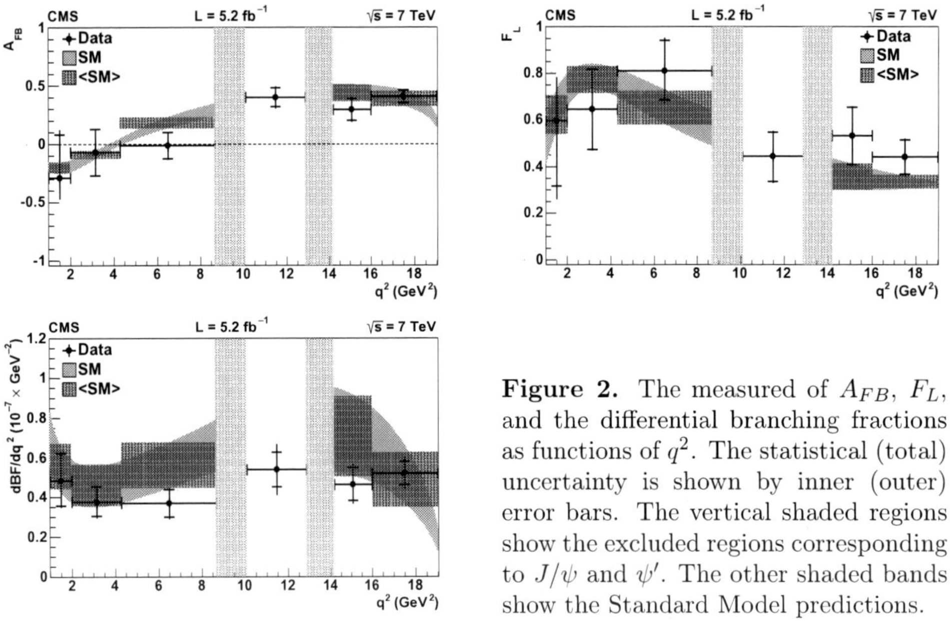

Figure 2. The measured of A_{FB}, F_L, and the differential branching fractions as functions of q^2. The statistical (total) uncertainty is shown by inner (outer) error bars. The vertical shaded regions show the excluded regions corresponding to J/ψ and ψ'. The other shaded bands show the Standard Model predictions.

3. Measurement of the CP-violating weak phase ϕ_s in $B_s \to J/\psi\phi$ decays

The weak phase ϕ_s is measured with $B_s \to J/\psi\phi$ decays, which is not a definite CP state and an angular analysis is required to disentangle the CP-odd and CP-even components of the final state. A time-dependent angular analysis is introduced by measuring the decay angles of the final state particles and the proper decay length of the B_s. In this analysis a dataset corresponding to an integrated luminosity of 20 fb^{-1} collected from pp collisions at $\sqrt{s} = 8$ TeV is used. The details of the analysis can be found in Ref. [15].

The $B_s \to J/\psi\phi$ signal events are triggered with a J/ψ candidate displaced from the primary vertex. In the offline section, the muons are required to have a minimum p_T of 4 GeV. Two oppositely charged muons are paired and form a common decay vertex. Dimuon pairs are selected if the invariant mass is within ±150 MeV of the average J/ψ mass. The $\phi \rightarrow K^+K^-$ candidates are reconstructed using pairs of oppositely charged tracks with nominal kaon mass assumed. The invariant mass of the K^+K^- pair is required to be within a ±10 MeV mass window around the nominal ϕ mass.

The B_s candidates are reconstructed from the J/ψ candidate and ϕ meson candidates. A common vertex position and its associated uncertainty are determined from a fit to all four final state charged tracks, constraining the dimuon invariant mass to the J/ψ mass. For events with multiple B_s candidates, the candidate with the best vertex fit quality is selected. The proper decay length, ct, is calculated based on the vertex position of B_s candidate. The proper decay length is required to be larger than 200 μm in order to avoid a bias due to the inefficiency from the trigger.

In this analysis, the angular distribution is defined in the transversity basis [16]. The definitions of the three angles $\Theta = (\theta_T, \psi_T, \phi_T)$ are illustrated in Fig. 3. The angle θ_T (ϕ_T) is the polar (azimuthal) angle defined of the μ^+ in the J/ψ rest frame, while the angle ψ_T is the helicity angle of ϕ meson defined by the direction of K^+. The differential decay rate for $B_s \rightarrow J/\psi\phi$ can be expressed as:

$$\frac{d^4\Gamma[B_s(t)]}{d\Theta dt} = \sum_{i=1}^{10} \mathcal{O}_i(\alpha, t) \cdot g_i(\Theta) \ , \tag{2}$$

where $\mathcal{O}_i$ are the time-dependent functions and g_i are the angular-dependent functions for i^{th} angular and CP states. The α are the physics parameters of interests; the Θ and t represent the measured angles and proper decay time of the B_s decay. The function $\mathcal{O}_i$ can be expanded as:

$$\mathcal{O}_i(\alpha, t) \propto e^{-\Gamma_s t} \left[a_i \cosh\left(\tfrac{1}{2}\Delta\Gamma_s t\right) + b_i \sinh\left(\tfrac{1}{2}\Delta\Gamma_s t\right) + c_i \cos\left(\Delta m_s t\right) + d_i \sin\left(\Delta m_s t\right)\right] \ , \tag{3}$$

where the parameters b_i and d_i contain the contribution from the weak phase ϕ_s, and are summarized in Tab. 1 together with the g_i functions. The details of the description of the time-dependent decay rate can be found in Ref. [17].

Table 1. Angular and time-dependent terms of the signal model for the $B_s \rightarrow J/\psi\phi$ decays, where the C, S, and D parameters are defined by $C = \frac{1-|\lambda|^2}{1+|\lambda|^2}$, $S = -\frac{2|\lambda|\sin\phi_s}{1+|\lambda|^2}$, $D = -\frac{2|\lambda|\cos\phi_s}{1+|\lambda|^2}$, respectively.

i	$g_i(\theta_T, \psi_T, \phi_T)$	a_i	b_i	c_i	d_i
1	$2\cos^2\psi_T(1-\sin^2\theta_T\cos^2\phi_T)$	1	D	C	$-S$
2	$\sin^2\psi_T(1-\sin^2\theta_T\sin^2\phi_T)$	1	D	C	$-S$
3	$\sin^2\psi_T\sin^2\theta_T$	1	$-D$	C	S
4	$-\sin^2\psi_T\sin 2\theta_T\sin\phi_T$	$C\sin(\delta_\perp-\delta_\parallel)$	$S\cos(\delta_\perp-\delta_\parallel)$	$\sin(\delta_\perp-\delta_\parallel)$	$D\cos(\delta_\perp-\delta_\parallel)$
5	$\frac{1}{\sqrt{2}}\sin 2\psi_T\sin^2\theta_T\sin 2\phi_T$	$\cos(\delta_\parallel-\delta_0)$	$D\cos(\delta_\parallel-\delta_0)$	$C\cos(\delta_\parallel-\delta_0)$	$-S\cos(\delta_\parallel-\delta_0)$
6	$\frac{1}{\sqrt{2}}\sin 2\psi_T\sin 2\theta_T\sin\phi_T$	$C\sin(\delta_\perp-\delta_0)$	$S\cos(\delta_\perp-\delta_0)$	$\sin(\delta_\perp-\delta_0)$	$D\cos(\delta_\perp-\delta_0)$
7	$\frac{2}{3}(1-\sin^2\theta_T\cos^2\phi_T)$	1	$-D$	C	S
8	$\frac{1}{3}\sqrt{6}\sin\psi_T\sin^2\theta_T\sin 2\phi_T$	$C\cos(\delta_\parallel-\delta_S)$	$S\sin(\delta_\parallel-\delta_S)$	$\cos(\delta_\parallel-\delta_S)$	$D\sin(\delta_\parallel-\delta_S)$
9	$\frac{1}{3}\sqrt{6}\sin\psi_T\sin 2\theta_T\cos\phi_T$	$\sin(\delta_\perp-\delta_S)$	$-D\sin(\delta_\perp-\delta_S)$	$C\sin(\delta_\perp-\delta_S)$	$S\sin(\delta_\perp-\delta_S)$
10	$\frac{4}{3}\sqrt{3}\cos\psi_T(1-\sin^2\theta_T\cos^2\phi_T)$	$C\cos(\delta_0-\delta_S)$	$S\sin(\delta_0-\delta_S)$	$\cos(\delta_0-\delta_S)$	$D\sin(\delta_0-\delta_S)$

Another important ingredient of this analysis is the capability of flavor tagging. If the other b-hadron opposite to the signal B_s candidate decays in one of the semileptonic channels, the

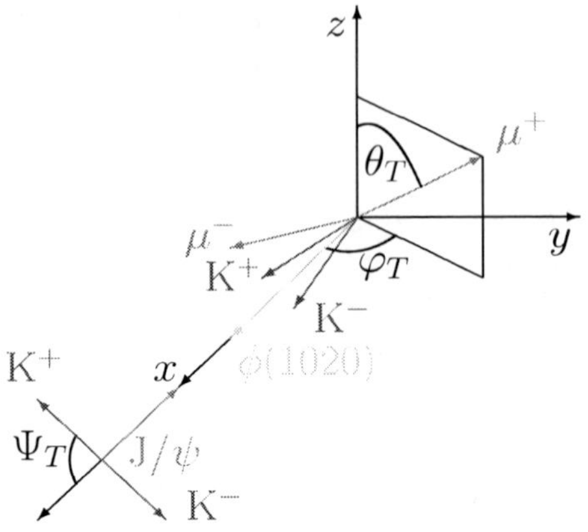

Figure 3. Definition of the three angular observables, θ_T, ψ_T, and ϕ_T for the $B_s \to J/\psi\phi$ decays.

charge of the opposite-side muon or electron is used to derive the flavor of the signal B_s meson at its production time. In practise the lepton with the highest p_T in the event is used as the tagging lepton. The tag lepton selection is optimized to reach the best tagging power, $P_{\text{tag}} = \epsilon_{\text{tag}}(1-2\omega)^2$, where the ϵ_{tag} is the tagging efficiency and ω is the mis-tag probability. The performance of the tagging is measured from the data with the self-tagging $B^+ \to J/\psi K^+$ decays. The measured average tagging performance with electron and muon taggers combined is $\omega = (32.3 \pm 0.3)\%$, $\epsilon_{\text{tag}} = (7.67 \pm 0.04)\%$, and resulting $P_{\text{tag}} = (0.97 \pm 0.03)\%$. The performance evaluated for simulated B^+ and B_s events are found to be consistent with each other.

An unbinned maximum likelihood fit is introduced to extract the physics parameters: the weak phase ϕ_s, the lifetime difference $\Delta\Gamma_s$, the mean lifetime of B_s meson, the decay amplitudes and the corresponding strong phases. The fit is performed by including information on the candidate B_s invariant mass, proper decay length, and the three decay angles. The per-event proper decay length uncertainty is included in the proper decay time resolution model. The fit sample contains around 49,000 signal events and 21,000 background events. The projections of the three decay angles are shown in Fig. 4, and the resulting physics parameters are summarized in Tab. 2. The value of the weak phase ϕ_s and the lifetime difference $\Delta\Gamma_s$ are in good agreement with the previous measurements and with the standard model predictions. Our measurements are the second best measurements to date and will improve the world averages. The uncertainties of the measurements are still dominated by statistical uncertainties and can be reduced further with upcoming new data from LHC run II.

Parameter	Result
$\lvert A_0\rvert^2$	$0.511 \pm 0.006 \pm 0.012$
$\lvert A_S\rvert^2$	$0.015 \pm 0.016 \pm 0.022$
$\lvert A_\perp\rvert^2$	$0.242 \pm 0.008 \pm 0.012$
$\delta_\parallel$	$3.48 \pm 0.09 \pm 0.68$ rad
$\delta_{S\perp}$	$0.34 \pm 0.24 \pm 1.12$ rad
$\delta_\perp$	$2.73 \pm 0.36 \pm 0.66$ rad
$c\tau$	$447.3 \pm 3.0 \pm 3.5$ μm
$\Delta\Gamma_s$	$0.096 \pm 0.014 \pm 0.007$ ps^{-1}
ϕ_s	$-0.03 \pm 0.11 \pm 0.03$ rad

Table 2. The measured physics parameters of interests from the unbinned maximum likelihood fit to the $B_s \to J/\psi\phi$ candidates.

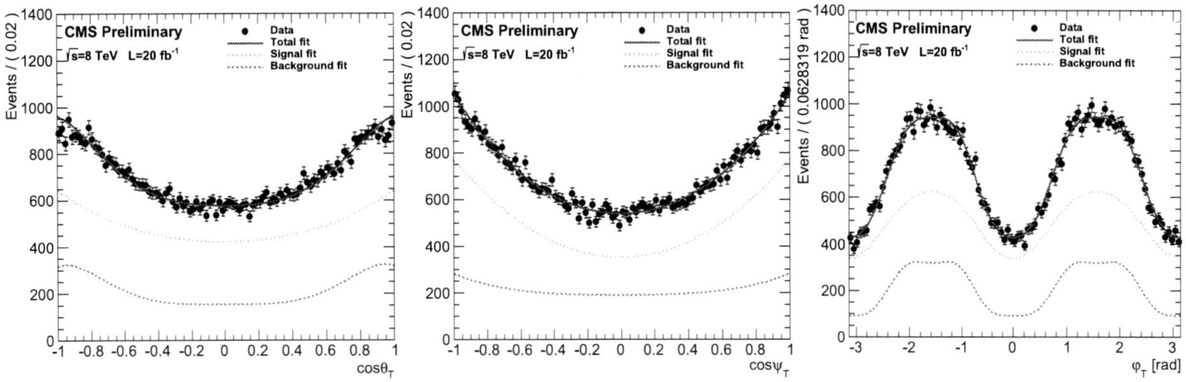

Figure 4. The measured angular distributions (θ_T, ψ_T, and ϕ_T) for the $B_s \to J/\psi\phi$ candidates. The solid lines are the result of the fit to the data, while the green (red) dashed lines represent the signal (background) components.

References

[1] Q. S. Yan, C. S. Huang, W. Liao and S. H. Zhu, "Exclusive semileptonic rare decays $B \to (K, K^{*})\ell^+\ell^-$ in supersymmetric theories," Phys. Rev. D **62**, 094023 (2000) [hep-ph/0004262].

[2] W. Altmannshofer, P. Ball, A. Bharucha, A. J. Buras, D. M. Straub and M. Wick, "Symmetries and Asymmetries of $B \to K^*\mu^+\mu^-$ Decays in the Standard Model and Beyond," JHEP **0901**, 019 (2009) [arXiv:0811.1214 [hep-ph]].

[3] A. Ali, P. Ball, L. T. Handoko and G. Hiller, "A Comparative study of the decays $B \to (K, K^{*})\ell^+\ell^-$ in standard model and supersymmetric theories," Phys. Rev. D **61**, 074024 (2000) [hep-ph/9910221].

[4] G. Buchalla, G. Hiller and G. Isidori, "Phenomenology of nonstandard Z couplings in exclusive semileptonic $b \to s$ transitions," Phys. Rev. D **63**, 014015 (2000) [hep-ph/0006136].

[5] T. Feldmann and J. Matias, "Forward backward and isospin asymmetry for $B \to K^*\ell^+\ell^-$ decay in the standard model and in supersymmetry," JHEP **0301**, 074 (2003) [hep-ph/0212158].

[6] J. Charles, O. Deschamps, S. Descotes-Genon, R. Itoh, H. Lacker, A. Menzel, S. Monteil and V. Niess *et al.*, "Predictions of selected flavour observables within the Standard Model," Phys. Rev. D **84**, 033005 (2011) [arXiv:1106.4041 [hep-ph]].

[7] A. J. Buras, "Flavour Theory: 2009," PoS EPS **-HEP2009**, 024 (2009) [arXiv:0910.1032 [hep-ph]].

[8] C. W. Chiang, A. Datta, M. Duraisamy, D. London, M. Nagashima and A. Szynkman, "New Physics in $B_s^0 \to J/\psi\phi$: A General Analysis," JHEP **1004**, 031 (2010) [arXiv:0910.2929 [hep-ph]].

[9] S. Chatrchyan *et al.* [CMS Collaboration], "The CMS experiment at the CERN LHC," JINST **3**, S08004 (2008).

[10] The CMS Collaboration, "Angular analysis and branching fraction measurement of the decay $B^0 \to K^{*0}\mu^+\mu^-$," Phys. Lett. B **727**, 77 (2013) [arXiv:1308.3409 [hep-ex]].

[11] C. Bobeth, G. Hiller and D. van Dyk, "General Analysis of $\bar{B} \to \bar{K}^{(*)}\ell^+\ell^-$ Decays at Low Recoil," Phys. Rev. D **87**, 034016 (2013) [arXiv:1212.2321 [hep-ph]].

[12] B. Aubert *et al.* [BaBar Collaboration], "Angular Distributions in the Decays $B \to K^*\ell^+\ell^-$," Phys. Rev. D **79**, 031102 (2009) [arXiv:0804.4412 [hep-ex]].

[13] J.-T. Wei *et al.* [BELLE Collaboration], "Measurement of the Differential Branching Fraction and Forward-Backward Asymmetry for $B \to K^{(*)}\ell^+\ell^-$," Phys. Rev. Lett. **103**, 171801 (2009) [arXiv:0904.0770 [hep-ex]].

[14] R. Aaij *et al.* [LHCb Collaboration], "Differential branching fraction and angular analysis of the decay $B^0 \to K^{*0}\mu^+\mu^-$," JHEP **1308**, 131 (2013) [arXiv:1304.6325, arXiv:1304.6325 [hep-ex]].

[15] The CMS Collaboration, "Measurement of the CP-violating weak phase ϕ_s and the decay width difference $\Delta\Gamma_s$ using the $B_s \to J/\psi\phi$ decay channel," CMS PAS BPH-13-012 (2014).

[16] A. S. Dighe, I. Dunietz and R. Fleischer, "Extracting CKM phases and $B_s - \bar{B}_s$ mixing parameters from angular distributions of nonleptonic B decays," Eur. Phys. J. C **6**, 647 (1999) [hep-ph/9804253].

[17] A. S. Dighe, I. Dunietz, H. J. Lipkin and J. L. Rosner, "Angular distributions and lifetime differences in $B_s \to J/\psi\phi$ decays," Phys. Lett. B **369**, 144 (1996) [hep-ph/9511363].

$\bar{B}_s \to K$ semileptonic decay from an Omnès improved nonrelativistic quark model

C Albertus[1], E Hernández[2], C Hidalgo-Duque[3] and J Nieves[3]

[1] Departamento de Física Atómica, Nuclear y Molecular e Instituto Carlos I de Física Teórica y Computacional, Universidad de Granada, Avenida de Fuentenueva s/n, E-18071 Granada, Spain

[2] Departamento de Física Fundamental e IUFFyM, Universidad de Salamanca, Plaza de la Merced s/n, E-37008 Salamanca, Spain

[3] Instituto de Física Corpuscular (IFIC), Centro Mixto CSIC-Universidad de Valencia, Institutos de Investigación de Paterna, Apartado 22085, E-46071 Valencia, Spain

E-mail: albertus@ugr.es,gajatee@usal.es,carloshd@ific.uv.es, jmnieves@ific.uv.es

Abstract. We study the f^+ form factor for the $\bar{B}_s \to K^+\ell^-\bar{\nu}_\ell$ semileptonic decay in a nonrelativistic quark model. The valence quark contribution is supplemented with a $\bar{B}^*$-pole term that dominates the high q^2 region. To extend the quark model predictions from its region of applicability near $q^2_{\rm max} = (M_{B_s} - M_K)^2$, we use a multiply-subtracted Omnès dispersion relation. We fit the subtraction constants to a combined input from previous light cone sum rule results in the low q^2 region and the quark model results (valence plus $\bar{B}^*$-pole) in the high q^2 region. From this analysis, we obtain $\Gamma(\bar{B}_s \to K^+\ell^-\bar{\nu}_\ell) = (5.47^{+0.54}_{-0.46})|V_{ub}|^2 \times 10^{-9}\,$MeV, which is about 10% and 20% higher than predictions based on Lattice QCD and QCD light cone sum rules respectively.

1. Introduction

Playing a critical role in testing the consistency of the Standard Model of particle physics and, in particular, the description of CP violation, V_{ub} is still the less well known element of the Cabibbo-Kobayashi-Maskawa (CKM) quark mixing matrix. Any new information that can be obtained from experimentally unexplored reactions is thus relevant. This is the case of the $\bar{B}_s \to K^+\ell^-\bar{\nu}_\ell$ semileptonic decay which is expected to be observed at LHCb and Belle and that it could be used to obtain an independent determination of $|V_{ub}|$. In this contribution we present a study of this reaction. All the details and further results to those presented here can be found in Ref. [1].

The hadronic matrix element for the reaction can be parameterized in terms of the $f^+(q^2)$ and $f^0(q^2)$ form factors, of which only $f^+(q^2)$ plays a significant role for the case of a light lepton in the final state ($l = e, \mu$). In fact, for zero lepton masses, the differential decay width is given solely in terms of $f^+(q^2)$ as

$$\frac{d\Gamma}{dq^2} = \frac{G_F^2}{192\pi^3}|V_{ub}|^2 \frac{\lambda^{3/2}(q^2, M_{B_s}^2, M_K^2)}{M_{B_s}^3}|f^+(q^2)|^2 \qquad (1)$$

with G_F the Fermi decay constant and λ the Källen function defined as $\lambda(a, b, c) = a^2 + b^2 + c^2 - 2ab - 2ac - 2bc$.

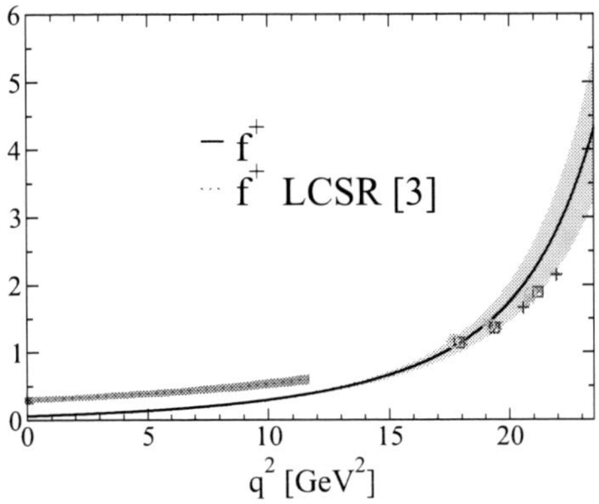

Figure 1. $f^+(q^2)$ form factor evaluated in the quark model (valence plus $\bar{B}^*$-pole contributions). We also show the results obtained in the LCSR calculation of Ref. [3] (dotted-line plus error band) and different lattice data in the high q^2 region reported in Ref. [4].

2. Results and discussion

To obtain the f^+ form factor we shall follow our earlier work in Ref. [2], where similar decays were analyzed, and then we use the quark model to evaluate the valence plus $\bar{B}^*$-pole contributions to the form factors. Calculational details can be found in [1] and references therein. Results are shown in figure 1. Taking into account theoretical uncertainties, shown as a band in the figure, we obtain a reasonable description of the form factor in the high q^2 region, as compared to the preliminary lattice data recently reported in Ref. [4]. For high q^2, the $\bar{B}^*$-pole term dominates but the valence contribution accounts for around 20% of the total. However, there is a large discrepancy in the low q^2 region between the quark model and the light cone sum rule (LCSR) results obtained in Ref. [3]. Since the latter are reliable for low q^2, it is clear that the non-relativistic quark model does not provide a good reproduction of the form factor in that region of q^2 where large relativistic effects are to be expected.

To obtain an $f^+(q^2)$ form factor valid for the whole q^2 region spanned by the decay, we adopt the scheme in Refs. [5, 6, 7], assuming a multiply subtracted Omnès functional ansatz that provides a parameterization of the form factor constrained by unitarity and analyticity properties. We take

$$f^+(q^2) \approx \frac{1}{M_{B^*}^2 - q^2} \prod_{j=0}^{n} \left[f^+(q_j^2) \left(M_{B^*}^2 - q_j^2 \right) \right]^{\alpha_j(q^2)} \quad , \quad \alpha_j(q^2) = \prod_{j \neq k=0}^{n} \frac{q^2 - q_K^2}{q_j^2 - q_k^2} \tag{2}$$

for $q^2 < s_{\rm th} = (M_{B_s} + M_K)^2$ and where $q_0, \cdots q_n^2 \in]-\infty, s_{\rm th}[$ are the $(n+1)$ subtraction points. Note that despite the factor $\frac{1}{M_{B^*}^2 - q^2}$, the functional form is not given by a single pole. The values of $f^+(q_j^2)$ are taken as free parameters and we fix them by making a combined fit to our quark model results in the high q^2 region and to the LCSR results, taken from Ref. [3], in the low q^2 part. As in Ref. [7] we only use four subtraction points corresponding to $q_j^2 = 0$, $q_{\rm max}^2/3$, $2q_{\rm max}^2/3$, $q_{\rm max}^2$.

Our final result for $f^+(q^2)$ together with its 68% confidence level band is displayed in figure 2. There, we show a comparison with different calculations using LCSR [3], LCSR+$\bar{B}^*$-pole fit [8], relativistic quark model (RQM) [9], light front quark model (LFQM) [10], perturbative QCD (PQCD) [11] and the extrapolation to the physical region done in Ref. [12] of the lattice QCD (LQCD) results obtained in Ref. [4] (also shown). In the LCSR calculation in Ref. [3] the results are only given up to $q^2 = 10\,{\rm GeV}^2$, whereas in Ref. [10] no $\bar{B}^*$-pole contribution is included as can be seen by the behavior of the predicted form factor in the high q^2 region. All other calculations include the $\bar{B}^*$-pole mechanism, but with different strengths. In Ref. [9], where a RQM is used, they obtain a form factor similar to ours for high q^2 values. However, their approach for low and intermediate values of q^2 should not be as appropriate as a LCSR one,

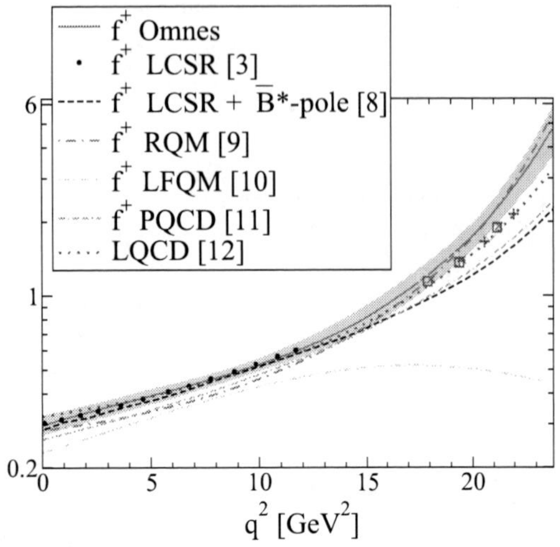

Figure 2. Global comparison of our final result (solid line plus 68% confidence level band) for the f^+ form factor with different calculations using LCSR [3], LCSR+$\bar{B}^*$-pole fit [8], RQM [9], LFQM [10], PQCD [11] and the extrapolation to the physical region done in Ref. [12] of the LQCD data from Ref. [4] which is also shown.

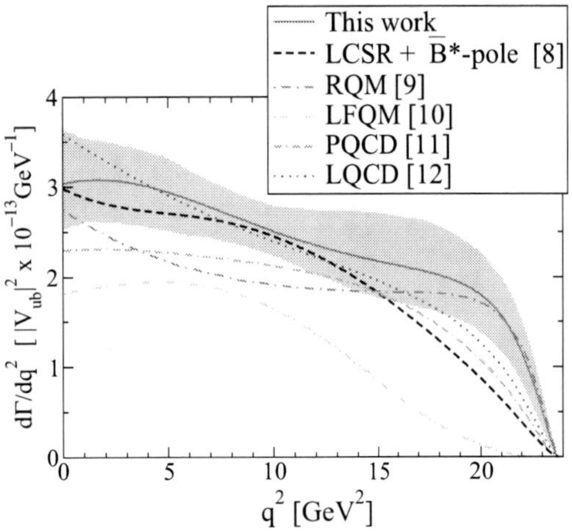

Figure 3. Differential decay width obtained in this work with the Omnès fit (solid line plus 68% confidence level band) and in LCSR+$\bar{B}^*$-pole fit [8], RQM [9], LFQM [10] and PQCD [11] and LQCD [12] approaches.

which we include in our combined analysis. Calculations in Refs. [11] and [8] give similar results at high q^2 but the one in Ref. [11] deviates from LCSR evaluations at small q^2 values. The high q^2 results obtained in LQCD [4, 12] are in between the results obtained in the approaches of Refs. [8, 11] and the quark model ones (both this work and the RQM calculation of Ref. [9]). For very low q^2 however, the central values of the LQCD extrapolation in Ref. [12] lie in the upper part of the LCSR band. Our combined approach should be more adequate in that region of q^2 since we use LCSR data to constraint our form factor.

The differential decay width, together with its 68% confidence level band, is displayed in figure 3. We also show the differential decay width from the calculations in Refs. [8, 9, 10, 11, 12]. For the integrated decay width we obtain

$$\Gamma(\bar{B}_s \to K^+ \ell^- \bar{\nu}_\ell) = (5.47^{+0.54}_{-0.46}) |V_{ub}|^2 \times 10^{-9}\,\text{MeV} \tag{3}$$

and a comparison with the results in other approaches is shown in table 1. The calculations in Refs. [8, 9] obtain results that are some 15% smaller than ours. The fact that their results are so similar when compared to each other seems to be a coincidence. As seen in figure 3, their differential decay widths deviate both for small and large q^2 values, but those differences compensate in the integrated width. The result of the PQCD calculation in Ref. [11] is also similar but with a larger uncertainty, around 50%. The LFQM calculation in Ref. [10] gives a much smaller result, in part because no $\bar{B}^*$-pole contribution seems to be included in that approach. The LQCD result in Ref. [12] is the one closest to ours. Its large uncertainty comes from the form factor extrapolation from high q^2, where the lattice points were obtained, to the low q^2 region. Our result is the largest although we are compatible within uncertainties with the predictions of Refs. [8, 9, 11, 12].

Table 1. Decay width in units of $|V_{ub}|^2 \times 10^{-9}$ MeV from several approaches. For the result of Ref. [8] we have propagated a 10% uncertainty in the form factor. Results for Refs. [9, 10, 11] have been adapted from Table IV in Ref. [13].

	This work	LCSR+$\bar{B}^*$-pole [8]	RQM [9]	LFQM [10]	PQCD [11]	LQCD [12]
$\Gamma\ [|V_{ub}|^2 \times 10^{-9}\,\text{MeV}]$	$5.47^{+0.54}_{-0.46}$	$4.63^{+0.97}_{-0.88}$	4.50 ± 0.45	2.75 ± 0.24	4.2 ± 2.2	5.1 ± 1.0

Acknowledgments

This research was supported by the Spanish Ministerio de Economía y Competitividad and European FEDER funds under Contracts Nos. FPA2010-21750-C02-02, FIS2011-28853-C02-02, and the Spanish Consolider-Ingenio 2010 Programme CPAN (CSD2007-00042), by Generalitat Valenciana under Contract No. PROMETEO/20090090, by Junta de Andalucia under Contract No. FQM-225, by the EU HadronPhysics3 project, Grant Agreement No. 283286, and by the University of Granada start-up Project for Young Researches contract No. PYR-2014-1. C.A. wishes to acknowledge a CPAN postdoctoral contract and C.H.-D. thanks the support of the JAE-CSIC Program.

References

[1] Albertus C, Hernández E, Hidalgo-Duque C and Nieves 2014 *Phys. Lett.* B **738** 144
[2] Albertus C, Flynn J M, Hernández E, Nieves J and Verde-Velasco J M 2005 *Phys. Rev.* D **72** 033002
[3] Duplancic G and Melic B 2008 *Phys. Rev.* D **78** 054015
[4] Bouchard C M, Lepage G P, Monahan C J, Na H and Shigemitsu J 2013 B and B_s semileptonic decay form factors with NRQCD/HISQ quarks *Preprint arXiv:1310.3207 [hep-lat]*
[5] Flynn J M and Nieves J 2007 *Phys. Rev.* D **75** 013008
[6] Flynn J M and Nieves J 2007 *Phys. Lett.* B **649** 269
[7] Flynn J M and Nieves J 2007 *Phys. Rev.* D **76** 031302
[8] Li Z H, Liang F Y, Wu X Y and Huang T 2001 *Phys. Rev.* D **64** 057901
[9] Faustov R N and Galkin V O 2013 *Phys. Rev.* D **87** 094028
[10] Verma R C 2012 *J. Phys.* G **39** 025005
[11] Wang W F and Xiao Z J 2012 *Phys. Rev.* D **86** 114025
[12] Bouchard C M, Lepage G P, Monahan C, Na H and Shigemitsu J 2014 *Phys. Rev.* D **90** 054506
[13] Meißner U G and Wang W 2014 *JHEP* **1401** 107

Top quark production in the ATLAS detector of the LHC

Minoru Hirose on behalf of the ATLAS Collaboration
Tokyo Institute of Technology, 2-12-1 Ookayama, Meguro, Tokyo, 152-8550, Japan
E-mail: hirose@hep.phys.titech.ac.jp

Abstract. Top quark has been re-discovered at LHC in 2010 by the CMS and the ATLAS experiments. The top quark is the heaviest known elementary particle, and its properties might offer a hint of physics beyond the Standard Model. Therefore, it is interesting to measure them as precisely as possible. We present a summary of a variety of measurements relating to the top quark production in pp collisions with $\sqrt{s} = 7$ and 8 TeV at LHC. Not only the inclusive production cross section, but also differential and fiducial cross sections are measured with the ATLAS detector. By comparing our results with various theoretical predictions, it is confirmed that Standard Model predictions are in good agreement with all measured results.

1. Introduction

The top quark was discovered at Tevatron in 1995. It was re-discovered at LHC in 2010. In the LHC operation from 2010 to 2012, the ATLAS experiment [1] has recorded a large number of top quark events. This allows us to measure properties of the top quark very precisely. These precise measurements might give us a clue to catch physics beyond the Standard Model.

Top quarks are produced mainly in pairs via the strong interaction. Predicted cross section of the pair production is calculated as $\sigma_{t\bar{t}} = 177^{+10}_{-11}$ pb for $\sqrt{s} = 7$ TeV and $\sigma_{t\bar{t}} = 253^{+13}_{-15}$ pb for $\sqrt{s} = 8$ TeV with the next-to-next-to-leading order precision [2]. In addition to the pair production, we can consider the single top production via the weak interaction. The dominant process for it happens via the t-channel W boson exchange. The next-to-next-to-leading order calculation predicts the cross section of this process as $\sigma_{tq+\bar{t}q} = 64.6^{+2.7}_{-2.0}$ pb for 7 TeV and $\sigma_{tq+\bar{t}q} = 87.8^{+3.4}_{-1.9}$ pb for 8 TeV [3]. Understanding of these production processes is an important topic because it proves the validity of the Standard Model at the highest energy region ever achieved.

Top quarks are predicted to decay via the electroweak interaction into a W boson and a bottom quark with nearly 100% branching ratio. Events are normally classified according to the number of charged leptons. For the $t\bar{t}$ final state, there are two W bosons. W boson can either decay into a pair of quarks or into a charged lepton and a neutrino. Single lepton and dilepton topologies (one and two charged leptons in the final state, respectively) were considered as the signal candidate event. Only electrons and muons, including those from tau decays, were considered in the analyses presented below.

2. Measurement of the inclusive cross section of $t\bar{t}$

The inclusive cross section of the top quark pair production $\sigma_{t\bar{t}}$ has been measured in pp collisions at $\sqrt{s} = 7$ and 8 TeV in the final state with oppositely charged $e\mu$ pair. The measurement was performed with the 7 TeV dataset taken in 2011 corresponding to an integrated luminosity of 4.6 fb^{-1}, and the 8 TeV dataset taken in 2012 of 20.3 fb^{-1}. The numbers of events with exactly one and exactly two b-tagged jets were counted and used to simultaneously determine $\sigma_{t\bar{t}}$, and efficiency to tag a jet from a top quark decay as a b-quark jet. This method allows us to minimize the associated systematic uncertainties. The cross section was measured to be

$$\begin{aligned} \sigma_{t\bar{t}} &= 182.9 \pm 3.1 \pm 4.2 \pm 3.6 \pm 3.3 \text{ pb } (\sqrt{s} = 7 \text{ TeV}) \text{ and} \\ \sigma_{t\bar{t}} &= 242.4 \pm 1.7 \pm 5.5 \pm 7.5 \pm 4.2 \text{ pb } (\sqrt{s} = 8 \text{ TeV}) \text{ [4]}, \end{aligned}$$

where four uncertainties arise from data statistics, systematic uncertainties, knowledge of the integrated luminosity and of the LHC beam energy. The results are consistent with the theoretical QCD calculations at next-to-next-to-leading order mentioned in Sec. 1.

3. Measurement of the differential cross section of $t\bar{t}$

Measurements of normalized differential cross sections for $t\bar{t}$ production are presented as a function of the top quark transverse momentum p_T^t, and of the mass $M_{t\bar{t}}$, the transverse momentum $p_T^{t\bar{t}}$, and the rapidity $y_{t\bar{t}}$ of the $t\bar{t}$ system, in proton-proton collisions at $\sqrt{s} = 7$ TeV. The dataset corresponds to an integrated luminosity of 4.6 fb^{-1}, recorded in 2011 [5]. Events are selected in the single lepton channel, requiring exactly one lepton and at least four jets with at least one of the jets tagged as b-quark jet. The measured kinematic variables are corrected for detector efficiency and resolution effects, and are compared to several Monte Carlo simulations and theoretical calculations. The results are in agreement with the predictions in a wide kinematic range. Nevertheless, data distributions are softer than predicted for higher values of the mass of the $t\bar{t}$ system and of the top quark transverse momentum as shown in Figure 1. The measurements can also shed a light on some features of different sets of parton distribution functions.

4. Jet multiplicities in the $t\bar{t}$ production

The $t\bar{t}$ production cross section as a function of jet multiplicities is reported for proton-proton collisions at a centre-of-mass energy of 7 TeV in the single lepton final state [6]. We used data collected with the ATLAS detector in 2011 corresponding to an integrated luminosity of 4.6 fb^{-1}. Differential cross sections are presented as a function of the jet multiplicity for up to eight jets using several jet transverse momentum thresholds between 25 and 80 GeV, and as a function of jet transverse momentum up to the fifth jet. The results are shown after background subtraction and corrections for all detector effects. Several QCD-based Monte Carlo models are compared with the results. Sensitivity to the parton shower modeling is found at the higher jet multiplicities and in the transverse momentum spectrum of the fifth leading jet as shown in Figure 2.

5. Top quark pair production associated with the vector boson

We have performed the analysis of the $t\bar{t}$ production associated with a W or Z boson [7]. Analyzed data corresponding to 20.3 fb^{-1} were taken in 2012 at $\sqrt{s} = 8$ TeV. Final states with two or three leptons are considered. Assuming $t\bar{t}W$ production at the Standard Model rate, predicted by next-to-leading-order QCD calculations, the ratio of the measured $t\bar{t}Z$ signal to the Standard Model expectation (signal strength) is found to be $\mu_{t\bar{t}Z} = 0.73^{+0.29}_{-0.26}$, corresponding to a 3.2σ excess over the background-only hypothesis. Assuming $t\bar{t}Z$ production at the Standard Model rate, the $t\bar{t}W$ signal strength is found to be $\mu_{t\bar{t}W} = 1.25^{+0.57}_{-0.47}$, corresponding to a

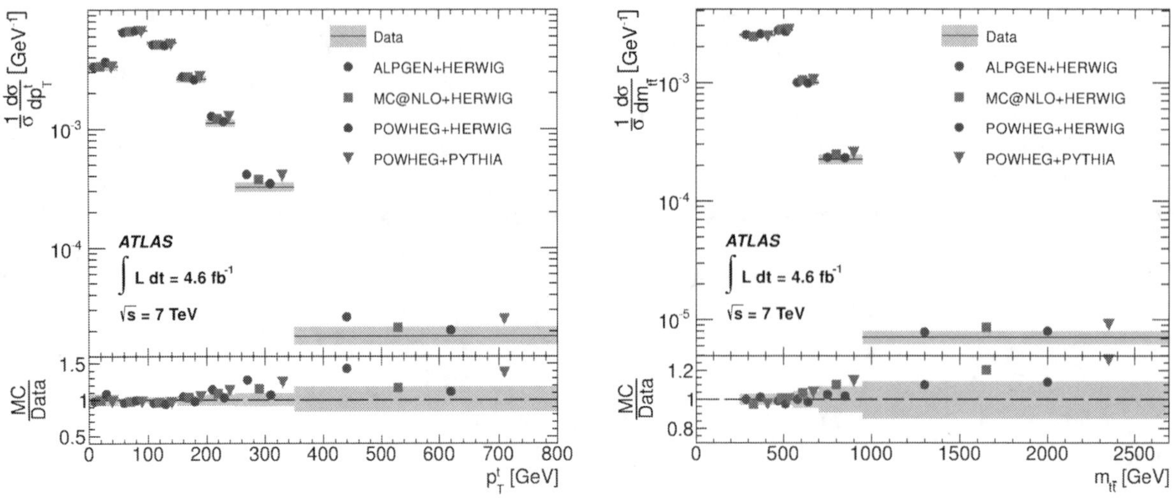

Figure 1. Normalized differential cross sections for the transverse momentum of the hadronically decaying top quark p_T^t (left), and the invariant mass of the $t\bar{t}$ system $m_{t\bar{t}}$ (right) [5]. Predictions by four Monte Carlo generators are shown as markers for ALPGEN+HERWIG (green circles), MC@NLO+HERWIG (squares), POWHEG+HERWIG (blue circles) and POWHEG+PYTHIA (inverted triangles). Shaded bands indicate the total uncertainty on data.

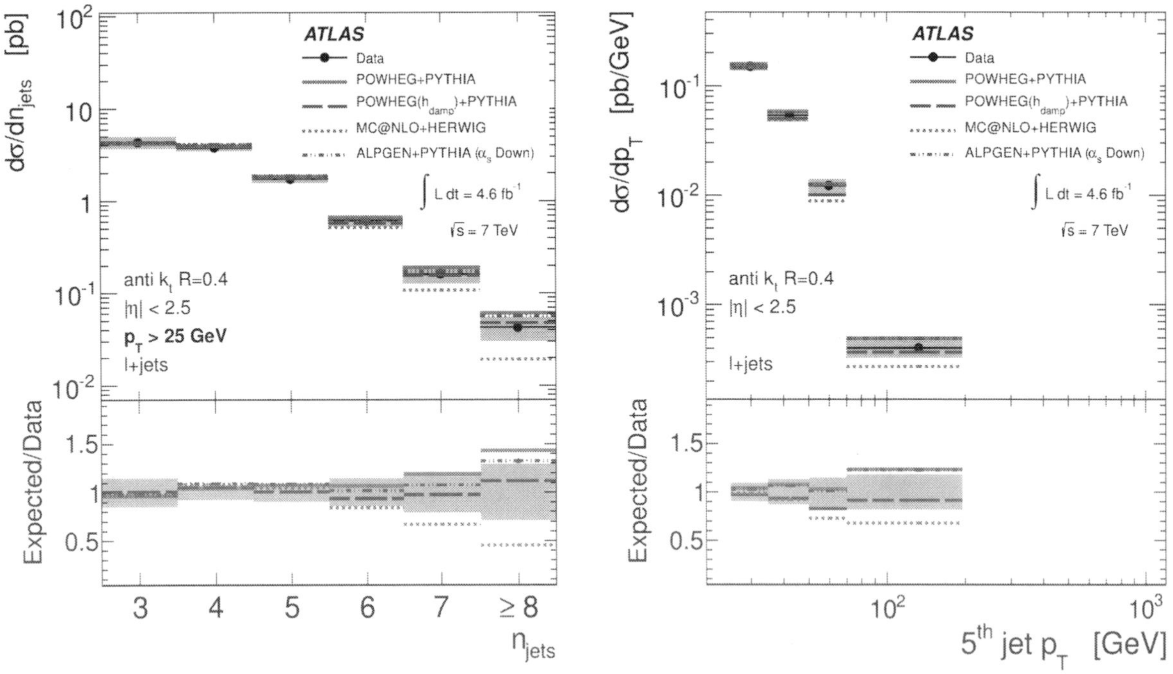

Figure 2. The $t\bar{t}$ cross section as a function of the jet multiplicities for the jet p_T threshold of 25 GeV (left), and of the transverse momentum of 5th leading jet (right) [6]. The data are shown in comparison to POWHEG+PYTHIA, POWHEG+PYTHIA with a parameter tuning for high-p_T radiation (referred to as POWHEG(h_{damp})+PYTHIA), MC@NLO+HERWIG and ALPGEN+PYTHIA(α_S Down) predictions. The data points and their corresponding total uncertainties are shown as a shaded band.

3.1σ excess over the background-only hypothesis. The combined $t\bar{t}Z$ and $t\bar{t}W$ signal strength, assuming the Standard Model cross section ratio between $t\bar{t}Z$ and $t\bar{t}W$ predicted by next-to-leading-order QCD calculations, is found to be $\mu_{t\bar{t}V} = 0.89^{+0.23}_{-0.22}$, corresponding to a 4.9σ excess over the background-only hypothesis. A simultaneous measurement of the $t\bar{t}Z$ and $t\bar{t}W$ signal strengths yields $\mu_{t\bar{t}Z} = 0.71^{+0.28}_{-0.26}$ and $\mu_{t\bar{t}W} = 1.30^{+0.59}_{-0.48}$, corresponding to a 4.4σ excess over the background-only hypothesis. Observed signal significance from the simultaneous measurement yields 3.1σ for both $t\bar{t}Z$ and $t\bar{t}W$ processes. All measurements are consistent with next-to-leading-order theoretical calculations for $t\bar{t}Z$ and $t\bar{t}W$ processes.

6. Measurement of the cross section of single top production

The cross section of the t-channel single top production has been measured with 4.6 fb^{-1} of $\sqrt{s} = 7$ TeV pp collision data recorded in 2011, and with 20.3 fb^{-1} of $\sqrt{s} = 8$ TeV in 2012 [8],[9]. Events are selected using the $t \to \ell\nu b$ decay mode of the top quark. The experimental signature of candidate events is given by one electron or muon, large missing transverse momentum from neutrino, and two jets with high transverse momentum. One of the jets is required to originate from a b-quark. To separate t-channel signal events from background, several kinematic variables are combined into one discriminant by employing a neural network. A large number of potential input variables have been studied, including kinematic variables of the reconstructed particles, but also variables obtained from the reconstruction of the W boson and the top quark. A binned maximum-likelihood fit to the neural network output distribution is performed to extract the cross section. Inclusive cross sections for each beam energy were obtained as

$$\begin{aligned} \sigma_t(7\text{ TeV}) &= 68 \pm 2(\text{stat.}) \pm 8(\text{syst.})\text{ pb and} \\ \sigma_t(8\text{ TeV}) &= 82.6 \pm 1.2(\text{stat.}) \pm 11.4(\text{syst.}) \pm 3.1(\text{PDF}) \pm 2.3(\text{lumi.})\text{ pb,} \end{aligned}$$

which is in excellent agreement with the Standard Model prediction.

7. Conclusions

The ATLAS experiment has performed various measurements relating to the top quark production with using whole datasets of pp collisions at LHC taken in 2011 and 2012. Thanks to the large amount of the top quark events, precision of measurements presented in this paper is not limited by the statistic uncertainty. We have compared our results with several theoretical predictions. No significant deviation has been observed so far.

References

[1] Aad G et al. (ATLAS Collaboration) 2008 *JINST* **3** S08001
[2] Czakon M, Fiedler P and Mitov A 2013 *Phys. Rev. Lett.* **110** 252004 (*arXiv:1303.6254*)
[3] Kidonakis N 2011 *Phys. Rev.* D **83** 091503 (*arXiv:1103.2792*)
[4] Aad G et al. (ATLAS Collaboration) 2014 submitted to *Eur. Phys. J.* C (*arXiv:1406.5375*)
[5] Aad G et al. (ATLAS Collaboration) 2014 submitted to *Phys. Rev.* D (*arXiv:1407.0371*)
[6] Aad G et al. (ATLAS Collaboration) 2014 submitted to *JHEP* (*arXiv:1407.0891*)
[7] Aad G et al. (ATLAS Collaboration) 2014 *ATLAS-CONF-2014-038* (http://cds.cern.ch/record/1735215)
[8] Aad G et al. (ATLAS Collaboration) 2014 submitted to *Phys. Rev.* D (*arXiv:1406.7844*)
[9] Aad G et al. (ATLAS Collaboration) 2014 *ATLAS-CONF-2014-007* (http://cds.cern.ch/record/1668960)

XI International Conference on Hyperons, Charm and Beauty Hadrons (BEACH 2014) IOP Publishing
Journal of Physics: Conference Series **556** (2014) 012028 doi:10.1088/1742-6596/556/1/012028

Recent Top Properties Measurements at CDF

Giorgio Chiarelli
INFN Sezione di Pisa, Largo Bruno Pontecorvo 3, I-56127 Pisa, Italy
E-mail: giorgio.chiarelli@pi.infn.it

Abstract. We present the most recent CDF results on the measurements of the decay and production vertex of the top-quark. New results on forward-backward asymmetry in top-antitop events are presented. Also, recent measurements of the branching fractions of top-quark are discussed. Finally, measurements in single top events, where top-quark is produced through electroweak processes, are presented. Despite the much larger number of top events collected at the LHC, due to the symmetric initial state and the better signal-to-background ratio in specific channels, some results will be lasting heritage of the Tevatron.

1. Introduction

Top quark was discovered at the Tevatron in 1994-1995. Since then its properties were extensively studied by the CDF and D0 experiments. The last Tevatron run (2001-2011) provided almost 10 fb^{-1} of data to each experiment. In this document I will concentrate on the most recent CDF results that exploit this wealth of data.

Thanks to its large mass, top is the only quark we can study before hadronization, therefore we have the unique opportunity to study a "bare" quark. From an experimental point of view, observables are (in most cases) perfectly well defined. The importance of this very peculiar quark cannot be underestimated. Its mass is strictly related, through loops, to the Higgs mass and to vacuum stability; its large Yukawa coupling is puzzingly close to one.

Top quark can be produced via electroweak mechanism (single top processes) or via strong interactions (in top-antitop pairs). Despite the very small signal/background ratio for single top production process, both Tevatron experiments detected (and studied) top quark in both cases. As top decays in a W and a quark (almost 100% of cases a b-quark), topologies of top-quark events are classified according to the charged boson decay. For strongly produced top events we call "dilepton" events in which both Ws decay leptonically, "l+jets" events in which one of the two Ws decays hadronically and "all-hadron" the case in which both bosons decay in quarks[1]. Therefore the typical physics objects present in a top-candidate event of interest are one or two charged leptons (e or μ), from 2 to 4 jets (two coming from b-quark hadronization), and large missing transverse energy signalling the presence of one or two neutrinos.

In the following analyses we use events with large (> 20 GeV) E_T electrons or muons, large (> 25 GeV) missing transverse energy (MET) and jets of hadrons (reconstructed by a fixed cone algorithm) with $E_T > 20$ GeV and pseudorapidity $|\eta| < 2.8$.

[1] In the following I will not present results related to the all-hadronic topology.

2. Study of top-quark properties in $t\bar{t}$ events

Top production vertex can be studied in top-antitop pair events and in events in which top-quark is produced singly. CDF studied all topologies and first measured production cross section. A summary of the Tevatron results for the top-pair production is shown in Figure 1. The final CDF result is $\sigma_{t\bar{t}} = 7.63 \pm 0.5$ pb, to be compared with the theoretical prediction of $7.35^{+0.11}_{-0.21}(scale)^{+0.17}_{-0.12}(PDF)$ pb. The Tevatron result is $\sigma_{t\bar{t}} = 7.6 \pm 0.41(stat)^{+0.2}_{-0.36}(syst.)$, and CDF contribution weights ≈ 60 % to the precision obtained with the combination.

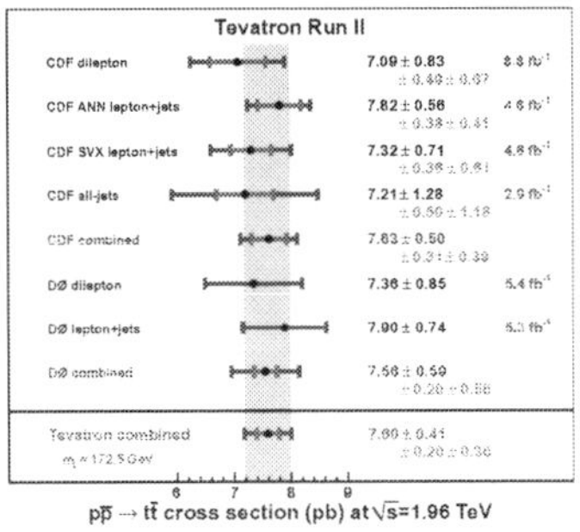

Figure 1. Measurements of $\sigma_{t\bar{t}}$.

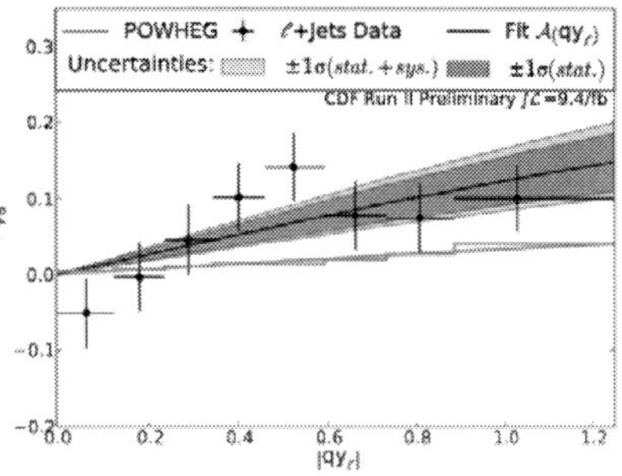

Figure 2. Fit to A_{FB} in l+jets data.

2.1. *Angular distributions in $t\bar{t}$ events*

CDF studied the distribution of the angle θ between the proton and the top direction in the $t\bar{t}$ reference frame looking for deviation from SM expectations. A description of $d\sigma/d\theta$ in terms of Legendre Polynomials shows that the first moment has some tension with the SM expected value (a difference at the 1.8 σ level), but there is not enough sensitivity to the other terms of the expansion to draw any conclusion.

Another important distribution to look at is the forward-backward asymmetry. Defined as $A_{FB} = (N_{\Delta Y>0} - N_{\Delta Y<0})/(N_{\Delta Y>0} + N_{\Delta Y<0})$, where ΔY is the difference in rapidity between top and antitop, this quantity is sensitive to new physics. Positive asymmetry is present in the production process which proceeds via $q\bar{q}$ annihilation at tree level or via a box diagram. Initial (ISR) and Final (FSR) state radiation can induce a negative asymmetry. Past CDF observation showed a tension with the SM with $A_{FB} = 0.066 \pm 0.02$ in the $l + jets$ channel. This value, brought back at the parton level, translates into an asymmetry of 16.4 ± 4.7 %, almost doubling the SM prediction [1].

This result prompted a number of studies. Here we present new measurements in the $l+$ jets and the dilepton channels and their combination.

The asymmetry in the $l + jets$ channel can also be studied in terms of direct observables, defining the leptonic asymmetry: $A^l_{FB} = (N_{q_l\eta_l>0} - N_{q_l\eta_l<0})/(N_{q_l\eta_l>0} + N_{q_l\eta_l>0})$, where l indicates the lepton. The final result is $A^l_{FB} = 9.32^{+3.2}_{-2.9}$ % against a SM expectation of 3.8 ± 0.3 %. Figure 2 shows the A^l_{FB} as a function of η [2]. This result shows again tension with the SM.

Recently CDF studied the leptonic asymmetry in the dilepton channel. This time one can separately study the symmetric and asymmetric part of the distribution of $q \times \eta$ and compare both to SM expectations. Figure 3 shows the distribution of $q \times \eta$ compared to POWHEG predictions, and Figure 4 the asymmetric part. The result, $A^l_{FB} = (7.2 \pm 5.2(stat) \pm 3(syst)) = (7.2 \pm 6.0)$ % is consistent with SM expectation of (3.8 ± 3) %. In order to improve the experiment sensitivity one can use both measurments to obtain the best estimate. Using 3864 events from the $l+jets$ channel (73% purity) and 569 from the dilepton channel (72% purity) , CDF combines

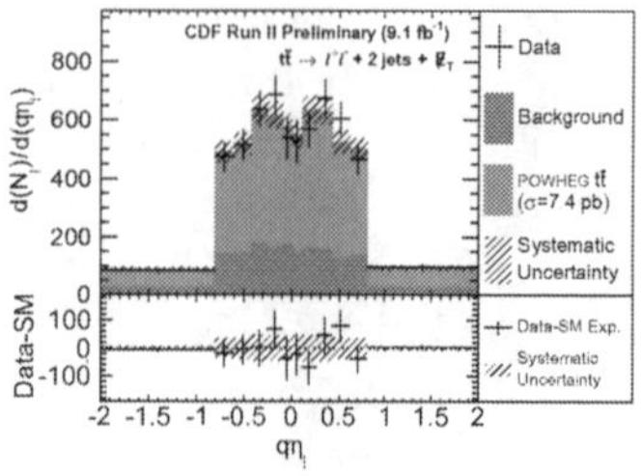

Figure 3. Distribution of $q \times \eta$ in dilepton events.

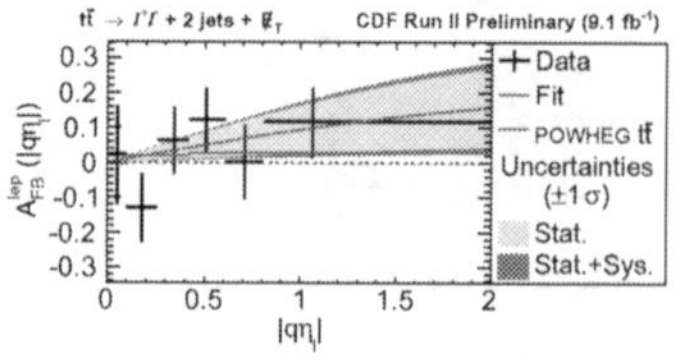

Figure 4. Asymmetric part of the $q \times \eta$ distribution compared to different theory predictions.

(using BLUE) the results and measures $A_{FB} = (9.0^{+2.8}_{-2.6})$ %. Figure 5 shows the CDF results compared to SM prediction. The final (combined) result is $\simeq 1.8\sigma$ away by the SM central value.

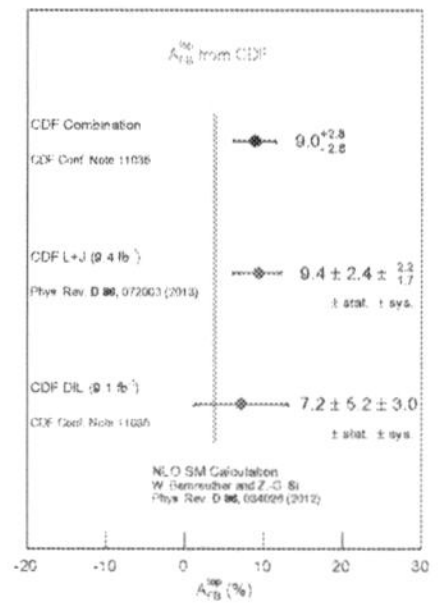

Figure 5. Summary of CDF results for A_{FB}.

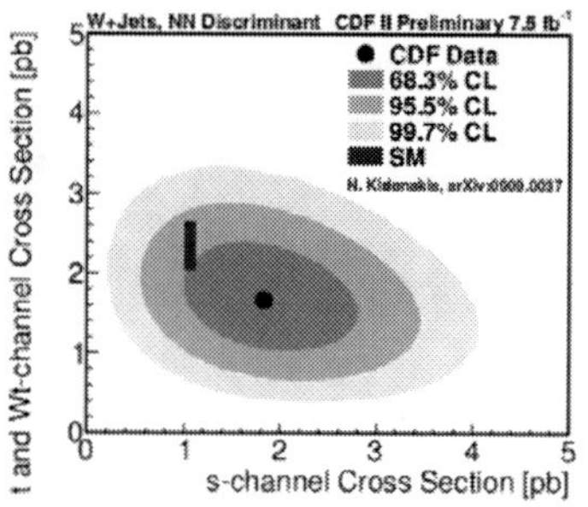

Figure 6. Simultaneous s and t cross section determination.

2.2. Indirect determination of the CKM element $|V_{tb}|$

The top-quark decay vertex is related to CKM matrix element V_{tb}. One can directly measure the branching fractions and there are two ways in which one can study V_{tb}: indirectly, via ratio of branching fractions, or directly, studying single top events.

In the SM, $\sum(|V_{td}|^2 + |V_{ts}|^2 + (|V_{tb}|^2) = 1$, therefore a measurement of the ratio of branching fractions $R = \frac{|V_{tb}|^2}{|V_{td}|^2+|V_{ts}|^2+|V_{tb}|^2}$ provides an estimate of $|V_{tb}|$.

From an experimental point of view, one proceeds by counting the number of top candidate events with 0,1,2 b-tagged jets, and comparing these figures with expectations as a function of R. Both CDF and D0 performed several measurements in this way. The most recent ones are due to CDF in the $l + jets$ and in the dilepton channels. In the l+jets by simultaneously fitting R and the $t\bar{t}$ production cross section, it measures $R = 0.94 \pm 0.09$ and $\sigma_{t\bar{t}} = 7.5 \pm 1.0$ pb, setting a 95 % C.L. limit of $|V_{tb}| > 0.89$ [3].

The strategy to extract R in the dilepton sample is different from the l+jets case as, thanks to the very good S/B ratio, it is possible to first extract the cross section, without using b-tagging techniques and then measure the b-content of the jets present in the events. Using an integrated luminosity of 8.7 fb^{-1} CDF measures $R = 0.87 \pm 0.07(stat + syst)$ [4] from which one extracts $|V_{tb}| = 0.93 \pm 0.04$ or $|V_{tb}| > 0.85$ at 95 % C.L. limit This result, together with the

the D0 result [5] in the dilepton channel ($R = 0.86 \pm 0.05$), is tantalizing away from the SM expectation.

3. Study of top properties in single top events

Single top production, is the generic definition that encompasses electroweak-produced events with only one top (or antitop) quark in the final state. Besides production through t or s channel (relevant at the Tevatron), single top-quark can also be produced in association with a W boson. Latter process is predicted to be negligible at the Tevatron, and evidence for its existence was recently obtained by CMS at the LHC.

Single top production is important, as its cross section is *directly* proportional to $|V_{tb}|^2$. Therefore it provides access to the $W-t-b$ vertex, probes the $V-A$ structure of the theory, gives access to the top spin and, of course, makes possible to directly measure $|V_{tb}|$ and explore possibilities of new physics. Besides, existence of FCNC currents or of heavy bosons, Kaluza Klein excitations, charged Higgs, may affect differently t and s channel. Although precision electroweak measurements rule out a fourth generation, there is still room for extensions [6]. Precise comparison of measurements of single top event properties, could open windows on New Physics (if any).

Single top production channels have final state topologies than can be mimicked by processes with much larger cross section. Indeed at the Tevatron, we are dealing with processes whose rate is smaller than the background fluctuations. In general, given a topology, we combine several channels, then we develop one (or more) Artificial Neural Network(s) (ANN) to identify the signal by fitting the appropriate distribution. CKM element is extracted by using the relation: $|V_{tb}|^2 = |V_{tb}^{SM}|^2 \times \frac{\sigma^{obs}}{\sigma^{SM}}$.

CDF recently submitted to PRL its result for the $s+t$ channel. Using 7.5 fb^{-1} we observe $\sigma_{s+t} = 3.04 \pm 0.55$ pb from which we extract $|V_{tb}| = 0.95 \pm 0.09(stat.+syst.) \pm 0.05(th.)$, and set a limit $|V_{tb}| > 0.78$ at 95 % C.L. Then, by simultaneously fitting s and t components[2], we extract $\sigma_{t+Wt} = 1.66^{+0.53}_{-0.47}$ pb and $\sigma_s = 1.81^{+0.63}_{-0.58}$ pb (see Figure 6), to be compared with SM prediction of $\sigma_{t+Wt} = 2.34 \pm 0.3$ pb, and $\sigma_s = 1.06 \pm 0.06$ pb [7]. If we combine this result with the previous analysis in the MET+$b\bar{b}$ channel, we set a limit $|V_{tb}| > 0.84$ at 95 % C.L.

Finally, as the s-channel is very difficult to observe at the LHC, both CDF and D0 optimized their selections to observe this mode of production. After this optimization, CDF re-analyzed its MET+$b\bar{b}$ and l+jets sample, also using a new b-tagger developed for the Higgs search. We find evidence for s-channel production at 4.2 σ, measuring $\sigma_s = 1.38^{+0.38}_{-0.37}$ pb [8]. By combining this result with D0 analysis, the Tevatron experiments reported the observation of the s channel production at 6.1 σ, measuring $\sigma_s = 1.29^{+0.26}_{-0.22}$ pb [9].

3.1. Acknowledgments

I would like to thank my colleagues D. Toback, J. Wilson, S. Leone, and Y. Peters.

References

[1] T. Aaltonen et al. *Phys. Rev.* D **87**, 89002 (2012).
[2] T. Aaltonen et al. *Phys. Rev.* D **88**, 072003 (2013).
[3] T. Aaltonen et al. *Phys. Rev.* D **87**, 111101 (2013).
[4] T. Aaltonen et al. *Phys. Rev. Lett.* **112**, 221801 (2014).
[5] V. M. Abazov et al. *Phys. Rev. Lett.* **107**, 121802 (2011).
[6] J. Alwall et al., *Eur. Phys J.* C **49** 791-801 (2007).
[7] T. Aaltonen et al. arXiv 14.07.4031.
[8] T. Aaltonen et al. *Phys. Rev. Lett.* **112**, 231805 (2013).
[9] T. Aaltonen et al. *Phys. Rev. Lett.* **112**, 231803 (2013).

[2] Wt component is included in the t channel as it shares the same final state topology.

Flavor Changing Neutral Currents in top production and decay at CMS

Tae Jeong Kim on behalf of the CMS collaboration
Chonbuk National University, Jeonju, South Korea
E-mail: tae.jeong.kim@cern.ch

Abstract. Searches for flavor changing neutral currents (FCNC) in top production and decay using data collected by the Compact Muon Solenoid (CMS) experiment at $\sqrt{s} = 7$ and 8 TeV corresponding to an integrated luminosity of around 5 fb^{-1} and 20 fb^{-1} are presented. FCNC searches are conducted to probe tqZ, $tq\gamma$, tqH, and tgq interactions in various channels. There is no excess of signal observed in the data. The upper limit of $\mathcal{B}(t \to u\gamma) < 0.0161\%$, $\mathcal{B}(t \to ug) < 0.0355\%$, $\mathcal{B}(t \to Zq) < 0.05\%$ and $\mathcal{B}(t \to cH) < 0.56\%$ at 95% confidence level are obtained. Future prospects of FCNC searches with the upgraded detector at 14 TeV are also presented.

1. Introduction

In 2012, a new boson with a mass around 125 GeV has been discovered. The discovery of this scalar type of particle opens a new window in our research. Within the standard model (SM) the new boson is expected to have a strong relation with the top quark, the heaviest known fundamental particle in nature. The top quark decays mostly to a bottom quark and a W boson. However, some extensions of the SM predict that a top quark can also decay through a neutral Z boson, a Higgs boson (H), a gamma (γ) or a gluon (g) changing its flavor, $t \to Zq$, $t \to Hq$, $t \to q\gamma$ or $t \to gq$, where q is either a u or c quark. The flavor changing neutral current (FCNC) interaction is suppressed in the SM by the Glashow-Iliopoulos-Maiani (GIM) mechanism and only occurs at the level of quantum loop corrections. The branching ratio B($t \to Xq$) where $X = Z, H, \gamma$ or gluon is predicted to be around $10^{-17} \sim 10^{-12}$ in the SM far below the experimental reach of the Large Hadron Collider (LHC). However, in several models beyond the SM such as R-parity-violation-supersymmetry models and topcolor-assisted technicolor models and, in particular, two scalar Higgs doublets (2HDMs), the branching ratio can reach up to 10^{-3}. Therefore, detection of this signal would be an indication of a large enhancement in the branching ratio and clear evidence for violations of the SM prediction.

The Compact Muon Solenoid (CMS) experiment [1] at the LHC has accumulated data corresponding to an integrated luminosity of 5 fb^{-1} in 2011 and 20 fb^{-1} in 2012. Searches for FCNC in top production and decay using data collected by the CMS experiment at $\sqrt{s} = 7$ and 8 TeV corresponding to an integrated luminosity of around 5 fb^{-1} and 20 fb^{-1} are presented. FCNC searches are conducted to probe tqZ, $tq\gamma$, tqH, and tgq interactions in various channels. Future prospects of FCNC searches with the upgraded detector at 14 TeV are also presented.

2. FCNC in a single top production

2.1. $tq\gamma$ coupling

The anomalous $tq\gamma$ coupling has been searched with events in association with a photon using the full data collected at 8 TeV in 2012 corresponding to 19.1 fb^{-1} [2]. In order to reduce the QCD multijet background, the top quark in the lepton decay mode is considered. Therefore, one isolated muon, one isolated photon and one b-tagged jet are required to select the final signature. Boosted Decision Tree (BDT) is used to separate the signal signature from the background contributions. There is no excess observed and the limits on the anomalous $\kappa_{tu\gamma}$ and $\kappa_{tc\gamma}$ coupling are calculated at 95% confidence level using the BDT distribution. Photon energy scale and the estimate of $W\gamma$ + jets process are the main uncertainty sources in this analysis. The limit on the coupling and on the corresponding branching ratio is shown in Fig 1 (left) and the limits on the branching ratios of $tq\gamma$ and $tc\gamma$ are shown in Fig 1 (right) for the case of coupling with c-quark and u-quark indicated by red vertical line. The observed limits on the anomalous coupling of $\kappa_{tu\gamma}$ and $\kappa_{tc\gamma}$ are 0.0279 and 0.0094, respectively. The corresponding branching ratios of $t \to u\gamma < 0.0161$ % and $t \to c\gamma < 0.182$ % are obtained. This result is the first measurement of the limit on these couplings at the LHC and the most stringent bounds on the anomalous FCNC $tc(u)\gamma$ coupling to date.

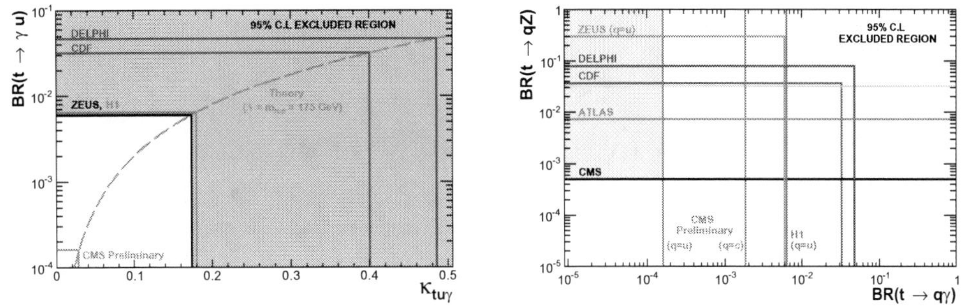

Figure 1. Relation between κ coupling and branching ratio (left). Limits on the branching ratio (right).

2.2. tqg coupling

The anomalous couplings of tcg and tug were searched in the t-channel single top-quark production using data collected at 7 TeV corresponding to an integrated luminosity of 5 fb^{-1} [3]. The final signature of the anomalous FCNC coupling is the same as the SM single top quark production processes with one isolated muon, 2 or 3 jets and one b-tagged jet. However, the signal has different kinematic distributions and is separated from the backgrounds using Bayesian Neural Network (BNN) based on their kinematical properties. Data and the SM predictions agree well within the uncertainties. In this analysis, the estimate of the W+jets process is one of the main uncertainty source. Taking into account the systematic uncertainties, the limit on the couplings of tcg and tug are calculated at 95% confidence level. The limits on the couplings of tcg and tug are shown in Fig. 2 (left) and the coupling can be converted to the limit on the branching ratio. The observed upper limits on $\mathcal{B}(t \to u + g) < 0.0355\%$ and $\mathcal{B}(t \to c + g) < 0.344\%$ are obtained.

2.3. $tqg(Z)$ coupling

The single top production in association with a Z boson has been also analyzed for the searches for the coupling of tqZ and tqg with data corresponding to an integrated luminosity of 4.9 fb^{-1}

collected at 7 TeV [4]. The BDT distribution for the coupling Zut is shown in Fig. 2 (right) which shows a good agreement between data and the SM background prediction. The observed limit on the branching ratios of $\mathcal{B}(t \to gu) < 0.56\%$ and $\mathcal{B}(t \to Zu) < 0.56\%$ are obtained. These results are not competitive with the one from the single top production 2.2. However, it is important to note that it gives interesting cross check in different physics process. For tqZ coupling, it is conceivable to combine the result with the one from the top quark decay mode in $t\bar{t}$ pair production to improve the sensitivity.

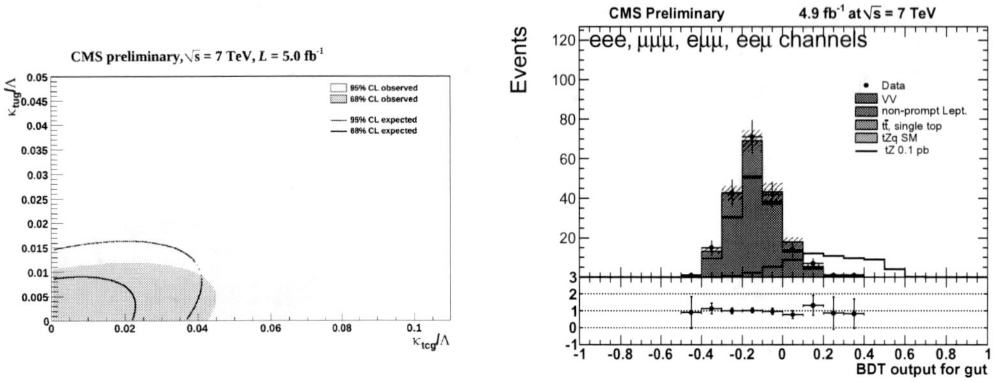

Figure 2. Exclusion upper limits in two-dimensions on FCNC tug and tcg couplings at 68% and 96% confidence level (left). BDT distributions for Zut (right).

3. FCNC in a top decay

A searching for $t\bar{t}$ events where one of top quarks decays into Zq has been pursued with the full data collected at 7 and 8 TeV [5]. The other top is assumed to decay in the SM leptonic decay mode. Events with opposite-sign, same-flavor, isolated leptons from a Z boson candidate and exclusive one extra lepton are selected. All three leptons must satisfy $p_T > 20$ GeV and $|\eta| < 2.5$ for electrons and $|\eta| < 2.4$ for muons. The large transverse momentum and the presence of at least two jets where one of these jets should be a b-tagged jet are also required to remove the background contributions from Z boson or diboson events. The top-quark mass decaying into Wb (m_{Wb}) in the SM decay is reconstructed with W boson and b-tagged jets and is required to be within 35 GeV of the top-quark mass, equal to 172.5 GeV in the simulation. (see Fig 3) Secondly, the top-quark decaying into Zq is reconstructed considering all possible pairings and the pair with the largest separation in azimuthal angle to the first top quark is selected. The second top-quark mass m_{Zj} is required to be within 25 GeV of the assumed top-quark mass. After the final selection, only 1 event remains in data while 3.3 ± 1.1 events are expected from the SM backgrounds. The dominant uncertainties are from the renormalization and factorization uncertainty, PDF and the cross section of $\sigma_{t\bar{t}}$. The observed upper limit on the branching ratio $\mathcal{B}(t \to Zq)$ at 95% confidence level is 0.06% with 8 TeV data alone. Combining the result with the statistically independent result with 7 TeV data, the observed upper limit on the $\mathcal{B}(t \to Zq)$ is 0.05%.

Projections in 3000 fb^{-1} The study of $t \to c$-quark and Z boson was simulated at 14 TeV for the upgraded detector and will be repeated at the new energy using data corresponding to 300 fb^{-1} and 3000 fb^{-1} [6]. Assuming that the systematic uncertainty is reduced by the $\sqrt{\mathcal{L}}$ where $\mathcal{L}$ is an integrated luminosity, the expected limits are predicted to be smaller than 0.010 % with 3000 fb^{-1}, this is 10 times better than the expected limit from the Run 1 period.

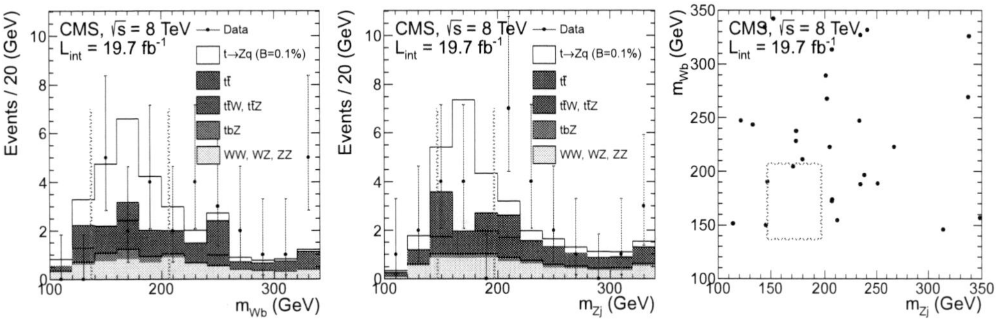

Figure 3. Reconstructed top-mass distributions in the SM decay (left). Reconstructed top-mass distributions in the FCNC decay (middle). The scatter distribution of reconstructed top-mass in 2D for both decay modes (right). The red dotted vertical lines (left and middle) and box (right) indicate the top-quark mass requirements.

4. Limit on $t \to cH$

The results of two CMS searches from heavy Higgs models with multileptons as well as diphotons final states [7] and the inclusive multilepton search [8] using data collected at 8 TeV corresponding to 19.5 fb^{-1} are reinterpreted for the rare flavor-changing decay of the top quark to a Higgs boson and a charm quark [9]. In this analysis, the flavor-changing decay in $t\bar{t}$ production followed by one of the top quarks decaying to cH where Higgs decays WW, ZZ, $\tau\tau$ and $\gamma\gamma$ are considered as the signal. No excess over the data is found in both analyses. Using the modified frequentist construction CL_S, the observed limit of $\mathcal{B}(t \to cH) < 0.56\%$ is calculated at 95% confidence level. This branching ratio is related to the left- and right-handed top flavor changing Yukawa couplings so that the observed limit can be converted to the limit on the coupling of $\sqrt{|\Lambda^H_{tc}|^2 + |\Lambda^H_{ct}|^2} < 0.14$. The result from the final state of multilepton search analysis alone is $\mathcal{B}(t \to cH) < 1.28\%$ and is 0.69% from two photon final state search analysis. The combination with the result from the two photon final state search analysis gives a significant improvement with respect to the one from multilepton search analysis alone.

5. Conclusions

The CMS collaboration has performed the FCNC searches in a single top-quark production and decay with data collected at 7 and 8 TeV and found no signature of FCNC. Limits of $\mathcal{B}(t \to u\gamma)$ $< 0.0161\%$, $\mathcal{B}(t \to ug) < 0.0355\%$, $\mathcal{B}(t \to Zq) < 0.05\%$ and $\mathcal{B}(t \to cH) < 0.56\%$ at 95% confidence level are obtained. More results from Run 1 are expected to come soon. Exciting time is ahead of us with data to be collected at 13 TeV in 2015 which would be suitable for the FCNC rare process searches.

References

[1] CMS Collaboration, "The CMS experiment at the CERN LHC", JINST 3:S08004,2008.
[2] CMS Collaboration, "Search for anomalous single top quark production in association with a photon in pp collisions at $\sqrt{s} = 8$ TeV", **PAS TOP-14-003**.
[3] CMS Collaboration, "Search for anomalous Wtb couplings and top FCNC in t-channel single-top-quark events", **PAS TOP-14-007**.
[4] CMS Collaboration, "Search for Flavor-Changing Neutral Currents in tZ events in proton-proton collisions at $\sqrt{s} = 7$ TeV", **PAS TOP-12-021**.
[5] CMS Collaboration, "Search for Flavor-Changing Neutral Currents in Top-Quark Decays $t \to Zq$ in pp Collisions at $\sqrt{s} = 8$ TeV", Phys. Rev. Lett., **112** (2014) 171802.
[6] CMS Collaboration, "Projections for Top FCNC Searches in 3000 fb^{-1} at the LHC" **PAS FTR-13-016**.

[7] CMS Collaboration, "Search for extended Higgs sectors in the $H \to hh$ and $A \to Zh$ channels in $\sqrt{s} = 8$ TeV pp collisions with multileptons and photons final states", **PAS HIG-13-025**.
[8] CMS Collaboration, "A search for anomalous production of events with three or more leptons using 19.5 fb^{-1} of $\sqrt{s} = 8$ TeV LHC data", **PAS SUS-13-002**.
[9] CMS Collaboration, "Combined multilepton and diphoton limit on t $\to$ cH", **PAS HIG-13-034**.

Effective approach to top-quark decay and FCNC processes at NLO accuracy

Cen Zhang
Centre for Cosmology, Particle Physics and Phenomenology, Université catholique de Louvain, 2 Chemin du Cyclotron, B-1348 Louvain-la-Neuve, Belgium
E-mail: cen.zhang@uclouvain.be

Abstract. The top quark is expected to be a probe to new physics beyond the standard model. Thanks to the large number of top quarks produced at the Tevatron and the LHC, various properties of the top quark can now be measured accurately. An effective field theory allows us to study the new physics effects in a model-independent way, and to this end accurate theoretical predictions are required. In this talk we will discuss some recent results on top-quark decay processes as well as flavor-changing processes, based on the effective field theory approach.

1. Introduction

Current strategies to search for new physics beyond the standard model (SM) can be broadly divided into two categories. In the first category we look for new resonant states. In the second category, new states are assumed to be heavy, and we look for their indirect effects in the interactions of known particles.

The top quark has been a natural probe to new physics in the first category, due to its large mass and strong coupling to the electroweak symmetry breaking sector. Searches for resonant states through decay processes involving the top quarks have been performed both at the Tevatron and at the LHC. Examples include $t\bar{t}$ resonance searches, top partner production, and so on. Unfortunately, until now no new states have been discovered, and exclusion limits have been placed, up to around several TeV scale.

On the other hand, the top quark physics has entered a precision era, thanks to the large number of top quarks produced at the Tevatron and LHC. Various properties of the top quark have been already measured with high precision, and the upcoming LHC Run-II will continue to increase the precision level. From the theory side, accurate SM predictions are also available. As a result the focus of top quark physics is now moving toward the second category, i.e. to measure accurately the known interactions and the rare processes of SM particles. Examples are measurements on W-helicity fractions in top-quark decay, and searches for processes involving flavor-changing neutral current (FCNC) of the top quark. These are the main topics of this talk.

2. The effective approach

When looking for deviations from the SM interactions, the standard approach is to utilize the effective field theory (EFT) framework, in which deviations from the SM are parameterized by including higher dimensional operators. The approach is valid when the new physics scale, Λ, is higher than the scale of the process. Assuming the full SM gauge symmetries, the leading new operators involving a top quark arise at dimension six. At this level the Lagrangian can be written as

$$\mathcal{L}_{EFT} = \mathcal{L}_{SM} + \sum_i \frac{C_i}{\Lambda^2} O_i + \text{h.c.} \tag{1}$$

A calculation in the EFT framework consists of three steps:

(i) Matching calculation. The EFT is matched to the full theory which lives at the high energy scale Λ. This determines the operator coefficients C_i, defined at the matching scale.

(ii) Renormalization group (RG) evolution. The coefficients will be evolved down to a lower scale E where the actual process takes place. This essentially resums the $\log(\Lambda/E)$ terms.

(iii) Matrix element calculation, to be performed at scale E.

The procedures are familiar from the RG-improved perturbation theory which plays an important role in flavor physics. Indeed there is a similarity between top quark physics and flavor physics. In flavor physics, the full theory is the SM, possibly with some new physics. The matching is performed at scale $\Lambda = m_W$, where the heavy SM particles are integrated out. The theory is then evolved down to lower scales where the matrix element can be computed either perturbatively or non-perturbatively. In the top quark physics, the full theory is the SM plus new physics, which is to be matched to the EFT at scale $\Lambda = \Lambda_{NP}$, where the new physics lives. The procedure integrates out heavy particles in the new physics, but the heavy particles of the SM remain in the effective theory, leading to 59 dimension-six operators in total for one generation of fermions [1]. We then select the ones that are relevant for top-quark processes and evolve them down to a scale of $\sim m_t$. Cross sections can then be computed in terms of these operator coefficients.

Recently the RG equations for all dimension-six operators have been derived at the one-loop level [2, 3, 4], completing the second step in this picture. There are also new developments in matrix element calculations. In particular, the top-quark decay and several FCNC processes have been computed at NLO in QCD, and the automation of the FCNC processes in the MG5_aMC@NLO framework [5] is now in progress. These will be discussed in the remainder of this talk.

3. Top-quark decay and W-helicity fractions

In this section we will discuss the main decay channel of the top quark, $t \to bW$, focusing on the W-helicity fractions. For SM couplings and unpolarized top-quark production, the helicity fractions are approximately 70% longitudinal and 30% left-handed. Accurate predictions at NNLO in QCD and NLO in electroweak are available.

In an EFT, dimension-six operators will modify the helicity fractions. First of all, there are operators that directly modify the standard Wtb vertex function. They are

$$\begin{aligned} O_{\varphi Q}^{(3)} &= i\tfrac{1}{2}y_t^2 \left(\varphi^\dagger \overleftrightarrow{D}^I_\mu \varphi\right)(\bar{Q}\gamma^\mu \tau^I Q) \quad & O_{tW} &= y_t g_W (\bar{Q}\sigma^{\mu\nu}\tau^I t)\tilde{\varphi} W^I_{\mu\nu} \\ O_{\varphi\varphi} &= i y_t^2 \left(\tilde{\varphi}^\dagger D_\mu \varphi\right)(\bar{t}\gamma^\mu b) \quad & O_{bW} &= y_t g_W (\bar{Q}\sigma^{\mu\nu}\tau^I b)\varphi W^I_{\mu\nu} \end{aligned} \tag{2}$$

where Q is the left-handed (t, b) doublet. The QCD corrections to these operators have been computed in Ref. [6]. Deviations from the SM predictions are expected to be dominated by contribution from O_{tW}.

A second possibility comes from the color-dipole operator,

$$O_{tG} = y_t g_s (\bar{Q}\sigma^{\mu\nu} T^A t)\tilde{\varphi} G^A_{\mu\nu} \tag{3}$$

which only contributes at NLO in QCD. This has been computed in Ref. [7]. There is a mixing effect from operator O_{tG} to O_{tW}, given by $\gamma_{C_{tW},C_{tG}} = 2\alpha_s/3\pi$.

Finally, four-fermion operators such as $O_{lQ}^{(3)} = \left(\bar{l}\gamma_\mu \tau^I l\right)\left(\bar{Q}\gamma^\mu \tau^I Q\right)$ should also be considered. These operators arise from for example a heavy W' particle exchange, and can directly contribute to the $bl\nu$ final state. Their contributions to the total decay rate are small. Nevertheless we can hope to set some bounds using differential decay rate, since they will affect not only the measured F_+ value (which would be tiny in the SM), but also the $m_{l\nu}$ spectrum.

The helicity fractions are measured both at the Tevatron and at the LHC. Using the results from Ref. [8], one can already set some constraints. For illustration, assuming other operators vanish, the limit on O_{tW} is

$$C_{tW} = -0.11 \pm 1.13\,(\Lambda/\mathrm{TeV})^2 \tag{4}$$

at 95% CL. However one should keep in mind that considering only one operator at a time is not natural, in particular due to the operator mixing effects. In principle, four-fermion operators should also be considered, however current strategy to top-quark reconstruction requires that $m_{l\nu}$ is equal to the W mass. This condition does not apply when four-fermion operators are present, so the current limits on the helicity fractions cannot be used to constrain four-fermion operators.

4. Top-quark decay via FCNC

Flavor-changing decay modes such as $t \to qX$ are suppressed by the Glashow-Iliopoulos-Maiani mechanism, and the SM predictions for the branching ratios are of order $10^{-12} - 10^{-15}$. Any evidence for these decay modes would be a definite signal for new physics. In the EFT framework the decay rates for $t \to ug$, $t \to u\gamma$, $t \to uh$ and $t \to ull$ are all computed at NLO in QCD [9, 10, 7]. Relevant operators can be found in Appendix A. On the experimental side, branching ratios for $u\gamma$, ull and uh final states have been measured both at the Tevatron and at the LHC. Single top production processes via FCNC couplings are also studied, via $e^+e^- \to t\bar{q}$ at LEP2, $e^-q \to e^-t$ at HERA, and $qg \to t(j)$ at the Tevatron and the LHC. Limits on FCNC couplings obtained from production processes are often converted to limits on branching ratios.

In general, a single process can depend on many operators, and so one measurement can only set bounds on a specific linear combination of operator coefficients. Once we combine all available measurements, a global analysis can be performed to determine the allowed region for all operator coefficients. However, limits obtained by experimental collaborations almost always assume only one FCNC interaction is present at a time. In addition, four-fermion operators are always neglected in both production and decay processes. For these two reasons it is difficult to perform a real global fit based on the EFT framework.

To illustrate the feasibility of such a global approach, here we present an "approximate" global fit for the top quark FCNC sector. We neglect all four-fermion operators, and only consider $t \to qZ$ [11], $t \to qh$ [12] [1] and $pp \to t$ [14] via FCNC couplings involving an up quark. Moreover we interpret the limit from $pp \to t\gamma$ [15] as a direct bound on the branching ratio of $t \to q\gamma$, which is much tighter than the current bound measured at the Tevatron. These four measurements suffice to constrain all two-quark FCNC operators. Unfortunately, the published information is not enough for a consistent combination of all four measurements at 95% CL, therefore we only require that all constraints at 95% CL are independently satisfied.

The result of the fit is shown in Figure 1. The blue lines are obtained by setting other coefficients to zero, while the red lines are obtained by allowing other coefficients to float. For some operators the blue and red lines are different, indicating that a correlation is present between the operators. This effect between O_{uB} and O_{uW} can be observed in Figure 2.

The situation can be more complicated if four-fermion operators are included. These operators may contribute to the decay channel $t \to qll$, which is the actual final state of the

[1] This limit applies for $t \to ch$, but limit for $t \to uh$ is expected to be better [13].

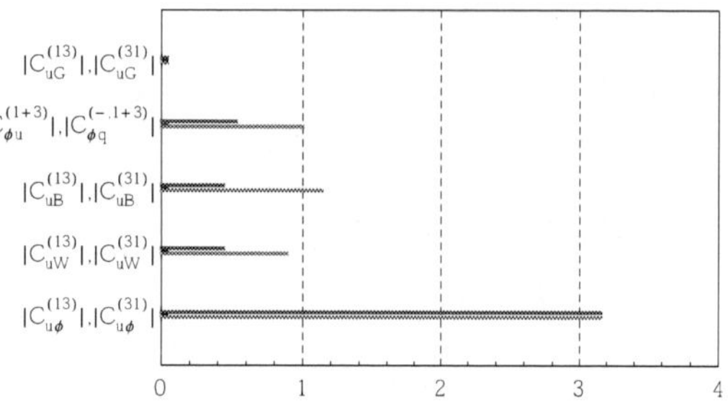

Figure 1. Limits on individual operator coefficients. Blue lines indicate limits obtained by setting other coefficients to zero. Red lines are obtained by allowing other coefficients to float. $\Lambda = 1$ TeV is assumed.

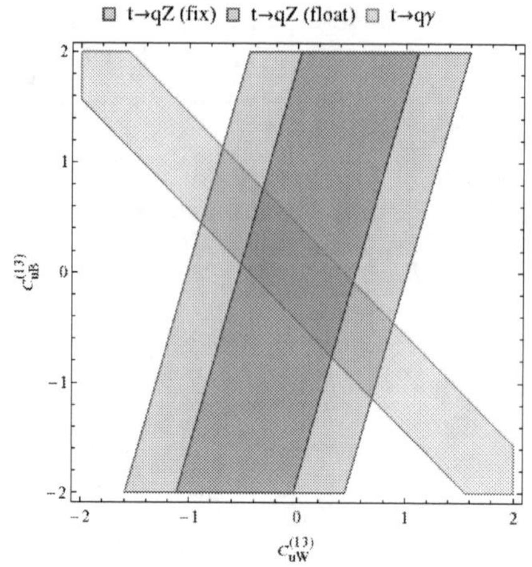

Figure 2. Allowed region in $C_{uW}^{(13)} - C_{uB}^{(13)}$ space, from $t \to uZ$ and $pp \to t\gamma$, by fixing other coefficients to zero and by allowing them to float. $\Lambda = 1$ TeV.

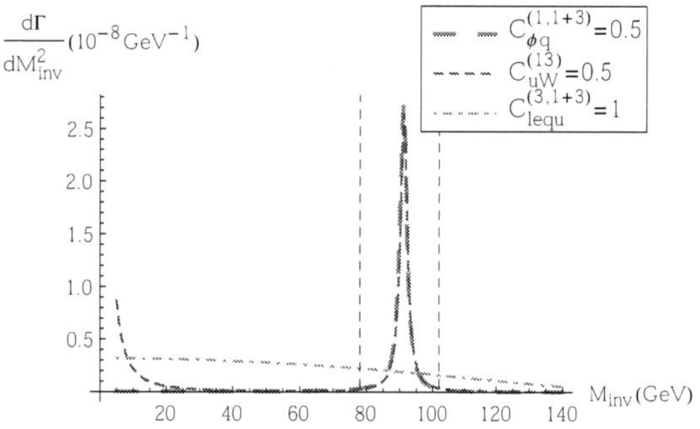

Figure 3. Invariant mass distribution of lepton pairs from FCNC top-quark decay. Contributions from two-fermion operators, $O_{\varphi q}^{(1,1+3)}$ and $O_{uW}^{(13)}$, and four-fermion operator $O_{lequ}^{(3,1+3)}$ are compared. Λ=1 TeV.

$t \to qZ$ measurements. Consider for example the following operator

$$O_{lequ}^{(3,13)} = \left(\bar{l}\sigma_{\mu\nu}e\right)\varepsilon\left(\bar{q}\sigma^{\mu\nu}t\right) , \tag{5}$$

its contribution is comparable with typical contributions from two-fermion operators. Therefore constraining four-fermion operator coefficients from $t \to qZ$ is possible.

The main difference between two- and four-fermion operators is reflected by the invariant mass distribution of the lepton pairs. Two-fermion operators involve a Z boson and show a peak near the Z mass, while four-fermion operators give a more flat spectrum. We illustrate this difference in Figure 3, where the four-fermion operator $O_{lequ}^{(3,1+3)}$ and two additional two-fermion operators $O_{\varphi q}^{(1,1+3)}$ and $O_{uW}^{(13)}$ are included. Current strategies to searching for FCNC top-quark decay impose some on-shell cuts on the invariant mass of the two leptons, e.g. $m_{ll} \in [78, 102]$ GeV, to select events with two leptons coming from an on-shell Z boson. These are displayed in Figure 3 as two dashed straight lines. After applying these cuts, the contributions from the three operators are:

$$\Gamma_{\text{on-shell}} = \left(7.0|C_{uW}^{(13)}|^2 + 7.3|C_{\varphi q}^{(1,1+3)}|^2 + 0.8|C_{lequ}^{(3,13)}|^2\right) \times 10^{-5}\ \text{GeV} . \tag{6}$$

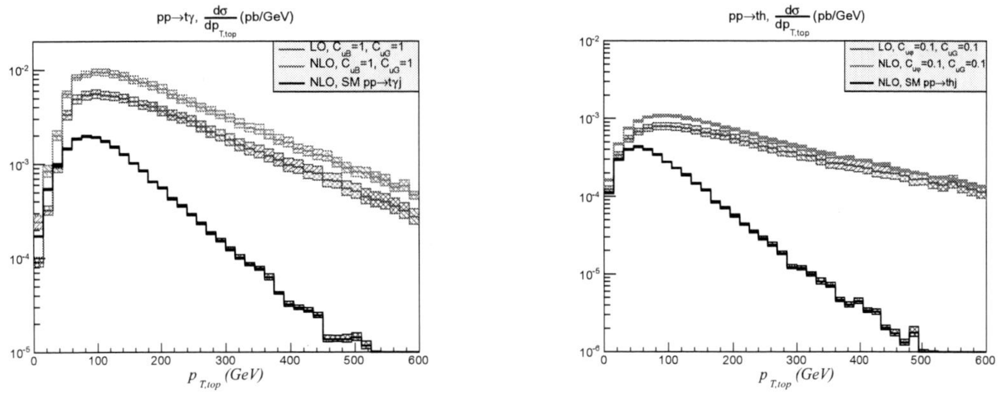

Figure 4. The p_T distribution of the top quark in $pp \to t\gamma$ (left), and in $pp \to th$ (right). The SM background is from $t\gamma j$ (left) and thj (right) production.

This implies that even with the on-shell cuts, the four-fermion operator is still not negligible. In fact, the above results will lead to limit on $C_{lequ}^{(3,13)}$ that is only 3 times larger than those of the two-fermion operator coefficients.

On the other hand, looking at the off-shell regions of the spectrum will give us more information. The decay rate in $m_{ll} \in [15, 78] \cup [102, \infty]$ GeV is:

$$\Gamma_{\text{off-shell}} = \left(0.6|C_{uW}^{(13)}|^2 + 0.4|C_{\varphi q}^{(1,1+3)}|^2 + 2.7|C_{lequ}^{(3,13)}|^2\right) \times 10^{-5}\ \text{GeV}. \quad (7)$$

We can see that two-fermion contributions are suppressed while the four-fermion one is enhanced, so by looking at both regions we can constrain two- and four-fermion operators separately. Due to less Drell-Yan background in the off-shell region, one might even get a better sensitivity on four-fermion operators.

5. Single Top Production via FCNC

We now turn to single top production via top-quark FCNC. Recently single top production associated with a photon and with a Z-boson have been searched for at the LHC [15, 16]. Limits are obtained on the FCNC couplings tqg, $tq\gamma$ and tqZ. Single top production associated with a Higgs boson has also been proposed as a probe of the tqg and tqh couplings. These analyses are important consistency checks for the FCNC searches already performed in top-decay processes. In a global fit based on an EFT approach, they can be combined together with the decay processes, to improve the limits on various FCNC couplings.

Theoretical predictions for $t\gamma$, tZ and th production processes have been computed in Refs. [17, 18, 19] at NLO in QCD. The NLO corrections are found to be at $\sim 40\%$ to 80% level. We have implemented the two-quark flavor-changing operators to the MG5_aMC@NLO framework [5], using FeynRules [20] and NLOCT [21]. As a result, processes such as $pp \to tX$ where $X = j, \gamma, Z, h$ can be computed automatically at NLO in QCD and are matched to parton shower simulation. For illustration, we present in Fig. 4 the p_T distribution of the top in $pp \to t\gamma$ and in $pp \to th$ at NLO. Because the calculation is automatic, a rich set of processes can be studied at NLO with parton shower, including single top production in not only pp collision but also e^+e^- collision, and top-quark decay in single top and $t\bar{t}$ production modes. In future four-fermion operators involving the top quark will also be available, allowing for more interesting processes to be studied. All these studies can give useful information about top-quark FCNC couplings.

6. Summary

The top quark physics has entered a precision era. Due to the scale separation between Λ_{NP} and m_t, the EFT approach is a natural tool to study the indirect effects from new physics. In this talk we have discussed theoretical results for top quark processes in the EFT framework. We have focused on top-quark decay and searches for top quark FCNC interactions, and discussed the need for a global analysis for top-quark properties based on an EFT approach. Moreover, NLO predictions are now available for these processes, and in particular, the top-quark FCNC operators are implemented to the MG5_aMC@NLO framework. These results provide information and tools needed to perform a global analysis.

Acknowledgments

I would like to thank my collaborators, Celine Degrande, Gauthier Durieux, Fabio Maltoni and Jian Wang. This work is supported by the IISN "Fundamental interactions" convention 4.4517.08.

Appendix A. Top-Quark FCNC Operators

Top FCNC interactions are characterized by the following two-quark operators:

$$O_{\varphi q}^{(3,i+3)} = i\left(\varphi^\dagger \overleftrightarrow{D}^I_\mu \varphi\right)(\bar{q}_i\gamma^\mu\tau^I Q) \quad O_{\varphi q}^{(1,i+3)} = i\left(\varphi^\dagger \overleftrightarrow{D}_\mu \varphi\right)(\bar{q}_i\gamma^\mu Q) \quad O_{\varphi u}^{(i+3)} = i\left(\varphi^\dagger \overleftrightarrow{D}_\mu \varphi\right)(\bar{u}_i\gamma^\mu t)$$

$$O_{uB}^{(i3)} = g_Y(\bar{q}_i\sigma^{\mu\nu}t)\tilde{\varphi}B_{\mu\nu} \quad O_{uW}^{(i3)} = g_W(\bar{q}_i\sigma^{\mu\nu}\tau^I t)\tilde{\varphi}W^I_{\mu\nu} \quad O_{uG}^{(i3)} = g_s(\bar{q}_i\sigma^{\mu\nu}T^A t)\tilde{\varphi}G^A_{\mu\nu}$$

$$O_{u\varphi}^{(i3)} = (\varphi^\dagger\varphi)(\bar{q}_i t)\tilde{\varphi}$$

where the subscript i is the generation of the quark field. Q is the third-generation doublet. For operators with $(i3)$ superscript, a similar set of operators with $(3i)$ flavor structure can be obtained by interchanging $(i3) \leftrightarrow (3i)$, $t \leftrightarrow u_i$ and $Q \leftrightarrow q_i$. We further define $O_{\varphi q}^{(-,i+3)} = (O_{\varphi q}^{(1,1+3)} - O_{\varphi q}^{(3,1+3)})/2$. There are also four-fermion operators involving two quarks and two leptons. A full list can be found elsewhere [7].

References

[1] Grzadkowski B, Iskrzynski M, Misiak M and Rosiek J 2010 *JHEP* **1010** 085
[2] Jenkins E E, Manohar A V and Trott M 2013 *JHEP* **1310** 087
[3] Jenkins E E, Manohar A V and Trott M 2014 *JHEP* **1401** 035
[4] Alonso R, Jenkins E E, Manohar A V and Trott M 2014 *JHEP* **1404** 159
[5] Alwall J, Frederix R, Frixione S, Hirschi V, Maltoni F, Mattelaer O, Shao H -S, Stelzer T, Torrielli P and Zaro M 2014 *JHEP* **1407** 079
[6] Drobnak J, Fajfer S and Kamenik J F 2010 *Phys. Rev.* D **82** 114008
[7] Zhang C 2014 *Phys. Rev.* D **90** 014008
[8] CMS Collaboration 2013 *JHEP* **1310** 167
[9] Drobnak J, Fajfer S and Kamenik J F 2010 *Phys. Rev. Lett.* **104** 252001
[10] Zhang J J, Li C S, Gao J, Zhu H X, Yuan C P and Yuan T C 2010 *Phys. Rev.* D **82** 073005
[11] CMS Collaboration 2014 *Phys. Rev. Lett.* **112** 171802
[12] CMS Collaboration 2014 Combined multilepton and diphoton limit on t to cH *CMS-PAS-HIG-13-034*
[13] Greljo A, Kamenik J F and Kopp J 2014 *JHEP* **1407** 046
[14] The ATLAS collaboration 2013 Search for single top-quark production via FCNC in strong interaction in $\sqrt{s} = 8$ TeV ATLAS data *ATLAS-CONF-2013-063*
[15] CMS Collaboration 2014 Search for anomalous single top quark production in association with a photon *CMS-PAS-TOP-14-003*
[16] CMS Collaboration 2013 Search for Flavour Changing Neutral Currents in single top events *CMS-PAS-TOP-12-021*
[17] Zhang Y, Li B H, Li C S, Gao J and Zhu H X 2011 *Phys. Rev.* D **83** 094003
[18] Li B H, Zhang Y, Li C S, Gao J and Zhu H X 2011 *Phys. Rev.* D **83** 114049
[19] Wang Y, Huang F P, Li C S, Li B H, Shao D Y and Wang J 2012 *Phys. Rev.* D **86** 094014
[20] Alloul A, Christensen N D, Degrande C, Duhr C and Fuks B 2014 *Comput. Phys. Commun.* **185** 2250
[21] C. Degrande 2014 Automatic evaluation of UV and R2 terms for beyond the Standard Model Lagrangians: a proof-of-principle *Preprint 1406.3030 [hep-ph]*

Top quark couplings and polarization

Giuseppe Codispoti on behalf of the CMS Collaboration
Dipartimento di Fisica e Astronomia, Viale Berti Pichat 6/2, 40127 Bologna (BO), Italy
E-mail: giuseppe.codispoti@bo.infn.it

Abstract.
Precise measurements of top quark couplings are presented. The measurements cover the single top tW cross section, the top-quark branching-fraction ratio R=B(t $\rightarrow$ Wb)/B(t $\rightarrow$ Wq) and the CKM matrix element V_{tb} from the single top cross sections. Top quark polarisation in the t-channel single-top quark production and dilepton $t\bar{t}$, and W-helicity fraction measurements along with searches of top quark anomalous couplings in lepton plus jet channel are also presented. Finally measurements of top quark pair production in association with a W or Z boson or a photon are presented.

1. Introduction

The top quark is the heaviest known fundamental particle. Because of its high mass, the top quark decays before hadronization and it can be thus studied as a free particle. The direct access to its bare properties provides us with a unique tool to probe the Standard Model (SM).

The top quark is produced in proton proton collisions in pairs (top, t, and anti-top, $\bar{t}$), through strong processes (gluon-gluon fusion or quark-antiquark annihilation), and as a single top, through electroweak processes.

The most recent results in top quark production, couplings and properties, performed by the CMS [1] Collaboration at the Large Hadron Collider (LHC) [2] are presented here. A detailed description of the techniques is out of scope for a short conference report: we will focus on the results, while addressing to the proper documentation for the technique details.

2. $|V_{tb}|$ measurements

The top quark decays almost exclusively to a W boson and a b quark. The SM prediction for the Cabibbo-Cobayashi-Maskawa (CKM) matrix element $|V_{tb}|$ is thus close to unity. The most precise measurement provided by the D0 Collaboration [3] shows some tension with the SM, raising more interest on the ongoing studies.

The CMS measurements of $|V_{tb}|$ come from the three analysis here described. The results are summarized in figure 1.

2.1. Measurement of the branching fraction R=B(t $\rightarrow$ Wb)/B(t $\rightarrow$ Wq)

The most sensitive analysis for a direct $|V_{tb}|$ estimation is given by the measurement of the ratio R of the branching fractions of the top quarks decaying in W plus a b quark, and the top quarks decaying in a W plus any down type quark (q=b,s,d). Under the assumption of the unitarity of the CKM matrix, it can indeed be shown that $R = |V_{tb}|^2$.

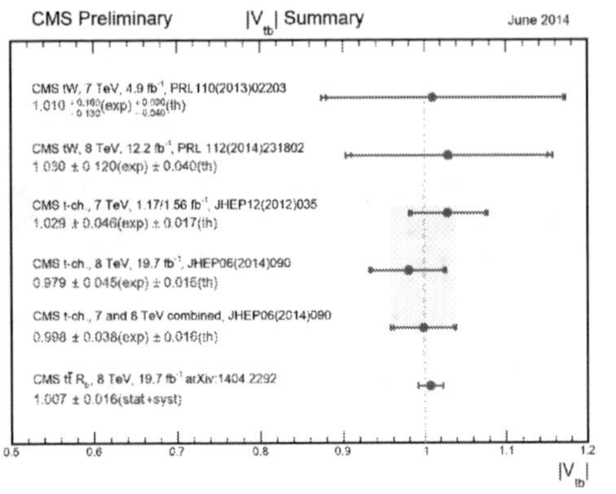

Figure 1. Summary of all the $\mid V_{\mathrm{tb}} \mid$ measurements performed by CMS.

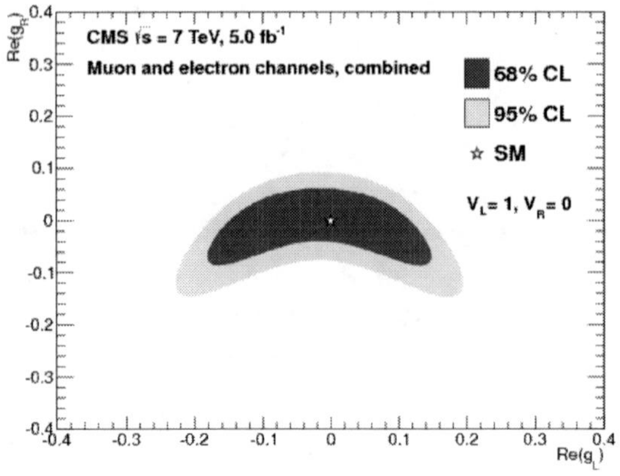

Figure 2. Limits on the real components of the anomalous couplings constants g_L, g_R at 68% and 95% CL.

The analysis [4] is performed in the dilepton channel only, given its channel-dependent purity of 70-90%. The selected events are used to measure the $\mathrm{t\bar{t}}$ cross section through a likelihood fit on the jet multiplicity constraining the signal and background contributions. The b-quark content is inferred from the distribution of the b-tagged jets per event as a function of the jet multiplicity. The model accounts for correction due to the single top contribution, the fraction of events with correct jet assignment, the b-tag efficiency and mis-identification.

The analysis leads to a measured value $R = 1.014 \pm 0.003(stat.) \pm 0.032(syst.)$ which translates to $\mid V_{\mathrm{tb}} \mid = 1.007 \pm 0.016(stat. + syst.)$. By requiring $R = \mid V_{\mathrm{tb}} \mid \leq 1$ we obtain $R > 0.955$ and $\mid V_{\mathrm{tb}} \mid > 0.975$ at 95% confidence level.

2.2. $\mid V_{\mathrm{tb}} \mid$ in single top production

Since the single top production involves a Wtb vertex, the $|\mathrm{V_{tb}}\mid$ matrix element can be determined from the cross section measurements of several production mechanisms.

In the t-channel, the single top is produced in association with a light quark emitted at high rapidity and a soft b quark. An event is selected if a muon or an electron that originated from the top quark is present. The signal is extracted from a maximum likelihood fit to the distribution of the absolute value of the pseudorapidity of the jet originating from the light quark. The t and $\bar{\mathrm{t}}$ production cross section are extracted independently. The analysis is performed both for 7 TeV and 8 TeV data. Here we report the combined results [5]. Assuming the CKM matrix elements $|\mathrm{V_{td}}\mid$ and $|\mathrm{V_{ts}}\mid$ to be negligible, we obtain:

$$\mid V_{\mathrm{tb}} \mid = \sqrt{\frac{\sigma_{t-ch}}{\sigma_{t-ch}^{th.}}} = 0.998 \pm 0.038(exp.) \pm 0.016(th.) \tag{1}$$

where the experimental uncertainty comes from the uncertainty on the measurement of σ_{t-ch} while the theoretical uncertainty comes from the uncertainty on $\sigma_{t-ch}^{th.}$; By requiring $\mid V_{\mathrm{tb}} \mid \leq 1$ we obtain $V_{\mathrm{tb}} > 0.92$ at 95% confidence level.

Single top can also be produced in association with a W boson (tW-channel). A clean signature is possible when both the top quark and the W have a leptonic decay: this leads to a final state with two opposite charge isolated leptons, a b-jet and missing transverse energy due to the neutrino. The CMS analysis [6], performed over 12.2 fb^{-1} of 8 TeV data, uses a multivariate tecnique exploiting the topological differences between the tW signal and the dileptonic $t\bar{t}$ background. Using the same assumptions as in t-channel, we measure $\mid V_{tb} \mid = 1.03 \pm 0.12(exp.) \pm 0.04(th.)$. By requiring $\mid V_{tb} \mid \leq 1$ we obtain $V_{tb} > 0.78$ at 95% confidence level.

3. $t\bar{t}$ and $W/Z/\gamma$ associated production

The associated production of $t\bar{t}$ pairs with W, Z and γ bosons can be disentangled at LHC and the corresponding cross sections measured in order to test the internal consistence of the SM. CMS performed the cross section measurements of the three processes using 8 TeV data.

Same sign dilepton events are used for the measurement of the $t\bar{t}W$ process, where one lepton originates from the leptonic decay of one of the two top quarks and the other from the decay of the W. Trilepton events, one lepton arising from the top quark decay and two opposite sign same flavour leptons from the vector boson, are used for the $t\bar{t}Z$ analysis, as well as four leptons events.

The different channels are then combined in order to provide the cumulative $t\bar{t}V$ cross section [7]. In order to solve the dependence of the measured cross section on the other $t\bar{t}V$ process, a simultaneous fit of the cross sections using all processes is also performed, resulting in the following:

$$\sigma(t\bar{t}W) = 170^{+110}_{-100}\,\mathrm{fb} \quad (2)$$

$$\sigma(t\bar{t}Z) = 170 \pm 90\,\mathrm{fb} \quad (3)$$

where the reported errors are 1σ standard deviation obtained from the combination of the different channels. The $t\bar{t}W$ and $t\bar{t}Z$ measurements have respectively a significance of 1.6σ and 3.1σ. The $t\bar{t}\gamma$ cross section [8] is instead evaluated in the context of a muon plus jets analysis, with the additional request of a high energy photon with extra angular distance requests from the b jets. The photon misidentification plays a big role in the analysis and is estimated using a template fit to the photon shower shape. The cross section is extracted from a ratio with the CMS measured cross section for the $t\bar{t}$ events. The resulting value is:

$$\sigma(t\bar{t}\gamma) = 2.4 \pm 0.2(\mathrm{stat.}) \pm 0.6(\mathrm{syst.})\,\mathrm{pb} \quad (4)$$

4. Top Properties

4.1. Polarization and spin correlation in $t\bar{t}$

In the SM, top quarks are produced in proton proton interactions with a small amount of polarization. Conversely the production mechanisms results in a full spin correlation between t and $\bar{t}$.

The polarization and spin correlation are studied at CMS in terms of angular asymmetries in the dilepton decay channel [9]. The polarization P is expressed in function of the asymmetry:

$$A_p = 2\frac{N[cos(\theta_l^*) - cos(\theta_l^*)]}{N[cos(\theta_l^*) + cos(\theta_l^*)]} = 0.005 \pm 0.013(stat.) \pm 0.014(syst.) \pm 0.008(p_T rw) \quad (5)$$

where $P = 2A_p$ and $cos(\theta_l^*)$ is the angle between the lepton and the top quark flight direction, in the top rest frame, and the major systematic uncertainties source is the top transverse momentum modeling ($p_T rw$). The result is in agreement with the expectation.

The best estimator found for the spin correlation is instead the $A_{\Delta\phi}$ asymmetry between the longitudinal angles ϕ of the two leptons.

The result for $A_{\Delta\phi}$ is rather precise, since it does not require any event reconstruction:

$$A_{\Delta\phi} = 0.113 \pm 0.010(stat.) \pm 0.006(syst.) \pm 0.012(p_T rw) \quad (6)$$

in good agreement with the fully correlated hypotesis $A^{corr.}_{\Delta\phi} = 0.115^{+0.014}_{-0.016}$.

4.2. Polarization in single top

In the single top case, the expected polarization is close to 100%, due to the electroweak nature of the process. As in the $t\bar{t}$ analysis, the polarization can be estimated using an asymmetry quantity. In the single top analysis the asymmetry is defined using the angle between the charged lepton and the recoiling jet in the t-channel [10], since the light quark recoiling against the top quark tends to have a direction parallel to the spin direction of the top. The measured polarization is in good agreement with the expectation:

$$P_t = 0.82 \pm 0.12(stat.) \pm 0.32(syst.) = 0.82 \pm 0.34 \quad (7)$$

4.3. W helicity in $t\bar{t}$

The W-helicity fraction in the top quark decays are directly related to the V-A structure of the SM and their measurement provides great sensitivity to physics beyond the SM.

The helicity fractions of the W boson are defined as partial decay rates for a given helicity state divided by the total decay rate. In the SM, the longitudinal helicity fraction is the dominat one, $F_0 = 0.0787 \pm 0.005$, while the right handed is suppressed at leading order, $F_R = 0.0017 \pm 0.0001$ [11] . The remaing left handed component is expected to be $F_L = 0.311 \pm 0.005$. The differential cross section connecting the three fractions can be written in terms of the $cos(\theta^*)$ variable, where θ^* is the angle between the charged lepton in the W rest frame with respect to the direction of the W in the top quark rest frame. The measurement of the relative contributions is extracted through a Poisson binned likelihood on the $cos(\theta^*)$ variable shape.

The results from the muon analysis with 8 TeV data [12] are the world most precise estimates:

$$F_0 = 0.659 \pm 0.015(stat.) \pm 0.0023(syst.) \quad (8)$$
$$F_L = 0.350 \pm 0.010(stat.) \pm 0.024(syst.) \quad (9)$$
$$F_R = -0.009 \pm 0.006(stat.) \pm 0.020(syst.) \quad (10)$$

We also report the exclusion limits calculated for anomalous coupling with 7 TeV data [13], using both electron and muon channels in figure 2.

5. Conclusions

The top quark physics has been deeply investigated in the CMS experiments using the data produced by two successful LHC data taking periods at 7 and 8 TeV center of mass energy. Several measurements were reported here, focusing on the most recent results. All the measurements show a good agreement with the SM expectations. New results will come from the next LHC runs at higher center of mass energy and luminosity.

References

[1] Chatrchyan, S et al (CMS Collaboration) *JINST* **3** S08004
[2] Evans L and Bryant P 2008 *JINST* **3** S08001
[3] Abazov et al (D0 Collaboration) 2011 *Phys. Rev. Lett.* **107** (12) 121802
[4] Chatrchyan S et al (CMS Collaboration) 2014 *Phys. Lett. B* **736** 33 57 ISSN 0370-2693
[5] Chatrchyan S et al (CMS Collaboration) 2014 *JHEP* **06** 90
[6] Chatrchyan S et al (CMS Collaboration) 2014 *Phys. Rev. Lett.* **112** (23) 231802

[7] Khachatryan V et al. (CMS Collaboration) 2014 Preprint CMS-TOP-12-036 and arXiv:1406.7830
[8] Chatrchyan S et al (CMS Collaboration) Preprint CMS-TOP-13-011
[9] Chatrchyan S et al (CMS Collaboration) 2014 *Phys. Rev. Lett.* **112** (18) 182001
[10] Chatrchyan S et al (CMS Collaboration) Preprint CMS-PAS-TOP-13-001
[11] Czarnecki A and Korner J G. and Piclum J 2010 *Phys.Rev.* **D81** 111503
[12] Chatrchyan S et al (CMS Collaboration) 2013 *JHEP* **10** 167
[13] Chatrchyan S et al (CMS Collaboration) Preprint CMS-PAS-TOP-13-008

Top Quark Studies at D0

Reinhild Yvonne Peters

University of Manchester, School of Physics and Astronomy, Oxford Road, Manchester M13 9PL, England; also at DESY, Hamburg, Germany

E-mail: `reinhild.peters@cern.ch`

Abstract. Years after its discovery in 1995 by CDF and D0, the top quark still undergoes intense investigations at the Tevatron. Using up to the full Run II data sample, new measurements of top quark production and properties by the D0 Collaboration are presented. In particular, the first observation of single top quark s-channel production, the measurement of differential $t\bar{t}$ distributions, forward-backward $t\bar{t}$ asymmetry, a new measurement of the top quark mass, and a measurement of the top quark charge are discussed.

1. Introduction

The top quark, discovered in 1995 by the CDF and D0 Collaborations at the Fermilab Tevatron collider, is the heaviest known elementary particle [1][2]. Due to its high mass and short lifetime, the top quark provides a unique environment to study a bare quark. It is believed to play a special role in electroweak symmetry breaking and provide a window to physics beyond the standard model (SM).

Despite the currently running LHC pp collider being a top quark factory, the study of the top quark using Run II Tevatron $p\bar{p}$ collision events provides complementary information. Tevatron Run II with a collision energy of 1.96 TeV (starting 2001 and ending September 30th, 2011) provided ≈ 10.5 fb^{-1} of integrated luminosity for each of the D0 and CDF experiments.

In this article, the latest studies in single top quark production and top quark pair production at the D0 experiment are presented. The discussed analyses are performed in dileptonic and semileptonic final states, where either both or one of the W bosons, coming from the decay of the top quark, decay into a charged lepton and associated neutrino.

2. Studies in Single Top Quark Production

Top quarks can be produced in pairs via the strong interaction, or singly via the electroweak interaction. The latter occurs via s-channel, t-channel and Wt-channel production. The Wt-channel has a negligible cross section at the Tevatron, and only became relevant and observable at the LHC [3]. In 2009, CDF and D0 reported first observation of single top quark production [4][5], where the s- and t-channel were measured together. For the observation, CDF and D0 were using up to 3.2 fb^{-1} and 2.3 fb^{-1} of data, respectively. The measurement of single top quark production is very challenging, as the main background from W+jets events looks very similar to the single top signature. Various multivariate techniques have been employed to distinguish the signal from the large background. The t-channel on its own has been first observed by D0 in 2011 [6] using 5.4 fb^{-1} of data.

Only recently, also the s-channel was observed by performing a combination of CDF and D0 measurements. The analyses employed for combination use up to the full Run II data sample. Semileptonic events are considered in the analyses by both collaborations, with the addition of an analysis by CDF, where events with a missing transverse energy plus jet signature are used, adding events to the sample in which the lepton is not directly reconstructed. The events are required to contain at least two jets, of which one or two have to be identified as b-jet. A multivariate discriminant is build to separate s-channel signal from background. In this analysis, the t-channel single top production cross section was set to its SM value. The combined analysis results in a cross section of $\sigma_s = 1.29^{+0.26}_{-0.24}$ pb, which deviates with more than 6.3 standard deviations (SD) from zero [7].

3. Studies in Top Quark Pair Production

The production of top quark pairs at the Tevatron is dominated by $q\bar{q}$ annihilation with a fraction of approximately 85%, and consists of $\approx$ 15% gluon gluon fusion. At the LHC, these fractions are roughly inverse. Besides the different cross section of $t\bar{t}$ production at the Tevatron relative to the LHC, these different fractions of production mechanisms are one of the main reasons why many analyses are complementary at the Tevatron relative to the LHC.

Recently, a new measurement of the $t\bar{t}$ cross section inclusively and differentially as function of the invariant $t\bar{t}$ mass, $m_{t\bar{t}}$, the rapidity of the top, $|y^{top}|$, and the transverse momentum of the top, p_T^{top}, has been performed by D0, using the full data sample of 9.7 fb^{-1} [8]. The measurement of the $t\bar{t}$ production provides a direct test of Quantum Chromo Dynamics (QCD) by confronting the measurement with SM predictions. The analysis uses lepton+jets events, where the $t\bar{t}$ event reconstruction is performed using a constrained kinematic fitter. The distributions are then corrected for detector and acceptance effects, using regularized unfolding, and are defined for parton-level top quarks including off-shell effects. Using events with at least four jets, the inclusive cross section has been measured as $\sigma_{t\bar{t}} = 8.3 \pm 0.7(\text{stat}) \pm 0.6(\text{syst}) \pm 0.5(\text{lumi})$ pb, in good agreement with the SM prediction. Figure 1 and Figure 2 show the unfolded $t\bar{t}$ distributions as function of $m_{t\bar{t}}$ and $|y^{top}|$, respectively, compared to approximate next-to-next-to-leading order (NNLO) calculations and different generator predictions. For these as well as the distribution as function of p_T^{top}, a good agreement between data and the NNLO calculations and generator predictions can be seen in general. In addition, a comparison to various axigluon models is performed. Using the differential distributions, these can be constrained.

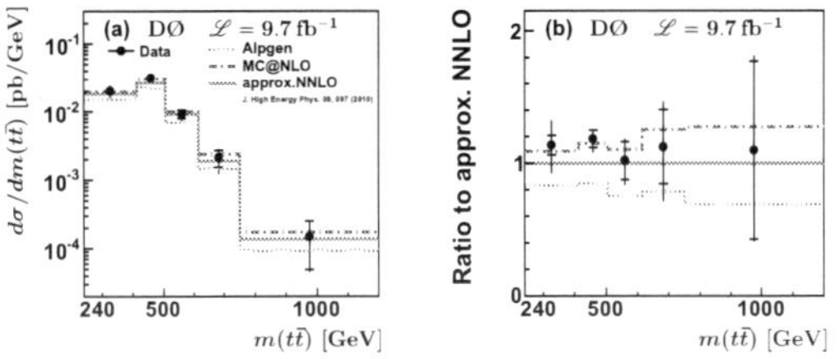

Figure 1. Differential $t\bar{t}$ distribution as function of $m_{t\bar{t}}$ [8].

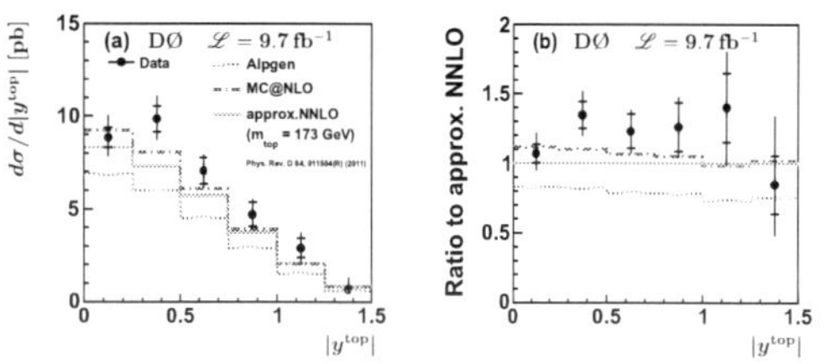

Figure 2. Differential $t\bar{t}$ distribution as function of $|\Delta y^{top}|$ [8].

An interesting feature of $t\bar{t}$ production is the forward-backward asymmetry. At next-to-leading order (NLO), interference between different $q\bar{q}$ diagrams causes a $t\bar{t}$ asymmetry, where the top quarks are more likely to go into the direction of the incoming quark. Various asymmetries can be studied, in particular the forward-backward asymmetry $A_{FB}^{t\bar{t}} = \frac{N(\Delta y>0)-N(\Delta y<0)}{N(\Delta y>0)+N(\Delta y<0)}$, with N being the number of events with the difference in top and antitop rapidity Δy smaller or larger than zero, and the leptonic asymmetry $A^{l}_{FB} = \frac{N(q_l y_l>0)-N(q_l y_l<0)}{N(q_l y_l>0)+N(q_l y_l<0)}$, with q_l and y_l being the charge and rapidity of the lepton from W boson decay, respectively. The asymmetries have been measured using dileptonic and semileptonic events, and unfolded to production level. The combined leptonic asymmetry in dileptonic and semileptonic events, using 9.7 fb^{-1} of D0 data, yields $A^{l}_{FB} = 4.7 \pm 2.3(\text{stat}) \pm 1.5(\text{syst})$ % [9][10], in good agreement with the SM prediction at NLO QCD, including electroweak (EW) corrections, of $A^{l}_{FB} = 3.8 \pm 0.2$ % [11]. The asymmetry is also measured as function of lepton p_T, showing good agreement with the prediction from MC@NLO [12]. The measurement of the forward-backward asymmetry in the semileptonic channel explores events with three or at least four jets. The measured asymmetry is $A^{t\bar{t}}_{FB} = 10.6 \pm 2.7(\text{stat}) \pm 1.3(\text{syst})$ % [13], in good agreement with the NLO+EW SM prediction of $A^{t\bar{t}}_{FB} = 8.8 \pm 0.6$ % [11]. The measurement of $A^{t\bar{t}}_{FB}$ as function of $m_{t\bar{t}}$ and $|\Delta y|$ is shown in Figure 3 and Figure 4, respectively. The differential $A^{t\bar{t}}_{FB}$ distributions as function of the two observables show good agreement with the prediction from MC@NLO, but are also consistent with the recent measurement by CDF [14].

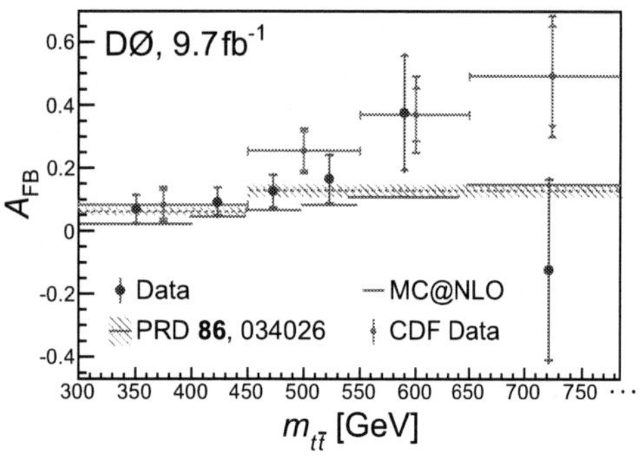

Figure 3. $A^{t\bar{t}}_{FB}$ as function of $m_{t\bar{t}}$ [13].

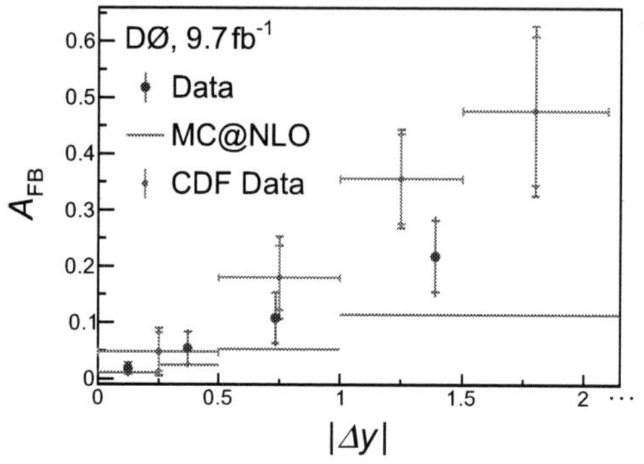

Figure 4. $A^{t\bar{t}}_{FB}$ as function of $|\Delta y^{top}|$ [13].

Another important analysis is the precise measurement of the top quark mass. The top quark mass is a free parameter in the SM. Together with the mass of the W boson, the top quark mass constrains the Higgs boson mass, and thus provides a self-consistency check. The D0 collaboration recently performed a new measurement of the top quark mass, using semileptonic events with at least one identified b-jet in the full Run II data sample of 9.7 fb^{-1}. For this, the matrix element technique is applied, which uses the full event kinematic and thus provides the most precise method. In this method, a probability is calculated for each event, where a phase space integration over leading order matrix elements, folded with parton distribution functions and transfer functions is performed. For the new analysis, the speed of the integration has been improved with respect to the previous implementation in D0. Furthermore, the handling of the systematic uncertainties has also been improved, allowing less conservative estimations of several sources of systematic uncertainties. To reduce the uncertainty from jet energy scale, jets from the hadronically decaying W boson are used as an in-situ constraint. The top quark mass is measured to be $m_t = 174.98 \pm 0.58(\text{stat} + \text{JES}) \pm 0.49(\text{syst})$ GeV [15], where JES is the jet energy scale. With a relative uncertainty of 0.43 %, this is the most precise single measurement of the top quark mass to date. The main systematic uncertainties originate from uncertainties on the residual jet energy scale, and hadronization and underlying event. Using this new measurement as well as updates on all-hadronic and dileptonic measurements by CDF, an updated Tevatron top quark mass combination has been performed. Using the BLUE method [16] for the combination of Run I and Run II results from CDF and D0 in various final states, the new Tevatron combination yields $m_t = 174.34 \pm 0.37(\text{stat}) \pm 0.52(\text{syst})$ GeV [17].

In the SM, the top quark has an electric charge of $+2/3$ of the electron charge (e). An exotic model could be possible, where the top quark carries a charge of $-4/3e$. Using 5.3 fb^{-1} of data, D0 performed a new measurement of the top quark charge in semileptonic $t\bar{t}$ events. In this analysis, at least four jets, of which at least two jets have to be identified as b jets, are required. To assign the final state objects to come from the top or antitop quark, a kinematic fitting algorithm is used. The top quark charge can then be determined by combining the charge of the lepton from the W boson decay with the charge of the b jet. The determination of the latter happens via a jet-charge algorithm, where a weighted sum of the charge of the tracks belonging to the b jet is calculated. The weight used in the sum is $p_T^{0.5}$ of the respective tracks. The factor 0.5 in the exponent has been optimized using $t\bar{t}$ events. The calibration of the jet charge algorithm is done using dijet events, where soft lepton and lifetime b-jet identification was applied. A SM and exotic top charge template is then constructed in $t\bar{t}$ events, and the fraction f of the events with SM top charge is extracted using a binned maximum likelihood fit. The measured fraction f yields $f = 0.88 \pm 0.13(\text{stat}) \pm 0.11(\text{syst})$ [18]. The dominant systematic uncertainty in this measurement is the statistics of the dijet sample. This value of f translates into the exclusion of the hypothesis that top quarks carry a charge of $-4/3$ of the electron charge at more than five SD, confirming earlier results by CDF and ATLAS. An alternative interpretation of the measurement can be done, assuming that the $t\bar{t}$ sample is a mixture of top quarks with charges $+2/3$ and $-4/3$. This measurement results in an upper limit on the admixture of exotic top quarks of $f < 0.46$ at 95% confidence level.

4. Conclusion

Despite the Tevatron having been switched off about three years ago, the exploitation of the data are still ongoing. Many new measurements of top quark production and properties have been released recently by the D0 collaboration. Several of these are Tevatron legacies, for example the most precise single measurement of the top quark mass, the final word from D0 on the forward-backward asymmetry, and the measurement of a variety of differential $t\bar{t}$ distributions.

Acknowledgments

I would like to thank my collaborators from the D0 collaboration for their help in preparing the presentation and this article. I also thank the staffs at Fermilab and collaborating institutions, and acknowledge the support from the Helmholtz association.

References

[1] F. Abe *et al.* [CDF Collaboration], *Phys. Rev. Lett.* **74**, 2626 (1995).
[2] S. Abachi *et al.* [D0 Collaboration], *Phys. Rev. Lett.* **74**, 2632 (1995).
[3] CMS Collaboration, *Phys. Rev. Lett.* **112**, 231802 (2014).
[4] T. Aaltonen *et al.* [CDF Collaboration], *Phys. Rev. Lett.* **103**, 092002 (2009).
[5] V. M. Abazov *et al.* [D0 Collaboration], *Phys. Rev. Lett.* **103**, 092001 (2009).
[6] V. M. Abazov et al. [D0 Collaboration], *Phys. Lett. B* **705**, 313 (2011).
[7] T. Aaltonen *et al.* [CDF and D0 Collaborations], *Phys. Rev. Lett.* **112** , 231803 (2014).
[8] V. M. Abazov *et al.* [D0 Collaboration], arXiv:1401.5785 [hep-ex] (submitted to *PRD*).
[9] V. M. Abazov et al. [D0 Collaboration], *Phys. Rev. D* **88**, 112002 (2013).
[10] V. M. Abazov *et al.* [D0 Collaboration], arXiv:1403.1294 [hep-ex] (submitted to *PRD*).
[11] W. Bernreuther and Z. -G. Si, *Phys. Rev. D* **86**, 034026 (2012).
[12] S. Frixione and B. R. Webber, *J. High Energy Phys.* **06**, 029 (2002).
[13] V. M. Abazov *et al.* [D0 Collaboration], arXiv:1405.0421 [hep-ex] (submitted to *PRD*).
[14] T. Aaltonen *et al.* [CDF Collaboration], *Phys. Rev. D* **87**, 092002 (2013).
[15] V. M. Abazov et al. [D0 Collaboration], *Phys. Rev. Lett.* **113**, 032002 (2014).
[16] L. Lyons, D. Gibaut, and P. Clifford, *Nucl. Instrum. Meth. A* **270**, 110 (1988); A. Valassi, *Nucl. Instrum. Meth. A* **500**, 391 (2003).
[17] T. Aaltonen *et al.* [CDF and D0 Collaborations], arXiv:1407.2682 [hep-ex].
[18] V. M. Abazov et al. [D0 Collaboration], arXiv:1407.4837 [hep-ex] (submitted to *PRD RC*).

XI International Conference on Hyperons, Charm and Beauty Hadrons (BEACH 2014) IOP Publishing
Journal of Physics: Conference Series **556** (2014) 012033 doi:10.1088/1742-6596/556/1/012033

Top quark mass measurements in ATLAS

S. Marti-Garcia

Institut de Física Corpuscular (IFIC). Centre mixte del CSIC i de la Universitat de València. Carrer Catedràtic José Beltrán 2, E-46980 Paterna, Spain.

E-mail: salvador.marti@ific.uv.es

Abstract. The top quark is a fundamental constituent of the Standard Model (SM). The properties of this quark are accurately predicted by this theory, except for its mass, which remains a fundamental parameter of the SM. With the advent of the Large Hadron Collider (LHC), many million of top-antitop quark pairs are available for study. With such statistics, the physics of the top quark has entered the precision era. In this note, the most recent experimental results by ATLAS concerning the top quark mass are reported.

1. Introduction

The top quark is the heaviest elementary particle known up to date. It was discovered at Tevatron in 1995 by the CDF [1] and DØ [2] experiments. It is an up-type quark, therefore it is expected to have the same quantum numbers as the u and c quarks. Only its mass is a free parameter. Thus, within the SM all production and decay properties of the top quark are fully defined. As the value of the top quark mass is close to the electroweak symmetry breaking scale, it has inspired speculations about the top quark having a special role in the symmetry breaking mechanism. For this reason, it is important to confirm experimentally at LHC [3] that the top quark has the properties predicted by the Standard Model. Any deviation from the expectations may be a hint to new physics. This is one of the reasons why the physics of the top quark is a major topic of the ATLAS [4] and CMS [5] experiments at LHC.

In the SM, the Higgs boson mass is related to the top quark and W boson masses through the radiative corrections. The most up to date electroweak fit uses NNLO theoretical predictions and compares the constraints from the electroweak fit on the Higgs with direct LHC measurements [6]. Current experimental results show this set of masses to be in agreement (Fig. 1).

The quark masses are fundamental parameters of the SM lagrangian which are not predicted by the theory. However, the definition of the quark mass is not unique as quarks are not observed as free particles. Therefore, the use of dynamical expressions is mandatory in order to determine unequivocally the masses of the quarks. The perturbative pole mass and the running mass in the $\overline{MS}$ scheme are two definitions among the most frequently used in theoretical QCD calculations. However, for the top quark, the Monte Carlo (MC) mass is also frequently used (as a kind of kinematic mass). According to references [7] [8] the difference between the kinematic mass and the pole mass is estimated to be ≤ 1 GeV.

1.1. Top quark production and decay

At LHC, top quarks are predominantly produced in quark-antiquark ($t\bar{t}$) pairs, although it is possible that a single (anti-)quark is being produced. The analyses reported in this note deal

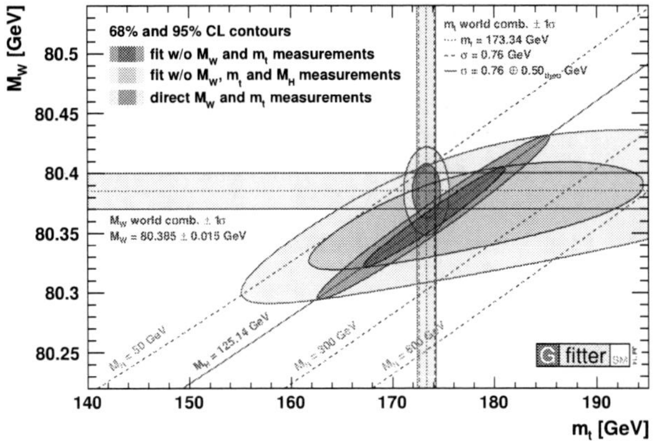

Figure 1. Contours of 68% and 95% confidence level of the experimental W boson mass versus the top quark mass including the Higgs boson mass (blue) and excluding it (grey). The values are in agreement as expected in the SM. Figure extracted from [6].

only with the $t\bar{t}$ production.

Once produced the top quark decays almost exclusively via the $t \to Wb$ vertex. Obviously a jet emanates from the b hadronization. In turn, the W boson decays either into leptons or into quarks. The W decay mode defines the topology of the event. When leptons are produced ($W \to \ell\nu$), the neutrinos escape detection and their transverse energy appears during reconstruction as an unbalance of the event transverse energy. When quarks stem from the $W \to q\bar{q}'$ decay, they hadronize giving rise to jets.

The usual classification of the events is: all-hadronic (when both Ws decay hadronically), lepton plus jets (ℓ+jets in short, when each W decays in either mode) and dilepton channel (both W decaying leptonically). The branching ratios are well established in the SM. The ℓ+jets channel has a large branching ratio[1]. Moreover the presence of the charged lepton and missing transverse energy carried away by the neutrino help to keep the SM physics background under control. The all-hadronic is the most probable channel (44% branching ratio) but the background is high. Besides, the reconstruction of many jets in the event is very challenging. The dilepton channel has a low background but its branching fraction is quite small. On the other hand, the presence of two neutrinos makes the event kinematic reconstruction more arduous.

2. Top quark mass measurements in ATLAS

ATLAS has measured the top quark mass (MC mass) in all event topologies [9, 10, 11] using the 7 TeV proton-proton collisions at LHC. The analyses use template methods to extract the top quark mass. They consist in characterizing the distribution of selected observables which are sensitive to the top quark mass. The shape of the distribution is parametrized as a function of the top quark MC mass in the simulations. Then a likelihood fit figures out which top quark mass better describes the real data distribution of those same observables. In the fit, the estimated physics background is also included.

In the case of the most recent ATLAS 7 TeV data analysis in the ℓ+jets, the template considers 3 variables (3D template) (Fig. 2). The first is the reconstructed invariant mass of the top quark (from a kinematic fit to the event data). Then comes the reconstructed invariant mass of the W boson from the hadronic decay part. This observable allows to calibrate the relative reconstructed jet energy scale factor (JSF) between real data and MC. Even a tiny difference in the JSF may bias irremediably the extracted value of the top quark mass. The third observable may control/detect/avoid possible (small) differences in the b-jet energy scale

[1] Its actual value depends to the treatment given to the $W \to \tau\nu$ decays. As the τ will decay later, some of the ATLAS analysis only consider as genuine signal the W decays into e or μ.

factor (bJSF) between real data and MC that may rise due to the different fragmentation of the b quark compared with respect to the light quarks.

The analysis uses either events with one or two b-tagged jets. Of course the templates of the observables are parametrized independently. Despite of the two b-tagged jets sample being more sensitive to the top quark mass, that with one b-tagged jet sample gets more statistics, resulting in a similar resolution power.

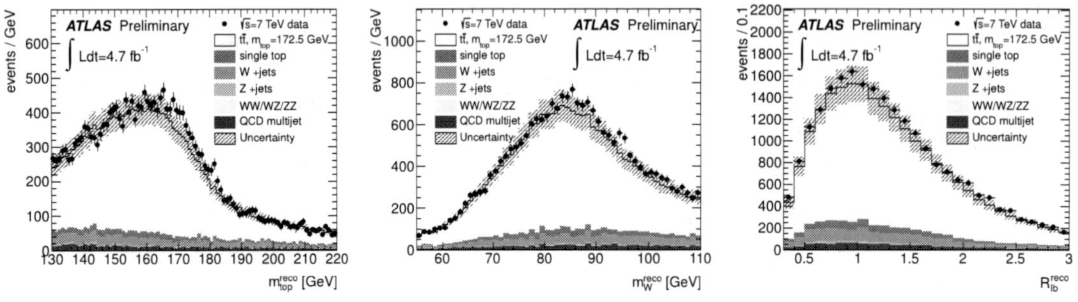

Figure 2. Distributions of the three observables used in the ℓ+jets channel top mass analysis with a 3D template. Real data (dots) are compared with MC expectation (using MC mass of 172.5 GeV, solid line). The contribution from different physics backgrounds is also presented. The sample includes events either with 1 b-tagged jet or 2 b-tagged jets. Taken from [9].

In the dilepton channel, the presence of two neutrinos in the final state is a challenge for the event reconstruction. The observable used is the invariant mass between the reconstructed charged lepton (ℓ) and the b-jet ($m_{\ell b}$). As there are two leptons and two b-jets, for each possible lepton-to-b-jet assignment, the average of the two invariant masses is calculated. Then the lowest average mass is taken as the $m_{\ell b}$ estimator for the event. Finally the top quark mass is extracted using also a template method.

Concerning the systematic uncertainties, both analysis are dominated by the Jet Energy Scale uncertainty (about 0.8 and 0.9 GeV in the ℓ+jets and dilepton analysis respectively). Of course the hadronization modeling and the ISR/FSR effects play a major role. The uncertainty arising from these effects is about 0.6 GeV for both analysis. The latests ATLAS results in all analysis are reported in Table 1.

Table 1. Summary of top mass (MC mass) values from ATLAS.

decay channel	reference	m_t (GeV)
ℓ+ jets	[9]	172.31 ± 0.75(stat.+JSF+bJSF) ± 1.35(syst.)
dilepton	[10]	173.09 ± 0.64(stat.) ± 1.50(syst.)
all-hadronic	[11]	174.9 ± 2.1(stat.) ± 3.8(syst.)

Another mass measurement comes from the $t\bar{t}$ cross section ($\sigma_{t\bar{t}}$). The $\sigma_{t\bar{t}}$ has been calculated at NNLO accuracy in α_s including the resummation of next-to-next-to-leading logarithmic (NNLL) soft gluon terms [12, 13]. These calculations are performed in a well defined scheme using the pole mass. ATLAS has already performed the measurements using the 7 and 8 TeV LHC data [14] and obtained: $m_t^{\rm pole} = 172.9\pm^{2.5}_{2.6}$ GeV. Unfortunately, the dependence of $\sigma_{t\bar{t}}$ on the mass is quite mild (Fig. 2). This is why the measurement of $m_t^{\rm pole}$ gets such a big uncertainty.

Finally, it is worth to mention that there are underway efforts to define a new observable sensitive to the $m_t^{\rm pole}$ that may provide a better accuracy than the $\sigma_{t\bar{t}}$ result. These studies consider the mass effects in the gluon radiation in $t\bar{t}$ events using NLO calculations [15].

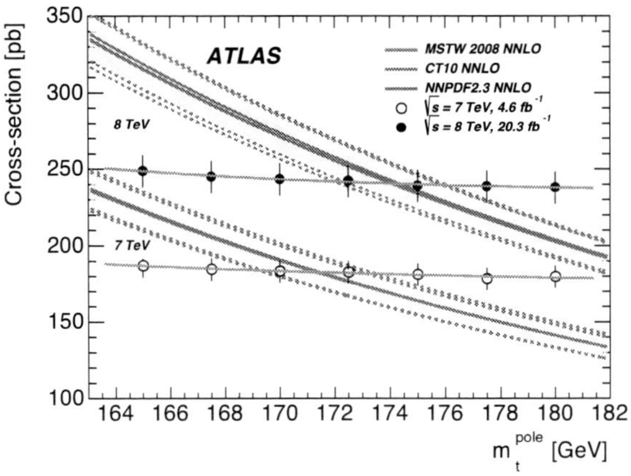

Figure 3. Predicted NNLO $\sigma_{t\bar{t}}$ at 7 and 8 TeV as a function of $m_t^{\rm pole}$, showing the central values and total uncertainty bands with several PDF sets. The yellow band shows the QCD scale uncertainty. The measurements of $\sigma_{t\bar{t}}$ are also shown, with their dependence on the assumed value of m_t. Figure taken from [14].

3. Summary

The most up to date measurements of the top quark mass using ATLAS data have been reported. ATLAS, CMS, CDF and DØ measurement have been used to get a m_t world combination [16].

$$m_t = 173.34 \pm 0.27(\text{stat.}) + 0.71(\text{syst.})\ \text{GeV}.$$

References

[1] CDF Collaboration, Abe F et al., Observation of top quark production in $p\bar{p}$ collisions, 1995 *Phys. Rev. Lett.* **74** 2626-2631

[2] DØ Collaboration, Abachi S et al., Observation of the top quark, 1995 *Phys. Rev. Lett.* **74** 2632-2637

[3] Evans L and Bryant Ph, *LHC Machine*, 2008 *JINST* **3** S08001

[4] ATLAS Collaboration, Aad G et al., The ATLAS Experiment at the CERN Large Hadron Collider, 2008 *JINST* **3** S08003

[5] CMS Collaboration, Chatrchyan S et al., The CMS Experiment at the CERN LHC, 2008 *JINST* **3** S08004

[6] Baak M et al., The global electroweak fit at NNLO and prospects for the LHC and ILC, 2014 *Preprint* arXiv:1407.3792

[7] Buckley A et al., General-purpose event generators for LHC physics. 2011 *Phys. Rept.* **504** 145-233

[8] Jegerlehner F et al., On the difference between the pole and the $\overline{MS}$ masses of the top quark and the electroweak scale, 2013 *Phys. Lett.* **B722** 123-129

[9] ATLAS Collaboration, Measurement of the Top Quark Mass from $\sqrt{s}$ = 7 TeV ATLAS Data using a 3-dimensional Template Fit. *Preprint* ATLAS-CONF-2013-046. https://cds.cern.ch/record/1547327

[10] ATLAS Collaboration, Measurement of the Top Quark Mass in Dileptonic Top Quark Pair Decays with $\sqrt{s}$=7 TeV ATLAS Data. *Preprint* ATLAS-CONF-2013-077. https://cds.cern.ch/record/1562935

[11] ATLAS Collaboration, Determination of the Top Quark Mass with a Template Method in the All-Hadronic Decay Channel using 2.04 fb^{-1}. *Preprint* ATLAS-CONF-2012-030. https://cds.cern.ch/record/1431895

[12] Cacciari M et al., Top-pair production at hadron colliders with next-to-next-to-leading logarithmic soft-gluon resummation, 2012 *Phys. Lett.* B **710** 612-622

[13] Bärnreuther P et al., Percent Level Precision Physics at the Tevatron: First Genuine NNLO QCD Corrections to $q\bar{q} \to t\bar{t} + X$, 2012 *Phys. Rev. Lett.* **109**, 132001

[14] ATLAS Collaboration, Measurement of the $t\bar{t}$ production cross-section using $e\mu$ events with b-tagged jets in pp collisions at $\sqrt{s}$=7 and 8 TeV with the ATLAS detector. *Preprint* arXiv:1406.5375

[15] Alioli S et al., A new observable to measure the top-quark mass at hadron colliders. 2013 *Eur. Phys. J.* C **73** 2438

[16] ATLAS, CDF, CMS and DØCollaborations. First combination of Tevatron and LHC measurements of the top-quark mass. *Preprint* arXiv:1403.4427

XI International Conference on Hyperons, Charm and Beauty Hadrons (BEACH 2014) IOP Publishing
Journal of Physics: Conference Series **556** (2014) 012034 doi:10.1088/1742-6596/556/1/012034

CKM angle γ measurements at LHCb

Alexis Vallier

Laboratoire de l'Accélérateur Linéaire, Université Paris-Sud 11, CNRS/IN2P3, Orsay, France.

E-mail: vallier@lal.in2p3.fr

Abstract. The CKM angle γ remains the least known parameter of the CKM mixing matrix. The precise measurement of this angle, as a Standard Model benchmark, is a key goal of the LHCb experiment. We present four recent CP violation studies related to the measurement of γ, including amplitude analysis of $B^{\pm} \to DK^{\pm}$ decays, the ADS/GLW analysis of $B^0 \to DK^{*0}$ decays and the time-dependent analysis of $B_s^0 \to D_s^{\mp} K^{\pm}$ decays.

1. Introduction

The CKM angle γ, defined as $\gamma \equiv \arg\left(-\frac{V_{ud}V_{ub}^*}{V_{cd}V_{cb}^*}\right)$, is the least known CKM parameter. B-factories and the LHCb experiment measured γ with an uncertainty larger than $10°$ [1, 2, 3]. To compare, global fits like CKMfitter [4] and UTfit [5] obtain an estimation of γ with an error about $2°$. This angle is directly measurable through tree processes, without significant loop contribution. Hence the extraction of its value is very clean and has a theoretical relative uncertainty lower than 10^{-7} [6]. Therefore a precise measurement of γ provides an excellent standard candle to check the consistency of the CKM paradigm in the Standard Model and to probe some new physics. The present paper summarises four measurements of γ performed by the LHCb collaboration and presented at the BEACH 2014 conference.

2. Time-integrated measurements of γ

Given its definition, γ is approximately the phase difference between the quark transitions $b \to c\bar{u}s$ and $b \to u\bar{c}s$ and the interference between these two transitions is sensitive to this angle. The interference is obtained by reconstructing the D^0 and $\bar{D}^0$ mesons produced in these decays in an identical final state (Fig. 1). In the case of a three body D meson decay a Dalitz plot analysis can be carried out. This method is called GGSZ [7, 8] and two recent LHCb results are reported in section 2.1. In the case of a two body D meson[1] decay a counting analysis is developed following the so called GLW [9, 10] or ADS [11, 12] methods. A recent ADS/GLW result from LHCb is presented in section 2.2. All of these methods can be applied to the channels $B^{\pm} \to DK^{\pm}$ and $B^0 \to DK^{*0}$. Since these B mesons decays are self-tagged, no time-dependent analysis is required.

[1] in the following D stands for either a D^0 or a $\bar{D}^0$ meson.

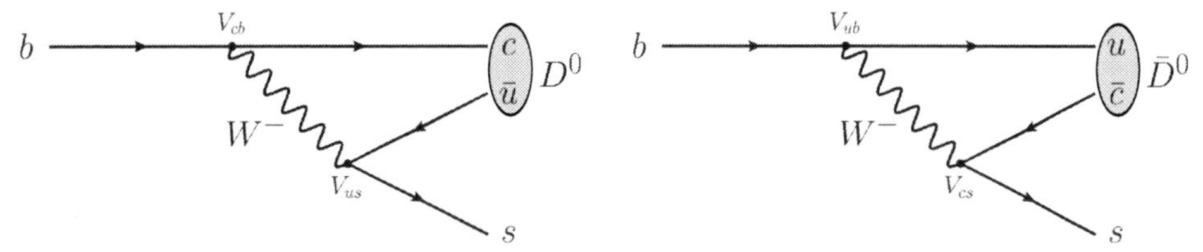

Figure 1. Feynman diagrams of $b \to c\bar{u}s$ (left) and $b \to u\bar{c}s$ (right) decays.

2.1. Measurements with 3-body D meson decays
The $B^{\pm} \to D(K^0_S h^+ h^-)K^{\pm}$decay amplitude — h stands for either a charged pion or a charged kaon — can be written as

$$A_{B^{\pm}}(\mathcal{D}) = A_D(\mathcal{D}) + e^{i(\delta_B \pm \gamma)} A_{\bar{D}}(\mathcal{D}),$$

where A_D ($A_{\bar{D}}$) is the D^0 ($\bar{D}^0$) decay amplitude, $\mathcal{D}$ represents the D meson phase space, δ_B is the strong phase difference and γ is the weak phase difference between the D^0 and $\bar{D}^0$ channels. The D meson phase space is parameterised by two squared invariant masses, for instance $m^2(K^0_S\pi^+)$ and $m^2(K^0_S\pi^-)$ on a D meson Dalitz plot. The sensitivity to γ arises from large asymmetries in some region of the Dalitz plot. In order to evaluate γ, the strong phase variation over the Dalitz plot must be known. This can be done in two different ways: with a model-dependent (MD) method using BaBar's amplitude model [13], or with a model-independent (MI) method using CLEO-c measurements as inputs [14]. Both methods involve fitting the D meson Dalitz plot to extract the polar coordinates $(r_{B^{\pm}}, \delta_B, \gamma)$. The parameter $r_{B^{\pm}}$ is the ratio of the magnitudes of the suppressed and favoured B decay amplitudes. The polar coordinates are not estimated directly from the Dalitz fit, but the cartesian coordinates $x_{\pm} = r_{B^{\pm}} \cos(\delta_B \pm \gamma)$ and $y_{\pm} = r_{B^{\pm}} \sin(\delta_B \pm \gamma)$.

2.1.1. Model-dependent analysis This section presents the model-dependent analysis of the $B^{\pm} \to D(K^0_S\pi^+\pi^-)K^{\pm}$ signal, using a sample of proton-proton collision data at a centre-of-mass energy of 7 TeV corresponding to an integrated luminosity of 1 fb^{-1}. Full details can be found in Ref. [15]. The analysis is carried out in two distinct stages. First a fit to the B meson reconstructed invariant mass is performed on the selected $B^{\pm} \to D(K^0_S\pi^+\pi^-)K^{\pm}$ and $B^{\pm} \to D(K^0_S\pi^+\pi^-)\pi^{\pm}$ candidates. This fit determines the signal and background fractions in the data sample. A total yield of 637 $B^{\pm} \to D(K^0_S\pi^+\pi^-)K^{\pm}$ signal events and 8866 events of the $B^{\pm} \to D(K^0_S\pi^+\pi^-)\pi^{\pm}$ control channel are found. Then a fit to the Daliz plot determines the CP violation observables $(x_{\pm}, y_{\pm})$. The signal and background yields and the parameters of the B invariant mass probability distribution function are fixed to the values obtained in the first stage. The model used to described the amplitude of the $D^0 \to K^0_S\pi^+\pi^-$ decay over the phase space is the one determined by the BaBar collaboration in Ref. [13]. Fitting simultaneously the distributions in the $D^0 \to K^0_S\pi^+\pi^-$ phase space for the $B^{\pm} \to D(K^0_S\pi^+\pi^-)K^{\pm}$ and the $B^{\pm} \to D(K^0_S\pi^+\pi^-)\pi^{\pm}$ candidates enables to take into account the variation of efficiency over the phase space. The $B^{\pm} \to D(K^0_S\pi^+\pi^-)\pi^{\pm}$ decay is a good proxy to get the efficiency variation, since it has a kinematic topology similar to the signal one and CP violation can be neglected in this channel. The resulting values of the cartesian coordinates are:

$$x_- = +0.027 \pm 0.044^{+0.010}_{-0.008} \pm 0.001, \qquad x_+ = -0.084 \pm 0.045 \pm 0.009 \pm 0.005,$$
$$y_- = +0.013 \pm 0.048^{+0.009}_{-0.007} \pm 0.003, \qquad y_+ = -0.032 \pm 0.048^{+0.010}_{-0.009} \pm 0.008,$$

where the first uncertainty is statistical, the second systematic and the third due to the amplitude model used to describe the $D^0 \to K^0_S\pi^+\pi^-$ decay. The leading experimental systematic errors are due to the efficiency and background description uncertainties. The constraints obtained on the polar coordinates are $r_{B^\pm} = 0.06 \pm 0.04$, $\delta_B = (115^{+41}_{-51})^\circ$ and $\gamma = (84^{+49}_{-42})^\circ$. These results are consistent with those of the LHCb model-independent analysis based on the same data set [16].

2.1.2. Model-independent analysis This section presents the model-independent analysis of the $B^\pm \to D(K^0_S h^+h^-)K^\pm$ decays, using a sample of proton-proton collision data at a centre-of-mass energy of 7 and 8 TeV corresponding to a total integrated luminosity of 3 fb^{-1}. Full details can be found in Ref. [17]. There is a significant improvement compared to the former results in Ref. [16], thanks to the increased statistics and a better analysis technique. To know the strong phase variation over the $D^0 \to K^0_S h^+h^-$ phase space the measurements made by CLEO-c, in a particular binning scheme, is used [14]. In this way the analysis is a counting experiment in bins of the Dalitz plot. The expected number of D from B^+ events falling in a particular bin labelled $\pm i$ (the $\pm$ sign comes from the phase symmetry with respect to the Dalitz diagonal) can be expressed as

$$N^+_{\pm i} = h^+_B \left[F_{\mp i} + (x^2_+ + y^2_+)F_{\pm i} + 2\sqrt{F_i F_{-i}}(x_+ c_{\pm i} \mp y_+ s_{\pm i})\right], \quad (1)$$

where c_i and s_i are the averaged cosine and sine of the strong phase difference in bin i (CLEO-c inputs), F_i is the expected fraction of pure D^0 events in bin i taking into account the efficiency profile over the phase space, and h^+_B is a normalisation factor. The F_i parameters are determined from the $B^0 \to D^{*\pm}\mu^\mp\nu_\mu$ control mode. This is an excellent proxy because the sample has a high purity, a high statistics and the D^0 meson is tagged thanks to the slow pion in the $D^{*+} \to D^0\pi^+$ decay. Some corrections are applied from simulated data to account for reconstruction and selection discrepancies between the $B^0 \to D^{*\pm}\mu^\mp\nu_\mu$ and the $B^\pm \to DK^\pm$ decays. The fit is performed in two steps. First the phase space integrated $B^\pm$ mass fit determines the total signal yields (around 2600) and fixes the model used in the second step. This last fit is made in each Dalitz bin with all the parameters in Eq. (1) fixed but the normalisation factor and the $(x_\pm, y_\pm)$ observables. The resulting values of the cartesian coordinates (Fig. 2) are the most precise to date:

$$x_- = (2.5 \pm 2.5 \pm 1.0 \pm 0.5) \times 10^{-2}, \qquad x_+ = (-7.7 \pm 2.4 \pm 1.0 \pm 0.4) \times 10^{-2},$$
$$y_- = (7.5 \pm 2.9 \pm 0.5 \pm 1.4) \times 10^{-2}, \qquad y_+ = (-2.2 \pm 2.5 \pm 0.4 \pm 1.0) \times 10^{-2},$$

where the first uncertainty is statistical, the second systematic and the third due to the experimental knowledge of the (c_i, s_i) parameters. Compared to the 1 fb^{-1} measurement [16], the statistical uncertainty is reduced thanks to the larger data sample, the experimental systematic is reduced by using the new control mode $B^0 \to D^{*\pm}\mu^\mp\nu_\mu$, and the (c_i, s_i) systematic is also improved by the increased LHCb sample size. The results are: $r_{B^\pm} = 0.080^{+0.019}_{-0.021}$, $\gamma = (62^{+15}_{-14})^\circ$ and $\delta_B = (62^{+14}_{-15})^\circ$.

2.2. Measurements with 2-body D meson decays

This section presents the ADS/GLW analysis of the $B^0 \to DK^{*0}$ decays, using a sample of proton-proton collision data at a centre-of-mass energy of 7 and 8 TeV corresponding to a total integrated luminosity of 3 fb^{-1}. Full details can be found in Ref. [18]. Compared to the $B^\pm \to DK^\pm$ decays, both Cabibbo favoured and suppressed diagrams are color suppressed, which brings about a higher interference amplitude (r_{B^0} is larger than $r_{B^\pm}$). Hence a better sensitivity to γ is expected. However this neutral channel is experimentally more challenging. For the GLW modes the D mesons are reconstructed in two CP eigenstate: K^+K^- and $\pi^+\pi^-$.

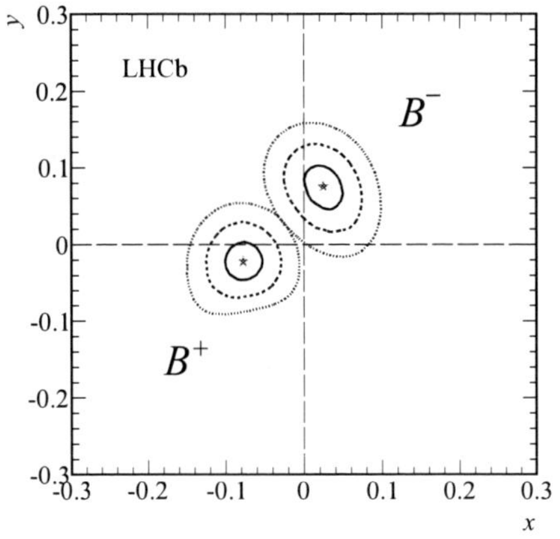

Figure 2. Confidence levels at 39.3%, 86.5% and 98.9% probability for (x_+, y_+) and (x_-, y_-) as measured in $B^\pm \to DK^\pm$ decays (statistical uncertainties only). The parameters (x_+, y_+) relate to B^+ decays and (x_-, y_-) refer to B^- decays. The stars represent the best fit central values.

For the ADS modes the D mesons are reconstructed in the $K^+\pi^-$ and $K^-\pi^+$ final states. From these decays several observables sensitive to γ can be built. For instance in the GLW modes the CP asymmetries

$$A_d^{hh} \equiv \frac{\Gamma(\bar{B}^0 \to D(h^+h^-)\bar{K}^{*0}) - \Gamma(B^0 \to D(h^+h^-)K^{*0})}{\Gamma(\bar{B}^0 \to D(h^+h^-)\bar{K}^{*0}) + \Gamma(B^0 \to D(h^+h^-)K^{*0})} = \frac{2r_{B^0}\kappa \sin\delta_B \sin\gamma}{1 + r_{B^0}^2 + 2r_{B^0}\kappa\cos\delta_B\cos\gamma}$$

are measured. The parameter κ is the coherence factor introduced to account for the effect of the non resonant $B^0 \to DK^+\pi^-$ contribution in the K^{*0} signal region. And in the ADS mode the ratio of suppressed $B^0 \to D(\pi^+K^-)K^{*0}$ to favoured $B^0 \to D(K^+\pi^-)K^{*0}$ partial widths are measured separately for B^0 and $\bar{B}^0$:

$$R_d^+ \equiv \frac{\Gamma(B^0 \to D(\pi^+K^-)K^{*0})}{\Gamma(B^0 \to D(K^+\pi^-)K^{*0})} = \frac{r_{B^0}^2 + r_D^2 + 2r_{B^0}r_D\kappa\cos(\delta_B + \delta_D + \gamma)}{1 + r_{B^0}^2 r_D^2 + 2r_{B^0}r_D\kappa\cos(\delta_B - \delta_D + \gamma)},$$

$$R_d^- \equiv \frac{\Gamma(\bar{B}^0 \to D(\pi^-K^+)\bar{K}^{*0})}{\Gamma(\bar{B}^0 \to D(K^-\pi^+)\bar{K}^{*0})} = \frac{r_{B^0}^2 + r_D^2 + 2r_{B^0}r_D\kappa\cos(\delta_B + \delta_D - \gamma)}{1 + r_{B^0}^2 r_D^2 + 2r_{B^0}r_D\kappa\cos(\delta_B - \delta_D - \gamma)}.$$

The parameters r_D and δ_D are the magnitude ratio and the phase difference, respectively, between the amplitudes of the $D^0 \to K^+\pi^-$ and $D^0 \to K^-\pi^+$ decays. The significances of the combined B^0 and $\bar{B}^0$ signals for the $B^0 \to D(K^+K^-)K^{*0}$, $B^0 \to D(\pi^+\pi^-)K^{*0}$ and $B^0 \to D(\pi^+K^-)K^{*0}$ decay modes are 8.6 σ, 5.8 σ and 2.9 σ respectively. Once the production and efficiency asymmetries are taken into account the results are:

$$A_d^{KK} = -0.20 \pm 0.15 \pm 0.02, \qquad R_d^+ = 0.06 \pm 0.03 \pm 0.01,$$
$$A_d^{\pi\pi} = -0.09 \pm 0.22 \pm 0.02, \qquad R_d^- = 0.06 \pm 0.03 \pm 0.01,$$

where the first uncertainties are statistical and the second systematic. $A_d^{\pi\pi}$, R_d^+ and R_d^- are first measurements and the A_d^{KK} result supersedes the former LHCb one [19]. From these measurements the value of r_{B^0} (proper to the $B^0 \to DK^{*0}$ channel) is found to be $r_{B^0} = 0.240^{+0.055}_{-0.048}$. This is the most precise measurement to date.

3. Time-dependent measurement of γ

This section presents the time-dependent analysis of the $B_s^0 \to D_s^\mp K^\pm$ decays, using a sample of proton-proton collision data at a centre-of-mass energy of 7 TeV corresponding to an integrated

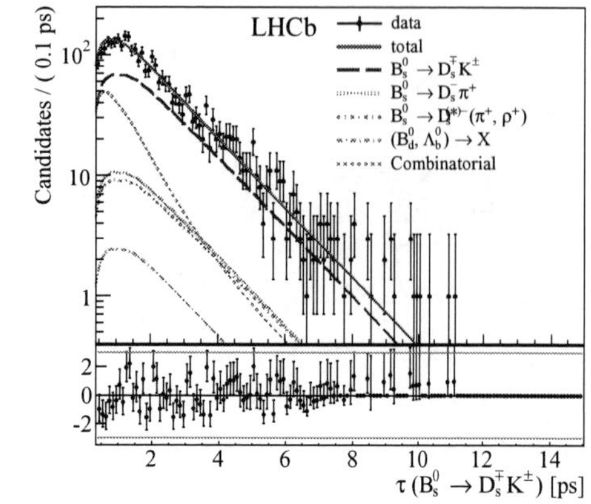

Figure 3. Result of the decay-time *cFit* to the $B_s^0 \to D_s^{\mp}K^{\pm}$ candidates.

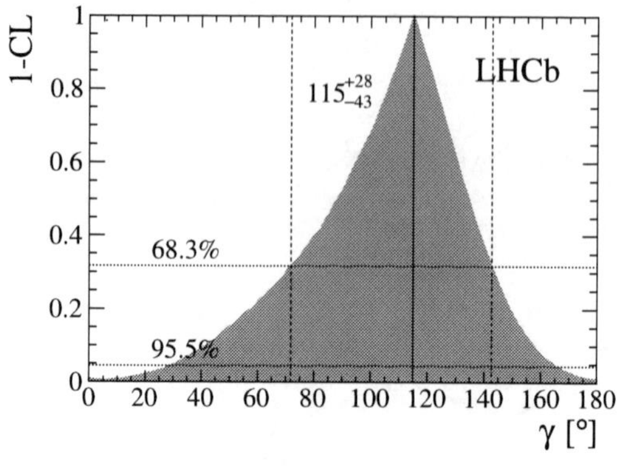

Figure 4. Frequentist interpretation of the γ measurement in term of confidence interval.

luminosity of 1 fb^{-1}. Full details can be found in Ref. [20]. In addition to the same tree level processes as in the time-integrated analysis (Fig. 1), the effect of the B_s mixing occurs. Hence the interference between mixing and decay amplitudes in $B_s^0 \to D_s^{\mp}K^{\pm}$ is sensitive to the CP violating phase $(\gamma - 2\beta_s)$ where $\beta_s \equiv \arg(-V_{ts}V_{tb}^*/V_{cs}V_{cb}^*)$. The time-dependent decay rates depend on the CP observables:

$$C_f = \frac{1-r_{D_sK}^2}{1+r_{D_sK}^2}, \quad A_f^{\Delta\Gamma} = \frac{-2r_{D_sK}\cos(\delta-(\gamma-2\beta_s))}{1+r_{D_sK}^2}, \quad A_{\bar{f}}^{\Delta\Gamma} = \frac{-2r_{D_sK}\cos(\delta+(\gamma-2\beta_s))}{1+r_{D_sK}^2},$$
$$S_f = \frac{2r_{D_sK}\sin(\delta-(\gamma-2\beta_s))}{1+r_{D_sK}^2}, \quad S_{\bar{f}} = \frac{-2r_{D_sK}\sin(\delta+(\gamma-2\beta_s))}{1+r_{D_sK}^2},$$

where r_{D_sK} is the magnitude of the amplitude ratio $|A(\bar{B}_s^0 \to D_s^- K^+)/A(B_s^0 \to D_s^- K^+)|$, δ the strong phase difference and $(\gamma - 2\beta_s)$ the weak phase difference. This analysis uses an independent measurement of ϕ_s [21] and assumes $\phi_s = -2\beta_s$ to interpret the results in terms of γ. To discriminate the signal and background components a 3D fit is performed on the B_s and D_s masses along with the log-likelihood difference between the kaon and pion hypothesis for the companion particle ($K^{\pm}$ for $B_s^0 \to D_s^{\mp}K^{\pm}$ signal and π^+ for $B_s^0 \to D_s^-\pi^+$ control mode). Then the output of this multivariate fit is used for the decay-time fit. Two fits are performed: a background subtracted fit, called *sFit*, using the *sWeights* [22, 23] determined by the multivariate fit; and a classical fit, called *cFit* (Fig. 3) where all signal and background time distributions are described. The results of these two fits are in excellent agreement:

Parameter	*sFit* fitted value	*cFit* fitted value
C_f	$0.52 \pm 0.25 \pm 0.04$	$0.53 \pm 0.25 \pm 0.04$
$A_f^{\Delta\Gamma}$	$0.29 \pm 0.42 \pm 0.17$	$0.37 \pm 0.42 \pm 0.20$
$A_{\bar{f}}^{\Delta\Gamma}$	$0.14 \pm 0.41 \pm 0.18$	$0.20 \pm 0.41 \pm 0.20$
S_f	$-0.09 \pm 0.31 \pm 0.06$	$-1.09 \pm 0.33 \pm 0.08$
$S_{\bar{f}}$	$-0.36 \pm 0.34 \pm 0.06$	$-0.36 \pm 0.34 \pm 0.08$

The first uncertainties are statistical and the second systematic. The main sources of systematic arise from the trigger-induced time-dependent efficiency, Γ_s and $\Delta\Gamma_s$. These results can be interpreted as a confidence interval $\gamma = (115^{+28}_{-43})^{\circ}$ at 68% CL (Fig. 4). This is the first measurement of γ with $B_s^0 \to D_s^{\mp}K^{\pm}$ decays.

4. Conclusion

The latest LHCb results on the CKM angle γ are reported: the model dependent GGSZ analysis of $B^\pm \to DK^\pm$ decays, the update to the full available data set of the model indepent GGSZ analysis of $B^\pm \to DK^\pm$ decays, the ADS/GLW analysis of $B^0 \to DK^{*0}$ decays and the first γ measurement with $B_s^0 \to D_s^\mp K^\pm$decays. Using these results and the corresponding improvements, the next combination of the LHCb γ measurements should yield a significant reduction of the uncertainty.

References

[1] BaBar collaboration, J. P. Lees *et al.*, *Observation of direct CP violation in the measurement of the Cabibbo-Kobayashi-Maskawa angle gamma with $B^\pm \to D^{(*)}K^{(*)\pm}$ decays*, Phys. Rev. **D87** (2013), no. 5 052015, arXiv:1301.1029.

[2] Belle collaboration, K. Trabelsi *et al.*, *Study of direct CP in charmed B decays and measurement of the CKM angle gamma at Belle*, arXiv:1301.2033.

[3] LHCb collaboration, *Improved constraints on γ from $B^\pm \to DK^\pm$ decays including first results on 2012 data*, , Linked to LHCb-ANA-2013-012.

[4] CKMfitter Group, J. Charles *et al.*, Eur. Phys. J. **C41** (2005) 1, arXiv:hep-ph/0406184, updated results and plots available at: http://ckmfitter.in2p3.fr.

[5] UTfit collaboration, M. Bona *et al.*, JHEP **0610** (2006) 081, arXiv:hep-ph/0606167, updated results and plots available at: http://www.utfit.org.

[6] J. Brod and J. Zupan, *The ultimate theoretical error on γ from $B \to DK$ decays*, JHEP **1401** (2014) 051, arXiv:1308.5663.

[7] A. Giri, Y. Grossman, A. Soffer, and J. Zupan, *Determining γ using $B^\pm \to DK^\pm$ with multibody D decays*, Phys. Rev. D **68** (2003) 054018.

[8] A. Bondar in *Proceedings of BINP special analysis meeting on Dalitz, unpublished*, 2002.

[9] M. Gronau and D. London, *How to determine all the angles of the unitarity triangle from $B_d^0 \to DK_S$ and $B_s^0 \to D\phi$*, Physics Letters B **253** (1991), no. 3–4 483 .

[10] M. Gronau and D. Wyler, *On determining a weak phase from charged b decay asymmetries*, Physics Letters B **265** (1991), no. 1–2 172 .

[11] D. Atwood, I. Dunietz, and A. Soni, *Enhanced CP Violation with $B \to KD^0(\bar{D}^0)$ Modes and Extraction of the Cabibbo-Kobayashi-Maskawa Angle γ*, Phys. Rev. Lett. **78** (1997) 3257.

[12] D. Atwood, I. Dunietz, and A. Soni, *Improved methods for observing* CP *violation in $B^\pm \to KD$ and measuring the CKM phase γ*, Phys. Rev. D **63** (2001) 036005.

[13] BaBar collaboration, P. del Amo Sanchez *et al.*, *Measurement of $D^0 - \bar{D}^0$ mixing parameters using $D^0 \to K_S^0\pi^+\pi^-$ and $D^0 \to K_S^0K^+K^-$ decays*, Phys. Rev. Lett. **105** (2010) 081803, arXiv:1004.5053.

[14] CLEO collaboration, J. Libby *et al.*, *Model-independent determination of the strong-phase difference between D^0 and $\bar{D}^0 \to K_{S,L}^0h^+h^-$ ($h = \pi, K$) and its impact on the measurement of the CKM angle γ/ϕ_3*, Phys. Rev. **D82** (2010) 112006, arXiv:1010.2817.

[15] LHCb collaboration, R. Aaij *et al.*, *Measurement of CP violation and constraints on the CKM angle γ in $B^\pm \to DK^\pm$ with $D \to K_S^0\pi^+\pi^-$ decays*, arXiv:1407.6211, submitted to Nucl. Phys. B.

[16] LHCb collaboration, R. Aaij *et al.*, *A model-independent Dalitz plot analysis of $B^\pm \to DK^\pm$ with $D \to K_S^0h^+h^-$ ($h = \pi, K$) decays and constraints on the CKM angle γ*, Phys. Lett. **B718** (2012) 43, arXiv:1209.5869.

[17] LHCb collaboration, R. Aaij *et al.*, *Measurement of the CKM angle γ using $B^\pm \to DK^\pm$ with $D \to K_S\pi^+\pi^-$, $K_SK^+K^-$ decays*, arXiv:1408.2748, submitted to JHEP.

[18] LHCb collaboration, R. Aaij *et al.*, *Measurements of CP violation parameters in $B^0 \to DK^{*0}$ decays*, arXiv:1407.8136, submitted to Phys. Rev. D.

[19] LHCb collaboration, R. Aaij *et al.*, *Measurement of CP observables in $B^0 \to DK^{*0}$ with $D \to K^+K^-$*, JHEP **03** (2013) 067, arXiv:1212.5205.

[20] LHCb collaboration, R. Aaij *et al.*, *Measurement of CP asymmetry in $B_s^0 \to D_s^\mp K^\pm$ decays*, arXiv:1407.6127, submitted to JHEP.

[21] LHCb collaboration, R. Aaij *et al.*, *Measurement of CP violation and the B_s^0 meson decay width difference with $B_s^0 \to J/\psi K^+K^-$ and $B_s^0 \to J/\psi\pi^+\pi^-$ decays*, Phys. Rev. **D87** (2013) 112010, arXiv:1304.2600.

[22] M. Pivk and F. R. Le Diberder, *sPlot: A statistical tool to unfold data distributions*, Nuclear Instruments and Methods in Physics Research A **555** (2005) 356, arXiv:physics/0402083.

[23] Y. Xie, *sFit: a method for background subtraction in maximum likelihood fit*, ArXiv e-prints (2009) arXiv:0905.0724.

Journal of Physics: Conference Series **556** (2014) 012035 doi:10.1088/1742-6596/556/1/012035

Towards new frontiers in CP violation in B decays

Robert Fleischer

Nikhef, Science Park 105, NL-1098 XG Amsterdam, Netherlands

Department of Physics and Astronomy, Vrije Universiteit Amsterdam, NL-1081 HV Amsterdam, Netherlands

E-mail: Robert.Fleischer@nikhef.nl

Abstract. CP-violating effects in decays of B mesons offer a wide spectrum of probes for testing the phase structure of the quark-flavour sector of the Standard Model. After a brief discussion of the picture emerging from the current LHC data, the focus will be put on two specific topics: hadronic uncertainties from penguin topologies on measurements of the B_q^0–$\bar{B}_q^0$ mixing phases ($q \in \{d, s\}$), and the U-spin-related decays $B_s^0 \to K^+K^-$ and $B_d^0 \to \pi^+\pi^-$. Valuable new insights are expected from future studies of CP violation in B decays. For the detection of possible new sources of CP violation, it will be crucial to match the experimental and theoretical precisions and to have a careful look at the underlying assumptions.

1. Introduction

In the Standard Model (SM), the rich phenomenology of quark-flavour physics and CP violation is related to the Cabibbo–Kobayashi–Maskawa (CKM) quark-mixing matrix [1, 2]. Information on the phase structure and elements of this matrix are encoded in weak decays of K, D and B mesons. Since the theory is formulated in terms of quarks while the mesons are bound states of strong interactions, we have to deal with process-dependent, non-perturbative "hadronic" parameters in the calculation of the relevant transition amplitudes. This feature gives rise to the main challenge in studies of CP violation in B decays: hadronic uncertainties.

In the presence of New Physics (NP), typically new sources of flavour and CP violation arise. Analyses of weak meson decays are facing an impressive hierarchy of scales:

$$\underbrace{\Lambda_{\rm NP} \sim 10^{(0\ldots?)}\,{\rm TeV} \gg \Lambda_{\rm EW} \sim 10^{-1}\,{\rm TeV}}_{\text{short distance physics}} \ggg \underbrace{\Lambda_{\rm QCD} \sim 10^{-4}\,{\rm TeV}}_{\text{long distance physics}} . \tag{1}$$

In order to deal with this situation, effective field theories offer the suitable theoretical tool. Within this framework, the heavy degrees of freedom (NP particles, top quark, Z and W bosons) are integrated out from appearing explicitly and are described in short-distance loop functions. Perturbative QCD corrections can be calculated in a systematic way, and renormalisation group techniques allow the summation of large logarithms. This machinery was applied to the SM and various popular NP scenarios, such as MSSM, models with universal and warped extra dimensions, little Higgs models, scenarios with extra Z' bosons, etc., as reviewed in [3].

Following these lines, low-energy effective Hamiltonians can be calculated for $\bar{B} \to \bar{f}$ processes, taking the following general form [4]:

$$\langle \bar{f}|\mathcal{H}_{\rm eff}|\bar{B}\rangle = \frac{G_{\rm F}}{\sqrt{2}}\sum_j \lambda^j_{\rm CKM}\sum_k C_k(\mu)\,\langle \bar{f}|Q^j_k(\mu)|\bar{B}\rangle, \tag{2}$$

where G_{F} is Fermi's constant and the $\lambda^{j}_{\mathrm{CKM}}$ denote combinations of CKM matrix elements. The short-distance contribution to the decay amplitude is described by the Wilson coefficient functions $C_k(\mu)$, which can be calculated in perturbation theory for the SM and its extensions. On the other hand, the hadronic matrix elements $\langle \bar{f}|Q^{j}_{k}(\mu)|\bar{B}\rangle$ are non-perturbative quantities, describing the long-distance contributions.

The key players for the exploration of CP violation in B decays are non-leptonic channels. In the previous decade, there were interesting developments for calculations of such processes within QCD: QCD factorisation, the perturbative hard-scattering (PQCD) approach, soft collinear effective theory (SCET) and QCD sum rules; the state of the art is discussed in [5]. There has recently been impressive progress in lattice QCD [6]. However, non-leptonic B decays generally remain a theoretical challenge, which is also indicated by experimental data.

The outstanding feature of analyses of CP violation in B decays is that the calculation of the hadronic matrix elements $\langle \bar{f}|Q^{j}_{k}(\mu)|\bar{B}\rangle$ can be circumvented in fortunate cases (for a detailed discussion, see [7]). The corresponding strategies play a key role to "over constrain" the unitarity triangle (UT) of the CKM matrix. Detailed analyses and continuous updates are performed by the CKMfitter [8] and UTfit [9] collaborations. The current picture of the UT shows impressive consistency with the CKM sector of the SM, despite a few tensions.

The previous run of the Large Hadron Collider (LHC) has resulted in the exciting discovery of the Higgs boson. On the other hand, the ATLAS and CMS experiments have so far not seen signals of NP particles, and the SM flavour sector has been confirmed by the LHCb data, apart from a few discrepancies which are unfortunately not yet conclusive. The implications for the general structure of physics beyond the SM are a large characteristic NP scale, i.e. not just $\sim$ TeV, or (and?) symmetries preventing large NP effects in the flavour sector, where models with "Minimal Flavour Violation" are the most prominent example.

Many more interesting results are expected from the next run of the LHC and its upgrade as well as high-precision flavour experiments. In view of the present situation, we have to prepare ourselves to deal with smallish NP effects. It will be crucial for resolving possible signals of NP in the data to have a careful look at the underlying theoretical assumptions and approximations. The challenge will be to match the experimental and theoretical precisions.

2. Penguin effects in benchmark probes of CP violation

Neutral B^0_q mesons ($q \in \{d, s\}$) show B^0_q–$\bar{B}^0_q$ mixing [10], which originates from box topologies in the SM but may well receive NP contributions. The CP-violating mixing phases are given by

$$\phi_d = \phi_d^{\mathrm{SM}} + \phi_d^{\mathrm{NP}} = 2\beta + \phi_d^{\mathrm{NP}}, \quad \phi_s = \phi_s^{\mathrm{SM}} + \phi_s^{\mathrm{NP}} = -2\lambda^2\eta + \phi_s^{\mathrm{NP}}, \tag{3}$$

where β is the usual angle of the UT, while $\lambda \equiv |V_{us}| \sim 0.22$ and η are parameters of the Wolfenstein parametrisation of the CKM matrix. The benchmark decays to measure the mixing phases ϕ_q through mixing-induced CP violation are given by $B^0_d \to J/\psi K_{\mathrm{S}}$, $B^0_s \to J/\psi\phi$ and $B^0_s \to J/\psi f_0(980)$. Decays of B_s mesons play the key role at the LHC [11].

These determinations are affected by uncertainties from doubly Cabibbo-suppressed penguin contributions [12–19], which cannot be calculated reliably and are usually neglected. In view of the current situation and the increasing experimental precision, the following questions arise: how important are the penguin contributions and how can they be controlled?

2.1. The $B^0_{d,s} \to J/\psi K_{\mathrm{S}}$ system

In the SM, the decay $B^0_d \to J/\psi K_{\mathrm{S}}$ originates from a colour-suppressed tree topology and penguin topologies with up, charm and top quarks running in the loops. Using the unitarity of the CKM matrix, the decay amplitude can be written as follows [12]:

$$A(B^0_d \to J/\psi\, K_{\mathrm{S}}) = \left(1 - \lambda^2/2\right) \mathcal{A}' \left[1 + \epsilon a' e^{i\theta'} e^{i\gamma}\right], \tag{4}$$

where $\mathcal{A}'$ and $a'e^{i\theta'}$ are CP-conserving parameters, involving the relevant hadronic matrix elements. Whereas the former quantity is governed by the colour-suppressed tree contribution, the latter measures the ratio of penguin to tree topologies. The key feature of (4) is that the penguin parameter a' enters with the tiny $\epsilon \equiv \lambda^2/(1-\lambda^2) = 0.05$; γ is the usual UT angle.

CP violation is probed through the time-dependent decay rate asymmetry

$$\frac{\Gamma(B_d^0(t) \to J/\psi K_S) - \Gamma(\bar{B}_d^0(t) \to J/\psi K_S)}{\Gamma(B_d^0(t) \to J/\psi K_S) + \Gamma(\bar{B}_d^0(t) \to J/\psi K_S)} = C_{J/\psi K_S} \cos(\Delta M_d t) - S_{J/\psi K_S} \sin(\Delta M_d t), \quad (5)$$

where the direct CP asymmetry $C_{J/\psi K_S}$ is proportional to $\epsilon\, a' \sin\theta' \sin\gamma$. On the other hand, the mixing-induced CP asymmetry can be written in the following form [14]:

$$S_{J/\psi K_S} = \sin(\phi_d + \Delta\phi_d), \quad (6)$$

where the hadronic phase shift $\Delta\phi_d$ is proportional to $\epsilon\, a' \cos\theta' \sin\gamma$ (and is usually neglected). In Eq. (5), ΔM_d is the mass difference of the B_d mass eigenstates.

The decay $B_s^0 \to J/\psi K_S$ is related to $B_d^0 \to J/\psi K_S$ through the U-spin symmetry of strong interactions [12]. In the SM, its decay amplitude can be written as

$$A(B_s^0 \to J/\psi\, K_S) = -\lambda\, \mathcal{A} \left[1 - a e^{i\theta} e^{i\gamma}\right]. \quad (7)$$

In contrast to (4), a does not enter with ϵ, i.e. is not doubly Cabibbo-suppressed. Consequently, the penguin effects are magnified in $B_s^0 \to J/\psi K_S$. It is useful to introduce a quantity $H \propto \mathrm{BR}(B_s \to J/\psi K_S)/\mathrm{BR}(B_d \to J/\psi K_S)$, which complements the direct and mixing-induced CP asymmetries of the time-dependent CP-violating rate asymmetry of $B_s^0 \to J/\psi K_S$.

The U-spin symmetry implies $a = a'$ and $\theta = \theta'$, thereby allowing the determination of γ, a and θ from H and the two CP-violating observables of $B_s^0 \to J/\psi K_S$ [12]. Since 1999, when this strategy was originally proposed, there has been a change of the main focus: a study [16] has shown that the extraction of γ will be feasible at LHCb but not competitive with other methods. As γ will be known by the time CP violation in $B_s^0 \to J/\psi K_S$ can be detected, the corresponding CP asymmetries allow a clean determination of the penguin parameters a and θ.

The $B_s^0 \to J/\psi K_S$ channel was observed by CDF [20] and LHCb [21] but its CP asymmetries have not yet been measured. Using currently available data for decays with a CKM structure similar to $B_s^0 \to J/\psi K_S$, i.e. $B_d^0 \to J/\psi \pi^0$ and $B^+ \to J/\psi \pi^+$, and complementing them with $B_d^0 \to J/\psi K^0$, $B^+ \to J/\psi K^+$ data, the size of the penguin parameters a and θ can be constrained. An analysis along these lines yields the following preliminary results [22]:

$$a = 0.17^{+0.13}_{-0.11}, \quad \theta = (182.4^{+4.7}_{-4.6})^\circ, \quad \Delta\phi_d = (-0.97^{+0.72}_{-0.65})^\circ. \quad (8)$$

2.2. CP violation in $B_s^0 \to J/\psi\phi$

The CKM structure of $B_s^0 \to J/\psi\phi$ is analogous to that of $B_d^0 \to J/\psi K_S$. However, as the final state is a mixture of CP-even and CP-odd linear polarisation states $f \in \{0, \parallel, \perp\}$, a time-dependent angular analysis of the $J/\psi[\to \mu^+\mu^-]\phi[\to K^+K^-]$ decay products has to be performed [23–25]. The impact of the SM penguin contributions is usually neglected. As in the case of $B_d^0 \to J/\psi K_S$, the expressions for the mixing-induced CP asymmetries are modified as follows [15]:

$$\mathcal{A}_{\mathrm{CP},f}^{\mathrm{mix}} = \sin\phi_s \to \sin(\phi_s + \Delta\phi_s^f), \quad (9)$$

where the hadronic phase shift $\Delta\phi_s^f$ depends on the final-state configuration f. The current average (neglecting the penguin effects) of the CDF, D0, ATLAS and LHCb data compiled by the Heavy Flavour Averaging Group is given by $\phi_s = (0.0 \pm 4.0)^\circ$ [26], which agrees with the SM value $\phi_s^{\mathrm{SM}} = -(2.11 \pm 0.08)^\circ$ [8] of the B_s^0–$\bar{B}_s^0$ mixing phase.

A tool to control the penguin effects is offered by $B_s^0 \to J/\psi \bar{K}^{*0}$ [15], which was observed by CDF [20] and LHCb [27]. Its branching ratio $(4.4^{+0.5}_{-0.4} \pm 0.8) \times 10^{-5}$ is found in agreement with the prediction $(4.6 \pm 0.4) \times 10^{-5}$ following from $B_d^0 \to J/\psi \rho^0$, and its polarisation fractions agree well with those of $B_d^0 \to J/\psi K^{*0}$. The $B_d^0 \to J/\psi \rho^0$ channel, which shows also mixing-induced CP violation, is another interesting decay to shed light on the hadronic penguin effects [15].

The experimental sensitivity from $B_s^0 \to J/\psi\phi$ at the LHCb upgrade ($50\,\mathrm{fb}^{-1}$) is expected as $\Delta\phi_s|_{\rm exp} \sim 0.008 = 0.46^\circ$ [28]. This impressive precision will make it mandatory to get a handle on the penguin effects, which may lead to phase shifts at the 1° level (as indicated by Eq. (8)).

2.3. CP violation in $B_s^0 \to J/\psi f_0(980)$

Another interesting probe to study CP violation is provided by $B_s^0 \to J/\psi f_0(980)$ [29]. In contrast to $B_s^0 \to J/\psi\phi$, as the $f_0(980)$ is a scalar state with quantum numbers $J^{PC} = 0^{++}$, the final state is present in a p wave and has the CP eigenvalue -1. Consequently, a time-dependent angular analysis is not needed. On the other hand, the hadronic structure of the $f_0(980)$ is still – after decades – not settled, with a variety of theoretical interpretations ranging from the quark–antiquark picture to tetraquarks. A detailed discussion of the implications of this feature for the extraction of ϕ_s was given in [18] (for $B_{s,d}^0 \to J/\psi\eta^{(\prime)}$ decays, see [30]), while recent LHCb measurements related to this topic are reported in [31].

2.4. Effective B_s^0 decay lifetimes

The measurement of effective lifetimes of $B_s^0 \to f$ decays, which are defined as

$$\tau_f \equiv \frac{\int_0^\infty t\ \langle\Gamma(B_s(t) \to f)\rangle\ dt}{\int_0^\infty \langle\Gamma(B_s(t) \to f)\rangle\ dt}, \tag{10}$$

offers yet another way to obtain insights into CP violation [32]. Here it is particularly interesting to compare B_s^0 decays into CP-odd final states, such as $B_s^0 \to J/\psi f_0(980)$, with those into CP-even final states, such as $B_s^0 \to K^+K^-$ and $B_s^0 \to D_s^+D_s^-$. The measured effective lifetimes can be converted into contours in the ϕ_s–$\Delta\Gamma_s$ plane, where $\Delta\Gamma_s$ is the decay width difference of the B_s-meson system (for an overview of the status of $\Delta\Gamma_s$, see [33]). The lifetime contours are very robust with respect to hadronic uncertainties [32]. For an update with the most recent LHCb data, see [34]. The B_s decay lifetimes result in a picture in agreement with the SM.

2.5. Comments for the LHCb upgrade era

In view of hadronic effects, it is important to give measurements of ϕ_s for the individual decay channels $B_s^0 \to f$, i.e. $B_s^0 \to J/\psi\phi$ and $B_s^0 \to J/\psi f_0(980)$. The pattern of the $(\phi_s)_f$ may provide insights into the hadronic effects: differences in the values of $(\phi_s)_f$ would indicate hadronic effects. On the other hand, should no differences between the individual ϕ_s emerge, there would be evidence for negligible hadronic effects (within the errors) or a universal hadronic phase shift.

The time-dependent analysis of CP violation in $B_s^0 \to J/\psi K_{\rm S}$ allows the clean determination of the corresponding penguin parameters (see Subsection 2.1). A sizeable penguin parameter a would indicate a potential problem in the measurement of ϕ_s from the $B_s^0 \to J/\psi\phi$ and $B_s^0 \to J/\psi f_0(980)$ channels. On the other hand, smallish penguin effects would give us confidence for the measurement of ϕ_s, although subtleties may arise due to the different final states.

3. CP Violation in $B_s^0 \to K^+K^-$ and $B_d^0 \to \pi^+\pi^-$

The decays $B_s^0 \to K^+K^-$ and $B_d^0 \to \pi^+\pi^-$ receive contributions from tree and penguin topologies. In the SM, their decay amplitudes can be written as

$$A(B_s^0 \to K^+K^-) \propto \mathcal{C}' \left[e^{i\gamma} + d' e^{i\theta'}/\epsilon\right], \quad A(B_d^0 \to \pi^+\pi^-) \propto \mathcal{C}\left[e^{i\gamma} - d\, e^{i\theta}\right], \tag{11}$$

where $d'e^{i\theta'}$, C' and their unprimed counterparts are CP-conserving strong quantities [35]. The direct and mixing-induced CP asymmetries of the $B_s^0 \to K^+K^-$ and $B_d^0 \to \pi^+\pi^-$ decays allow the determination of theoretically clean contours in the γ–d' and γ–d planes, respectively. Since these decays are related to each other through the interchange of all down and strange quarks, the U-spin symmetry of strong interactions implies $d' = d$ and $\theta' = \theta$; the former relation allows the extraction of γ and $d(= d')$ from the contours [35, 36]. Moreover, the strong phases θ and θ' can be determined, allowing an internal consistency check of the U-spin symmetry. Further insights into the hadronisation dynamics are provided by $|C|$ and $|C'|$, which can be extracted from the ratio $K \propto \mathrm{BR}(B_s \to K^+K^-)/\mathrm{BR}(B_d \to \pi^+\pi^-)$. This strategy is promising for the LHCb physics programme [28]. It will be particularly interesting to compare the resulting value of γ with those following from methods using only tree-diagram-like $B_{(s)}$-meson decays.

The picture resulting from the current data was explored in detail in [36, 37]; the numerical results given below refer to the update by Rob Knegjens in [38]. An interesting variant of the method was proposed in [39], as discussed in detail by Marco Ciuchini in [40].

Using information on K, CP violation in $B_d^0 \to \pi^+\pi^-$ and $B_d^0 \to \pi^\mp K^\pm$, and allowing for U-spin-breaking corrections $\xi \equiv d'/d = 1 \pm 0.15$, $\Delta\theta \equiv \theta' - \theta = \pm 20^\circ$ results in

$$\gamma = (67.7^{+4.5}_{-5.0}|_{\rm input}{}^{+5.0}_{-3.7}|_{\xi}{}^{+0.1}_{-0.2}|_{\Delta\theta})^\circ, \tag{12}$$

which agrees with the "tree-level" results $\gamma = (70.0^{+7.7}_{-9.0})^\circ$ [8] and $(69.4 \pm 7.1)^\circ$ [9] within the uncertainties. There are no indications for sizeable non-factorisable $SU(3)$-breaking corrections in the corresponding data. In the SM, the mixing-induced CP asymmetry is predicted as

$$\mathcal{A}_{\rm CP}^{\rm mix}(B_s \to K^+K^-)|_{\rm SM} = -0.220^{+0.042}_{-0.054}, \tag{13}$$

while $\mathcal{A}_{\rm CP}^{\rm dir}(B_s \to K^+K^-) \approx \mathcal{A}_{\rm CP}^{\rm dir}(B_d \to \pi^\pm K^\pm) = 0.082 \pm 0.04$.

The first LHCb measurement [41] of the CP-violating $B_s \to K^+K^-$ observables yields

$$\mathcal{A}_{\rm CP}^{\rm mix}(B_s \to K^+K^-) = -0.30 \pm 0.12 \pm 0.04, \quad \mathcal{A}_{\rm CP}^{\rm dir}(B_s \to K^+K^-) = 0.14 \pm 0.11 \pm 0.03, \tag{14}$$

and agrees with the SM predictions given above. In the future, once the experimental precision for the CP asymmetries improves, γ can be extracted exclusively from the γ–$d^{(\prime)}$ contours. The observable K, which is affected by form factors and non-factorisable effects, will then yield insights into hadronic physics. The current data point towards a fortunate situation for the determination of γ which is very robust with respect to U-spin-breaking corrections [36–38].

4. Outlook

The exploration of CP violation in B decays is a very broad field, with many other interesting topics complementing those discussed above. I would like to briefly give two more examples:

- The penguin decay $B_s^0 \to \phi\phi$ (for a recent theoretical discussion, see [42]). This summer, LHCb announced the first time-dependent angular analysis of CP violation in this channel [43]. The result $\phi_s = -0.17 \pm 0.15(\mathrm{stat}) \pm 0.03(\mathrm{syst}) = -(9.7 \pm 8.8)^\circ$ is consistent with the SM at the present level of precision. This channel has a lot of potential for the future.
- A promising decay for Belle II at SuperKEKB is the decay $B_d^0 \to \pi^0 K_{\rm S}$. A correlation between the direct and mixing-induced CP asymmetries of this channel can be predicted in the SM, with current data showing an intriguing discrepancy [44]. It will be interesting to monitor the future measurements of the corresponding observables at Belle II.

We are moving towards new frontiers in particle physics. There are still no unambiguous signals for NP at the LHC, and it is impressive – also frustrating – to see how the SM stands more and more stringent tests, both at the high-energy and at the high-precision frontier. Much more is yet to come with the future running of the LHC and dedicated studies of flavour physics, including CP violation in B decays. However, we have to prepare ourselves to deal with smallish NP effects in the data. In view of the increasing experimental precision, we have to be careful with respect to theoretical assumptions and approximations. The challenge will be the matching of the experimental and theoretical uncertainties in the future high-precision era. Interesting and fruitful years for the further testing of the SM and the search of NP are ahead of us!

Acknowledgements
I would like to thank Cristina Lazzeroni and her co-organizers for inviting me to this most interesting and enjoyable conference and their kind hospitality in Birmingham.

References

[1] Cabibbo N 1963 *Phys. Rev. Lett.* **10** 531
[2] Kobayashi M and Maskawa T 1973 *Prog. Theor. Phys.* **49** 652
[3] Buras A J and Girrbach J 2013 arXiv:1306.3775 [hep-ph]
[4] Buchalla G, Buras A J and Lautenbacher M E 1996 *Rev. Mod. Phys.* **68** 1125
[5] Bell G 2014 *these proceedings*
[6] Jüttner A 2014 *these proceedings*
[7] Fleischer R 2002 *Phys. Rept.* **370** 537
[8] Charles J *et al.* [CKMfitter Group] `http://ckmfitter.in2p3.fr`
[9] Bevan A *et al.* [UTfit Collaboration] `http://www.utfit.org/UTfit/`
[10] Bobeth C 2014 *these proceedings*
[11] Borissov G, Fleischer R and Schune M-H 2013 *Ann. Rev. Nucl. Part. Sci.* **63** 205
[12] Fleischer R 1999 *Eur. Phys. J.* C **10** 299
[13] Ciuchini M, Pierini M and Silvestrini L 2005 *Phys. Rev. Lett.* **95** 221804
[14] Faller S, Fleischer R, Jung M and Mannel T 2009 *Phys. Rev.* D **79** 014030
[15] Faller S, Fleischer R and Mannel T 2009 *Phys. Rev.* D **79** 014005
[16] De Bruyn K, Fleischer R and Koppenburg P 2010 *Eur. Phys. J.* C **70** 1025
[17] Ciuchini M, Pierini M and Silvestrini L 2011 arXiv:1102.0392 [hep-ph]
[18] Fleischer R, Knegjens R and Ricciardi G 2011 *Eur. Phys. J.* C **71** 1832
[19] Jung M 2012 *Phys. Rev.* D **86** 053008
[20] Aaltonen T *et al.* [CDF Collaboration] 2011 *Phys. Rev.* D **83** 052012
[21] Aaij R *et al.* [LHCb Collaboration] 2012 *Phys. Lett.* B **713** 172
[22] De Bruyn K and Fleischer R 2014 *in preparation.*
[23] Dighe A S , Dunietz I , Lipkin H J and Rosner J L 1996 *Phys. Lett.* B **369** 144
[24] Dighe A S, Dunietz I and Fleischer R 1999 *Eur. Phys. J.* C **6** 647
[25] Dunietz I, Fleischer R and Nierste U 2001 *Phys. Rev.* D **63** 114015
[26] Amhis Y *et al.* [Heavy Flavour Averaging Group] `http://www.slac.stanford.edu/xorg/hfag/`
[27] Aaij R *et al.* [LHCb Collaboration] 2012 *Phys. Rev.* D **86** 071102
[28] Aaij R *et al.* [LHCb Collaboration] 2013 *Eur. Phys. J.* C **73** 2373
[29] Stone S and Zhang L 2009 *Phys. Rev.* D **79** 074024; arXiv:0909.5442 [hep-ex].
[30] Fleischer R, Knegjens R and Ricciardi G 2011 *Eur. Phys. J.* C **71** 1798
[31] Aaij R *et al.* [LHCb Collaboration] 2014 *Phys. Rev.* D **90** 012003
[32] Fleischer R and Knegjens R 2011 *Eur. Phys. J.* C **71** 1789
[33] Lenz A 2012 arXiv:1205.1444 [hep-ph].
[34] Gandini P *talk at Beauty 2014*, 14–18 July 2014, Edinburgh, UK, `http://www.ph.ed.ac.uk/beauty2014`
[35] Fleischer R 1999 *Phys. Lett.* B **459** 306
[36] Fleischer R 2007 *Eur. Phys. J.* C **52** 267
[37] Fleischer R and Knegjens R 2011 *Eur. Phys. J.* C **71** 1532
[38] Knegjens R 2014 *PhD thesis* Vrije Universiteit Amsterdam
[39] Ciuchini M, Franco E, Mishima S and Silvestrini L 2012 *JHEP* **1210** 029
[40] Ciuchini M *talk at Beauty 2014*, 14–18 July 2014, Edinburgh, UK, `http://www.ph.ed.ac.uk/beauty2014`
[41] Aaij R *et al.* [LHCb Collaboration] 2013 *JHEP* **1310** 183
[42] Bhattacharya B, Datta A, Duraisamy M and London D 2013 *Phys. Rev.* D **88** 016007
[43] De Bruyn K 2014 *these proceedings*; Aaij R *et al.* [LHCb Collaboration] 2014 arXiv:1407.2222 [hep-ex]
[44] Fleischer R, Jäger S, Pirjol D and Zupan J 2008 *Phys. Rev.* D **78** 111501

Measurement of CP violation in the B_s^0 system

Kristof De Bruyn on behalf of the LHCb collaboration
Nikhef National Institute for Subatomic Physics,
Science Park 105, 1098 XG Amsterdam, The Netherlands
E-mail: debkr@nikhef.nl

Abstract. Studies of CP violation play a key role in the search for physics beyond the Standard Model. In the B_s^0 system these mainly consist of measurements of mixing-induced CP violation, which arises from the interference between the B_s^0–$\bar{B}_s^0$ mixing process and the subsequent decay. This report focusses on two such asymmetry measurements recently published by the LHCb collaboration. The first analysis presents the measurement of a CP violating phase in $B_s^0 \to \phi\phi$, which is found to be $\phi_{\phi\phi} = -0.17 \pm 0.15\,(\text{stat}) \pm 0.03\,(\text{syst})\,\text{rad}$. The second provides an updated measurement of the B_s^0–$\bar{B}_s^0$ mixing phase ϕ_s using the $B_s^0 \to J/\psi\pi^+\pi^-$ channel. It yields $\phi_s = 0.070 \pm 0.068\,(\text{stat}) \pm 0.008\,(\text{syst})\,\text{rad}$. Both results are consistent with the Standard Model expectations and no indication for direct CP violation was found in either decay mode.

1. Introduction

In the Standard Model (SM), CP violation arises through a single phase in the CKM quark mixing matrix. In decays of neutral B_s^0 mesons to a final state accessible to both B_s^0 and $\bar{B}_s^0$, the interference between the amplitude for the direct decay and the amplitude for decay via oscillation leads to time-dependent CP violation. The resulting CP asymmetry can be parametrised in terms of the convention-independent quantity

$$\lambda_f \equiv \frac{q}{p}\frac{A(\bar{B}_s^0 \to f)}{A(B_s^0 \to f)} = |\lambda_f|\, e^{i\phi_f} , \tag{1}$$

where q and p are complex parameters that relate the B_s^0 flavour and mass eigenstates to each other, and A is the decay amplitude. Where accurate SM predictions for λ_f are available, these CP asymmetries provide sensitive tests of the SM flavour sector and thus are excellent probes to search for signs of new physics, well beyond the reach of direct detection methods. Under the assumption that $|q/p| = 1$, as is consistent with current data [1], a deviation of $|\lambda_f|$ from unity would be a measure of direct CP violation in the $B_s^0 \to f$ mode. As direct CP violation is in general expected to be small, the primary target for mixing-induced CP violation measurements is the decay-mode-dependent complex phase ϕ_f.

This report presents two such asymmetry measurements: the first measurement of a CP violating phase in $B_s^0 \to \phi\phi$ [2] and an updated measurement of the B_s^0–$\bar{B}_s^0$ mixing phase ϕ_s from $B_s^0 \to J/\psi\pi^+\pi^-$ [3]. Both analyses are performed on a data sample corresponding to an integrated luminosity of $3.0\,\text{fb}^{-1}$ of pp collisions, recorded by the LHCb experiment in 2011 and 2012. The LHCb detector [4] is a single-arm forward spectrometer covering the pseudo-rapidity range $2 < \eta < 5$, designed for the study of particles containing b or c quarks. Its vertex locator and tracking system allow for the identification of the secondary vertices associated with the decay of B_s^0 mesons, and achieve a decay time resolution of about 40 fs. This is small enough to resolve the rapid B_s^0–$\bar{B}_s^0$ oscillations and allow for a time-dependent analysis of the above-mentioned modes.

2. Measurement of a CP violating phase in $B_s^0 \to \phi\phi$

The $B_s^0 \to \phi\phi$ decay originates from a $\bar{b} \to \bar{s}s\bar{s}$ transition. In the SM, such a transition is forbidden at tree level and can only proceed via higher order loop diagrams, where it is dominated by the so-called gluonic penguin topologies. The channel is therefore sensitive to possible new heavy degrees of freedom showing up in such loops. The presence of these new physics contributions can be constrained through a measurement of the complex phase $\phi_{\phi\phi}$, defined in Eq. (1). Due to a difference in the structure of the decay amplitude, this weak phase should not directly be compared to the CKM angle ϕ_s measured in $B_s^0 \to J/\psi h^+h^-$ decays. Instead, its SM value is limited to be smaller than [5]

$$\phi_{\phi\phi} \approx \left(S_{\phi\phi} \equiv \frac{2\,\mathrm{Im}[\lambda_{\phi\phi}]}{1 + |\lambda_{\phi\phi}|^2} \right) \leq 0.02\,. \tag{2}$$

The experimental measurement at LHCb [2] uses a topological b-trigger to identify potential $B_s^0 \to \phi\phi$ candidates from a four kaon final state. The subsequent event selection is based on a multivariate analysis trained on a combination of simulated data for the signal and data events from the B_s^0 mass sidebands for the background. Due to differences in the run conditions between 2011 and 2012 and in particular the pile-up, which affects the particle identification performance, the selection procedure is optimised for both years individually and yields 1185 ± 35 (2011) plus 2765 ± 57 (2012) reconstructed $B_s^0 \to \phi\phi$ decays.

$B_s^0 \to \phi\phi$ is a vector-vector final state and thus requires an angular analysis to disentangle the CP-even and CP-odd contributions to its decay amplitude. In addition, also contributions from non-resonant final states, split in a vector-scalar type (S-wave) and a scalar-scalar type (double S-wave) component, are included in the analysis. The angular analysis is performed in the helicity frame, illustrated in Fig. 1.

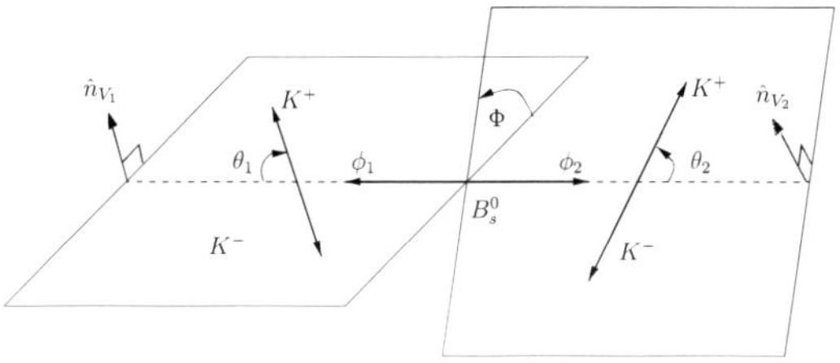

Figure 1. Helicity framework for the angular analysis of the $B_s^0 \to \phi\phi$ decay. The angle Φ, used in the construction of the two *triple product asymmetries*, is the angle between the decay planes of the two ϕ mesons.

For the time-dependent analysis, knowledge of the initial flavour of the B_s^0 meson is required. This analysis uses both information from the other b-quark in the event, referred to as opposite side tagging, as well as information from kaons that can be associated with the hadronisation of the signal b-quark, referred to as same side kaon tagging. The effective tagging power is measured to be $(5.33 \pm 0.37)\%$ for the 2011 data and $(5.44 \pm 0.30)\%$ for the 2012 data. The reader is referred to Ref. [2] for further details regarding the analysis and the construction of the used likelihood. The results from the time-dependent analysis of the data are

$$\begin{aligned} \phi_{\phi\phi} &= -0.17 \pm 0.15\,(\mathrm{stat}) \pm 0.03\,(\mathrm{syst})\ \mathrm{rad}\,, \\ |\lambda_{\phi\phi}| &= 1.04 \pm 0.07\,(\mathrm{stat}) \pm 0.03\,(\mathrm{syst})\,. \end{aligned} \tag{3}$$

The value of $\phi_{\phi\phi}$ is found to be compatible with the SM expectation.

In addition, also the T-odd *triple product asymmetries* can be used — in conjunction with the CPT theorem — to deduce information about CP violation in the $B_s^0 \to \phi\phi$ mode. These time-integrated quantities are defined in terms of the angular observables only. Starting from the trigonometric functions

$$U \equiv \sin(\Phi)\cos(\Phi)\,, \qquad V \equiv \sin(\pm\Phi)\,, \tag{4}$$

where Φ is the angle between the decay planes of the two ϕ mesons, as shown in Fig. 1, they are defined as

$$\begin{aligned} \mathcal{A}_U &\equiv \frac{\Gamma(U>0)-\Gamma(U<0)}{\Gamma(U>0)+\Gamma(U<0)}\,, \\ \mathcal{A}_V &\equiv \frac{\Gamma(V>0)-\Gamma(V<0)}{\Gamma(V>0)+\Gamma(V<0)}\,, \end{aligned} \tag{5}$$

and measured to be

$$\begin{aligned} \mathcal{A}_U &= -0.003 \pm 0.017\,(\text{stat}) \pm 0.006\,(\text{syst})\,, \\ \mathcal{A}_V &= -0.017 \pm 0.017\,(\text{stat}) \pm 0.006\,(\text{syst})\,. \end{aligned} \tag{6}$$

Also these results are compatible with the SM expectation.

3. Update on the measurement of ϕ_s in $B_s^0 \to J/\psi\pi^+\pi^-$

The B_s^0–$\bar{B}_s^0$ mixing phase ϕ_s is a gold plated probe to search for new physics. Its SM value is precisely known [6],

$$\phi_s^{\text{SM}} = -0.0364 \pm 0.0016\,\text{rad}\,, \tag{7}$$

and it is experimentally well accessible through the $B_s^0 \to J/\psi K^+K^-$ and $B_s^0 \to J/\psi\pi^+\pi^-$ decay modes. LHCb has recently updated its 1 fb^{-1} analysis [7] for the $B_s^0 \to J/\psi\pi^+\pi^-$ mode, which now includes the full LHC run 1 data [3].

The measurement sample uses high momentum muons to identify potential $B_s^0 \to J/\psi\pi^+\pi^-$ candidates from a $\mu^+\mu^-\pi^+\pi^-$ final state. The subsequent event selection is based on a multivariate analysis trained on a combination of simulated data for the signal and data events from the B_s^0 mass sidebands for the background. The final selection achieves a signal purity of 79.6% in the ± 20 MeV mass window around the B_s^0 peak, with a total signal yield of 27100 ± 200 events.

A important prerequisite for the time-dependent CP analysis of the $B_s^0 \to J/\psi\pi^+\pi^-$ mode is a precise determination of its CP content. An amplitude analysis in combination with information from the Dalitz plane was used [8] to identify the resonant contributions in the three-body final state. It was found that five interfering $\pi^+\pi^-$ states are required to describe the data. These include the dominant $f_0(980)$ as well as the $f_0(1500)$, $f_0(1790)$, $f_2(1270)$, and $f_2(1525)$. The resulting decomposition of the $\pi^+\pi^-$ invariant mass spectrum is shown in Fig. 2. The $B_s^0 \to J/\psi\pi^+\pi^-$ mode is consistent with being fully CP odd, where the CP even contribution is smaller than 2.3% at 90% C.L. Consequently, the full $\pi^+\pi^-$ mass spectrum can be used to perform the CP analysis.

For the time-dependent analysis knowledge of the initial flavour of the B_s^0 meson is required. Also here, both opposite side tagging and same side kaon tagging are used. The effective tagging power is measured to be $(3.89 \pm 0.25)\%$. The reader is referred to Ref. [3] for further details regarding the analysis and the construction of the used likelihood. Contrary to its $B_s \to J/\psi K^+K^-$ counterpart, the $B_s \to J/\psi\pi^+\pi^-$ analysis cannot determine $\Delta\Gamma_s$. Instead, the 1 fb^{-1} results from $B_s \to J/\psi K^+K^-$ [7],

$$\begin{aligned} \Gamma_s &= 0.663 \pm 0.005\,(\text{stat}) \pm 0.006\,(\text{syst})\,\text{ps}^{-1}\,, \\ \Delta\Gamma_s &= 0.100 \pm 0.016\,(\text{stat}) \pm 0.003\,(\text{syst})\,\text{ps}^{-1}\,, \end{aligned} \tag{8}$$

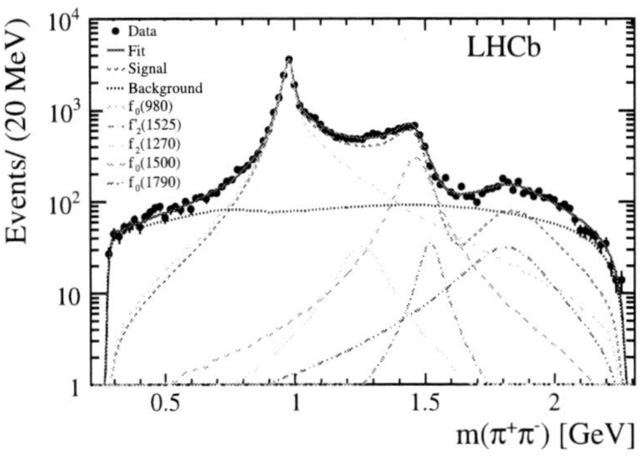

Figure 2. Decomposition of the $\pi^+\pi$ invariant mass spectrum for candidates in the ± 20 MeV mass window around the B_s^0 peak.

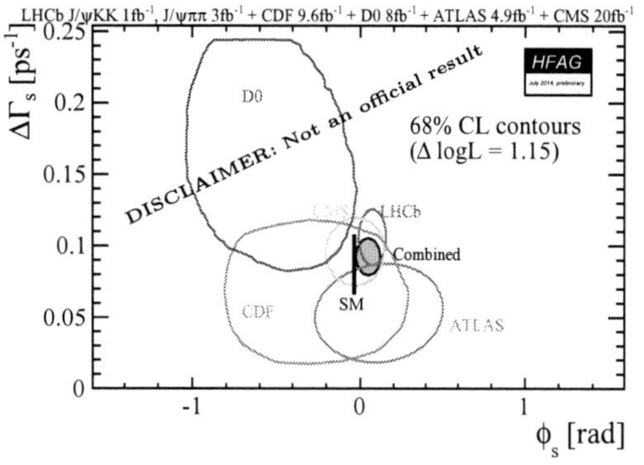

Figure 3. Unofficial overview of the current experimental constraints in the ϕ_s–$\Delta\Gamma_s$ plane, modified compared to the original [9] to include these results as well as the summer 2014 updates from Atlas [10] and CMS [11].

are used as Gaussian constraints in the likelihood. The results from the time-dependent analysis of the data are

$$\begin{aligned} \phi_s &= 0.070 \pm 0.068\,(\text{stat}) \pm 0.008\,(\text{syst})\ \text{rad}\,, \\ |\lambda| &= 0.89 \ \pm 0.05 \ \ (\text{stat}) \pm 0.01 \ \ (\text{syst})\,. \end{aligned} \tag{9}$$

The value of ϕ_s is found to be compatible with the SM expectation. An unofficial overview of the current experimental constraints in the ϕ_s–$\Delta\Gamma_s$ plane, modified compared to the original [9] to include these results as well as the summer 2014 updates from Atlas [10] and CMS [11], is shown in Fig. 3. This shows that new physics contributions, if present, are necessarily small and will thus be challenging to find. Further experimental improvements should help settle this matter.

References

[1] Beringer J et al [PDG] 2012 Review of particle physics *Phys. Rev.* D**86** 010001
[2] Aaij R et al [LHCb collaboration] 2014 Measurement of CP violation in $B_s^0 \to \phi\phi$ decays *Preprint* 1407.2222 [hep-ex]
[3] Aaij R et al [LHCb collaboration] 2014 Measurement of the CP-violating phase ϕ_s in $\bar{B}_s^0 \to J/\psi\pi^+\pi^-$ decays *Phys. Lett.* B**736** 186
[4] Alves Jr A et al [LHCb collaboration] 2008 The LHCb detector at the LHC *JINST* **3** S08005
[5] Bartsch M, Buchalla G and Kraus C 2008 $B \to V_L V_L$ Decays at Next-to-Leading Order in QCD *Preprint* 0810.0249 [hep-ph]
[6] Charles J et al [CKMfitter collaboration] 2011 Predictions of selected flavour observables within the Standard Model *Phys. Rev.* D**84** 033005
[7] Aaij R et al [LHCb collaboration] 2013 Measurement of CP-violation and the B_s^0-meson decay width difference with $B_s^0 \to J/\psi K^+K^-$ and $B_s^0 \to J/\psi\pi^+\pi^-$ decays *Phys. Rev.* D**87** 112010
[8] Aaij R et al [LHCb collaboration] 2014 Measurement of resonant and CP components in $\bar{B}_s^0 \to J/\psi\pi^+\pi^-$ decays *Phys. Rev.* D**89** 092006
[9] Amhis Y [HFAG collaboration] 2012 Averages of b-hadron, c-hadron, and τ-lepton properties as of early 2012 *Preprint* 1207.1158 [hep-ex]
[10] Aad G [Atlas collaboration] 2014 Flavour tagged time dependent angular analysis of the $B_s^0 \to J/\psi\phi$ decay and extraction of $\Delta\Gamma_s$ and the weak phase ϕ_s in ATLAS *Preprint* 1407.1796 [hep-ex]
[11] [CMS collaboration] 2014 Measurement of the CP-violating weak phase ϕ_s and the decay width difference $\Delta\Gamma_s$ using the $B_s \to J/\psi\phi(1020)$ decay channel *Preprint* CMS PAS BPH-13-012

CP violation measurements in B mesons at Belle

Zbyněk Drásal (on behalf of the Belle Collaboration)
Charles University in Prague, 147 00 Praha, Czech Republic
E-mail: Zbynek.Drasal@mff.cuni.cz

Abstract. We present recent measurements of time-dependent CP asymmetries in two neutral B decay modes: $B^0 \to \eta' K^0$ and $B^0 \to \omega K_S^0$. The measured values of mixing-induced ($\sin 2\phi_1^{\rm eff}$) and direct CP violation parameters are: $\mathcal{S}_{\eta' K^0} = +0.68 \pm 0.07_{\rm stat} \pm 0.03_{\rm syst}$, $\mathcal{A}_{\eta' K^0} = +0.03 \pm 0.05_{\rm stat} \pm 0.04_{\rm syst}$ and $\mathcal{S}_{\omega K_S^0} = +0.91 \pm 0.32_{\rm stat} \pm 0.05_{\rm syst}$, $\mathcal{A}_{\omega K_S^0} = -0.36 \pm 0.19_{\rm stat} \pm 0.05_{\rm syst}$, respectively. These results are fully consistent with the current world average of $\sin 2\phi_1$ measured in $b \to c\bar{c}s$ transitions, and represent the most accurate measurements in these modes up-to-date. In addition, we present a new preliminary update on branching ratio measurement in $B^0 \to \pi^0\pi^0$ decay mode, the result of which: $\mathcal{B}(B^0 \to \pi^0\pi^0) = (0.90 \pm 0.12_{\rm stat} \pm 0.10_{\rm syst}) \times 10^{-6}$ represents an important ingredient in a $\pi\pi$ SU(2) isospin analysis, used to experimentally extract ϕ_2 angle from $B \to \pi\pi$ measurements. To obtain CP violation parameters ($\mathcal{B}(B^0 \to \pi^0\pi^0)$) the final data set of 772×10^6 (751.5×10^6) $B\bar{B}$ pairs collected at the $\Upsilon(4S)$ resonance with the Belle detector at the KEKB asymmetric e^+e^- collider in Japan ($3.5 \times 8.0\,$GeV) is used.

1. Introduction
Within the Standard Model (SM) the CP violation is established via charged weak currents and three generation Cabbibo-Kobayashi-Maskawa quark mixing matrix (CKM) [1], namely its single irreducible complex phase. Requirement on CKM unitarity gives rise to six triangles in the complex plane, one of which naturally describes the b-quark transitions and is characterized by three measurable quantities (angles): ϕ_1, ϕ_2 and ϕ_3. The B-factories with their experimental apparatuses, Belle [2] and BaBar, were designed to test the CKM mechanism. Particularly, one of the main tasks was an experimental determination of the angles together with the sides of the unitarity triangle. In the past decade, both Belle and BaBar clearly observed the CP violation in all statistically significant $b \to c\bar{c}s$ induced decays [3, 4]. The experimental outcome is a precise determination of $\sin 2\phi_1 = 0.667 \pm 0.023_{\rm stat} \pm 0.012_{\rm syst}$ (Belle result) [3] via time-dependent CP asymmetries in B^0 decays. Despite the fact that these measurements proved CKM predictions to be extremely successful, the matter-antimatter asymmetry (as incorporated in SM via CP violation) is well below what Universe observations require to solve the baryon asymmetry puzzle [5]. To search for new sources of CP violation, experimental effort has gradually focused on other quark transitions, e.g. $b \to q\bar{q}s$. They dominantly proceed via loop Feynman diagram, and thus can be regarded highly sensitive to a possible new heavy particle(s) contribution.

2. Recent ϕ_1 related Belle measurements
In B factories the B mesons are produced coherently from a $\Upsilon(4S)$ resonance. The $\Upsilon(4S)$ decays almost exclusively to a $B\bar{B}$ pair, where one $B^0_{\rm rec}$ may be reconstructed in a studied decay mode while the other $\bar{B}^0_{\rm tag}$ may reveal its flavour. Using a technique called time-dependent CP

analysis, one can extract CP violation parameters by measuring a CP asymmetry evolving in time. For a B decaying into a given CP final state f the CP asymmetry follows as:

$$a_{CP}(\Delta t) = \mathcal{A}_{\rm f}\cos(\Delta m_{\rm d}\Delta t) + \mathcal{S}_{\rm f}\sin(\Delta m_{\rm d}\Delta t), \tag{1}$$

where Δt stands for the proper time interval between $B^0_{\rm rec}$ and $\bar{B}^0_{\rm tag}$, $\Delta m_{\rm d}$ ($\Delta\Gamma_{\rm d}$) is the difference between the masses (decay rates) of heavy and light B^0 mass eigenstates, and $\mathcal{A}_{\rm f}$ and $\mathcal{S}_{\rm f}$ represent direct and mixing-induced CP violation parameters, respectively. This relation assumes CPT invariance, perfect tagging and that $\Delta\Gamma_{\rm d} \sim 0$. For $b \to c\bar{c}s$ transitions, the SM predicts that $\mathcal{S}_{\rm f} = -\xi\sin 2\phi_1$ (with a very small theoretical uncertainty), where $\xi = -1(+1)$ for CP-odd (CP-even) final state. The ϕ_1 angle is related to the CKM elements as: $\phi_1 \equiv \arg[-V_{\rm cd}V^*_{\rm cb}/V_{\rm td}V^*_{\rm tb}]$.

In contrast to $c\bar{c}s$ transitions, $B^0 \to \eta' K^0$ and $B^0 \to \omega K^0_S$ represent physics decays induced by $b \to u\bar{u}s$ and $b \to d\bar{d}s$ transitions, where a penguin graph plays a dominant role. Nevertheless, a small pollution by CKM and colour-suppressed tree processes occurs and the CP violation parameters $\mathcal{S}_{\rm f}$ and $\mathcal{A}_{\rm f}$ may slightly deviate from the nominal values measured in $c\bar{c}s$ transitions. We expect $\mathcal{S}_{\rm f} \simeq -\xi\sin 2\phi_1$, $\mathcal{A}_{\rm f} \sim 0$. Therefore, observing a large deviation (further denoted as $\Delta\mathcal{S}_{\rm f}$) would be a clear sign for New Physics (NP). For the $B^0 \to \eta' K^0$ the SM predicts $\Delta\mathcal{S}_{\rm f}$ to be in the range: $[-0.05; +0.09]$ ($SU(3)_{\rm F}$ [6]) or $[-0.03; +0.03]$ (QCD factorization scheme [7]). For the $B^0 \to \omega K^0_S$ the SM predicts $\Delta\mathcal{S}_{\rm f}$ to be within $[+0.1; +0.2]$ (pQCD, QCD$_{\rm F}$, ... [8]).

2.1. *Measurement of CP violation parameters in $B^0 \to \eta' K^0$*

The overall reconstruction is performed separately in two CP eigenstates: $B^0 \to \eta' K^0_S$ ($CP = -1$) and $B^0 \to \eta' K^0_L$ ($CP = +1$) [9]. After a signal/background shape is determined, both channels are combined together and a common CP fit is performed. In order to increase the signal yield, several η' decay chains are used to reconstruct B^0 candidates: $\eta' \to \eta\pi^+\pi^-$ ($\eta \to \gamma\gamma$ or $\eta \to \pi^+\pi^-\pi^0$) and $\eta' \to \rho^0\gamma$ ($\rho^0 \to \pi^+\pi^-$). The last decay chain is not used in $B^0 \to \eta' K^0_L$ reconstruction due to its low signal-to-background ratio.

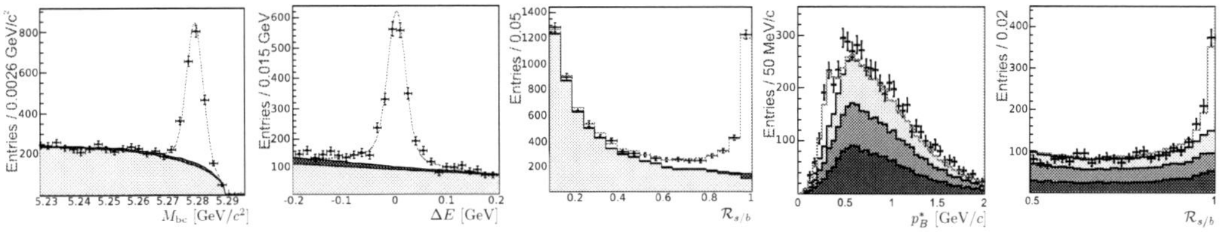

Figure 1. From left to right: Signal-enhanced $M_{\rm bc}$, ΔE and $\mathcal{R}_{\rm S/B}$ distributions of $B^0 \to \eta' K^0_S$ candidates (red - total PDF, yellow - $q\bar{q}$ background, blue - $B\bar{B}$ background); signal-enhanced p^*_B and $\mathcal{R}_{\rm S/B}$ distributions of $B^0 \to \eta' K^0_L$ candidates (red - total PDF, blue, green and yellow - combinatorial background with fake η', real η' with fake K^0_L, and both real η' with real K^0_L, respectively). All analysed η' decay modes are combined in the plots.

In order to find the signal/background fraction, several event-distributions are simultaneously fitted (see Fig. 1). In $\eta' K^0_S$ analysis the following variables are used: a beam-constraint mass, $M_{\rm bc} = \sqrt{(E^{\rm CMS}_{\rm beam})^2 - (p^{\rm CMS}_B)^2}$, difference between the B meson energy and beam energy, $\Delta E = E^{\rm CMS}_B - E^{\rm CMS}_{\rm beam}$, and $q\bar{q}$ (continuum) suppressing variable ($q = u, d, s, c$) - likelihood ratio, $\mathcal{R}_{\rm S/B}$. Similarly, in $\eta' K^0_L$ decay the yield is extracted by $\mathcal{R}_{\rm S/B}$ and the B^0 momentum in the CMS, $p{*}_B$, which is calculated using η' momentum and the K^0_L flight direction, assuming $\Delta E = 0$. In total, $3541 \pm 90_{\rm stat}$ signal candidates are reconstructed. Afterwards, the values of $\sin 2\phi^{\rm eff}_1$ and

$\mathcal{A}_{\eta' K^0}$ are obtained by performing an unbinned, maximum-likelihood (uML) fit to the $(\Delta t; q)$ distributions of the signal region events (see Fig. 2). In the plots $q = +1$ and $q = -1$ stand for the B flavour on the tag side B^0 and $\bar{B}^0$, respectively. The new result, superseding previous Belle

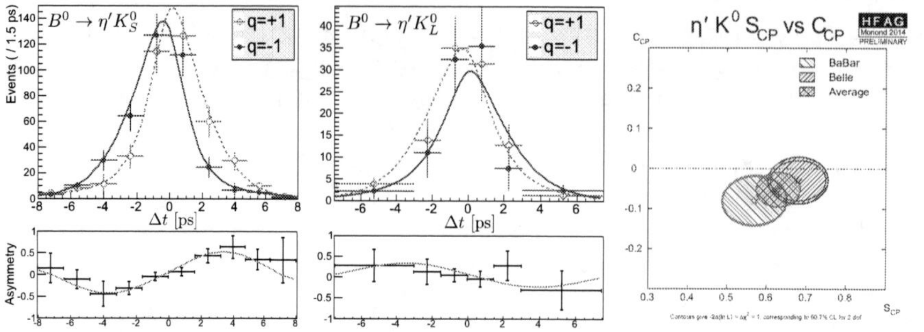

Figure 2. Data $(\Delta t; q)$ fit distributions for $\eta' K^0_S$ (left) and $\eta' K^0_L$ (middle). Plot combining results from Belle and BaBar by Heavy Flavour Averaging Group (HFAG) is on the right [10].

measurements [11], is: $\mathcal{S}_{\eta' K^0} = +0.68 \pm 0.07_{\text{stat}} \pm 0.03_{\text{syst}}$ and $\mathcal{A}_{\eta' K^0} = +0.03 \pm 0.05_{\text{stat}} \pm 0.04_{\text{syst}}$, which is fully consistent with the older Belle measurement [11], BaBar result [12] and the SM predictions, see the HFAG plot in Fig. 2.

2.2. Measurement of CP violation parameters in $B^0 \to \omega K^0_S$

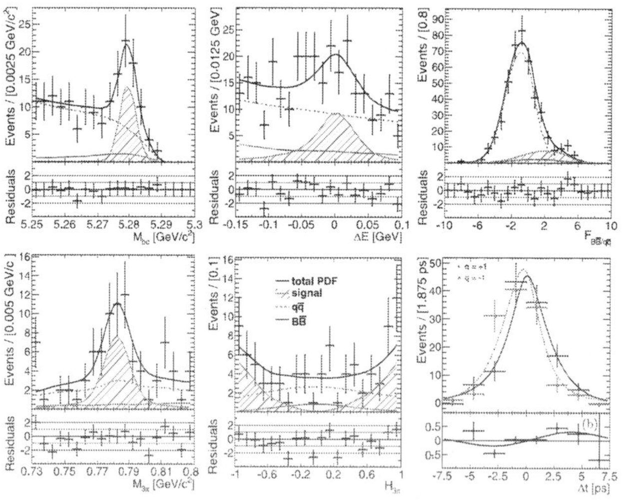

Figure 3. From left to right: Signal-enhanced fit projections in: M_{bc}, ΔE, Fisher $(F_{B\bar{B}/q\bar{q}})$, ω invariant mass $(M_{3\pi})$, ω helicity $(H_{3\pi})$ and $(\Delta t; q)$. The black curve corresponds to the total PDF, green to signal PDF, dashed red to $q\bar{q}$ background, blue dotted to $B\bar{B}$ background.

In $B^0 \to \omega K^0_S$, the CP parameters are directly obtained from a simultaneous seven-dimensional uML fit to signal/background distributions: M_{bc}, ΔE, invariant mass of $\omega \to \pi^+\pi^-\pi^0$, ω helicity angle, $\mathcal{H}_{3\pi}$, Fisher discriminant, $\mathcal{F}_{B\bar{B}/q\bar{q}}$ (used to suppress continuum), and fit to $(\Delta t; q)$ distribution (Fig. 3). In total, $234 \pm 22_{\text{stat}}$ signal candidates are reconstructed [13].

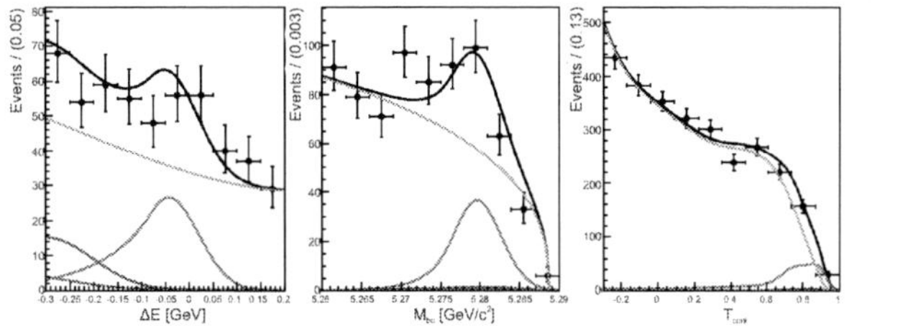

Figure 4. Signal-enhanced fit projections: ΔE (left), $M_{\rm bc}$ (middle), $T_{\rm cont}$ (right). Total PDF is displayed in black, signal in red, $q\bar{q}$ in green, rare B decays in brown and $B \to \rho\pi$ in blue.

The final result can be summarized as: $\mathcal{S}_{\omega K^0_S} = +0.91 \pm 0.32_{\rm stat} \pm 0.05_{\rm syst}$ and $\mathcal{A}_{\omega K^0_S} = -0.36 \pm 0.19_{\rm stat} \pm 0.05_{\rm syst}$. By performing a likelihood scan, a $\mathcal{S}_{\omega K^0_S}$ deviation of $3.1\,\sigma$ from the CP conservation hypothesis is found. That makes this result the first evidence of CP violation in this mode. In addition, the values are in a good agreement with previous Belle [14] and BaBar [12] results, so no evidence of NP is observed at the current level of exp. sensitivity.

3. New preliminary update on branching fraction measurement in $B^0 \to \pi^0\pi^0$

This measurement is a preliminary update on a previous Belle result: $\mathcal{B} = (2.3^{+0.4+0.2}_{-0.5-0.3}) \times 10^{-6}$ [15], and represents an important ingredient in a $\pi\pi$ SU(2) isospin analysis [16]. This analysis is applied to disentangle contribution of $\Delta\phi_2$ in $S_{\rm f} \sim \sin(2\phi_2 + 2\Delta\phi_2)$ relation, and thus to experimentally extract ϕ_2 angle ($\phi_2 \equiv \arg[-V_{\rm td}V^*_{\rm tb}/V_{\rm ud}V^*_{\rm ub}]$) from $B \to \pi\pi$ measurements [17]. The theoretical prediction based on QCD factorization scheme is $\mathcal{B} < 1.0 \times 10^{-6}$ [18].

The analysis strategy consisted in a sequential reconstruction of four gammas, two neutral pions and B^0 candidates. The event-distributions used in a final uML fit are: $M_{\rm bc}$, ΔE and a $q\bar{q}$ suppression variable, $\rm T_{cont}$. $\rm T_{cont}$ is the output of a Fisher discriminant which combines event shape variables: $\cos\vartheta$ of the $\pi^0\pi^0$ system to the B^0-flight direction and $\cos\vartheta_{\rm thrust}$ of the reconstructed B^0 compared to the thrust of the rest of the event. In the distributions (see Fig. 4) the continuum component forms the largest background contribution. The Belle new result is: $\mathcal{B}(B^0 \to \pi^0\pi^0) = (0.90 \pm 0.12_{\rm stat} \pm 0.10_{\rm syst}) \times 10^{-6}$, measured with $6.7\,\sigma$ significance.

References

[1] Cabibbo N 1963, *Phys. Rev. Letters* **10**, 531; Kobayashi M, Maskawa T 1973, *Prog. Theor. Phys.* **49**, 652
[2] Abashian A *et al.* (Belle Collab.) 2002, *Nucl. Instrum. Methods Phys. Res.* A **479**, 117
[3] Adachi I *et al.* (Belle Collab.) 2012, *Phys. Rev. Letters* **108**, 171802
[4] Aubert B *et al.* (BaBar Collab.) 2009, *Phys. Rev.* D **79**, 072009
[5] Gavela M B, Hernández P, Orloff J and Pène O 1994, *Mod. Phys. Lett.* A **9**, 795
[6] Gronau M, Rosner J L and Zupan J 2006, *Phys. Rev.* D **74**, 093003
[7] Beneke M 2005, *Phys. Lett.* B **620**, 143; Williamson A R and Zupan J 2006, *Phys. Rev.* D **74**, 014003; Cheng H-Y, Chua C-K and Soni A 2005, *Phys. Rev.* D **72**, 014006
[8] Akeroyd A G, Chen C H and Geng C Q 2007, *Phys. Rev.* D **75**, 054003; Li H-N and Mishima S 2006, *Phys. Rev.* D **74**, 094020; Cheng H-Y and Chua C-K 2009, *Phys. Rev.* D **80**, 114008
[9] Santelj L *et al.* (Belle Collab.) 2014, arXiv:1408.5991 (submitted to *JHEP*)
[10] Amhis Y *et al.* Averages of b-hadron, c-hadron, and tau-lepton properties as of early 2012, arXiv:1207.1158
[11] Chen K-F *et al.* (Belle Collab.) 2007, *Phys. Rev. Lett.* **98**, 031802
[12] Aubert B *et al.* (BaBar Collab.) 2009, *Phys. Rev.* D **79**, 052003
[13] Chobanova V *et al.* (Belle Collab.) 2014, *Phys. Rev.* D **90**, 012002
[14] Chao Y *et al.* (Belle Collab.) 2007, *Phys. Rev.* D **76**, 091103(R)
[15] Chao Y *et al.* (Belle Collab.) 2005, *Phys. Rev. Lett.* **94**, 181803
[16] Gronau M and London D 1990, *Phys. Rev. Lett* **65**, 3381
[17] Dalseno J *et al.* (Belle Collab.) 2013, *Phys. Rev.* D **88**, 092003
[18] Beneke M, Neubert M 2003, *Nucl. Phys.* B **675**, 333; Keum Y-Y, Sanda A I 2003, *Phys. Rev.* D **67**, 54009

Measurements of B^0 and B_s^0 mixing frequencies at LHCb

Giulia Tellarini

INFN and University of Ferrara, Department of Physics, via Saragat 1, Ferrara, IT

E-mail: giulia.tellarini@cern.ch

Abstract. This document describes the measurements of the B^0 and B_s^0 mixing frequencies performed at LHCb using the data sample corresponding to an integrated luminosity of 1 fb^{-1} collected during 2011. The knowledge of the mixing frequencies provides a constraint on the apex of the CKM unitarity triangle and is a key ingredient for several CP-violation measurements. Analyses of the flavour oscillations allow also to measure the B^0 and the B_s^0 production asymmetries.

1. Introduction

The $B^0_{(q)}$-$\bar{B}^0_{(q)}$ mixing (q=d,s), described in the SM by box Feynman diagrams, are due to the misalignment of the flavour eigenstates and the mass eigenstates. The latter, denoted as B_{H^q} and B_{L^q}, have a mass difference $\Delta m_q = m_{H^q} - m_{L^q}$, which corresponds to the mixing frequency. The theoretical predictions on Δm_q are affected by uncertainties due to hadronic parameters entering the calculations, which partially cancel in the ratio $\frac{\Delta m_s}{\Delta m_d}$, allowing to constraint the apex of the CKM unitarity triangle. The mixing frequencies are measured through the time-dependent mixing asymmetry:

$$A(t) = \frac{N_{unmix}(t) - N_{mix}(t)}{N_{unmix}(t) - N_{mix}(t)} \approx \cos(\Delta m t)$$

where $N_{unmix}(t)$ is the number of events for which the flavour at the decay time and the production time is the same, while $N_{mix}(t)$ is the number of events that have a different flavour at decay time and production time. The flavour of the B meson at the decay time is known from the charge of the final state particles in flavour specific decays; in LHCb, measurements of the mixing frequencies are performed with the following decay modes: $B_s^0 \to D_s^- \pi^+$, $B^0 \to D^- \pi^+$, $B^0 \to J/\psi K^{*0}$, and $B^0_{(s)} \to D^-_{(s)} \mu^+ \nu$. The knowledge of the flavour of the B meson at the production time is inferred by means of flavour tagging algorithms, which are classified in two categories: the Same Side (SS) and the Opposite Side (OS) taggers [?]. The SS taggers exploit the charge of the fragmentation particles accompanying the B signal meson (SSK for B_s^0 and SSπ for B^0). The OS tagging algorithms use an inclusive reconstruction of the other b-hadron produced in the pp interaction (OSe, OSμ, OSK, OSVtxCharge) where the charge of the tagging particle is correlated with the flavour of the B meson. For each B candidate the tagging algorithms provide a decision on the flavour and a probability of such decision being wrong, namely mistag probability. If more than one tagger is available, it is possible to combine the tagging decisions and the mistag probabilities to obtain a final decision and a final mistag. In the following analyses the OS and the SS informations are combined in order to increase the tagging performance. The figure of merit of the tagging algorithms is called the effective tagging power $\epsilon_{eff} = \epsilon_{tag}(1 - 2\omega)^2$, where ϵ_{tag} is the tagging efficiency and ω is the mistag probability. It gives the factor by which the statistical power is reduced due to imperfect tagging decisions. The sensitivity to the mixing frequency is reduced by the

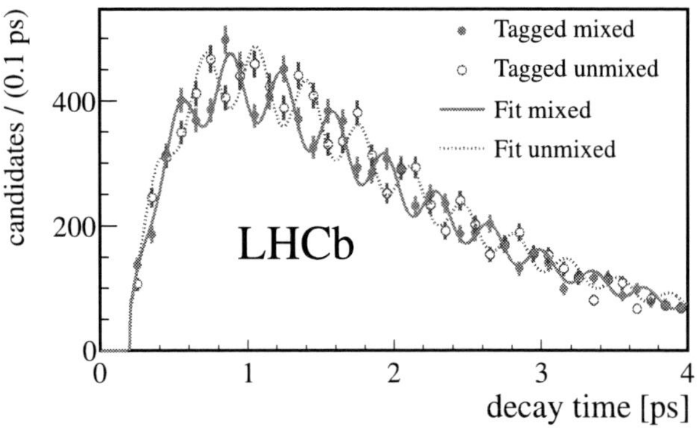

Figure 1. Decay time distribution fit for mixed and unmixed events in the $B_s^0 \to D_s^- \pi^+$ decay [?].

mistag probability which dilutes the mixing frequency by O(60-80%), depending on the decay mode. Another source of dilution is due to the resolution of the decay time measurement; the good capabilities of the LHCb tracking system allow to achieve a decay time resolution of O(50) fs for fully reconstructed decay, to be compared with the fast oscillation of the B_s^0 mesons, which have a period of 350 fs.

2. Mixing Frequency Measurements

In what follows, I report the measurements of the mixing frequencies obtained in a data set corresponding to an integrated luminosity of 1 fb^{-1} collected by the LHCb detector in proton-proton collisions at 7 TeV center-of-mass energy.

2.1. Δm_s in $B_s^0 \to D_s^- \pi^+$

The measurement of Δm_s in the $B_s^0 \to D_s^- \pi^+$ decay mode is described in Ref.[?]. The measurement is obtained through a maximum likelihood fit to the B_s^0 invariant mass, decay time and tagging decision for subsamples corresponding to 5 D_s decay modes ($D_s \to \phi\pi$, $D_s \to K^*K$, $D_s \to K\pi\pi$ non resonant, $D_s \to KK\pi$, $D_s \to \pi\pi\pi$). The total probability density function (PDF) is $PDF(m)PDF(t, q|\sigma_t, \eta) \cdot P(\sigma_t)P(\eta)$, where q is the tagging decision and η the mistag probability for each event from the combination of the OS and SSK taggers. The time resolution, σ_t, is estimated event by event and calibrated on a large sample of D_s decays. The corresponding measured average effective time resolution $< \sigma_t >$ is 44 fs. Some of the selection requirements on variables correlated with the decay time introduce an efficiency as a function of the decay time (decay time acceptance), which is parametrized from simulations. The measured signal yield is 34 000 events. The effective tagging power ϵ_{eff} is 2.6% for the OS and 1.2% for the SSK. In figure **??** the fit to the decay time distribution for mixed and unmixed events is shown in the signal mass range ($5320 < m_{B_s} < 5550$). The B_s^0 fast oscillations are well resolved and allow to determine the value of $\Delta m_s = 17.768 \pm 0.023$ (stat.) $\pm$ 0.006 (syst.) ps^{-1}. The main systematic sources are due to the uncertainty on the detector alignment: the longitudinal scale (0.004 ps^{-1}) and the overall momentum scale (0.004 ps^{-1}). This is the best Δm_s measurement to date.

2.2. Combined measurement of Δm_d in $B^0 \to D^- \pi^+$ and in $B^0 \to J/\psi K^{*0}$

The measurements of Δm_d in the $B^0 \to D^- \pi^+$ and in the $B^0 \to J/\psi K^{*0}$ decay modes are described in [?]. The two analyses exploit a maximum likelihood fit to the B^0 invariant mass, decay time and tagging decision (combining the OS and the SSπ taggers). A single gaussian with a width of 50 fs is

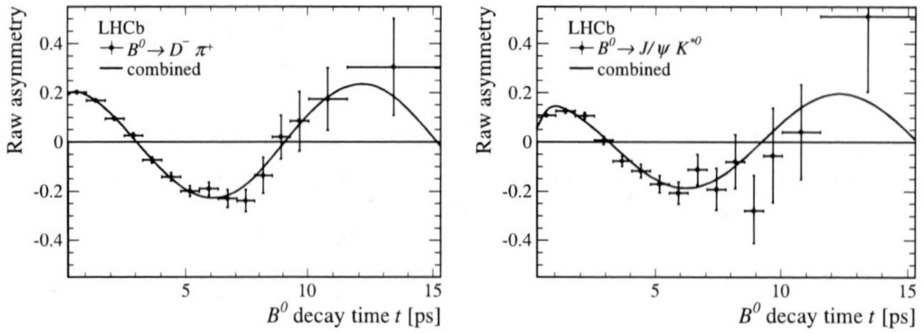

Figure 2. Mixing asymmetry fit in $B^0 \to D^-\pi^+$ on the left and in $B^0 \to J/\psi K^{*0}$ on the right [**?**].

used to model the time resolution effects. For the decay time acceptance the $B^0 \to D^-\pi^+$ analysis extracts the parametrization using simulated events while the $B^0 \to J/\psi K^{*0}$ uses a specific data sample of $B^0 \to J/\psi K^{*0}$ in which the events are collected without applying any of the decay time biasing selections. In the $B^0 \to D^-\pi^+$ analysis the D^- meson is reconstructed in $K^+\pi^-\pi^-$ and the measured signal yield is 87 724 $\pm$ 321 events in a mass window of 5200 $< m_B <$ 5450. On the left of figure **??** the fit to the mixing asymmetry is reported from which a mixing frequency of Δm_d = 0.5178 $\pm$ 0.0061 (stat.) $\pm$ 0.0037 (syst.) ps^{-1} is extracted. In the $B^0 \to J/\psi K^{*0}$ analysis the J/ψ is reconstructed in the di-muon final state and the K^{*0} in the $K^+\pi^-$. The corresponding statistics is 39 148 $\pm$ 316 signal events and the OS and SSπ tagging performances are a slightly worse ($\epsilon_{eff}(\%)$ in $B^0 \to J/\psi K^{*0}$ is 2.0 for OS and 0.6 for SSπ while in $B^0 \to D^-\pi^+$ is 3.0 for OS and 1.32 for SSπ); in addition the background fraction in the signal region is higher. The measured Δm_d value is Δm_d = 0.5096 $\pm$ 0.0114 (stat.)$\pm$ 0.0022 (syst.) ps^{-1}. The larger statistical error in $B^0 \to J/\psi K^{*0}$ is due to the lower statistics, the worse tagging performance and a higher background fraction in the signal region. On the right side of figure **??** the mixing asymmetry fit is shown. For both measurements the most relevant systematic effect is due to the fit model: 0.0037 ps^{-1} for $B^0 \to D^-\pi^+$ and 0.0022 ps^{-1} for $B^0 \to J/\psi K^{*0}$. The combined result is Δm_d = 0.5156 $\pm$ 0.0114 (stat.) $\pm$ 0.0022 (syst.) ps^{-1}. This is the most precise combination of Δm_d results from a single experiment.

2.3. Δm_d and Δm_s in semi-leptonic decays

An analysis of $B^0_{(s)} \to D^-_{(s)}\mu^+\nu_\mu X$ decays where $D^-_{(s)} \to KK\pi$ is presented in [**?**]. The measurements of Δm_d and Δm_s are closely related and represent the first observation of B^0_s mixing using only semi-leptonic decays. The analysis proceeds through a simultaneous fit to the D^-_s measured mass, the $B^0_{(s)}$ decay time and the tagging decision (only from OS for the B^0 while for B^0_s both the OS and SSK information is used). In semi-leptonic decays the reconstructed momentum is systematically low due to the escape of the neutrino which is not detected. Consequently the estimation of the decay time is biased. A correction based on simulation ("k-factor") is applied event by event. As a consequence the time resolution is worse compared to fully reconstructed modes and it depends on the decay time. For the $B^0_s \to D^-_s\mu^+X$ the measured yield is $\approx$ 201 200 events and the B^0_s mixing frequency is found to be Δm_s = 17.93 $\pm$ 0.22 (stat.) $\pm$ 0.15 (syst.) ps^{-1}. The left of figure **??** shows the fit to the mixing asymmetry for the B^0_s events. The oscillations are visible at small time while at larger time they are dumped by the time resolution effect. For this reason even if the yield is larger than the previous $B^0_s \to D^-_s\pi^+$ analysis, the statistical precision is worse. For the $B^0 \to D^-\mu^+X$ the yield is $\approx$ 50 700 events and the B^0 mixing frequency is equal to: Δm_d = 0.503 $\pm$ 0.011 (stat.) $\pm$ 0.014 (syst.) ps^{-1}. On the right of figure **??** the fit to the B^0 mixing asymmetry is shown. Several systematic effects are studied for both the measurements. The most relevant ones for Δm_s are connected to the time resolution model (0.09 ps^{-1}),

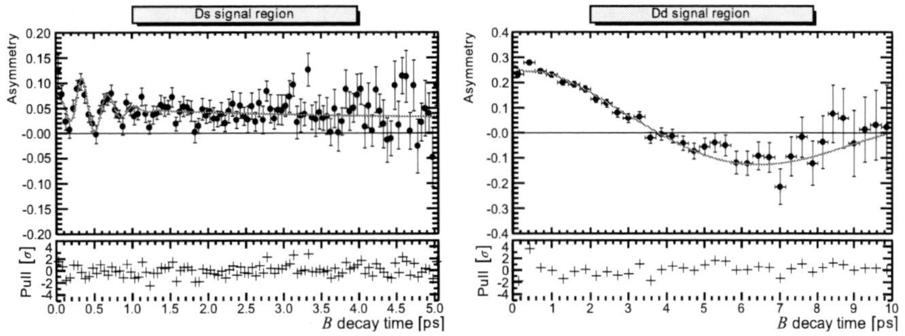

Figure 3. Fit to the mixing asymmetry for B_s^0 on the left and for B^0 on the right

the k-factor correction (0.09 ps^{-1}) and the simulation input for the k-factor evaluation (0.06 ps^{-1}). The Δm_d measurement is mostly affected by the parametrization of the background (0.008 ps^{-1}), the time resolution model (0.007 ps^{-1}) and the k-factor correction (0.0055 ps^{-1}).

3. B^0 and B_s^0 Production Asymmetries

The $B^0_{(s)}$-$\bar{B}^0_{(s)}$ production asymmetry in proton-proton collisions represents an important ingredient for CP violation analyses since CP asymmetries must be disentangled from other sources. The measurements of the B^0 and B_s^0 production asymmetry, A_P, are reported in [?]. LHCb has measured these quantities by using data collected at 7 TeV during the 2011 run and by analyzing the flavour specific decays: $B^0 \to D^-\pi^+$ and $B^0 \to J/\psi K^{*0}$ for $A_P(B^0)$ and $B_s^0 \to D_s^-\pi^+$ for $A_P(B_s^0)$. No flavour tagging is necessary. The analysis is based on a simultaneous fit to the B invariant mass and decay time in which the decay rates depend on the production asymmetry:

$$\Gamma_{B^0_{(s)}}(t) \approx (1 - f(A_{CP} + A_{det}))e^{-\Gamma_{(s)}t} \cdot [\cosh(\frac{\Delta\Gamma_{(s)}t}{2}) - f \cdot A_P \cos(\Delta m_{(s)}t)]$$

where A_{CP} is the CP asymmetry and it is assumed to be zero in these specific decays. A_{det} is the detection asymmetry and is a free parameter of the fit. The mixing frequencies are constrained: for Δm_d the world average value Δm_d = 0.510 $\pm$ 0.004 ps^{-1} [?] is used and for Δm_s the LHCb measurement Δm_s = 17.768 $\pm$ 0.024 ps^{-1} [?] is used. The production asymmetries are measured in bins of transverse momentum of the B meson and pseudorapidity; no dependence is observed. An integrated measurement of A_P is performed in the ranges $4 < p_T < 30$ GeV/c, $2.5 < \eta < 4.5$. The measured production asymmetries are $A_P(B^0)$ = -0.35 $\pm$ 0.76 (stat.) $\pm$ 0.28 (syst.) % and $A_P(B_s^0)$ = 1.09 $\pm$ 2.61 (stat.) $\pm$ 0.61 (syst.)%. Both the measurements are compatible with zero.

[1] LHCb Collaboration, *Performance of flavour tagging algorithms optimised for the analysis of* $B_s^0 \to J/\psi\phi$., Eur. Phys. J. C 72 (2012) 2022
[2] LHCb Collaboration, New J. Phys. 15 (2013) 053021.
[3] LHCb Collaboration, Phys. Lett. B 719 (2013) 318-325.
[4] LHCb Collaboration, Eur. Phys. J. C 73 (2013) 2655.
[5] LHCb Collaboration, *arXiv:1408.0275* Submitted to Phys. Lett. B
[6] Heavy Flavour Averaging Group, Y. Amhis et al., *arXiv:1207.1158, update at http://www.slac.stanford.edu/xorg/hfag*

XI International Conference on Hyperons, Charm and Beauty Hadrons (BEACH 2014) IOP Publishing
Journal of Physics: Conference Series **556** (2014) 012039 doi:10.1088/1742-6596/556/1/012039

FLAVOUR(267104)-ERC-79

Status of B-mixing in the standard model and beyond

Christoph Bobeth
Technische Universität München,
Institute for Advanced Study,
D-85748 Garching, Germany

E-mail: christoph.bobeth@ph.tum.de

Abstract. The theoretical methods underlying predictions of mixing quantities ΔM_q and $\Delta\Gamma_q$ are well understood nowadays. Currently the uncertainties are dominated by nonperturbative matrix elements of local $|\Delta B| = 2$ operators of dimension 6 and 7, which require preciser calculations, hopefully with lattice methods in the future. Global fits show good support of the standard model, restricting potential effects of new physics to be small. An improved measurement of $\Delta\Gamma_d$ will be helpful to further analyse the dimuon charge asymmetry as measured by DØ.

1. Introduction

The phenomenon of neutral B_q meson mixing ($q = d, s$) proceeds in the standard model (SM) only at the loop level and provides important constraints in the determination of elements of the Cabibbo-Kobayashi-Maskawa (CKM) quark mixing matrix, $V_{tb}V_{tq}^*$. Moreover, the loop suppression makes it sensitive to new sources of CP and flavour violation. It affects CP violation in all neutral B_q meson decays in the form of mixing-induced CP asymmetries. In the absence of reliable methods for the calculation of strong (CP-conserving) phases of decay amplitudes, mixing thus provides an important additional access to CP violating parameters.

The time evolution of the two flavour eigenstates B_q and $\overline{B}_q$ is described by

$$i\frac{d}{dt}\begin{pmatrix} |B_q(t)\rangle \\ |\overline{B}_q(t)\rangle \end{pmatrix} = \left(M^q - \frac{\Gamma^q}{2}\right)\begin{pmatrix} |B_q(t)\rangle \\ |\overline{B}_q(t)\rangle \end{pmatrix}. \tag{1}$$

Under the assumption of CPT-invariance, the elements of the 2×2 mass- and decay-matrices, M^q and Γ^q, fulfil the relations $M_{11}^q = M_{22}^q \equiv M_{B_q}$, $M_{12}^q = (M_{21}^q)^*$ and similarly for Γ^q. The diagonalisation yields the light and heavy mass eigenstates $|B_{L,H}\rangle = p|B\rangle \pm q|\overline{B}\rangle$ and their averaged masses, M_{B_q} and decay widths Γ_{B_q}. The according differences, ΔM_q and $\Delta\Gamma_q$, as well as the flavour-specific CP-asymmetries, $a_{\rm sl}^q$

$$\begin{aligned} \Delta M_q &= M_H^q - M_L^q = 2|M_{12}^q| + \ldots \geq 0\,, \\ \Delta\Gamma_q &= \Gamma_L^q - \Gamma_H^q = 2|\Gamma_{12}^q|\cos(\zeta_q) + \ldots \gtrless 0\,, \\ a_{\rm sl}^q &= \frac{\Gamma(\overline{B}_q \to f) - \Gamma(B_q \to \bar{f})}{\Gamma(\overline{B}_q \to f) + \Gamma(B_q \to \bar{f})} = \left|\frac{\Gamma_{12}^q}{M_{12}^q}\right|\sin(\zeta_q) + \ldots\,, \end{aligned} \tag{2}$$

Table 1. Compilation of B_q decay constants (left) and bag factors (right) from $N_f = 2+1$ lattice averages (see text for details) and the relative errors they induces in predictions of ΔM_q.

	[MeV]	$\delta(\Delta M_q)$
f_{B_s}	227.7 ± 4.5	4.0%
f_{B_d}	190.5 ± 4.2	4.4%

		$\delta(\Delta M_q)$
$\hat{B}_{B_s}$	1.33 ± 0.06	4.5%
$\hat{B}_{B_d}$	1.27 ± 0.10	7.9%

are determined by the off-diagonal elements M_{12}^q and Γ_{12}^q and their relative phase $\zeta_q = \arg(-M_{12}^q/\Gamma_{12}^q)$. Above the dots stand for terms of $\mathcal{O}(1/8|\Gamma_{12}^q/M_{12}^q|^2 \sin^2\zeta_q)$, which is smaller than 10^{-7} in the SM for both systems $q = d, s$ [1].

The structure of electroweak (EW) interactions in the SM allows for the construction of an effective field theory, in which the off-diagonal element M_{12}^q is given by $|\Delta B| = 2$ operators, whereas Γ_{12}^q by $|\Delta B| = 1$ current-current operators, which again can be matched on local $|\Delta B| = 2$ operators by means of heavy quark expansion (HQE).

2. Standard model

The prediction of M_{12}^q is well understood in the SM,

$$M_{12}^q = \frac{G_F^2 M_{B_q}}{12\pi^2}(V_{tb}V_{tq}^*)^2 m_W^2 S_0(x_t)\hat{\eta} B_{B_q} f_{B_q}^2 \,. \tag{3}$$

Contributions of the EW interaction at short-distance scales of the order of the W-boson mass, m_W, are decoupled and are contained at leading order (LO) in $S_0(x_t = m_t^2/m_W^2)$ [2] depending also on the top-quark mass m_t, at next-to-LO (NLO) QCD $\hat{\eta}$ [3], whereas tiny NLO EW corrections [4] are usually neglected. The accuracy of the determination of CKM elements is strongly limited by the uncertainties of the matrix element of a single local operator $\langle \overline{B}_q|(\bar{b}q)_{V-A}(\bar{b}q)_{V-A}|B_q\rangle = 8/3\, f_{B_q}^2 B_{B_q} M_{B_q}$ in terms of the B_q-meson decay constant, f_{B_q} and the associated bag parameter B_{B_q}. Both are nowadays calculated with the help of lattice QCD techniques — see [5] — and the current precision of latest world averages [6] ($N_f = 2+1$) are summarised in table 1, together with the arising uncertainty on ΔM_q. Since the small uncertainty of the decay constant is currently dominated by a single lattice calculation [7, 8], it would be reassuring to have confirmations from other lattice groups in the future. Beyond the SM, NP scenarios can give rise to additional $|\Delta B| = 2$ operators. The renormalisation group evolution of their Wilson coefficients from m_W to m_b are known up to NLO [9, 10] in QCD, whereas the according bag parameters are currently only available from the single unquenched lattice ($N_f = 2$) calculation [11] — for quenched results see [12] — and some preliminary ($N_f = 2+1$) results are reported in [13].

The calculation of Γ_{12}^q is based on HQE [14] giving rise to a double series in $\lambda = \Lambda_{\rm QCD}/m_b \sim 0.15$ and the strong coupling, α_s,

$$\Gamma_{12}^q = \lambda^3\left(\Gamma_3^{(0)} + \frac{\alpha_s}{4\pi}\Gamma_3^{(1)} + \ldots\right) + \lambda^4\left(\Gamma_4^{(0)} + \ldots\right) + \lambda^5\left(\Gamma_5^{(0)} + \ldots\right), \tag{4}$$

which is also applied to calculate of the total and partial inclusive decay widths [15]. The individual contributions shown in (4) can be found in [16, 17, 18, 19, 20, 21]. The expansion of $\Delta\Gamma^q$ (and also Γ^q) shows a convergent behaviour and predictions are in satisfactory agreement with measurements — see figure 1 for the B_s system. Especially lifetime ratios can be predicted quite precisely $\tau_{B_s}/\tau_{B_d} \in [0.996, 1.000]$ [1], since they are free of hadronic uncertainties, and agree well with the measured value 0.995 ± 0.006. Concerning $\Delta\Gamma_q$, current measurements [22] and the according SM predictions [1] are

$$\begin{aligned} \Delta\Gamma_d|_{\text{Exp}} &= (0.0059 \pm 0.0079)\,\text{ps}^{-1}\,, & \Delta\Gamma_s|_{\text{Exp}} &= (0.091 \pm 0.008)\,\text{ps}^{-1}\,, \\ \Delta\Gamma_d|_{\text{SM}} &= (0.0029 \pm 0.0007)\,\text{ps}^{-1}\,, & \Delta\Gamma_s|_{\text{SM}} &= (0.087 \pm 0.021)\,\text{ps}^{-1}\,. \end{aligned} \tag{5}$$

A comprehensive summary of theoretical uncertainties [1] shows that the largest sources are due to matrix elements of local $|\Delta B| = 2$ operators of dimension 6 and 7, about 5% and 17%, respectively. Especially bag factors of dim-7 operators are currently known only with low precision from sum rule calculations [23], but no dedicated lattice studies are pursued at the moment. Further, with most recent values of f_{B_q} the according relative uncertainty is the same as for ΔM_q given in table 1. Another large source of uncertainty is the factorisation scale dependence on $\mu_b \sim m_b$ with about 8%.

Whereas experimental uncertainties are below theoretical ones for $\Delta\Gamma_s$, $\Delta\Gamma_d$ is not well known yet. A better measurement of the latter can also help to further investigate the deviation of the dimuon charge asymmetry measured by DØ [24]. Although the interpretation of this measurement within the SM has been refined lately [25]

$$A_{\text{CP}} = C_d\, a_{\text{sl}}^d + C_s\, a_{\text{sl}}^s + C_{\Gamma_d}\frac{\Delta\Gamma_d}{\Gamma_d} + C_{\Gamma_s}\frac{\Delta\Gamma_s}{\Gamma_s}\,, \tag{6}$$

it still points towards large deviations of about 3.6σ from the SM predictions of a_{sl}^q and/or $\Delta\Gamma_d$ [24, 26]. The numerical values of the coefficients C_{d,s,Γ_d} can be extracted from [25, 24] and C_{Γ_s} turns out to be negligible.

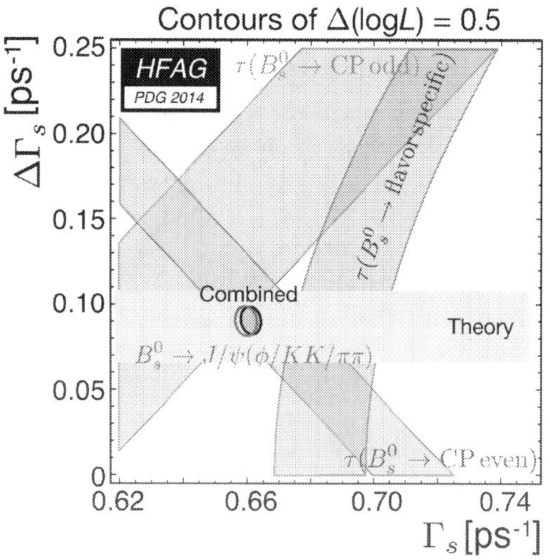

Figure 1. Combined region of $\Delta\Gamma_s$ vs. Γ_s from different channels. The ($B_s \to$ CP-odd) decays allow to measure directly Γ_H^s, whereas ($B_s \to$ CP-even) decays Γ_L^s. Also shown the SM prediction of $\Delta\Gamma_s$ [1]. For comparison, a "prediction" of $\Gamma_s = (0.660 \pm 0.004)$ ps^{-1} might be obtained when using the prediction of τ_{B_s}/τ_{B_d} [1], and the current measured world average of $\tau_{B_d} = (1.519 \pm 0.005)$ ps [22, 26].

3. Beyond the standard model

Effects of NP can be model-independently parametrised with two complex-valued parameters Δ_q and $\tilde{\Delta}_q$ as

$$\begin{aligned} M_{12}^q &= M_{12}^{q,\text{SM}}\,\Delta_q\,, & \Delta_q &= |\Delta_q|\, e^{i\phi_q^\Delta}\,, \\ \Gamma_{12}^q &= \Gamma_{12}^{q,\text{SM}}\,\tilde{\Delta}_q\,, & \tilde{\Delta}_q &= |\tilde{\Delta}_q|\, e^{i\tilde{\phi}_q^\Delta}\,. \end{aligned} \tag{7}$$

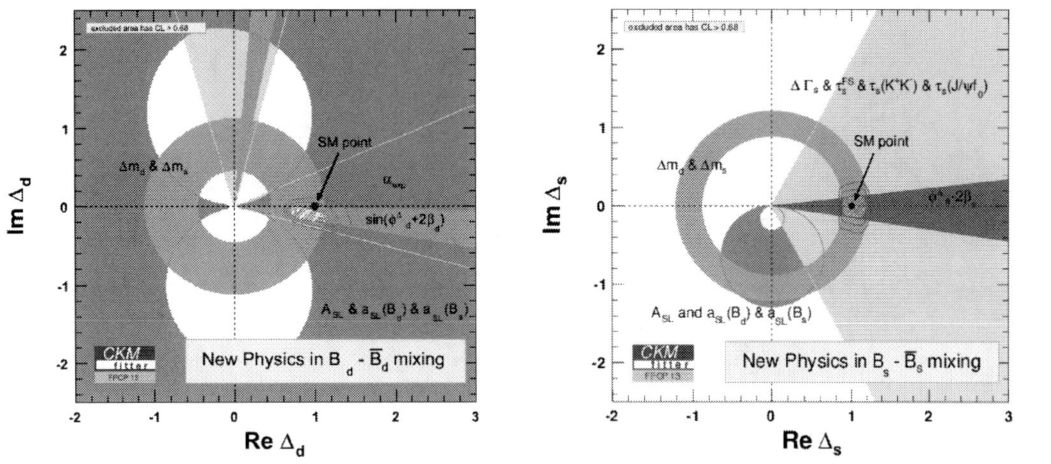

Figure 2. The allowed regions of the NP parameters Δ_d (left) and Δ_s (right), as determined in a global CKM-fit [27].

The global fits of the CKM elements in the framework of the SM take into account also ΔM_q. Beyond the SM, recent constraints on the scenario of NP only in M_{12}^q, i.e. $\Delta_d \neq \Delta_s \neq 0$ and also permitting NP independently in the $K\bar{K}$-system, are shown in figure 2. A 1.5σ deviation is obtained for the 2-dimensional SM hypothesis $\Delta_d = 1$ and 0.0σ for $\Delta_s = 1$. The SM point $\Delta_d = \Delta_s = 1$ is disfavoured by 1σ, compared to the 3.6σ in the year 2010 [28]. The highest pull value of 3.4σ arises for $A_{\rm CP} = C_d\, a_{\rm sl}^d + C_s\, a_{\rm sl}^s$, i.e. setting $C_{\Gamma_d} = 0$. Large deviations from the SM are by now excluded, whereas at 1σ CL deviations of $\mathcal{O}(40\%)$ and $\mathcal{O}(20\%)$ are still allowed

$$\begin{aligned} |\Delta_d| &= 0.81^{+0.27}_{-0.10}\,, \qquad & \phi_d^\Delta &= \left(-7.9^{+5.0}_{-2.3}\right)^\circ, \\ |\Delta_s| &= 0.97^{+0.20}_{-0.08}\,, \qquad & \phi_s^\Delta &= \left(-0.3^{+5.1}_{-5.3}\right)^\circ. \end{aligned} \tag{8}$$

An important role in this analysis plays the LHCb measurement of mixing-induced CP asymmetry $S(B \to J/\psi\phi)$ [29, 30], which does not permit an explanation of the DØ result for $A_{\rm CP}$ in terms of NP in M_{12}^s. Instead, from the best fit point of M_{12}^d follows the prediction $a_{\rm sl}^d = (-2.46^{+0.63}_{-0.45}) \cdot 10^{-3}$, which is enhanced by a factor of almost 8 over the SM prediction [1]. Improved determinations of $a_{\rm sl}^d$ at LHCb and Belle II are needed to further elucidate this issue.

In a more general NP scenario, also Γ_{12}^q might be modified. Whereas Γ_{12}^s is determined in the SM by Cabibbo-favoured tree-level exchange $b \to c\bar{c}s$ that gives rise to a large inclusive decay rate $Br(b \to c\bar{c}s) = (23.7 \pm 1.3)\%$, Γ_{12}^d is dominated by $b \to c\bar{c}d$, which leads only to small $Br(b \to c\bar{c}d) = (1.31 \pm 0.07)\%$ [31]. Therefore larger deviations from the SM prediction in Γ_{12}^d are at present less constrained, compared to Γ_{12}^s [32, 33]. Modifications of Γ_{12}^d affect both $\Delta\Gamma_d$ and $a_{\rm sl}^d$, being presently not well measured and thus potentially alleviating current tensions in $A_{\rm CP}$.

Some possible NP scenarios include violations of CKM unitarity [34, 33], large effects in almost unconstrained channels like $b \to \tau\bar{\tau}d$ or modifications of the current-current operators $b \to c\bar{c}d$, $c\bar{u}d$, $u\bar{u}d$ [33]. The latter two scenarios have been investigated in a model-independent fashion based on $|\Delta B| = 1$ dim-6 operators [33]. Concerning $b \to \tau\bar{\tau}d$, model-independent constraints on the complete set of dim-6 operators $O_i \sim [\bar{d}\Gamma_A b][\bar{\tau}\Gamma_B \tau]$ ($\Gamma_A = P_A$, $\gamma_\mu P_A, \sigma_{\mu\nu} P_A$

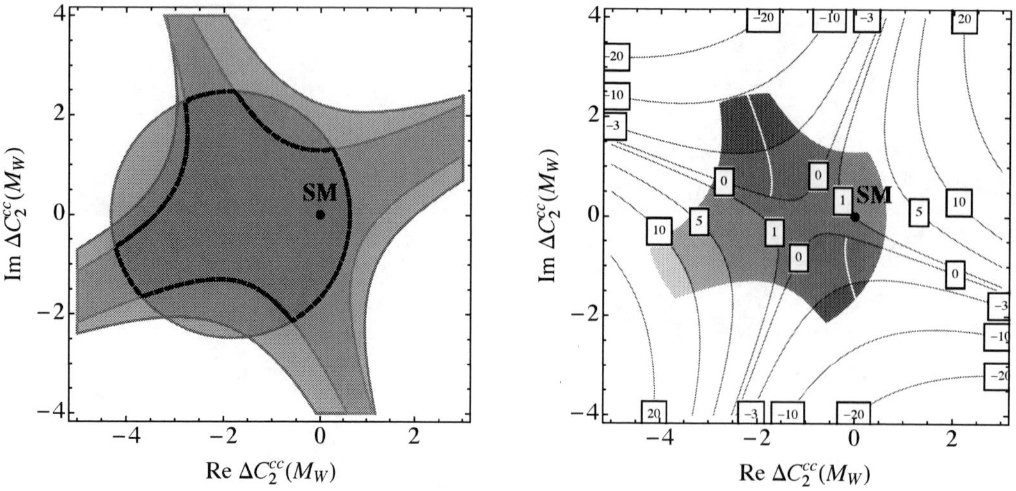

Figure 3. The allowed regions for the NP contributions of the colour-allowed current-current $b \to \tau\bar{\tau}d$ operator [left] from $Br(B \to X_d\gamma)$ (circle-shaped), $a_{\rm sl}^d$ (inner star-shaped) and dim-8 contributions to $\sin(2\beta)$ (outer star-shaped) [33]. Contours in $\Delta\Gamma_d/\Delta\Gamma_d^{\rm SM}$ [right].

with $P_A = (1 \pm \gamma_5)/2)$ have been derived on the according Wilson coefficients. In the case of scalar Wilson coefficients direct bounds from $Br(B_d \to \bar{\tau}\tau)$ cannot preclude enhancements of about 60% of $\Delta\Gamma_d$. Only indirect constraints from $Br(B \to X_d\gamma)$ and $Br(B^+ \to \pi^+\bar{\tau}\tau)$ are available on vector Wilson coefficients, which can enhance $\Delta\Gamma_d$ by about 270% over the SM prediction. In the case of current-current operators, NP in colour-suppressed $O_1^{pp'} = [\bar{d}^\alpha\gamma^\mu P_L p^\beta][\bar{p}'^\beta\gamma_\mu P_L b^\alpha]$ and colour-allowed $O_2^{pp'} = [\bar{d}\gamma^\mu P_L p][\bar{p}'\gamma_\mu P_L b]$ operators have been assumed, where $pp' = uu, cu, cc$. In the SM the Wilson coefficients $C_i^{pp'} = V_{pd}^* V_{p'b} C_i$ are universal, once factoring out CKM elements. For $b \to (u\bar{u},\ c\bar{u})d$, the experimental constraints from $B \to \pi\pi,\ \rho\pi,\ \rho\rho$ and $B \to D^*\pi$, respectively, allow for $\Delta\Gamma_d/\Delta\Gamma_d^{\rm SM} \in [-1.0,\ 1.4]$, however for $b \to c\bar{c}d$ huge effects of several 100% cannot be ruled out, see figure 3.

4. Conclusion

The mixing of neutral B_q mesons ($q = d, s$) plays an important role in CP violating phenomena, tests of the picture of quark mixing in the standard model (SM) and search for new physics (NP). Currently the largest uncertainties in predictions of ΔM_q and $\Delta\Gamma_q$ are due to hadronic parameters: the decay constants f_{B_q} and bag factors of various $|\Delta B| = 2$ operators. There is encouraging progress in lattice predictions concerning ΔM_q in the standard model, dominated by f_{B_q} and a single bag factor. However, NP searches require the knowledge of the bag factors of all $|\Delta B| = 2$ operators of dimension 6, which receives currently much less attention. Moreover, $\Delta\Gamma_q$ also depends strongly on bag factors of $|\Delta B| = 2$ operators of dimension 7 that cause at present the largest uncertainties, but for which so far exist only sum rule estimates.

Recent global CKM fits that account for NP in M_{12}^q only, exclude now larger deviations then $\mathcal{O}(20\%)$ and $\mathcal{O}(40\%)$ in the B_s and B_d system from the SM, respectively. Only the DØ measurement of the like-sign dimuon charge asymmetry has a pull value of about 3σ. This tension could be caused by non-standard contributions to Γ_{12}^d for which at present, large NP contributions cannot be excluded. A direct measurement of $\Delta\Gamma_d$, more precise data on the individual semi-leptonic CP asymmetries $a_{\rm sl}^{d,s}$, but also improved experimental determinations of rare and radiative $b \to d$ decays would shed light on this issue and should be pursued with vigour in the future.

Acknowledgments

I am indebted to the organisers of the XI International Conference on Hyperons, Charm and Beauty Hadrons (BEACH 2014) in Birmingham for the invitation and the kind hospitality. I thank my collaborators for our fruitful work. This work received support from the ERC Advanced Grant project "FLAVOUR" (267104).

References

[1] Lenz A and Nierste U 2011 (*Preprint* 1102.4274)
[2] Inami T and Lim C 1981 *Prog.Theor.Phys.* **65** 297
[3] Buras A J, Jamin M and Weisz P H 1990 *Nucl.Phys.* **B347** 491–536
[4] Gambino P, Kwiatkowski A and Pott N 1999 *Nucl.Phys.* **B544** 532–556 (*Preprint* hep-ph/9810400)
[5] Jüttner A (*Preprint* this proceedings)
[6] Aoki S, Aoki Y, Bernard C, Blum T, Colangelo G *et al.* 2013 (*Preprint* 1310.8555)
[7] McNeile C, Davies C, Follana E, Hornbostel K and Lepage G P 2012 *Phys.Rev.* **D85** 031503 (*Preprint* 1110.4510)
[8] Na H, Monahan C J, Davies C T, Horgan R, Lepage G P *et al.* 2012 *Phys.Rev.* **D86** 034506 (*Preprint* 1202.4914)
[9] Buras A J, Misiak M and Urban J 2000 *Nucl.Phys.* **B586** 397–426 (*Preprint* hep-ph/0005183)
[10] Buras A J, Jager S and Urban J 2001 *Nucl.Phys.* **B605** 600–624 (*Preprint* hep-ph/0102316)
[11] Carrasco N *et al.* (ETM Collaboration) 2014 *JHEP* **1403** 016 (*Preprint* 1308.1851)
[12] Becirevic D, Gimenez V, Martinelli G, Papinutto M and Reyes J 2002 *JHEP* **0204** 025 (*Preprint* hep-lat/0110091)
[13] Chang C, Bernard C, Bouchard C, El-Khadra A, Freeland E *et al.* 2013 (*Preprint* 1311.6820)
[14] Shifman M A and Voloshin M 1985 *Sov.J.Nucl.Phys.* **41** 120
[15] Bigi I I, Uraltsev N and Vainshtein A 1992 *Phys.Lett.* **B293** 430–436 (*Preprint* hep-ph/9207214)
[16] Beneke M, Buchalla G and Dunietz I 1996 *Phys.Rev.* **D54** 4419–4431 (*Preprint* hep-ph/9605259)
[17] Beneke M, Buchalla G, Greub C, Lenz A and Nierste U 1999 *Phys.Lett.* **B459** 631–640 (*Preprint* hep-ph/9808385)
[18] Beneke M, Buchalla G, Lenz A and Nierste U 2003 *Phys.Lett.* **B576** 173–183 (*Preprint* hep-ph/0307344)
[19] Ciuchini M, Franco E, Lubicz V, Mescia F and Tarantino C 2003 *JHEP* **0308** 031 (*Preprint* hep-ph/0308029)
[20] Lenz A and Nierste U 2007 *JHEP* **0706** 072 (*Preprint* hep-ph/0612167)
[21] Badin A, Gabbiani F and Petrov A A 2007 *Phys.Lett.* **B653** 230–240 (*Preprint* 0707.0294)
[22] Amhis Y *et al.* (Heavy Flavor Averaging Group) 2012 (*Preprint* 1207.1158)
[23] Mannel T, Pecjak B and Pivovarov A 2007 (*Preprint* hep-ph/0703244)
[24] Abazov V M *et al.* (D0 Collaboration) 2014 *Phys.Rev.* **D89** 012002 (*Preprint* 1310.0447)
[25] Borissov G and Hoeneisen B 2013 *Phys.Rev.* **D87** 074020 (*Preprint* 1303.0175)
[26] Bertram I (*Preprint* this proceedings)
[27] Lenz A, Nierste U, Charles J, Descotes-Genon S, Lacker H *et al.* 2012 *Phys.Rev.* **D86** 033008 (*Preprint* 1203.0238)
[28] Lenz A, Nierste U, Charles J, Descotes-Genon S, Jantsch A *et al.* 2011 *Phys.Rev.* **D83** 036004 (*Preprint* 1008.1593)
[29] Aaij R *et al.* (LHCb collaboration) 2013 *Phys.Rev.* **D87** 112010 (*Preprint* 1304.2600)
[30] De Bruyn K (*Preprint* this proceedings)
[31] Krinner F, Lenz A and Rauh T 2013 *Nucl.Phys.* **B876** 31–54 (*Preprint* 1305.5390)
[32] Bobeth C and Haisch U 2013 *Acta Phys.Polon.* **B44** 127–176 (*Preprint* 1109.1826)
[33] Bobeth C, Haisch U, Lenz A, Pecjak B and Tetlalmatzi-Xolocotzi G 2014 *JHEP* **1406** 040 (*Preprint* 1404.2531)
[34] Botella F, Branco G, Nebot M and Sanchez A 2014 (*Preprint* 1402.1181)

Charmless *B* decays at *BABAR*

Thomas Latham (on behalf of the *BABAR* experiment)
Department of Physics, University of Warwick, Coventry CV4 7AL, UK
E-mail: T.Latham@warwick.ac.uk

Abstract. Recent results from the *BABAR* experiment in charmless B decays are presented, in particular the results of a search for the decays $B^0 \to \omega\omega$ and $B^0 \to \omega\phi$ and preliminary results of a Dalitz-plot analysis of $B^+ \to K^0_S\pi^+\pi^0$. The latter includes first evidence for CP violation in the intermediate decay $B^+ \to K^*(892)^+\pi^0$.

1. Introduction

Charmless hadronic decays of B mesons can typically proceed via two topologies: $b \to u$ tree and $b \to d, s$ loop (or "penguin") diagrams. The amplitudes for these two processes, which have a relative weak phase (equal to the CKM angle γ in the Standard Model (SM)) and may have a similar magnitude, can interfere with one another, potentially giving rise to large CP violation effects. Recent analyses from LHCb of $B^+ \to h^+h^+h^-$ decays[1] have observed unexpectedly large CP asymmetries in regions of the Dalitz plots [1, 2, 3]. In addition, the observed CP asymmetries in $B \to K\pi$ decays do not follow the expected behaviour, which is that $\Delta\mathcal{A}_{CP}(K\pi) \equiv \mathcal{A}_{CP}(K^+\pi^0) - \mathcal{A}_{CP}(K^+\pi^-) = 0$. The current world average experimental results are $\mathcal{A}_{CP}(K^+\pi^0) = 0.040 \pm 0.021$ and $\mathcal{A}_{CP}(K^+\pi^-) = -0.082 \pm 0.006$ [4], leading to a value $\Delta\mathcal{A}_{CP}(K\pi) = 0.122 \pm 0.022$, which is significantly different from zero. It has been pointed out (e.g. in Ref. [5]) that additional information can be obtained from $B \to K^*\pi$ and $B \to \rho K$ decays, giving incentive to improve the experimental precision on the asymmetry measurements in these modes. In the $K^*\pi$ system, the least well measured decay is $B^+ \to K^*(892)^+\pi^0$, where the only existing measurement comes from a *BABAR* analysis of $B^+ \to K^+\pi^0\pi^0$ [6]: $\mathcal{A}_{CP}(K^*(892)^+\pi^0) = -0.06 \pm 0.24$.

Another puzzle from charmless B decays is the lower than expected longitudinal polarisation fraction in $B \to \phi K^*$ decays [7, 8, 9]. It is not clear if the large transverse component arises from SM contributions or beyond-SM effects. The study of other vector-vector decay modes is therefore strongly motivated. In particular, constraints on the SM contributions can be made using information on the branching fractions of $B^0 \to \omega\omega$ and $B^0 \to \omega\phi$ [10].

Both analyses described here use the full *BABAR* dataset collected at the $\Upsilon(4S)$ resonance. It corresponds to an integrated luminosity of $429\,\text{fb}^{-1}$ and contains 470.9 ± 2.8 million $B\overline{B}$ events.

2. First evidence for $B^0 \to \omega\omega$ and search for $B^0 \to \omega\phi$

The SM expectations for the branching fractions of the decays $B^0 \to \omega\omega$ and $B^0 \to \omega\phi$ are of order 10^{-6} and 10^{-7}, respectively. The existing experimental limits come from a previous

[1] The inclusion of charge conjugate processes is implied throughout, except in definitions of CP asymmetries.

BABAR analysis [11] that used approximately half the data sample used here. Measurements of the mixing-induced *CP* asymmetry and the longitudinal polarisation fraction made in these loop-dominated decays will be sensitive to possible new physics contributions. The ω (ϕ) candidates are reconstructed in their decay to $\pi^+\pi^-\pi^0$ (K^+K^-), with their invariant masses required to be in the range $740 < m_{\pi^+\pi^-\pi^0} < 820\,\mathrm{MeV}/c^2$ ($1009 < m_{K^+K^-} < 1029\,\mathrm{MeV}/c^2$). Backgrounds are considered from both continuum light-quark production and from B decays.

A maximum likelihood fit is used to separate signal from background and to obtain the signal yields. The likelihood fit is performed to following variables: the beam energy substituted mass of the B candidate (m_{ES}), the difference between the beam energy and that of the reconstructed B candidate (ΔE), a Fisher discriminant of topological variables, the invariant mass and helicity angle of each ω and ϕ resonance, and the "internal" helicity angle of the ω resonance (defined as the polar angle of the π^0 with respect to the ω flight direction in the $\pi^+\pi^-$ rest frame). More details of the analysis can be found in Ref. [12].

The results of the fit can be seen in Figure 1 and the fitted yields of the two decays are found to be 69^{+16}_{-15} for $B^0 \to \omega\omega$ and $-2.9^{+5.7}_{-4.0}$ for $B^0 \to \omega\phi$, where the quoted uncertainties are statistical only. The significance of the $B^0 \to \omega\omega$ signal is determined to be 4.4σ (including systematic uncertainties) and the branching fraction is calculated to be $(1.2 \pm 0.3\,[\mathrm{stat.}]\,^{+0.3}_{-0.2}\,[\mathrm{syst.}]) \times 10^{-6}$. The upper limit on the $B^0 \to \omega\phi$ branching fraction is determined to be 0.7×10^{-6} at 90% confidence level. The largest sources of systematic uncertainty arise form the uncertainty on the longitudinal polarisation fraction (which is assumed to be 0.88 in the nominal fit and then varied between 0.58 and 1.0 to determine the systematic uncertainty) and from possible fit bias.

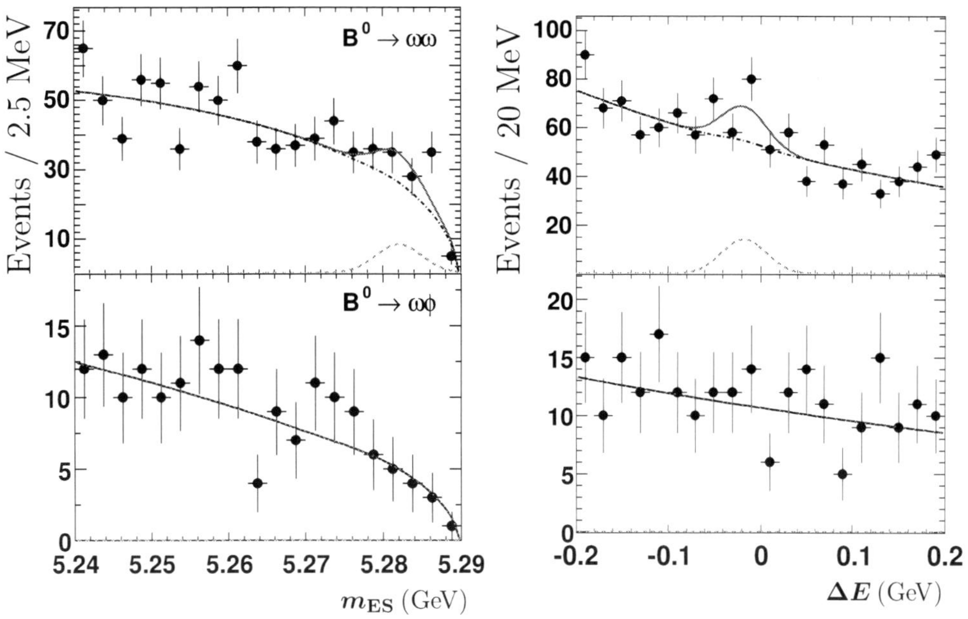

Figure 1. Projections of the fit results on the (left) m_{ES} and (right) ΔE variables, for (top) $B^0 \to \omega\omega$ and (bottom) $B^0 \to \omega\phi$ decays. The points are the data, the solid curve is the total fit, the dashed curve is the signal, and the dot-dashed curve is the background.

3. Amplitude analysis of $B^+ \to K^0_S\pi^+\pi^0$

The only previous measurement of the inclusive branching fraction of $B^+ \to K^0_S\pi^+\pi^0$ is an upper limit from CLEO: $\mathrm{BF}(B^+ \to K^0\pi^+\pi^0) < 66\times10^{-6}$ [13]. However, the intermediate decay mode $B^+ \to \rho(770)^+K^0_S$ has been observed by the *BABAR* collaboration and its branching fraction measured to be $\mathrm{BF}(B^+ \to \rho(770)^+K^0) = (8.0^{+1.4}_{-1.3} \pm 0.6)\times10^{-6}$ [14]. The other expected intermediate decays $B^+ \to K^*(892)^+\pi^0$ and $B^+ \to K^*(892)^0\pi^+$ have both been observed, although not in their decays to this final state. In addition to providing measurements of the branching fractions and *CP* asymmetries of the $K^{*+}\pi^0$, $K^{*0}\pi^+$ and $\rho^+K^0_S$ intermediate states, a Dalitz-plot analysis of $B^+ \to K^0_S\pi^+\pi^0$ will measure the relative phases of these various components. Of particular interest are the relative phases of the two $K^*\pi$ components, which can be used to determine the CKM angle γ [15, 16].

The K^0_S candidates are reconstructed in their decay to $\pi^+\pi^-$ and are required to have an invariant mass within $15\,\mathrm{MeV}/c^2$ of the known K^0_S mass. The π^0 candidates are are required to be in the invariant mass range $110 < m_{\gamma\gamma} < 160\,\mathrm{MeV}/c^2$. The largest backgrounds from B decays contain D^0 mesons that subsequently decay to $K^0_S\pi^0$. These are removed with a veto on the $K^0_S\pi^0$ invariant mass: $1804 < m_{K^0_S\pi^0} < 1924\,\mathrm{MeV}/c^2$. Approximately 32,000 candidates remain after all selection criteria are applied.

A maximum likelihood fit is performed to the variables m_{ES}, ΔE, the output of a Boosted Decision Tree (BDT) of topological variables and the Dalitz plot (DP). Since the kinematic variables, m_{ES} and ΔE, are strongly correlated with the Dalitz-plot position for signal events, the signal probability density functions are parametrised as a function of the DP position. The fitted signal yield is 1014 ± 63, where the uncertainty is statistical only.

The signal Dalitz-plot model follows the isobar model formalism, where the total amplitude is formed from the sum of the amplitudes for the various intermediate states

$$\overset{(-)}{A} = \sum_i \overset{(-)}{A_i} = \sum_i \overset{(-)}{c_i} F_i(m^2_{K^0_S\pi^+}, m^2_{\pi^+\pi^0}), \tag{1}$$

where the F_i describe the dependence of the amplitude on the Dalitz-plot position (e.g., a Breit–Wigner function) and the c_i are the complex coefficients that are determined from the fit. However, since only relative amplitudes can be determined, one component ($K^*(892)^0\pi^+$ in the nominal fit) is fixed to be real and to have unit magnitude. Quantities such as the branching fractions, *CP* asymmetries and relative phases are derived from these fitted coefficients. The signal model includes contributions from both the charged and neutral $K^*(892)$ resonances plus the corresponding $K\pi$ S-wave as well as the $\rho(770)^+$ resonance. The $K\pi$ S-wave is modelled using the LASS parametrisation [17], which consists of the sum of the $K^*_0(1430)$ resonance and an effective-range nonresonant term. No other nonresonant contributions were found to be needed to give a satisfactory fit to the data.

The preliminary results for the branching fractions, *CP* asymmetries and relative phases are given in Table 1. The *CP* asymmetry of $B^+ \to K^*(892)^+\pi^0$ has a very large, negative central value (-52%) and is found to have a significance of 3.4σ, corresponding to first evidence of *CP* violation in this decay. The projection of the fit onto the $K^0_S\pi^+$ invariant mass can be seen in Figure 2, separated by the charge of the B candidate, where the asymmetry in the $K^*(892)^+$ region can be clearly seen. The uncertainty on this quantity is much improved but is still large with respect to $K^{*+}\pi^-$ and as such still dominates the uncertainty on the updated value of $\Delta\mathcal{A}_{CP}(K^*\pi) = -0.16 \pm 0.14$. An analysis of this decay mode by the Belle experiment would help to further improve this uncertainty. A journal paper describing the *BABAR* analysis and its results in detail is in preparation.

Acknowledgments

Work supported by the European Research Council under FP7.

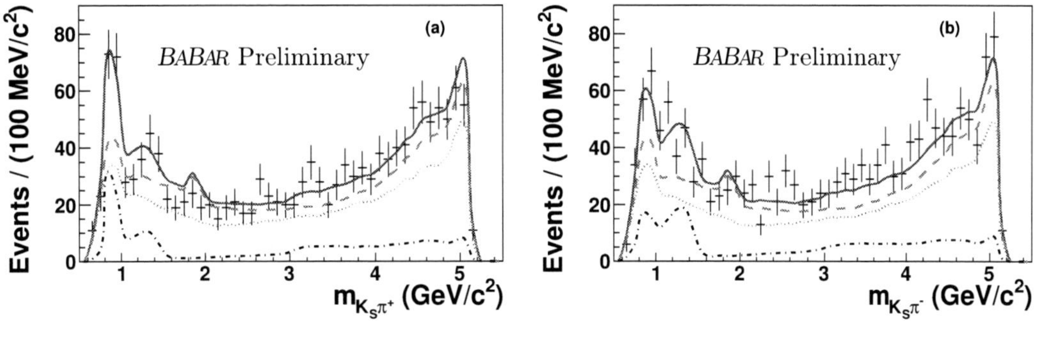

Figure 2. Data distributions of $m_{K_S^0\pi^+}$ and the corresponding fit projections for (a) B^+, and (b) B^- candidates. Points with error bars are the data, the solid (blue) lines are the total fit result, the dashed (green) lines are the total background contribution, and the dotted (red) lines are the $q\bar{q}$ component. The dash-dotted lines represent the signal contribution. The signal to background ratio has been increased by applying a tighter selection on $m_{\rm ES}$, ΔE, and BDT.

Table 1. Measured *CP*-averaged branching fractions (BF), *CP* asymmetries ($\mathcal{A}_{CP}$), and *CP*-averaged phases measured relative to the $K^*(892)^0\pi^+$ reference amplitude. The first uncertainty is statistical, the second is systematic, and the third is due to the signal Dalitz-plot model.

Isobar	BF $(\times 10^{-6})$	$\mathcal{A}_{CP}$	ϕ (°)
$K^0\pi^+\pi^0$	$45.9 \pm 2.6 \pm 3.0 \pm 8.6$	$0.07 \pm 0.05 \pm 0.03 \pm 0.04$	–
$K^*(892)^0\pi^+$	$14.6 \pm 2.4 \pm 1.3 \pm 0.5$	$-0.12 \pm 0.21 \pm 0.08 \pm 0.11$	–
$K^*(892)^+\pi^0$	$9.2 \pm 1.3 \pm 0.6 \pm 0.5$	$-0.52 \pm 0.14 \pm 0.04 \pm 0.04$	$-95 \pm 43^{+48}_{-36} \pm 70$
$K_0^*(1430)^0\pi^+$	$50.0 \pm 4.8 \pm 6.0 \pm 4.0$	$0.14 \pm 0.10 \pm 0.04 \pm 0.14$	$174 \pm 11 \pm 11 \pm 6$
$K_0^*(1430)^+\pi^0$	$17.2 \pm 2.4 \pm 1.5 \pm 1.8$	$0.26 \pm 0.12 \pm 0.08 \pm 0.12$	$-89 \pm 43^{+53}_{-40} \pm 18$
$\rho(770)^+K^0$	$9.4 \pm 1.6 \pm 1.0 \pm 2.6$	$0.21 \pm 0.19 \pm 0.07 \pm 0.30$	$-122 \pm 43^{+55}_{-47} \pm 68$

References

[1] Aaij R *et al.* (LHCb collaboration) 2013 *Phys.Rev.Lett.* **111** 101801 (*Preprint* 1306.1246)
[2] Aaij R *et al.* (LHCb collaboration) 2014 *Phys.Rev.Lett.* **112** 011801 (*Preprint* 1310.4740)
[3] Aaij R *et al.* (LHCb collaboration) 2014 (*Preprint* 1408.5373)
[4] Amhis Y *et al.* (Heavy Flavor Averaging Group) 2012 (*Preprint* 1207.1158)
[5] Gronau M, Pirjol D and Zupan J 2010 *Phys.Rev.* **D81** 094011 (*Preprint* 1001.0702)
[6] Lees J *et al.* (*BABAR* Collaboration) 2011 *Phys.Rev.* **D84** 092007 (*Preprint* 1109.0143)
[7] Aubert B *et al.* (*BABAR* Collaboration) 2007 *Phys.Rev.Lett.* **98** 051801 (*Preprint* hep-ex/0610073)
[8] Aubert B *et al.* (*BABAR* Collaboration) 2007 *Phys.Rev.Lett.* **99** 201802 (*Preprint* 0705.1798)
[9] Prim M *et al.* (Belle Collaboration) 2013 *Phys.Rev.* **D88** 072004 (*Preprint* 1308.1830)
[10] Oh S 1999 *Phys.Rev.* **D60** 034006 (*Preprint* hep-ph/9812530)
[11] Aubert B *et al.* (*BABAR* Collaboration) 2006 *Phys.Rev.* **D74** 051102 (*Preprint* hep-ex/0605017)
[12] Lees J *et al.* (*BABAR* Collaboration) 2014 *Phys.Rev.* **D89** 051101 (*Preprint* 1312.0056)
[13] Eckhart E *et al.* (CLEO Collaboration) 2002 *Phys.Rev.Lett.* **89** 251801 (*Preprint* hep-ex/0206024)
[14] Aubert B *et al.* (*BABAR* Collaboration) 2007 *Phys.Rev.* **D76** 011103 (*Preprint* hep-ex/0702043)
[15] Ciuchini M, Pierini M and Silvestrini L 2006 *Phys.Rev.* **D74** 051301 (*Preprint* hep-ph/0601233)
[16] Gronau M, Pirjol D, Soni A and Zupan J 2007 *Phys.Rev.* **D75** 014002 (*Preprint* hep-ph/0608243)
[17] Aston D, Awaji N, Bienz T, Bird F, D'Amore J *et al.* 1988 *Nucl.Phys.* **B296** 493

Journal of Physics: Conference Series **556** (2014) 012041 doi:10.1088/1742-6596/556/1/012041

Charmless B decays at the LHCb experiment

David Dossett on behalf of the LHCb Collaboration
University of Warwick, Coventry, CV4 7AL, UK
E-mail: d.a.dossett@warwick.ac.uk

Abstract. Recent measurements of charmless B meson decays using proton-proton collision data collected by the LHCb experiment are presented here. Branching fraction measurements of the $B^0_{d,s} \to K^{*\pm}K^{\mp}$ decays have been made, as well as measurements of CP asymmetry in both the $B^{\pm} \to h^{\pm}h^{+}h^{-}$ and $B^{\pm} \to h^{\pm}p\bar{p}$ decays. In addition, the forward-backward asymmetry in the $B^{\pm} \to h^{\pm}p\bar{p}$ decays was measured.

1. Introduction

One of the most important areas in quark flavour physics is the study of the so-called "charmless" B meson decays. These are decays of B mesons to hadronic final states which do not contain charm quarks or antiquarks. As shown in Fig. 1, these decays will in general have contributions from both "tree" and "loop" diagrams. When there is a phase difference between the CP even ("strong") and CP odd ("weak") phases of the contributing diagrams, CP violation may occur.

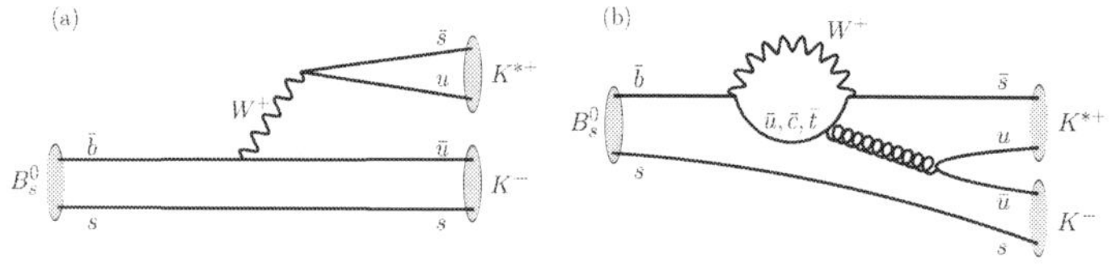

Figure 1: (a) Tree and (b) loop diagrams for the decay $B^0_s \to K^{*+}K^{-}$.

It is possible that beyond Standard Model (SM) particles could participate in the loop contribution, causing a deviation from the SM expectation of branching fractions or CP asymmetry. For decays containing three or more final state particles, a Dalitz plot analysis [1] allows the study of the weak and strong phases and the variation of the CP asymmetries across the phase space of the decay.

The analyses presented here are based on two data samples recorded at the LHCb detector at CERN [2]. The 2011 data sample corresponding to an integrated luminosity of $1.0\,\text{fb}^{-1}$ of proton-proton collisions at a centre-of-mass energy of $7\,\text{TeV}$, and the 2012 data sample with an integrated luminosity of $2.0\,\text{fb}^{-1}$ at $8\,\text{TeV}$.

2. $B^0_{d,s} \to K^{*\pm}h^{\mp}$ branching fractions [3]

Previous measurements have been made at LHCb of the inclusive branching fractions of $B^0_{d,s} \to K^0_S h^{\pm}h'^{\mp}$ decays [4]. The $B^0_{d,s} \to K^{*\pm}h^{\mp}$ decays are intermediate resonances of the

$B^0_{d,s} \to K^0_S h^\pm h'^\mp$ ones, where the $K^{*\pm}$ decays to a K^0_S and a $\pi^\pm$. This analysis aimed to measure the branching fractions of the three unobserved $B^0_{d,s} \to K^{*\pm} h^\mp$ modes with respect to the previously measured $B^0 \to K^{*+}\pi^-$ normalisation mode.

Using the 2011 data sample $B^0_{d,s} \to K^{*\pm} h^\mp$ events were selected using a method based on the $B^0_{d,s} \to K^0_S h^\pm h'^\mp$ analysis. Each B candidate was formed from three tracks, a K^0_S candidate and two charged tracks. The K^0_S candidates were reconstructed by combining $\pi^+\pi^-$ tracks and assigning them (and the B candidate they form) one of two categories, "long" and "downstream". Long K^0_S candidates come from pion tracks which have hits in both the VELO subdetector [5] and the tracking systems downstream of the magnet. Downstream candidates do not have tracks in the VELO and consequently have worse mass, momentum and vertex resolution. Due to this there were slightly different selection requirements on the long and downstream candidates, and until the final calculation of branching fractions the analyses of the two categories were completely separate.

To extract the yields of the different decays separate 2D fits for the $B^0_{d,s} \to K^{*\pm} K^\mp$ and $B^0_{d,s} \to K^{*\pm}\pi^\mp$ final state were created. These fits used the m_B and $m_{K^0_S\pi}$ mass combination variables and fit simultaneously between the two K^0_S categories. Figs. 2-3 show the results of the these fits.

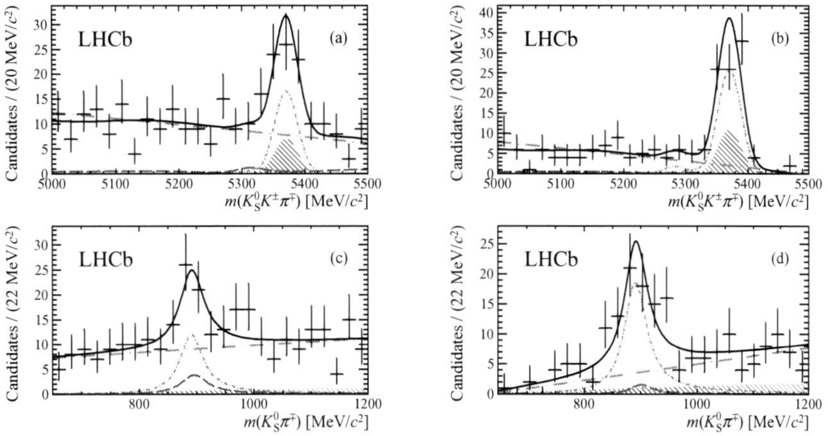

Figure 2: Results of the fit to $K^{*\pm}K^\mp$ candidates projected onto (a,b) B candidate and (c,d) K^* candidate mass distributions, for (a,c) long and (b,d) downstream candidates. The total fit result (solid black line) is shown together with the data points and multiple background components.

The total significances of the $B^0_s \to K^{*\pm}K^\mp$ and $B^0_s \to K^{*-}\pi^+$ yields were 7.8σ and 3.4σ respectively, the first observation and evidence for these decays. While the significance of the $B^0 \to K^{*\pm}K^\mp$ was below 2σ. The yields were used to find the branching fractions relative to the $B^0 \to K^{*+}\pi^-$ mode. Multiplying the relative branching fractions by the world average $\mathcal{B}\left(B^0 \to K^{*+}\pi^-\right) = (8.5 \pm 0.7) \times 10^{-6}$ [6] gave the absolute branching fractions and limits

$$\begin{aligned}
\mathcal{B}\left(B^0_s \to K^{*\pm}K^\mp\right) &= (12.7 \pm 1.9(\text{stat}) \pm 1.9(\text{syst})) \times 10^{-6}\,, \\
\mathcal{B}\left(B^0 \to K^{*\pm}K^\mp\right) &= (0.17 \pm 0.15(\text{stat}) \pm 0.05(\text{syst})) \times 10^{-6}\,, \\
&< 0.4\ (0.5) \times 10^{-6} \text{ at } 90\,\%\ (95\,\%)\ \text{CL}\,, \\
\mathcal{B}\left(B^0_s \to K^{*-}\pi^+\right) &= (3.3 \pm 1.1(\text{stat}) \pm 0.5(\text{syst})) \times 10^{-6}\,.
\end{aligned}$$

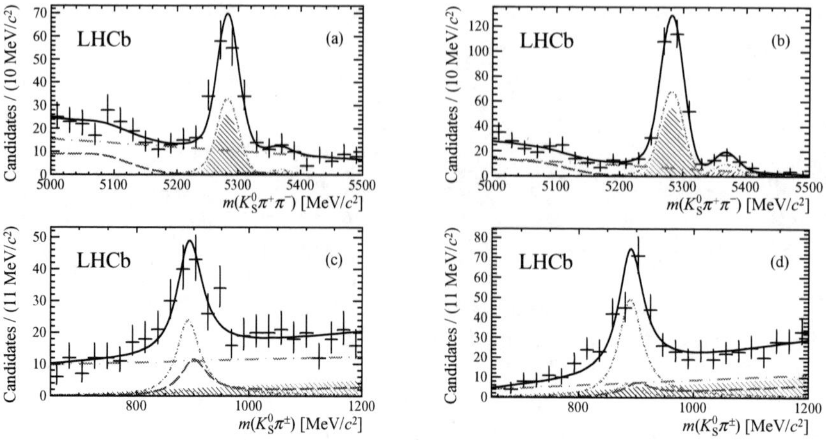

Figure 3: Results of the fit to $K^{*\pm}\pi^\mp$ candidates projected onto (a,b) B candidate and (c,d) K^* candidate mass distributions, for (a,c) long and (b,d) downstream candidates. The total fit result (black solid line) is shown together with the data points and multiple background components.

3. $\mathbf{B^\pm \to h^\pm h^+ h^-}$ CP asymmetries [7]

This analysis supersedes those of Ref. [8] and Ref. [9] using the full data sample of 2011 and 2012 and an updated selection of events. The inclusive *CP* asymmetries were measured for all four $B^\pm \to h^\pm h^+ h^-$ decay modes. Raw asymmetries between the B^+ and B^- candidates were extracted from fits to the mass spectra of the selected B candidates.These were then corrected by the detector and production asymmetries to find the final *CP* asymmetry ($\mathcal{A}^{\mathrm{CP}}$) using

$$\mathcal{A}^{\mathrm{CP}} = \mathcal{A}^{\mathrm{raw}} - \mathcal{A}^{\mathrm{D}}(h) - \mathcal{A}^{\mathrm{P}}(B^\pm). \tag{1}$$

The production asymmetry, $\mathcal{A}^{\mathrm{P}}(B^\pm)$, was calculated from $B^\pm \to J/\psi K^\pm$ decays and the world average of $\mathcal{A}^{\mathrm{CP}}(B^\pm \to J/\psi K^\pm)$. The detector asymmetries are different for the $B^\pm \to K^\pm h^+ h^-$ and $B^\pm \to \pi^\pm h^+ h^-$ modes. The $\mathcal{A}^{\mathrm{D}}(\pi)$ has been previously measured [10] and the value of $\mathcal{A}^{\mathrm{D}}(K)$ was determined from $D^{*+} \to \pi^+ D^0$ decays.

The final values and significances for the inclusive *CP* asymmetries were

$$\begin{aligned}
\mathcal{A}^{\mathrm{CP}}(B^\pm \to K^\pm \pi^+ \pi^-) &= +0.025 \pm 0.004 \pm 0.004 \pm 0.007 \quad (2.8\sigma),\\
\mathcal{A}^{\mathrm{CP}}(B^\pm \to K^\pm K^+ K^-) &= -0.036 \pm 0.004 \pm 0.002 \pm 0.007 \quad (4.3\sigma),\\
\mathcal{A}^{\mathrm{CP}}(B^\pm \to \pi^\pm \pi^+ \pi^-) &= +0.058 \pm 0.008 \pm 0.009 \pm 0.007 \quad (4.2\sigma),\\
\mathcal{A}^{\mathrm{CP}}(B^\pm \to \pi^\pm K^+ K^-) &= -0.123 \pm 0.017 \pm 0.012 \pm 0.007 \quad (5.6\sigma),
\end{aligned}$$

where the first uncertainty is statistical, the second systematic and the third is from the uncertainty on $\mathcal{A}^{\mathrm{CP}}(B^\pm \to J/\psi K^\pm)$.

An investigation was also performed into the *CP* asymmetries in different regions of the Dalitz plot. Specifically in the mass regions where $KK \leftrightarrow \pi\pi$ rescattering and intermediate resonances are favoured. The results suggested that both rescattering and interference between the P-wave and S-wave contributions around the $\rho^0(770)$ could be important sources to the strong phase difference in these decays. However, a full Dalitz plot amplitude analysis is required to gain a full understanding of these results.

4. $\mathbf{B^\pm \to h^\pm p\overline{p}}$ asymmetries [11]

This analysis is an update to Ref. [12] with the full data sample, focused on the $B^\pm \to h^\pm p\overline{p}$ decays. These are similar to the previous $B^\pm \to h^\pm h^+ h^-$ decays but with a proton-antiproton pair rather than a $\pi^+\pi^-$ or KK pair in the final state.

The first measurement was of the forward-backward asymmetry in the $m_{p\bar{p}} < 2.85\,\mathrm{GeV}/c^2$ region of the Dalitz plot. The forward-backward asymmetry is defined as

$$\mathcal{A}^{\mathrm{FB}} = \frac{N_{\mathrm{pos}} - N_{\mathrm{neg}}}{N_{\mathrm{pos}} + N_{\mathrm{neg}}}, \tag{2}$$

where N_{pos} and N_{neg} are the efficiency corrected yields for the $\cos\theta_p > 0$ and $\cos\theta_p < 0$ regions respectively. Here θ_p is the angle between the charged meson and the oppositely charged baryon in the rest frame of the $p\bar{p}$ system. The obtained values of $\mathcal{A}^{\mathrm{FB}}$ for both decay modes were

$$\begin{aligned}\mathcal{A}^{\mathrm{FB}}(B^{\pm}\to K^{\pm}p\bar{p}) &= +0.495 \pm 0.012(\mathrm{stat}) \pm 0.007(\mathrm{syst}),\\ \mathcal{A}^{\mathrm{FB}}(B^{\pm}\to \pi^{\pm}p\bar{p}) &= -0.409 \pm 0.033(\mathrm{stat}) \pm 0.006(\mathrm{syst}).\end{aligned}$$

Additionally, Fig. 4a shows the dependence of $\mathcal{A}^{\mathrm{FB}}$ on $m_{p\bar{p}}$. These results indicate a strong dependence on $m_{p\bar{p}}$ and a sign flip between the kaon and pion decay modes.

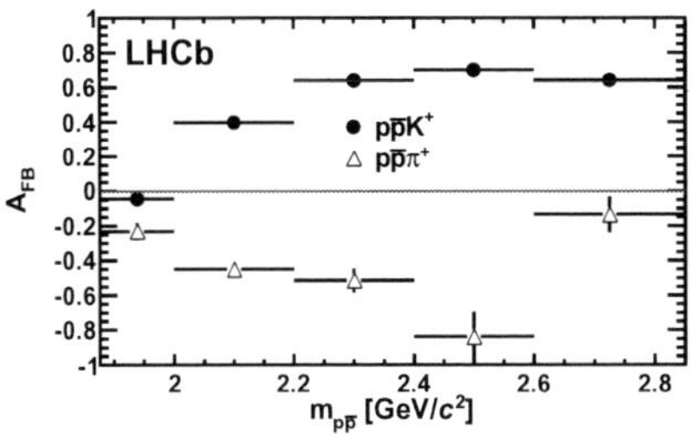

(a) Forward-backward asymmetry in bins of m_{pp} for $B^{\pm}\to K^{\pm}p\bar{p}$ and $B^{\pm}\to \pi^{\pm}p\bar{p}$ decays. The data points are shown with their total uncertainties.

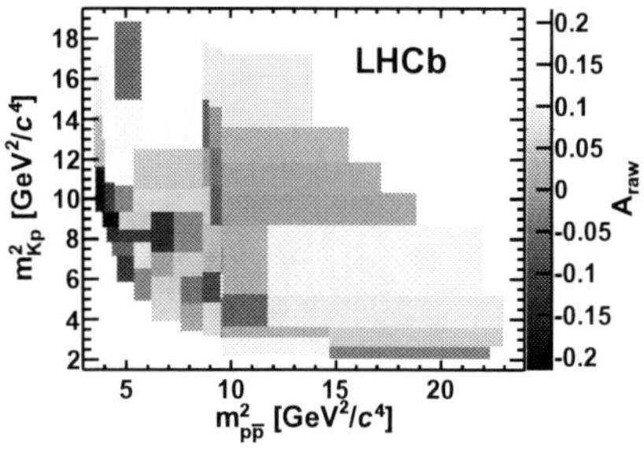

(b) Asymmetries of the number of signal events in bins of the Dalitz plot variables for $B^{\pm}\to K^{\pm}p\bar{p}$. The number of events in each bin is approximately 300.

The $\mathcal{A}^{\mathrm{CP}}$ was also measured for various regions of the Dalitz plot for the $B^{\pm}\to K^{\pm}p\bar{p}$ decay mode. The method to obtain $\mathcal{A}^{\mathrm{CP}}$ was very similar to that of the $B^{\pm}\to h^{\pm}h^{+}h^{-}$ analysis above and the raw asymmetries across the Dalitz plot can be seen in Fig. 4b. A positive $\mathcal{A}^{\mathrm{CP}}$ of nearly 4σ was found in the region $m_{p\bar{p}} < 2.85\,\mathrm{GeV}/c^2$, $m^2_{Kp} > 10\,\mathrm{GeV}^2/c^4$ with a value of $0.096 \pm 0.024(\mathrm{stat}) \pm 0.004(\mathrm{syst})$. This is the first evidence of *CP* violation in B decays with baryons in the final state.

References

[1] Dalitz R H 1953 *Phil. Mag.* **44**
[2] J Alves A A *et al.* (LHCb Collaboration) 2008 *Journal of Instrumentation* **3** S08005 ISSN 1748-0221
[3] Aaij R *et al.* (LHCb Collaboration) 2014 (*Preprint* 1407.7704)
[4] Aaij R *et al.* (LHCb Collaboration) 2013 *J. High Energy Phys.* **10** 143. 18 p
[5] Aaij R, Affolder A, Akiba K, Alexander M, Ali S *et al.* 2014 *JINST* **9** 09007 (*Preprint* 1405.7808)
[6] Heavy Flavor Averaging Group http://www.slac.stanford.edu/xorg/hfag/
[7] Aaij R *et al.* (LHCb Collaboration) 2014 (*Preprint* 1408.5373)
[8] Aaij R *et al.* (LHCb Collaboration) 2013 *Phys. Rev. Lett.* **111**(10) 101801
[9] Aaij R *et al.* (LHCb Collaboration) 2014 *Phys. Rev. Lett.* **112**(1) 011801
[10] Aaij R *et al.* (LHCb Collaboration) 2012 *Phys.Lett.* **B713** 186–195 (*Preprint* 1205.0897)
[11] Aaij R *et al.* (LHCb Collaboration) 2014 (*Preprint* 1407.5907)
[12] Aaij R *et al.* (LHCb Collaboration) 2013 *Phys.Rev.* **D88** 052015 (*Preprint* 1307.6165)

Symmetry violations at *BABAR*

Adrian John Bevan
Queen Mary University of London, Mile End Road, E1 4NS, United Kingdom
E-mail: a.j.bevan@qmul.ac.uk

Abstract. Following a brief introduction I report the current status of symmetry violation tests from the *BABAR* experiment, including recent results on the measurement of T violation, and searches for CP and T violation in mixing.

1. Introduction

The study of the discrete symmetries C (charge conjugation), P (parity), CP, T (motion reversal), and CPT has provided many insights as to the underlying structure of the weak interaction. Noether's theorem reminds us that in general we expect that exact symmetries in nature relate to conservation laws. Hence, any conserved symmetry should lead to a corresponding conservation law that underpins some physical phenomenon. Likewise symmetry violation would lead to the consequence that the conservation law did not work exactly, but was at best a rule of thumb that may be used to make estimates with some degree of uncertainty regarding the accuracy of any conclusion inferred; for example SU(3) falls into this category of symmetry. The symmetries C and P are known to be broken, and the violation of parity discovered by Wu in 1957 [1] was the first experimental signal of how important it was to understand violation of these symmetries for weak interactions. Parity violation led inextricably to the $V - A$ structure of the weak interaction in the Standard Model of particle physics. The fact that there are weak decays to flavour specific final states also highlights the presence of C violation. However a remarkable thing at that time was that CP was found to be conserved. Hence the large violation of parity and charge conjugation was such that it balanced in the combination CP. This assertion remained true until 1964, when Cronin, Fitch, Christenson and Turlay discovered CP violation in neutral kaon decay as a consequence of trying to improve our understanding of regeneration [2]. It was quickly noted by Sakharov as to how important CP violation was for cosmology [3]. Without C and CP violation the matter and antimatter manifest in the big bang would annihilate each other and the residual Universe would have been devoid of the large imbalance that we observe today. It is interesting to note that since that discovery we have found several other manifestations of CP violation, and also tested the level of T violation. However it took 45 years from the initial discovery for direct CP violation to be found in kaons [4][5]. A few years after this, large (10% level) CP violation was found studying the longitudinal polarisation basis of a rare K_L^0 decay to a four body final state [6][7]. On the theoretical side a model of CP violation was proposed by Kobayashi and Maskawa, building on earlier work concerning quark mixing by Cabibbo [8][9]. With hindsight, after several decades of detailed experimentation, it turns out that this model is the correct leading order description of CP violation in the SM. It is worth noting that just as in the case of C and P violations balancing to conserve *CP* (most of the time), that the measured levels of *CP* and T violation balance such that overall *CPT* is conserved. The ramifications of *CPT* conservation include Lorentz invariance, which underpins our understanding of modern physics. At some energy scale it is expected that quantum effects will become important when trying to describe space-time, and a consequence of this would be Lorentz violation, and hence *CPT* violation. Thus far however, experimental evidence continues to support both Lorentz and *CPT* invariance. The remainder of these proceedings focus on the B Factory tests of these discrete symmetries from the perspective of *BABAR* results, with a brief historical interlude followed by more recent results. The avid reader should refer to the Physics of the B Factories for more details [10].

2. *CP* violation in B meson decay

The B Factories, the *BABAR* experiment and PEP-II collider at the SLAC National Accelerator Laboratory and the Belle experiment and KEKB collider at KEK, were built to discover *CP* violation in the decay of neutral B mesons. This primary goal drove the teams building those experimental facilities to build in safety margins in order to be sure that the mission would be accomplished, even if *CP* violation turned out to be much smaller than expected by many theorists of the day. Nature was kind, and in 2001 *BABAR* unveiled results showing *CP* violation in the decay of a b quark to a $c\bar{c}s$ final state (modes such as $B \to J/\psi K^0_S$, and $J/\psi K^0_L$ were included in this discovery). That was quickly confirmed by Belle, firmly establishing that indeed *CP* violation was exhibited in B decay, and that it was large. These two results were published as back-to-back articles in Phys. Rev. Lett. [11][12]. In contrast with kaons, it only took a further two years before direct *CP* violation was established in B decays via the decay $B \to K^{\pm}\pi^{\mp}$. Both *CP* violation in the interference between mixing and decay amplitudes and direct *CP* violation turned out to be order one effects in B decays, in contrast to the $10^{-3} - 10^{-6}$ levels seen in kaons. The foresight that led to including safety margins in the design of the B Factories has resulted in a richer and broader physics programme than originally envisaged.

3. Other symmetry tests in B meson decay

Banuls and Bernabeu proposed using entangled pairs of B mesons to test T, *CP*, and *CPT* asymmetries [13] as a generalisation of the Kabir asymmetry measurement proposed in 1970 [14] that used only a flavour filter basis. Using two pairs of orthonormal states, one based on b-quark flavour and one based on *CP* eigenvalue, *BABAR* are able to measure 12 asymmetries of which there are four distinct tests of each of T, *CP*, and *CPT* [15]. The methodology relies on comparing some reference process with that of the symmetry conjugated one as a function of proper time between two events for the entangled pair of mesons evolving into their flavour and *CP* decay filters. The first event is marked by the collapse of the entangled wave function (at some time t_1) which coincides with the decay of one of the B mesons in the entangled pair, and the second event is marked by the decay of the remaining B (at time t_2). For example one of the four tests of T involves comparing the time-dependent ($t_2 - t_1 > 0$) rate of $\overline{B}^0 \to B_-$ with that of the conjugate transition $B_- \to \overline{B}^0$.[1] The standard *BABAR* flavour tagging techniques developed for the *CP* violation discovery measurement are used in order to select B^0 and $\overline{B}^0$ mesons, and the selection of *CP* even and *CP* odd filter basis pairs, for $b \to c\bar{c}s$ transitions differ by a K^0_L vs a K^0_S in the final state (see for example [16]). As reported in [15], the *BABAR* data are consistent with both *CP* and T violation, whilst *CPT* is conserved. The significance reported for the observation of T violation (assuming Gaussian uncertainties) exceeds 14σ. Thus the level of *CP* and T violation balance each other. A number of related measurements have recently been

[1] Here the $-(1)$ subscript refers to the *CP* eigenvalue of the B decay filter; in this case it is *CP* odd, corresponding to $J/\psi K^0_S$. The other pairings of reference and transformed transitions can be found in Refs. [13][15].

proposed [17][18] and Ref. [19] is a recent review on T violation measurements in mesons, which may be of interest to the reader.

Using only a flavour filter basis pair one can define two additional asymmetries. One of these is the Kabir asymmetry, which tests CP and T (in mixing), and the other tests CP and CPT. The corresponding asymmetry is given by

$$A_{CP,T} = \frac{\Gamma(\overline{B}^0 \to B^0) - \Gamma(B^0 \to \overline{B}^0)}{\Gamma(\overline{B}^0 \to B^0) + \Gamma(B^0 \to \overline{B}^0)} = \frac{1 - |q/p|^4}{1 + |q/p|^4} = \frac{N^{++} - N^{--}}{N^{++} + N^{--}}, \tag{1}$$

where q and p are parameters related to mixing and N^{++} and N^{--} are event yields. The SM expectation for $A_{CP,T}$ is $(-4.0 \pm 0.6) \times 10^{-4}$ [20][21][22], which is beyond current experimental reach, however large enhancements are possible in the presence of physics beyond the SM (c.f. the D0 measurement of this quantity for B_s mesons [23][24]). Traditional measurements searching for CP (and T) violation in mixing use semileptonic decays of B mesons (see for example Section 17.5 of [10]). The experimental signature typically used for this measurement is a pair of semi-leptonic decays reconstructed with same sign leptons, as the first meson decays to fix the initial flavour of the un-decayed B, and that meson subsequently mixes before decay. However those results are systematically limited by the size of control samples to estimate the level of wrong sign events. As a consequence it is interesting to attempt other ways of performing the same test. *BABAR* have recently performed a measurement using a combination of hadronic and semi-leptonic tags in order to explore an experimentally distinct final state with the hope of bypassing the traditional systematic limitations. The semi-leptonic decay $B^0 \to D^{*-} X \ell^+ \nu$ with a partially reconstructed $D^{*-} \to \pi^- \overline{D}^0$ is used for this measurement where the asymmetry between ℓ^+ and ℓ^- observed in data includes the effect of B mixing, and of charge reconstruction bias in the detector, both of which have to be corrected for in order to determine $A_{CP,T}$. Hadronic B decays used in this analysis are "kaon tagged" events, where a neutral B meson decays into a final state with a charged kaon. The charge asymmetry observed in data for these events, in addition to including an overall detector bias, has a correction from the asymmetry in reconstructing the charged kaon, arising from the difference in nuclear cross sections of K^+ and K^- in detector material. The measured value of $A_{CP,T}$ is $(0.06 \pm 0.17^{+0.38}_{-0.32})\%$, which is compatible with expectations. However, the uncertainties are considerably larger than the SM level and dominated by systematic uncertainties associated with peaking background contributions and the Δt (the proper time difference between the decay of the semi-leptonic and hadronic B decays in the event) resolution function model. Hence these uncertainties can be further reduced with input from larger data samples in the future.

BABAR has also measured CP asymmetries for a number of rare decays, and the contribution by Simon Akar to these proceedings gives a review of some of those recent measurements.

4. Symmetry violation searches in charm decay

BABAR has also searched for CP violation in charm decays. The details can be found in [10], where time-integrated and time-dependent analyses have been performed. In recent years there has been a resurgence of interest in this area for two reasons (i) charm mixing is now firmly established, which sets the scale for mixing-dependent effects to be manifest and opens the door for such measurements, and (ii) in 2011 LHCb reported time-integrated CP asymmetry difference between $D \to K^+K^-$ and $\pi^+\pi^-$ modes that was non trivial. The latest results are compatible with no evidence for CP violation [25]. There is also interest in the study of triple product asymmetries that probe CP violation using the longitudinal polarisation basis of a decay (c.f. kaon decays), in particular the decay modes $D^0 \to K^+K^-\pi^+\pi^-$, $D^+ \to K^0_S K^+\pi^+\pi^-$, and $D^+_s \to K^0_S K^+\pi^+\pi^-$ have been studied [26][27]. The results are consistent with CP conservation, in line with expectations that the charm system exhibits small weak phase differences in the

SM, hence small levels of *CP* violation. A recent re-analysis of the *BABAR* data was presented at CKM '14 [28] using the methodology introduced in Ref. [29]. This shows non-trivial C and P violations for the D^0 and D_s^+ decay, but not for D^+. These results may ultimately yield deeper insights on the weak interaction and hadronisation.

5. The search for *CP* violation in τ decay

Searches for *CP* violation are not confined to quark interactions and following CLEO and Belle's lead, *BABAR* has measured the direct *CP* asymmetry of $\tau^+ \to K_S^0\pi^+\nu$. A non-trivial level of *CP* violation is expected as a result of the neutral kaon in the final state, as well as K_S^0-K_L^0 interference and regeneration effects. The result obtained $A_{CP} = (-0.36 \pm 0.23 \pm 0.11)\%$ is compatible with no evidence for *CP* violation, however this lies 2.8σ from the SM expectation of $+0.36\%$ [30].

6. Summary

In summary *BABAR* has performed many interesting tests of discrete symmetries for B, D, and τ decays, and has observed several symmetry violations in meson decay. While *BABAR* stopped taking data several years ago, it is still producing results that are competitive and able to teach us something new about the behaviour of the weak interaction and of these discrete symmetries.

References

[1] Wu C S, Ambler E, Hayward R W, Hoppes D D and Hudson R P 1957 *Phys. Rev.* **105**, 1413.
[2] Christenson J H, Cronin J W, Fitch V L and Turlay R 1964 *Phys. Rev. Lett.* **13**, 138.
[3] Sakharov A D Pisma Zh. Eksp. Teor. Fiz. **5** (1967) 32 [1967 *JETP Lett.* **5** 24] [1991 *Sov. Phys. Usp.* **34** 392] [1991 *Usp. Fiz. Nauk* **161** 61].
[4] Alavi-Harati A *et al.* [KTeV Collaboration] 1999 *Phys. Rev. Lett.* **83** 22 [hep-ex/9905060].
[5] Fanti V *et al.* [NA48 Collaboration] 1999 *Phys. Lett.* B **465** 335 [hep-ex/9909022].
[6] Abouzaid E *et al.* [KTeV Collaboration] 2006 *Phys. Rev. Lett.* **96** 101801 [hep-ex/0508010].
[7] Lai A *et al.* [NA48 Collaboration] 2003 *Eur. Phys. J.* C **30** 33.
[8] Cabibbo N 1963 *Phys. Rev. Lett.* **10**, 531-533.
[9] Kobayashi M and Maskawa T 1972 *Prog. Theor. Phys.* **49**, 652-657.
[10] Bevan A J, Golob B, Mannel T, Prell S, Yabsley B D, *et al.*, arXiv:1406.6311 [hep-ex].
[11] Aubert B *et al.* [*BABAR* Collaboration] 2001 *Phys. Rev. Lett.* **87** 091801 [hep-ex/0107013].
[12] Abe K *et al.* [Belle Collaboration] 2001 *Phys. Rev. Lett.* **87** 091802 [hep-ex/0107061].
[13] Banuls M C and Bernabeu J 2000 *Nucl. Phys.* B **590**, 19 [hep-ph/0005323].
[14] Kabir P K 1970 *Phys. Rev.* D **2**, 540.
[15] Lees J P *et al.* [*BABAR* Collaboration] 2012 *Phys. Rev. Lett.* **109**, 211801 [arXiv:1207.5832 [hep-ex]].
[16] Aubert B *et al.* [BaBar Collaboration] 2009 *Phys. Rev.* D **79** 072009 [arXiv:0902.1708 [hep-ex]].
[17] Bevan A, Inguglia G and Zoccali M 2013 arXiv:1302.4191 [hep-ph].
[18] Dadisman R, Gardner S and Yan X 2014 arXiv:1409.6801 [hep-ph].
[19] Schubert K R 2014 arXiv:1409.5998 [hep-ex].
[20] Lenz A and Nierste U 2007 *JHEP* **0706** 072 [hep-ph/0612167].
[21] Charles J, Deschamps O, Descotes-Genon S, Itoh R, Lacker H, Menzel A, Monteil S and Niess V *et al.* 2011 *Phys. Rev.* D **84** 033005 [arXiv:1106.4041 [hep-ph]].
[22] Lenz A, Nierste U, Charles J, Descotes-Genon S, Lacker H, Monteil S, Niess V and T'Jampens S 2012 *Phys. Rev.* D **86** 033008 [arXiv:1203.0238 [hep-ph]].
[23] Abazov V M *et al.* [D0 Collaboration] 2011 *Phys. Rev.* D **84** 052007 [arXiv:1106.6308 [hep-ex]].
[24] Abazov V M *et al.* [D0 Collaboration] 2012 *Phys. Rev.* D **86** 072009 [arXiv:1208.5813 [hep-ex]].
[25] Heavy Flavor Averaging Group, `http://www.slac.stanford.edu/xorg/hfag`.
[26] del Amo Sanchez P *et al.* [*BABAR* Collaboration] 2010 *Phys. Rev.* D **81** 111103 [arXiv:1003.3397 [hep-ex]].
[27] Lees J P *et al.* [*BABAR* Collaboration] 2011 *Phys. Rev.* D **84** 031103 [arXiv:1105.4410 [hep-ex]].
[28] Martinelli M 2014 contribution to the 8th international workshop on the CKM Unitarity Triangle, Vienna.
[29] Bevan A J 2014 arXiv:1408.3813 [hep-ph].
[30] Lees J P *et al.* [BaBar Collaboration] 2012 *Phys. Rev.* D **85** 031102 [Erratum-ibid. D **85** (2012) 099904] [arXiv:1109.1527 [hep-ex]].

Journal of Physics: Conference Series **556** (2014) 012043 doi:10.1088/1742-6596/556/1/012043

Semileptonic B decays at Belle

Alexei Sibidanov[1]
School of Physics, University of Sydney, NSW 2006, Australia
E-mail: sibid@physics.usyd.edu.au

Abstract. Recent results from the Belle experiment on semileptonic decays of B-mesons are reviewed, including their effect on the determination of the CKM matrix elements $|V_{cb}|$ and $|V_{ub}|$.

1. Introduction

Precise experimental determinations of the values of the Cabibbo-Kobayashi-Maskawa (CKM) matrix elements is the way to test unitarity of the matrix and thus to probe phenomena beyond the Standard Model (SM) of Particle Physics, which can potentially violate this unitarity.

In the most commonly studied Unitarity Triangle (UT) based on the CKM matrix, the well measured angle ϕ_1 (also known as β in the literature) is opposite to the side whose length is proportional to the ratio $|V_{ub}|/|V_{cb}|$, which is currently known much less precisely, as shown in Fig. 1 made by the CKMfitter group [1]. In this ratio, the main contribution to the uncertainty comes from the value of $|V_{ub}|$. The usual way to obtain the values of $|V_{cb}|$ and $|V_{ub}|$ is to extract them from semileptonic decays of B-mesons, in which the decay rate is directly proportional to, to first order, the matrix element squared, and where QCD uncertainties due to hadronic recoil are under control.

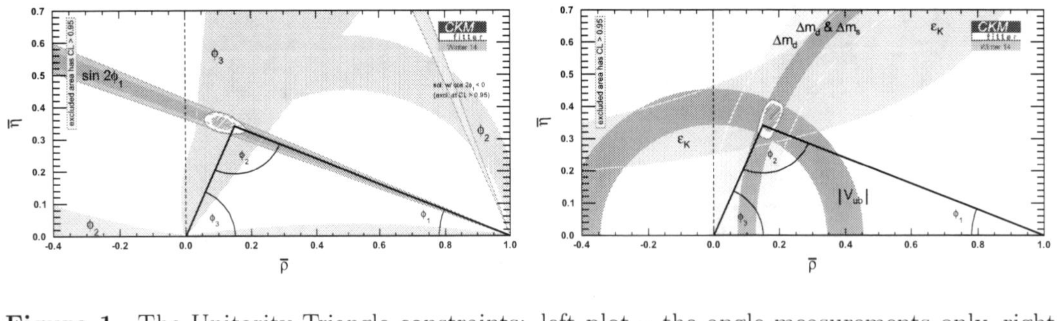

Figure 1. The Unitarity Triangle constraints: left plot – the angle measurements only, right plot – the angle measurements are excluded from the global fit [1].

Vast numbers of B_u^-, B_d^0 and B_s^0 mesons were collected by the Belle experiment at the e^+e^- asymmetric-energy KEKB collider, operating at the $\Upsilon(4S)$ and $\Upsilon(5S)$ resonances. This provides

[1] On behalf of the Belle collaboration

the opportunity to study their properties in great detail using various analysis techniques and theoretical approaches.

2. Semileptonic decays of B_s mesons

2.1. $\bar{B}^0_s \to X^+\ell^-\bar{\nu}_\ell$ inclusive branching fraction

The inclusive branching fraction measurement of $\bar{B}^0_s \to X^+\ell^-\bar{\nu}_\ell$ decay, described in more detail elsewhere [2], is based on 121 fb^{-1} of data collected at the $\Upsilon(5S)$ resonance ($\sqrt{s} = 10.87$ GeV) which contains $(7.1 \pm 1.3) \times 10^6$ $B_s^{0(*)}\bar{B}_s^{0(*)}$ pairs. The extracted branching fractions separated by lepton flavor are $\mathcal{B}(\bar{B}^0_s \to X^+e^-\bar{\nu}_e) = (9.1 \pm 0.5_{\rm stat} \pm 0.6_{\rm syst})\%$ and $\mathcal{B}(\bar{B}^0_s \to X^+\mu^-\bar{\nu}_\mu) = (10.2 \pm 0.6_{\rm stat} \pm 0.8_{\rm syst})\%$. The combined branching fraction is $\mathcal{B}(\bar{B}^0_s \to X^+\ell^-\bar{\nu}_\ell) = (9.6 \pm 0.4_{\rm stat} \pm 0.7_{\rm syst})\%$ which matches the theoretical expectations [3, 4] and agrees, and is more precise than, the previous *BABAR* measurement $\mathcal{B}(\bar{B}^0_s \to X^+\ell^-\bar{\nu}_\ell) = (9.5^{+2.5}_{-2.0}({\rm stat})^{+1.1}_{-1.9}({\rm syst}))\%$ [5].

2.2. Preliminary results of $\bar{B}^0_s \to D_s^{(*)+}X\ell^-\bar{\nu}$ study

Using the same data sample as in the previous section 2.1 Belle has studied the semi-inclusive semileptonic decays $\bar{B}^0_s \to D_s^{(*)+}X\ell^-\bar{\nu}$. To select signal decays the D_s^+ meson is reconstructed in the mode $D_s^+ \to \phi(K^+K^-)\pi^+$ and the D_s^{*+} meson in $D_s^{*+} \to D_s^+\gamma$. The numbers of D_s^+ and D_s^{*+} mesons in data are determined from a fit to the $M_{KK\pi}$ invariant mass and $\Delta M = M_{KK\pi\gamma} - M_{KK\pi}$ distributions, respectively. The preliminary branching fractions are $\mathcal{B}(\bar{B}^0_s \to D_s^+X\ell^-\bar{\nu}) = (8.2 \pm 0.2_{\rm stat} \pm 0.8_{\rm syst} \pm 1.5_{N_{B^0_s}})\%$ and $\mathcal{B}(\bar{B}^0_s \to D_s^{*+}X\ell^-\bar{\nu}) = (5.4 \pm 0.4_{\rm stat} \pm 0.5_{\rm syst} \pm 1.0_{N_{B^0_s}})\%$, where the main uncertainty comes from the estimate $N_{B^0_s}$ of the number of B^0_s mesons.

3. Semileptonic decays with a fully reconstructed tag at Belle

In e^+e^- collisions at the $\Upsilon(4S)$ resonance it is possible to fully reconstruct one B-meson decay in a known hadronic "tagging" mode where all decay products are registered by a detector, and by using energy-momentum conservation the kinematic variables of the other B meson can be calculated. This is extremely useful for the exclusive semileptonic decays $B \to X\ell\bar{\nu}_\ell$, where a particular hadronic final state X is reconstructed in the detector and the kinematic properties of the missing neutrino are reconstructed using tag-side information, providing a clean signal sample with very little background.

Recently a new reconstruction procedure for B hadronic decays based on the NeuroBayes neural net package has been introduced in Belle [6]. This procedure tries to reconstruct B mesons in more than 1100 exclusive hadronic decay channels. Compared to the previous cut-based algorithm it offers roughly a factor of two efficiency gain and about 2.1×10^6 (1.4×10^6) fully reconstructed $B^\pm$ (B^0) decays with 711 fb^{-1} collected at the $\Upsilon(4S)$ resonance.

Using this new tagging method Belle has studied the exclusive semileptonic decays $B \to D\ell\bar{\nu}_\ell$, $B^- \to p\bar{p}\ell^-\bar{\nu}_\ell$ [8], $B \to \pi\ell\bar{\nu}_\ell$, $B \to \rho\ell\bar{\nu}_\ell$ and $B^+ \to \omega\ell\bar{\nu}_\ell$ [9]. The study is based on the full data set of 711 fb^{-1} at the $\Upsilon(4S)$ resonance.

3.1. Preliminary study of $B \to D\ell\bar{\nu}_\ell$ decays

With the new hadronic tag the Belle study the $B \to D\ell\bar{\nu}_\ell$ decays compliments previous $|V_{cb}|$ evaluation from the $B \to D^*\ell\bar{\nu}_\ell$ decays [7]. To increase the statistical significance of the result D mesons are reconstructed multiple hadronic states. The signal yields in 10 bins of the $w = P_B \cdot P_D/(m_B m_D)$ variable, where m_B and m_D are the masses of the B and D mesons, respectively, and P_B and P_D are their four-momenta, are extracted from binned maximum likelihood fits to the missing mass squared ($M^2_{\rm miss}$) distributions. The preliminary branching fractions are $\mathcal{B}(\bar{B}^0 \to D^+\ell\bar{\nu}) = (2.49 \pm 0.17) \times 10^{-2}$ and $\mathcal{B}(B^- \to \bar{D}^0\ell\bar{\nu}) = (2.70 \pm 0.19) \times 10^{-2}$.

The fit to the w distribution yields $\mathcal{G}(1)|V_{cb}| = (42.6 \pm 1.0 \pm 1.4) \times 10^{-3}$ and the form factor slope $\rho_D^2 = 1.00 \pm 0.05 \pm 0.02$, which are the most precise single evaluations of these parameters to date. This result can be compared with the world average value $\mathcal{G}(1)|V_{cb}| = (42.64 \pm 0.72 \pm 1.35) \times 10^{-3}$ and $\rho_D^2 = 1.186 \pm 0.036 \pm 0.041$. Currently the exclusive value $|V_{cb}| = (39.5 \pm 0.8_{\text{tot}}) \times 10^{-3}$ and inclusive one $|V_{cb}| = (42.2 \pm 0.7_{\text{tot}}) \times 10^{-3}$ are marginally consistent [17].

3.2. Evidence for $B^- \to p\bar{p}\ell^-\bar{\nu}_\ell$ decay

The direct experimental search for $B^- \to p\bar{p}\ell^-\bar{\nu}_\ell$ decay at Belle has been triggered by a recent paper [10] which predicts an unexpectedly large branching fraction, $\mathcal{B}(B^- \to p\bar{p}\ell^-\bar{\nu}_\ell) = (1.04 \pm 0.38) \times 10^{-4}$. Previous phenomenological estimations had suggested that the branching fractions of semileptonic B decays with a baryon-antibaryon pair in the final state are at the level of $10^{-5} - 10^{-6}$ [11], which would lead to only a marginal possibility for observing such decays in the data collected by the Belle or *BABAR* experiments.

The measured branching fractions obtained [8], separated by lepton flavor are $\mathcal{B}(B^- \to p\bar{p}e^-\bar{\nu}_e) = (8.2^{+3.7}_{-3.2}(\text{stat}) \pm 0.6(\text{syst})) \times 10^{-6}$ and $\mathcal{B}(B^- \to p\bar{p}\mu^-\bar{\nu}_\mu) = (3.1^{+3.1}_{-2.4}(\text{stat}) \pm 0.7(\text{syst})) \times 10^{-6}$. The combined branching fraction, assuming lepton universality is $\mathcal{B}(B^- \to p\bar{p}\ell^-\bar{\nu}_\ell) = (5.8^{+2.4}_{-2.1}(\text{stat}) \pm 0.9(\text{syst})) \times 10^{-6}$ and has 3.2σ significance. Upper limits at the 90% confidence level are also evaluated: $\mathcal{B}(B^- \to p\bar{p}e^-\bar{\nu}_e) < 14 \times 10^{-6}$, $\mathcal{B}(B^- \to p\bar{p}\mu^-\bar{\nu}_\mu) < 8.5 \times 10^{-6}$ and for the combined mode $\mathcal{B}(B^- \to p\bar{p}\ell^-\bar{\nu}_\ell) < 9.6 \times 10^{-6}$. This result clearly contradicts the prediction from [10] and contributes to a deeper understanding of the baryonic transition form factors in B-meson decays.

3.3. Results for $B \to \pi\ell\bar{\nu}_\ell$ decays

Among charmless semileptonic decays of B mesons, the $B \to \pi\ell\bar{\nu}_\ell$ decay has the most developed theoretical apparatus to describe the $B \to \pi$ hadronic transition form factors. The number of signal events is extracted from an extended binned maximum likelihood fit to the M^2_{miss} distribution. The fit result for the $B^0 \to \pi^+\ell\bar{\nu}_\ell$ decay is shown in Fig. 2 where the signal peak is clearly visible and the signal-to-background ratio is much better in comparison with untagged measurements. The extracted branching fractions are $\mathcal{B}(B^0 \to \pi^+\ell\bar{\nu}_\ell) = (1.49 \pm 0.09(\text{stat}) \pm 0.07(\text{syst})) \times 10^{-4}$ and $\mathcal{B}(B^+ \to \pi^0\ell\bar{\nu}_\ell) = (0.80 \pm 0.08(\text{stat}) \pm 0.04(\text{syst})) \times 10^{-4}$ [9].

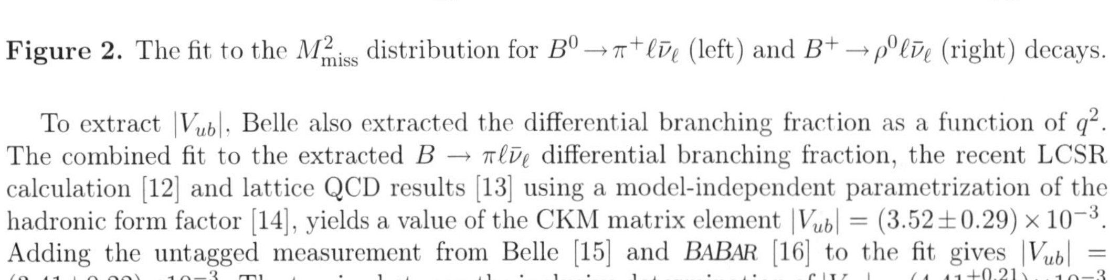

Figure 2. The fit to the M^2_{miss} distribution for $B^0 \to \pi^+\ell\bar{\nu}_\ell$ (left) and $B^+ \to \rho^0\ell\bar{\nu}_\ell$ (right) decays.

To extract $|V_{ub}|$, Belle also extracted the differential branching fraction as a function of q^2. The combined fit to the extracted $B \to \pi\ell\bar{\nu}_\ell$ differential branching fraction, the recent LCSR calculation [12] and lattice QCD results [13] using a model-independent parametrization of the hadronic form factor [14], yields a value of the CKM matrix element $|V_{ub}| = (3.52 \pm 0.29) \times 10^{-3}$. Adding the untagged measurement from Belle [15] and *BABAR* [16] to the fit gives $|V_{ub}| = (3.41 \pm 0.22) \times 10^{-3}$. The tension between the inclusive determination of $|V_{ub}| = (4.41^{+0.21}_{-0.23}) \times 10^{-3}$ quoted by PDG [17] remains significant at the $\sim 3\sigma$ level.

3.4. Results for $B \to \rho\ell\bar{\nu}_\ell$ decays

The maximum likelihood fit result for $B^+ \to \rho^0\ell\bar{\nu}_\ell$ decays is also shown in Fig. 2, where again the signal peak is clearly visible and the signal-to-background ratio is excellent. The extracted branching fractions are $\mathcal{B}(B^0 \to \rho^+\ell\bar{\nu}_\ell) = (3.22 \pm 0.27_{\text{stat}} \pm 0.24_{\text{syst}}) \times 10^{-4}$ and $\mathcal{B}(B^+ \to \rho^0\ell\bar{\nu}_\ell) = (1.83 \pm 0.10_{\text{stat}} \pm 0.10_{\text{syst}}) \times 10^{-4}$. The isospin average of measured branching fractions is 43% (2.7σ) higher than the previous PDG [17] value and its precision is almost a factor of two better.

The projection onto the $M_{\pi\pi}$ axis in the region $|M^2_{\text{miss}}| < 0.25$ GeV/c^2 of the two-dimensional binned likelihood fit in the M^2_{miss}-$M_{\pi\pi}$ plane is shown in Fig. 3. The extracted number of $B^- \to f_2\ell^-\bar{\nu}_\ell$ decays from the fit is $N(B^- \to f_2\ell^-\bar{\nu}_\ell) = 154 \pm 22$ which is more than 5σ away from zero and almost 3 times larger than the ISGW2 model prediction, whereas the main anticipated background from the HQE model and the PYTHIA package, the non-resonant process $\bar{B} \to X_u(\pi\pi)\ell^-\bar{\nu}_\ell$, is compatible with zero. This study can help to better estimate the uncertainty in inclusive determinations of $|V_{ub}|$ which comes from modeling of charmless semileptonic decays, and might decrease the tension with the exclusive measurements.

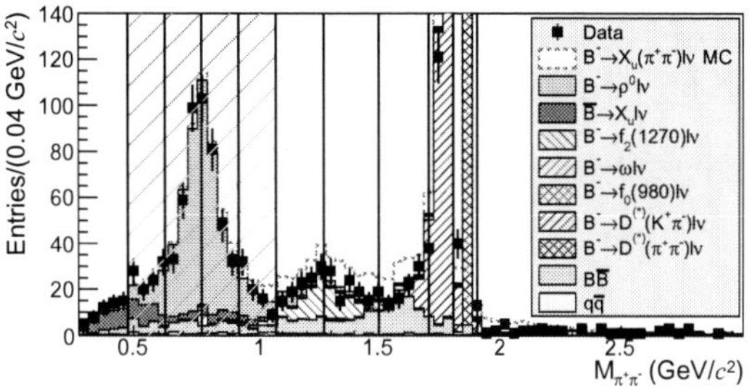

Figure 3. Projection of the fitted distribution to data for the $B^+ \to \rho^0\ell\bar{\nu}_\ell$ decay onto the $M_{\pi\pi}$ axis.

4. Conclusions

The Belle detector was decommissioned in 2010 but analysis of the collected data is not finished and is still producing outstanding scientific results. The clean environment of e^+e^- colliders is especially useful for studying semileptonic decays of B mesons in order to derive fundamental parameters of the SM such as the elements $|V_{cb}|$ and $|V_{ub}|$ of the CKM matrix.

Belle has significant progress in the (semi-)inclusive semileptonic branching fractions measurements of the B^0_s meson.

Belle has also recently developed a new procedure for B-meson full reconstruction in hadronic decay modes. This allows a number of semileptonic modes: $B \to D\ell\bar{\nu}_\ell$, $B^- \to p\bar{p}\ell^-\bar{\nu}_\ell$, $B \to \pi\ell\bar{\nu}_\ell$, $B \to \rho\ell\bar{\nu}_\ell$ and $B^+ \to \omega\ell\bar{\nu}_\ell$ to be studied.

There is much progress in the determination of $|V_{cb}|$ and $|V_{ub}|$ from B-meson semileptonic decays, where recent high-statistic measurements allow the form factor shapes to be constrained from the experimental data. Together with theory calculations this allows $|V_{cb}|$ and $|V_{ub}|$ values to be determined with high precision.

Despite all of this progress, there are still tensions at the 3σ level between exclusive and inclusive measurements of $|V_{cb}|$ and $|V_{ub}|$. This might yet be solved by improved theoretical calculations of hadronic form factors, better description of charmless semileptonic decays, and more sophisticated analysis of the existing data.

References

[1] Phys. Rev. D **84**, 033005 (2011).
[2] Phys. Rev. D **87**, 072008 (2013).
[3] Phys. Rev. D **83**, 034025 (2011).
[4] JHEP **1109**, 012 (2011).
[5] Phys. Rev. D **85**, 011101 (2012).
[6] Nucl. Instrum. Meth. A **654**, 432 (2011).
[7] Phys. Rev. D **82**, 112007 (2010).
[8] Phys. Rev. D **89**, no. 1, 011101 (2014).
[9] Phys. Rev. D **88**, 032005 (2013).
[10] Phys. Lett. B **704**, 495 (2011).
[11] Phys. Rev. Lett. **86**, 4247 (2001).
[12] JHEP **1205**, 092 (2012).
[13] Phys. Rev. D **79**, 054507 (2009).
[14] Phys. Rev. D **79**, 013008 (2009).
[15] Phys. Rev. D **83**, 071101 (2011).
[16] Phys. Rev. D **86**, 092004 (2012).
[17] Chin. Phys. C **38**, 090001 (2014).

Next-to-next-to-leading order QCD and next-to-leading order electroweak corrections to $B_s \to \mu^+\mu^-$ within the Standard Model

Matthias Steinhauser

Institut für Theoretische Teilchenphysik, Karlsruhe Institute of Technology (KIT), D-76128 Karlsruhe, Germany

E-mail: matthias.steinhauser@kit.edu

Abstract. In this contribution, the recent calculations of next-to-next-to-leading order (NNLO) QCD and next-to-leading order (NLO) electroweak (EW) corrections to the decay $B_s \to \mu^+\mu^-$ are reviewed.

1. Introduction

The decay of a B_s meson to two muons is a very promising process to look for deviations from the Standard Model (SM). The reason for this is that, on the one hand, there is both a loop and a helicity suppression which leads to a branching ratio of the order 10^{-9}. On the other hand, there is a quite clean experimental signal and thus, despite its rare nature, it can be measured experimentally. Hence, there was the hope that "New Physics" would have appeared immediately once a measurement was going to be available. However, in real life this is not the case. In 2013 the CMS and LHCb experiments at the LHC have provided measurements for the averaged time-integrated branching ratios for $B_s \to \mu^+\mu^-$ with the preliminary combination [1]

$$\overline{\mathcal{B}}_{s\mu} = (2.9 \pm 0.7) \times 10^{-9}, \tag{1}$$

which is based on Refs. [2] and [3]. This number has to be compared with the theory prediction [4] given by

$$\overline{\mathcal{B}}_{s\mu} = (3.65 \pm 0.23) \times 10^{-9}, \tag{2}$$

which shows good agreement within the uncertainty. The experimental uncertainty is statistically dominated and will be reduced in the coming years (see discussion in Ref. [5]). On the other hand, the uncertainty of the theory prediction is dominated by the B_s meson decay constant f_{B_s} and CKM matrix elements (see below). The uncertainty which comes from truncation of the perturbative series is sub-leading, thanks to the calculations performed in Refs. [6, 7] as we will briefly discuss in the next sections.

2. NNLO QCD and NLO EW corrections to $B_s \to \mu^+\mu^-$

There are several energy scales involved in decays of B mesons: first of all there is the typical scale of the decay process, μ_b, which is of the order of the bottom quark mass. Furthermore,

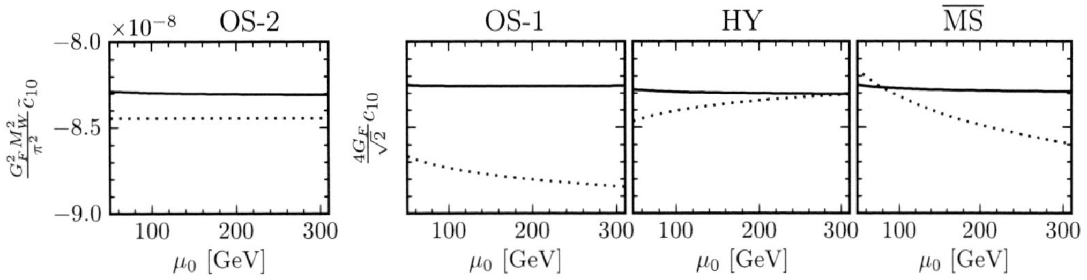

Figure 1. EW corrections adopting different renormalization schemes. In each panel the LO (dotted) and NLO (solid) results are shown as a function of the renormalization scale μ_0.

there are masses of the virtual particles in the loops, which in the SM are essentially given by the W boson and the top quark. In the following, the latter scale is denoted by μ_0. The framework which can be used to perform calculations involving widely separated scales is based on an effective theory where the heavy degrees of freedom are integrated out from the underlying theory. In the case at hand, this leads to an effective Lagrange density with only one relevant effective operator[1] $Q_A = (\bar{b}\gamma_\alpha\gamma_5 s)(\bar{\mu}\gamma^\alpha\gamma_5\mu)$. In order to arrive at predictions for the decay rate, the corresponding matching coefficient between the full and effective theory, C_A, has to be computed in a first step at the high scale μ_0. Afterwards, it has to be evolved to μ_b using renormalization group techniques.

The one-loop calculation of C_A has been performed for the first time in Ref. [8], and NLO QCD corrections have been considered in Refs. [9, 10, 11, 12]. Recently, the three-loop corrections have been computed in Ref. [6]. The calculation can be reduced to vacuum integrals involving the two masses M_W and M_t, which have been solved with the help of expansions for $M_t \gg M_W$ and $M_t \approx M_W$. In this way, only one-scale integrals have to be computed and simple analytic results are obtained, which are of polynomial form with at most logarithmic coefficients. Thus, the numerical evaluation is simple and straightforward. The inclusion of the three-loop QCD corrections to C_A reduces the uncertainties to the branching ratio from scale variation of μ_0 in the interval $M_t/2$ to $2M_t$ from 1.8% to below 0.2% [6].

The complete NLO EW corrections have been computed in Ref. [7]. Before, only the leading M_t^2 terms were known from Ref. [13] and the corresponding uncertainties have been estimated to be of the order of about 7%. In contrast to QCD, there is a non-trivial running of C_A from μ_0 to μ_b after including $\mathcal{O}(\alpha_{em})$ terms which originate from mixing of operators once QED corrections are turned on [14, 15, 16]. In Ref. [7] this effect has been taken into account together with a detailed study of the renormalization scheme dependence. The latter is shown in Fig. 1 where $\tilde{c}_{10} = -2C_A$ is plotted for four different scheme choices (see Ref. [7] for details) at LO and NLO. One observes huge differences at LO (dotted line) which basically disappear at NLO (solid line). After including NLO EW corrections the uncertainty is reduced to below 1%.

3. Prediction for $\overline{\mathcal{B}}(B_s \to \mu^+\mu^-)$

In Ref. [4] the values in Eq. (1) have been confronted with theory predictions including NNLO QCD and NLO EW corrections, as discussed in Section 2. In the following discussion, we concentrate on $B_s \to \mu^+\mu^-$; the arguments hold analogously also for $B_d \to \mu^+\mu^-$ and, thus, we

[1] In beyond-SM theories sizable contributions are also obtained from operators with scalar and pseudo-scalar currents.

		f_{B_q}	CKM	τ_H^q	M_t	α_s	other param.	non-param.	Σ
$\overline{\mathcal{B}}_{s\mu}$	$(3.65 \pm 0.23) \times 10^{-9}$	4.0%	4.3%	1.3%	1.6%	0.1%	< 0.1%	1.5%	6.4%
$\overline{\mathcal{B}}_{d\mu}$	$(1.06 \pm 0.09) \times 10^{-10}$	4.5%	6.9%	0.5%	1.6%	0.1%	< 0.1%	1.5%	8.5%

Table 1. Central value and relative uncertainties from various sources for $\overline{\mathcal{B}}_{s\mu}$ and $\overline{\mathcal{B}}_{d\mu}$. In the last column they are added in quadrature.

also present results for $\overline{\mathcal{B}}_{d\mu}$.

Within the effective-theory framework the branching ratio $\overline{\mathcal{B}}_{s\mu}$ can be written in the form (see also Ref. [17])

$$\overline{\mathcal{B}}_{s\mu} = \frac{|N|^2 M_{B_s}^3 f_{B_s}^2}{8\pi\,\Gamma_H^s}\,\beta_{s\mu}\,r_{s\mu}^2\,|C_A(\mu_b)|^2 + \mathcal{O}(\alpha_{em})\,, \tag{3}$$

with $N = V_{tb}^\star V_{ts}\, G_F^2 M_W^2/\pi^2$, $r_{s\mu} = 2m_\mu/M_{B_s}$, $\beta_{s\mu} = \sqrt{1-r_{s\mu}^2}$ and Γ_H^s denoting the heavier mass-eigenstate total width. M_{B_s} is the B_s-meson mass, and f_{B_s} its decay constant which is defined by the QCD matrix element $\langle 0|\bar{b}\gamma^\alpha\gamma_5 s|B_s(p)\rangle = ip^\alpha f_{B_s}$. The term "$\mathcal{O}(\alpha_{em})$" originates from the fact that NLO QED corrections to $\overline{\mathcal{B}}_{s\mu}$ are not complete as virtual NLO corrections to the matrix element $\langle\mu^+\mu^-|Q_A|B_s\rangle$ are missing. The $\mathcal{O}(\alpha_{em})$ term has been estimated to be of the order of 0.3% [4] (see also Ref. [18] for more details). It is so small because higher-order QED contributions either remain helicity suppressed or undergo phase-space suppression at the endpoint of the invariant mass distribution.

At that point it is straightforward to evaluate the branching ratio. We refrain from repeating the discussion about the input parameters which can be found in Ref. [4] but want to mention a few important issues in connection to the uncertainty of the branching ratio. A summary is given in Table 1.

- The largest contributions to the uncertainties arise from the decay constants and the CKM matrix elements. The former is based on lattice determinations, and the values for f_{B_s} and f_{B_d} are taken from a compilation of the Flavour Lattice Averaging Group (FLAG) [19].
- In the case of B_s, we write the CKM factors $|V_{tb}^\star V_{ts}|$ as $|V_{cb}| \times |V_{tb}^\star V_{ts}/V_{cb}|$, which allows us to use numerical results for the accurately known ratio $|V_{tb}^\star V_{ts}/V_{cb}|$. Furthermore, a precise result for $|V_{cb}|$ has recently been obtained in Ref. [20] taking into account both the semileptonic data and the precise quark mass determinations from flavor-conserving processes.
- The parameters M_t (on-shell top quark mass) and α_s enter the matching coefficient $C_A(\mu_b)$ in a non-trivial way. In Ref. [4] formulae for the branching ratio are provided which allow for a convenient change of these parameters.
- The column "other param." in Table 1 shows that the uncertainties originating from parameters that are not explicitly listed (like the Higgs or gauge boson masses or the Fermi constant) are negligible.
- The contributions to the non-parametric uncertainties, which are estimated to 1.5% both for $\overline{\mathcal{B}}_{s\mu}$ and $\overline{\mathcal{B}}_{d\mu}$ include

$\mathcal{O}(\alpha_{em})$ in Eq. (3)	0.3%
NNLO QCD μ_0 dependence	0.2%
NLO EW μ_0 dependence	0.2%
NLO EW renormalization scheme dependence	0.6%
Higher-order $M_{B_q}^2/M_W^2$ power corrections	0.4%
$\overline{\text{MS}}$−OS top quark mass conversion	0.3%

Combining these uncertainties in quadrature would give 0.9%. Our overall estimate of 1.5% is somewhat more conservative.

- In total relative uncertainties of 6.4% and 8.5% are obtained for $\overline{\mathcal{B}}_{s\mu}$ and $\overline{\mathcal{B}}_{d\mu}$, respectively.

4. Conclusions

This contribution shows that the inclusion of higher order QCD and EW corrections into the theoretical prediction of the decays $B_q \to \mu^+\mu^-$ $(q = d, s)$ significantly reduces the uncertainty, which is now dominated by CKM matrix elements and the decay constant f_{B_q}. Assuming the validity of the SM these parameters can be replaced by the mass difference of the $B_q\bar{B}_q$ system and the bag parameter B_{B_q} which might be a way to further reduce the uncertainties (see, e.g., Refs. [21, 4, 5]).

Acknowledgements

This work was supported by the DFG through the SFB/TR 9 "Computational Particle Physics". I would like to thank Christoph Bobeth, Martin Gorbahn, Thomas Hermann, Mikołaj Misiak, and Emmanuel Stamou for a fruitful collaboration on the subject of this contribution.

References

[1] CMS and LHCb Collaborations, EPS-HEP 2013 European Physical Society Conference on High Energy Physics, Stockholm, Sweden, 2013, Conference Report No. CMS-PAS-BPH-13-007, LHCb-CONF-2013-012, `http://cds.cern.ch/record/1564324` .
[2] S. Chatrchyan *et al.* (CMS Collaboration), Phys. Rev. Lett. **111**, 101804 (2013) [arXiv:1307.5025].
[3] R. Aaij *et al.* (LHCb Collaboration), Phys. Rev. Lett. **111** 101805, (2013) [arXiv:1307.5024].
[4] C. Bobeth, M. Gorbahn, T. Hermann, M. Misiak, E. Stamou and M. Steinhauser, Phys. Rev. Lett. **112** (2014) 101801 [arXiv:1311.0903 [hep-ph]].
[5] C. Bobeth, arXiv:1405.4907 [hep-ph].
[6] T. Hermann, M. Misiak and M. Steinhauser, JHEP **1312**, 097 (2013) [arXiv:1311.1347].
[7] C. Bobeth, M. Gorbahn and E. Stamou, Phys. Rev. D **89**, 034023 (2014) [arXiv:1311.1348].
[8] T. Inami and C. S. Lim, Prog. Theor. Phys. **65** (1981) 297 [Erratum-ibid. **65** (1981) 1772].
[9] G. Buchalla and A. J. Buras, Nucl. Phys. B **398** (1993) 285.
[10] G. Buchalla and A. J. Buras, Nucl. Phys. B **400** (1993) 225.
[11] M. Misiak and J. Urban, Phys. Lett. B **451** (1999) 161 [hep-ph/9901278].
[12] G. Buchalla and A. J. Buras, Nucl. Phys. B **548** (1999) 309 [hep-ph/9901288].
[13] G. Buchalla and A. J. Buras, Phys. Rev. D **57** (1998) 216 [hep-ph/9707243].
[14] C. Bobeth, P. Gambino, M. Gorbahn and U. Haisch, JHEP **0404**, 071 (2004) [hep-ph/0312090].
[15] T. Huber, E. Lunghi, M. Misiak and D. Wyler, Nucl. Phys. B **740**, 105 (2006) [hep-ph/0512066].
[16] M. Misiak, in Proceedings of 15th Lomonosov Conference on Elementary Particle Physics (Moscow, Russia, August 18-24, 2011), 301, [arXiv:1112.5978].
[17] K. De Bruyn *et al.*, Phys. Rev. Lett. **109**, 041801 (2012) [arXiv:1204.1737].
[18] M. Steinhauser, [arXiv:1406.6787 [hep-ph]].
[19] S. Aoki *et al.* (Flavour Lattice Averaging Group), [arXiv:1310.8555]; updates at `http://itpwiki.unibe.ch/flag` .
[20] P. Gambino and C. Schwanda, Phys. Rev. D **89**, 014022 (2014) [arXiv:1307.4551].
[21] A. J. Buras, Phys. Lett. B **566**, 115 (2003) [hep-ph/0303060].

Theoretical overview of kaon decays

Antonio Pich

IFIC, Universitat de València – CSIC, Apt. Correus 22085, E-46071 València, Spain

E-mail: Antonio.Pich@ific.uv.es

Abstract. Kaon decays are an important testing ground of the electroweak flavour theory. They can provide new signals of CP violation and, perhaps, a window into physics beyond the Standard Model. At the same time, they exhibit an interesting interplay of long-distance QCD effects in flavour-changing transitions. A brief overview is presented, focusing on a few selected topics of particular interest. A more detailed and comprehensive review can be found in Ref. [1].

1. Effective Field Theory

Kaon physics has been at the origin of many fundamental ingredients of the Standard Model (SM), such as flavour quantum numbers, parity violation, meson-antimeson mixing, quark mixing, CP violation and the GIM mechanism [1]. Rare kaon decays provide sensitivity to short-distance scales (c, t, $W^\pm$, Z) and offer the exciting possibility of unravelling new physics beyond the SM. Searching for forbidden lepton-flavour-violating processes ($K_L \to e^\pm\mu^\mp$, $K_L \to e^\pm e^\pm\mu^\mp\mu^\mp$, $K^+ \to \pi^+\mu^\pm e^\mp \ldots$) beyond the 10^{-10} level, one is actually exploring energy scales above the 10 TeV region. The study of allowed decay modes provides, at the same time, very interesting tests of the SM itself, improving our understanding of the interplay among electromagnetic, weak and strong interactions. In addition, new signals of CP violation, which would help to elucidate the source of CP-violating phenomena, can be looked for.

Owing to the presence of very different mass scales ($m_\pi < m_K \ll M_W$), the QCD corrections are amplified by large logarithms. The short-distance logarithmic corrections can be summed up, using the operator product expansion (OPE) and the renormalization group, all the way down from M_W to scales $\mu < m_c$ [2]. One gets in this way an effective Lagrangian, defined in the three-flavour theory, which is a sum of local four-fermion operators Q_i, constructed with the light degrees of freedom ($u, d, s; e, \mu, \nu_\ell$), modulated by Wilson coefficients $C_i(\mu)$ which are functions of the heavy (Z, W, t, b, c, τ) masses:

$$\mathcal{L}_{\mathrm{eff}}^{\Delta S=1} = -\frac{G_F}{\sqrt{2}} V_{ud}V_{us}^* \sum_i C_i(\mu)\, Q_i\,, \qquad\qquad C_i(\mu) = z_i(\mu) - y_i(\mu)\,\frac{V_{td}V_{ts}^*}{V_{ud}V_{us}^*}\,. \tag{1}$$

The CP-violating decay amplitudes are proportional to the components $y_i(\mu)$. The overall renormalization scale μ separates the short- ($M > \mu$) and long-distance ($m < \mu$) contributions, which are contained in $C_i(\mu)$ and Q_i, respectively. The Wilson coefficients are fully known at the next-to-leading order (NLO) [3, 4]; this includes all corrections of $\mathcal{O}(\alpha_s^n t^n)$ and $\mathcal{O}(\alpha_s^{n+1} t^n)$, where $t \equiv \log(M_1/M_2)$ refers to the logarithm of any ratio of heavy mass scales ($M_{1,2} \geq \mu$). In order to calculate the kaon decay amplitudes, we also need to know the non-perturbative matrix elements of the operators Q_i between the initial and final hadronic states.

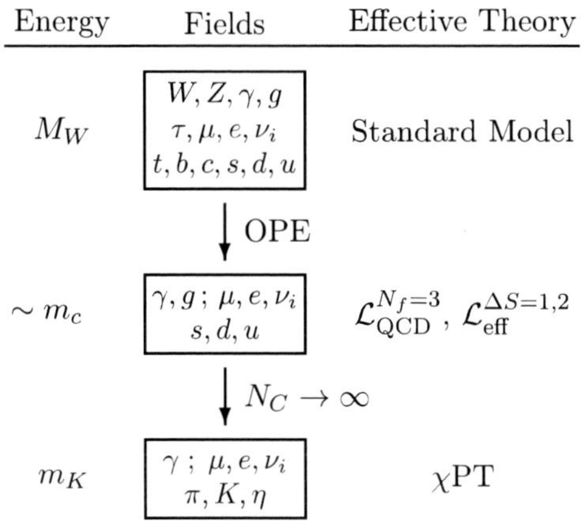

Figure 1. Evolution from M_W to m_K.

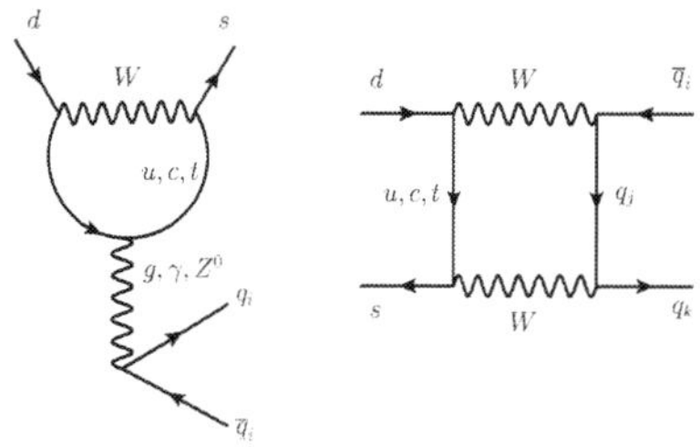

Figure 2. Short-distance diagrams.

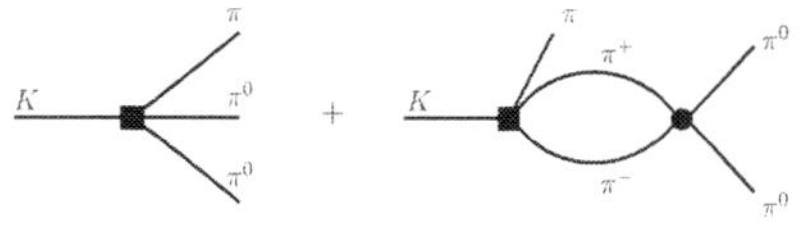

Figure 3. Long-distance diagrams.

At low energies, one can use symmetry considerations to define another effective field theory in terms of the QCD Goldstone bosons (π, K, η). Chiral Perturbation Theory (χPT) [5,6] describes the pseudoscalar-octet dynamics through a perturbative expansion in powers of momenta and quark masses over the chiral symmetry-breaking scale $\Lambda_\chi \sim 1$ GeV. Chiral symmetry fixes the allowed operators, while all short-distance information is encoded in their low-energy couplings (LECs) [7,8]. At LO the most general effective Lagrangian, with the same $SU(3)_L \otimes SU(3)_R$ transformation properties as the short-distance Lagrangian (1), contains three terms [1]:

$$\mathcal{L}_2^{\Delta S=1} = -\frac{G_F}{\sqrt{2}} V_{ud} V_{us}^* \left\{ g_8 \, \langle \lambda L_\mu L^\mu \rangle + g_{27} \left(L_{\mu 23} L_{11}^\mu + \frac{2}{3} L_{\mu 21} L_{13}^\mu \right) + e^2 g_8 g_{\rm ew} F^6 \, \langle \lambda U^\dagger Q U \rangle \right\}, \quad (2)$$

where $U \equiv \exp(i\vec{\lambda}\vec{\phi}/F)$ parameterizes the Goldstone fields, $L_\mu = iF^2 U^\dagger D_\mu U$ represents the octet of $V - A$ currents, $\lambda \equiv (\lambda_6 - i\lambda_7)/2$ projects onto the $\bar{s} \to \bar{d}$ transition, $Q = \frac{1}{3}\,\mathrm{diag}(2, -1, -1)$ is the quark charge matrix and $\langle\,\rangle$ denotes a 3-dimensional flavour trace. The LECs g_8 and g_{27} measure the strength of the two parts of $\mathcal{L}_{\rm eff}^{\Delta S=1}$ transforming as $(8_L, 1_R)$ and $(27_L, 1_R)$, respectively, under chiral rotations, while $g_{\rm ew}$ accounts for the electromagnetic penguin operators.

The χPT framework determines the most general form of the K decay amplitudes, compatible with chiral symmetry, in terms of the LECs multiplying the relevant chiral operators. These LECs, which encode the short-distance dynamics, can be determined phenomenologically and/or calculated in the limit of a large number of QCD colours N_C. Chiral loops generate non-polynomial contributions, with logarithms and threshold factors as required by unitarity. Fig. 1 shows schematically the procedure used to evolve down from M_W to m_K. While the OPE resums the short-distance logarithmic corrections $\log(M/\mu)$, the χPT loops take care of the large infrared logarithms $\log(\mu/m_\pi)$ associated with unitarity corrections (final-state interactions).

2. Leptonic and Semileptonic Decays

In (semi)leptonic decays strong interactions only appear through the hadronic matrix elements of the left-handed current, which can be precisely studied within χPT and with lattice simulations.

The ratio $\Gamma[K^- \to e^-\bar{\nu}_e(\gamma)]/\Gamma[K^- \to \mu^-\bar{\nu}_\mu(\gamma)]$ has been calculated [9, 10] and measured [11, 12] with high accuracy, allowing for a precise test of charged-current lepton universality: $|g_\mu/g_e| = 0.9978\,(20)$. Similar precisions have been achieved in $K \to \pi\ell\nu_\ell$ [13], $\pi \to \ell\nu_\ell$ and $\tau \to \nu_\tau\ell\nu_\ell$ [14] decays. The ratio $\Gamma[K^- \to \mu^-\bar{\nu}_\mu(\gamma)]/\Gamma[\pi^- \to \mu^-\bar{\nu}_\mu(\gamma)]$ provides information on the quark mixing matrix [13, 15]. With a careful treatment of electromagnetic and isospin-violating corrections, one extracts $|V_{us}/V_{ud}||F_K/F_\pi| = 0.2763 \pm 0.0005$ [16]. Taking for the ratio of decay constants the ($N_f = 2+1+1$) lattice average $F_K/F_\pi = 1.196 \pm 0.005$ [17], this gives

$$|V_{us}/V_{ud}| \;=\; 0.2310 \pm 0.0011 \,. \tag{3}$$

Including electromagnetic and isospin-breaking corrections [18, 19], the most recent $K_{\ell 3}$ data leads to $|V_{us}\, f_+(0)| = 0.2163 \pm 0.0005$ [13], with $f_+(0) = 1 + \mathcal{O}[(m_s - m_u)^2]$ the $K^0 \to \pi^- \ell^+ \nu_\ell$ vector form factor. The exact value of $f_+(0)$ has been thoroughly investigated since the first precise estimate by Leutwyler and Roos, $f_+(0) = 0.961 \pm 0.008$ [20]. While analytical calculations based on χPT obtain higher values [21, 22], owing to the large ($\sim +0.01$) 2-loop corrections [23], lattice results [17] used to agree with the Leutwyler–Roos estimate. The most recent lattice analyses [24, 25] give, however, larger values in better agreement with the χPT expectations. The FLAG $N_f = 2+1$ average $f_+(0) = 0.9661\,(32)$ [17] implies $|V_{us}| = 0.2239\,(9)$, while a recent $N_f = 2+1+1$ determination [24], $f_+(0) = 0.9704\,(32)$, results in

$$|V_{us}| \;=\; 0.2229 \pm 0.0005_{\rm exp} \pm 0.0007_{\rm th} \,. \tag{4}$$

Together with $|V_{ud}| = 0.97425 \pm 0.00022$ and the tiny $|V_{ub}|$ contribution [26], Eqs. (3) and (4) imply an stringent test of the unitarity of the quark mixing matrix:

$$\Delta_{\rm CKM} = |V_{ud}|^2 + |V_{us}|^2 + |V_{ub}|^2 - 1 \;=\; -0.0007 \pm 0.0007 \,. \tag{5}$$

3. Non-leptonic Decays

The measured $A(K \to \pi\pi)_I$ decay amplitudes show a strong enhancement of the octet $\Delta I = \frac{1}{2}$ transition amplitude into a 2π final state with isospin $I = 0$: $|A_0/A_2| \approx 22$. In the χPT framework this manifests as a huge difference between the LECs in Eq. (2). A LO fit to the data gives $|g_8| \simeq 5.0$ and $|g_{27}| \simeq 0.285$. Part of the enhancement originates in the strong rescattering of the final pions, which at one loop increases A_0 by roughly 35%. Taking the χPT 1-loop contributions into account, one finds a sizeably smaller octet coupling (the central values change slightly to 3.61 and 0.297, respectively, if isospin violation is included) [1, 27, 28]:

$$|g_8| \;=\; 3.62 \pm 0.28 \,, \qquad\qquad |g_{27}| \;=\; 0.286 \pm 0.028 \,. \tag{6}$$

In the absence of QCD, the SM (W exchange) prediction $g_8 = g_{27} = \frac{3}{5}$ would disagree with (6). The short-distance QCD corrections show the needed qualitative trend to understand the data. The matching of the effective descriptions $\mathcal{L}_{\rm eff}^{\Delta S=1}$ (short-distance) and $\mathcal{L}_2^{\Delta S=1}$ (χPT) can be done in the large-N_C limit; including the large short-distance logarithmic corrections $\sim \frac{1}{N_C} \log{(M_W/\mu)}$, this gives the results $g_8^\infty = 1.13 \pm 0.18$ and $g_{27}^\infty = 0.46 \pm 0.01$ [1], which show the relevance of missing NLO corrections in $1/N_C$.

A dynamical understanding of the $\Delta I = \frac{1}{2}$ enhancement was achieved long time ago [29], through a combined expansion in powers of momenta (χPT) and $1/N_C$. With one virtual $W^\pm$ field emitted and reabsorbed, and to LO in the chiral expansion (two L_μ insertions at most), there are three possible chiral invariant configurations which give rise to the effective Lagrangian

$$\mathcal{L}_{\rm eff} \;=\; -\frac{G_F}{\sqrt{2}} \left\{ a\, \langle Q_L^{(-)} L_\mu \rangle \langle Q_L^{(+)} L^\mu \rangle \,+\, b\, \langle Q_L^{(-)} L_\mu Q_L^{(+)} L^\mu \rangle \,+\, c\, \langle Q_L^{(-)} Q_L^{(+)} L_\mu L^\mu \rangle \right\} . \tag{7}$$

The operators $Q_L^{(+)} = Q_L^{(-)\dagger}$ represent the emission and absorption of the virtual $W^\pm$ field; thus, $[Q_L^{(+)}]_{ij} = \delta_{i1}\delta_{j2} V_{ud} + \delta_{i1}\delta_{j3} V_{us}$. The underlying functional integral over quark and gluon fields

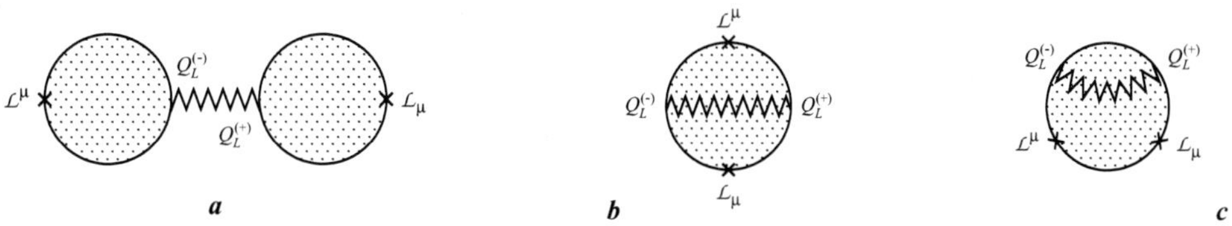

Figure 4. $K \to \pi\pi$ topologies. The solid lines represent quark fields propagating in a gluon background simulated by the dotted lines. The $W^{\pm}$ is exchanged between two $Q_L^{(\pm)}$ sources [29].

which gives rise to the effective couplings in (7) is represented diagrammatically in Fig. 4. When further restricted to $\Delta S = 1$ transitions, the effective Lagrangian (7) reduces to (2) with [29]

$$g_8 = \frac{3}{5}(a+b) - b + c\,, \qquad\qquad g_{27} = \frac{3}{5}(a+b)\,. \tag{8}$$

The topology which leads to the a-type coupling is $\mathcal{O}(N_C^2)$, while those generating b and c are $\mathcal{O}(N_C)$. In the large-N_C limit the coupling a can be calculated because the four-quark operators factorize into QCD currents with well-known χPT realizations. This factorization is only broken by (at least two) gluonic exchanges which are of NNLO in the $1/N_C$ expansion. The c-type configuration corresponds to the so-called penguin-like diagrams which can also be calculated at LO in terms of known phenomenological parameters. Taking into account the factor $F^2 \sim \mathcal{O}(N_C)$, included in the definition of the currents L_μ, one easily finds [29]:

$$\begin{aligned} a &= 1 + \mathcal{O}(1/N_C^2)\,, \qquad\qquad b = \mathcal{O}(1/N_C)\,, \\ c &= \mathrm{Re}\, C_4 - 16\, L_5\, \mathrm{Re}\, C_6(\mu)\, \left(\langle 0|\bar{q}q|0\rangle(\mu)/F^3\right)^2 + \mathcal{O}(1/N_C^2) \simeq 0.3 \pm 0.2\,, \end{aligned} \tag{9}$$

where L_5 is an $\mathcal{O}(p^4)$ coupling of the strong χPT Lagrangian. To $\mathcal{O}(1/N_C)$ the scale dependence in $C_6(\mu)$ cancels with the one in the quark condensate $\langle 0|\bar{q}q|0\rangle(\mu)$, while C_4 is scale-independent.

The non-leading topology b cannot be evaluated in a model-independent way. The important observation made in Ref. [29] is that it contributes with opposite signs to g_8 and g_{27}. Taking Eq. (9) into account, the experimental value of A_2, *i.e.* $|g_{27}| \approx 0.29$, implies [29]:

$$b \approx -0.52 + \mathcal{O}(1/N_C^2)\,, \qquad\qquad g_8 \approx 1.1 + \mathcal{O}(1/N_C^2)\,. \tag{10}$$

Thus, the measured ratio $|g_8/g_{27}|$ requires a large and negative value of b, generating a significant cancellation in g_{27} and a sizeable enhancement of g_8. This is precisely what was previously predicted through model-dependent calculations [30–33] and confirmed through a rigorous inclusive NLO analysis of the two-point correlators associated with the short-distance Lagrangians $\mathcal{L}_{\mathrm{eff}}^{\Delta S=1,2}$ [29, 30, 34]. Recently, the predicted cancellation in g_{27} has been observed by a lattice calculation of A_2 which finds $b/a \approx -0.7$ [35]; the corresponding enhancement of g_8 is also seen in A_0, although the present lattice results are still obtained at unphysical kinematics.

A similar cancellation is observed in lattice simulations of the $\Delta S = 2$ transition amplitude [36, 37]. In the chiral limit, the so-called B_K parameter which regulates the K^0–$\bar{K}^0$ hadronic matrix element is given by $B_K = \frac{3}{4}(a+b)$ [29], which behaves as g_{27}. This result agrees with explicit model calculations [30–33, 38, 39] and QCD sum-rule determinations [40, 41].

4. Direct CP Violation: ε'/ε

The measured CP-violating ratio [42–45] $[\eta_{ab} \equiv A(K_L \to \pi^a\pi^b)/A(K_S \to \pi^a\pi^b)]$

$$\mathrm{Re}\,(\varepsilon'/\varepsilon) = \frac{1}{3}\left(1 - \left|\frac{\eta_{00}}{\eta_{+-}}\right|\right) = (16.8 \pm 2.0) \times 10^{-4} \tag{11}$$

demonstrates the existence of direct CP violation in the $K \to 2\pi$ decay amplitudes. When CP violation is turned on, the amplitudes A_I acquire imaginary parts. To first order in CP violation,

$$\varepsilon' = -\frac{i}{\sqrt{2}}\, e^{i(\chi_2-\chi_0)}\, \frac{\mathrm{Re}A_2}{\mathrm{Re}A_0}\left[\frac{\mathrm{Im}A_0}{\mathrm{Re}A_0} - \frac{\mathrm{Im}A_2}{\mathrm{Re}A_2}\right], \tag{12}$$

where the strong phases χ_I can be identified with the S-wave $\pi\pi$ scattering phase shifts at $\sqrt{s} = m_K$, up to isospin-breaking effects [27,28]. The phase $\phi_{\varepsilon'} = \chi_2 - \chi_0 + \pi/2 = (42.5 \pm 0.9)^\circ$ is very close to the so-called superweak phase, $\phi_\varepsilon \approx \tan^{-1}[2(m_{K_L} - m_{K_S})/(\Gamma_{K_S} - \Gamma_{K_L})] = (43.52 \pm 0.05)^\circ$, implying that $\cos(\phi_{\varepsilon'} - \phi_\varepsilon) \approx 1$. The CP-conserving amplitudes $\mathrm{Re}A_I$ can be set to their experimentally determined values, avoiding in this way the large uncertainties associated with the hadronic matrix elements of the four-quark operators in $\mathcal{L}_{\mathrm{eff}}^{\Delta S=1}$. Thus, one only needs a first-principle calculation of the CP-odd amplitudes $\mathrm{Im}A_0$ and $\mathrm{Im}A_2$; the first one is completely dominated by the strong penguin operator Q_6, while the leading contribution to the second one comes from the electromagnetic penguin Q_8. Fortunately, those are precisely the only operators that are well approximated through a large-N_C estimate of LECs, because their anomalous dimensions are leading in $1/N_C$. Owing to the large ratio $\mathrm{Re}A_0/\mathrm{Re}A_2$, isospin violation plays also an important role in ε'/ε [28]. The one-loop χPT enhancement of the isoscalar amplitude [46,47] destroys an accidental LO cancellation of the two terms in (12) [48–50], bringing the SM prediction of ε'/ε in good agreement with the experimental measurement in Eq. (11) [1,46,47]:

$$\mathrm{Re}\,(\epsilon'/\epsilon) = \left(19 \pm 2_\mu\, {}^{+9}_{-6}{}_{m_s} \pm 6_{1/N_C}\right) \times 10^{-4}. \tag{13}$$

5. Rare and Radiative Decays

Kaon decays mediated by flavour-changing neutral currents are suppressed in the SM and their main interest, other than their own understanding, relies on the possible observation of new physics effects. Most of these processes ($K \to \gamma^{(*)}\gamma^{(*)}$, $K \to \ell^+\ell^-$, $K \to \pi\gamma^{(*)}\gamma^{(*)}$, $K \to \pi\ell^+\ell^- \ldots$) are dominated by long-distance contributions which can be analyzed with χPT techniques [51–64]. However, there are also processes governed by short-distance amplitudes, such as $K \to \pi\nu\bar{\nu}$ [48, 65, 66]. The decays $K_L \to \pi^0 e^+ e^-$ and $K_L \to \pi^0\nu\bar{\nu}$ provide very interesting measures of CP violation. The first one receives contributions from both direct and indirect CP violation which dominate the decay amplitude; the CP-conserving contribution being suppressed by an additional power of α [54]. The decay $K_L \to \pi^0\nu\bar{\nu}$ violates CP and it is completely dominated by the direct-CP contribution. A detailed overview of rare K decays with a more complete reference list can be found in Refs. [1,67].

Acknowledgments

Work supported in part by the Spanish Government [grants FPA2011-23778 and CSD2007-00042 (Consolider Project CPAN)] and the Generalitat Valenciana [PrometeoII/2013/007].

References

[1] V. Cirigliano, G. Ecker, H. Neufeld, A. Pich and J. Portolés, *Rev. Mod. Phys.* **84** (2012) 399.
[2] F. J. Gilman and M. B. Wise, *Phys. Rev.* D **20** (1979) 2392, **21** (1980) 3150.
[3] A. J. Buras *et al.*, *Nucl. Phys.* B **400** (1993) 37, 75, **408** (1993) 209; *Phys. Lett.* B **389** (1996) 749.
[4] M. Ciuchini *et al.*, *Z. Phys.* C **68** (1995) 239; *Phys. Lett.* B **301** (1993) 263; *Nucl. Phys.* B **415** (1994) 403.
[5] S. Weinberg, *Physica* A **96** (1979) 327.
[6] J. Gasser and H. Leutwyler, *Nucl. Phys.* B **250** (1985) 465, 517; *Annals Phys.* **158** (1984) 142.
[7] G. Ecker, *Prog. Part. Nucl. Phys.* **35** (1995) 1.
[8] A. Pich, *Rept. Prog. Phys.* **58** (1995) 563.
[9] W. J. Marciano and A. Sirlin, *Phys. Rev. Lett.* **71** (1993) 3629.
[10] V. Cirigliano and I. Rosell, *Phys. Rev. Lett.* **99** (2007) 231801; *JHEP* **0710** (2007) 005.
[11] C. Lazzeroni *et al.* [NA62 Collaboration], *Phys. Lett.* B **719** (2013) 326, **698** (2011) 105.

[12] F. Ambrosino *et al.* [KLOE Collaboration], *Eur. Phys. J.* C **64** (2009) 627 [*Erratum-ibid.* **65** (2010) 703].
[13] M. Antonelli *et al.*, *Eur. Phys. J.* C **69** (2010) 399.
[14] A. Pich, *Prog. Part. Nucl. Phys.* **75** (2014) 41.
[15] W. J. Marciano, *Phys. Rev. Lett.* **93** (2004) 231803.
[16] V. Cirigliano and H. Neufeld, *Phys. Lett.* B **700** (2011) 7.
[17] S. Aoki *et al.* [FLAG Working Group], arXiv:1310.8555v4 [hep-lat].
[18] V. Cirigliano, M. Giannotti and H. Neufeld, *JHEP* **0811** (2008) 006.
[19] A. Kastner and H. Neufeld, *Eur. Phys. J.* C **57** (2008) 541.
[20] H. Leutwyler and M. Roos, *Z. Phys.* C **25** (1984) 91.
[21] M. Jamin, J. A. Oller and A. Pich, *JHEP* **0402** (2004) 047.
[22] V. Cirigliano, G. Ecker, M. Eidemuller, R. Kaiser, A. Pich and J. Portolés, *JHEP* **0504** (2005) 006.
[23] J. Bijnens and P. Talavera, *Nucl. Phys.* B **669** (2003) 341.
[24] A. Bazavov *et al.* [FNAL–MILC Collab.], *Phys. Rev. Lett.* **112** (2014) 112001; *Phys. Rev.* D **87** (2013) 073012.
[25] P. A. Boyle *et al.* [RBC/UKQCD Collaboration], *JHEP* **1308** (2013) 132.
[26] K. A. Olive *et al.* [Particle Data Group Collaboration], *Chin. Phys.* C **38** (2014) 090001.
[27] V. Cirigliano, G. Ecker and A. Pich, *Phys. Lett.* B **679** (2009) 445.
[28] V. Cirigliano *et al.*, *Eur. Phys. J.* C **33** (2004) 369; *Phys. Rev. Lett.* **91** (2003) 162001.
[29] A. Pich and E. de Rafael, *Phys. Lett.* B **374** (1996) 186.
[30] A. Pich and E. de Rafael, *Nucl. Phys.* B **358** (1991) 311.
[31] S. Bertolini *et al.*, *Nucl. Phys.* B **514** (1998) 63, **469** (1996) 143.
[32] W. A. Bardeen *et al.*, *Phys. Lett.* B **192** (1987) 138, **211** (1988) 343; *Eur. Phys. J.* C **74** (2014) 2871.
[33] J. Bijnens and J. Prades, *JHEP* **9901** (1999) 023; *Nucl. Phys.* B **444** (1995) 523.
[34] M. Jamin and A. Pich, *Nucl. Phys.* B **425** (1994) 15.
[35] P. A. Boyle *et al.* [RBC and UKQCD Collaborations], *Phys. Rev. Lett.* **110** (2013) 152001, **108** (2012) 141601; *Phys. Rev.* D **86** (2012) 074513.
[36] L. Lellouch, arXiv:1104.5484 [hep-lat].
[37] N. Carrasco, V. Lubicz and L. Silvestrini, *Phys. Lett.* B **736** (2014) 174.
[38] J. Bijnens, E. Gámiz and J. Prades, *JHEP* **0603** (2006) 048.
[39] S. Peris and E. de Rafael, *Phys. Lett.* B **490** (2000) 213.
[40] A. Pich and E. De Rafael, *Phys. Lett.* B **158** (1985) 477.
[41] J. Prades *et al.*, *Z. Phys.* C **51** (1991) 287.
[42] J. R. Batley *et al.* [NA48 Collaboration], *Phys. Lett.* B **544** (2002) 97, **465** (1999) 335; *Eur. Phys. J.* C **22** (2001) 231.
[43] G. D. Barr *et al.* [NA31 Collaboration], *Phys. Lett.* B **317** (1993) 233, **206** (1988) 169.
[44] E. Abouzaid *et al.* [KTeV Collaboration], *Phys. Rev.* D **83** (2011) 092001, **67** (2003) 012005 [*Erratum-ibid.* D **70** (2004) 079904]; *Phys. Rev. Lett.* **83** (1999) 22.
[45] L. K. Gibbons *et al.*, *Phys. Rev. Lett.* **70** (1993) 1203.
[46] E. Pallante and A. Pich, *Phys. Rev. Lett.* **84** (2000) 2568; *Nucl. Phys.* B **592** (2001) 294.
[47] E. Pallante, A. Pich and I. Scimemi, *Nucl. Phys.* B **617** (2001) 441.
[48] G. Buchalla, A. J. Buras and M. E. Lautenbacher, *Rev. Mod. Phys.* **68** (1996) 1125.
[49] S. Bertolini, M. Fabbrichesi and J. O. Eeg, *Rev. Mod. Phys.* **72** (2000) 65.
[50] T. Hambye *et al.*, *Nucl. Phys.* B **564** (2000) 391.
[51] G. D'Ambrosio and D. Espriu, *Phys. Lett.* B **175** (1986) 237.
[52] J. L. Goity, *Z. Phys.* C **34** (1987) 341.
[53] G. Ecker, A. Pich and E. de Rafael, *Phys. Lett.* B **189** (1987) 363, **237** (1990) 481.
[54] G. Ecker, A. Pich and E. de Rafael, *Nucl. Phys.* B **303** (1988) 665.
[55] L. Cappiello and G. D'Ambrosio, *Nuovo Cim.* A **99** (1988) 155.
[56] G. Ecker and A. Pich, *Nucl. Phys.* B **366** (1991) 189.
[57] A. G. Cohen, G. Ecker and A. Pich, *Phys. Lett.* B **304** (1993) 347.
[58] J. Kambor and B. R. Holstein, *Phys. Rev.* D **49** (1994) 2346.
[59] L. Cappiello, G. D'Ambrosio and M. Miragliuolo, *Phys. Lett.* B **298** (1993) 423.
[60] G. D'Ambrosio and J. Portolés, *Phys. Lett.* B **386** (1996) 403; *Nucl. Phys.* B **492** (1997) 417.
[61] G. D'Ambrosio, G. Ecker, G. Isidori and J. Portolés, *JHEP* **9808** (1998) 004.
[62] J. F. Donoghue and F. Gabbiani, *Phys. Rev.* D **51** (1995) 2187.
[63] G. Ecker, H. Neufeld and A. Pich, *Nucl. Phys.* B **413** (1994) 321; *Phys. Lett.* B **278** (1992) 337.
[64] J. Bijnens, G. Ecker and A. Pich, *Phys. Lett.* B **286** (1992) 341.
[65] A. J. Buras *et al.*, *Phys. Rev. Lett.* **95** (2005) 261805; *JHEP* **0611** (2006) 002.
[66] J. Brod, M. Gorbahn and E. Stamou, *Phys. Rev.* D **83** (2011) 034030.
[67] A. Pich, *PoS HQL* **2012** (2012) 019.

Journal of Physics: Conference Series **556** (2014) 012046 doi:10.1088/1742-6596/556/1/012046

Status of KOTO Experiment at J-PARC

Manabu Togawa

Department of Physics, Osaka University, Toyonaka 560-0043, Osaka, Japan

E-mail: togawa@champ.hep.sci.osaka-u.ac.jp

Abstract. The KOTO experiment at the J-PARC laboratory seeks to obtain the first observation of the $K_L \to \pi^0\nu\bar{\nu}$ decay. The branching ratio is calculated in the Standard Model to be 2.4×10^{-11}. This decay is a good probe to explore new physics beyond the Standard Model with its small theoretical uncertainty (~2%). In 2012 - 2013, KOTO performed commissioning runs with most of the detectors, and the first physics run started in May 2013. We accumulated 100 hours with the 24 kW beam power, and the analysis is on going. In this paper, the status of the KOTO experiment, including the data analysis and prospects, is reported.

1. Introduction

The very rare decay $K_L^0 \to \pi^0\nu\bar{\nu}$ is a sensitive probe for direct CP violation in the quark sector. The decay is a Flavor Changing Neutral Current process and is induced through electroweak loop diagrams. The branching ratio for $K_L^0 \to \pi^0\nu\bar{\nu}$ is predicted to be 2.4×10^{-11} in the Standard Model [1], and the theoretical uncertainty is estimated to be only a few percent. The decay is also sensitive to new physics scenarios beyond the Standard Model such as Supersymmetric theories.

A model-independent upper bound on the branching ratio, called the Grossman-Nir bound [2], is derived as $BR(K_L^0 \to \pi^0\nu\bar{\nu}) < 4.4 \times BR(K^+ \to \pi^+\nu\bar{\nu})$ from isospin symmetry arguments. The measured $BR(K^+ \to \pi^+\nu\bar{\nu})$ value of $(1.7 \pm 1.1) \times 10^{-10}$ from the BNL E787 and E949 experiments [3] yields an upper limit on $BR(K_L^0 \to \pi^0\nu\bar{\nu})$ of 1.2×10^{-9} (68 % C.L.).

The latest experimental results were obtained from the E391a experiment at KEK 12 GeV PS, and the upper limit on the branching ratio was 2.6×10^{-8} [4]. The goal of the KOTO (K0 at TOkai) experiment is to study $K_L^0 \to \pi^0\nu\bar{\nu}$ at the Standard Model sensitivity [5].

2. Experiment

A schematic cross-sectional view of the KOTO detector is shown in Figure 1. The KOTO detector subsystems are categorized into two parts. One is the CsI calorimeter, which detects two photons from π^0 decay, and the other is a set of hermetic veto counters, which demand no energy deposition in the $K_L^0 \to \pi^0\nu\bar{\nu}$ events. With two photon detection in the CsI calorimeter, the decay vertex can be reconstructed assuming the nominal π^0 mass using the knowledge that the beam generated by the upstream collimator system is very narrow.

The detector design follows the same concepts as those in the E391a experiment, and most part of counters used in the E391a are reused. A major upgrade was performed to the detector subsystems that are placed around the beam hole to suppress the halo-neutron background reported at the E391a experiment [4].

We finished the construction of the K_L beam line in 2009, and most of the detectors were installed by 2013.

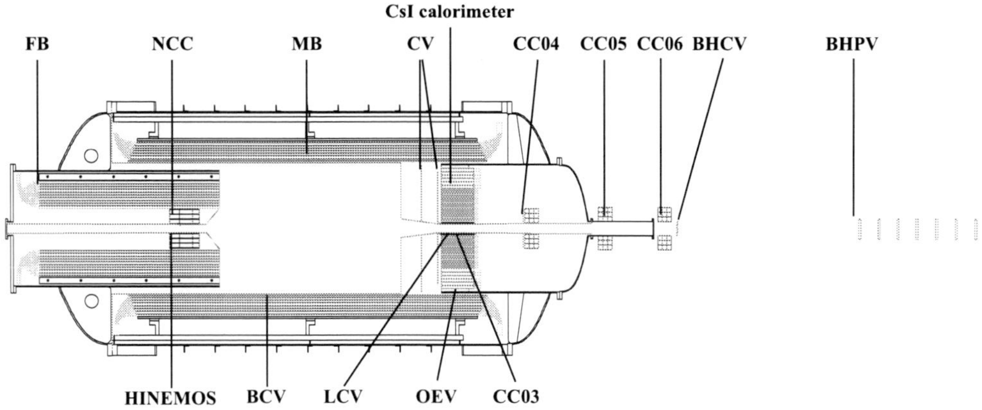

Figure 1. Schematic cross-sectional view of the KOTO experiment.

3. First physics run in 2013

In January 2013, we conducted engineering runs for studying the response of all detectors and confirming the K_L yield using $K_L \to 3\pi^0, 2\pi^0, 2\gamma$ decay modes [7]. We took 5-day data across March and April for tuning the detectors and DAQ for physics data taking. Finally, we started the first physics run in May; unfortunately the beam was terminated due to the hadron hall accident. We accumulated data for 100 hours with the 24 kW beam power, and the data sample is expected to be close to the sensitivity of the E391a experiment.

The performance of the detectors was checked by the $K_L \to 3\pi^0$ and $2\pi^0$ decays. Figure 2 shows the K_L invariant mass spectrum and the reconstructed vertex distribution using the events of 6 photons in the CsI calorimeter. These events are from $K_L \to 3\pi^0$ decay and are useful to check the performance of the calorimeter. The Monte Carlo simulation, considering the detector response and accidental hits which were accumulated by a beam-related trigger, reproduces the data well.

Figure 3 shows the K_L invariant mass spectrum using the events of 4 photons in the CsI calorimeter. The K_L mass peak from $K_L \to 2\pi^0$ and a large amount of continuum from $K_L \to 3\pi^0$ due to missing photons in the calorimeter are observed. Most of such missing photons are going to the veto detectors. Thus, after imposing extra-particle veto for all the hermetic detectors, $K_L \to 3\pi^0$ event is highly suppressed and the K_L mass peak from $K_L \to 2\pi^0$ is very clean as shown in Figure 3 (right). The Monte Carlo simulation reproduces the data well. We concluded that the performance of the CsI calorimeter and the veto detectors were well understood.

The analysis of $K_L^0 \to \pi^0\nu\bar{\nu}$ is on going with the signal region masked.

Journal of Physics: Conference Series **556** (2014) 012046 doi:10.1088/1742-6596/556/1/012046

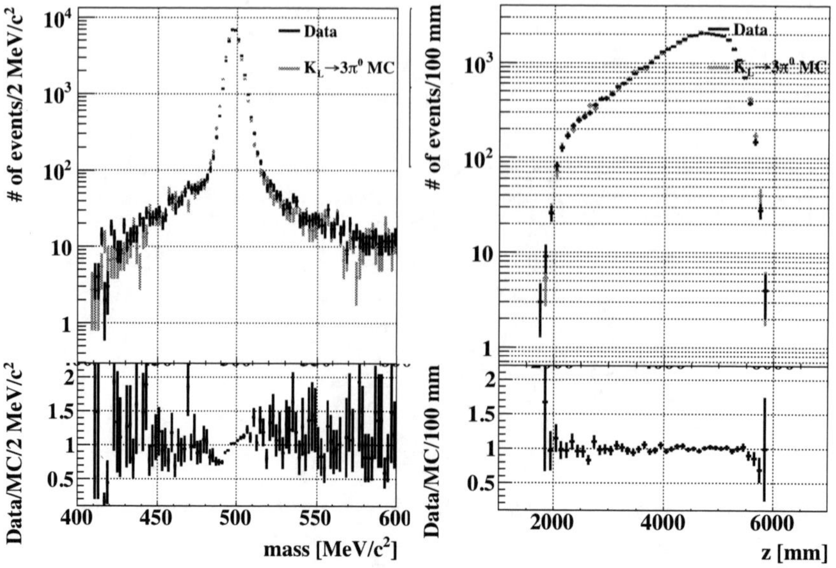

Figure 2. Left) K_L invariant-mass spectrum and Right) reconstructed vertex distribution using the events of 6 photons in the CsI calorimeter. Black and red colors indicate events of real data and of the Monte Carlo simulation, respectively; their ratio is shown in the bottom.

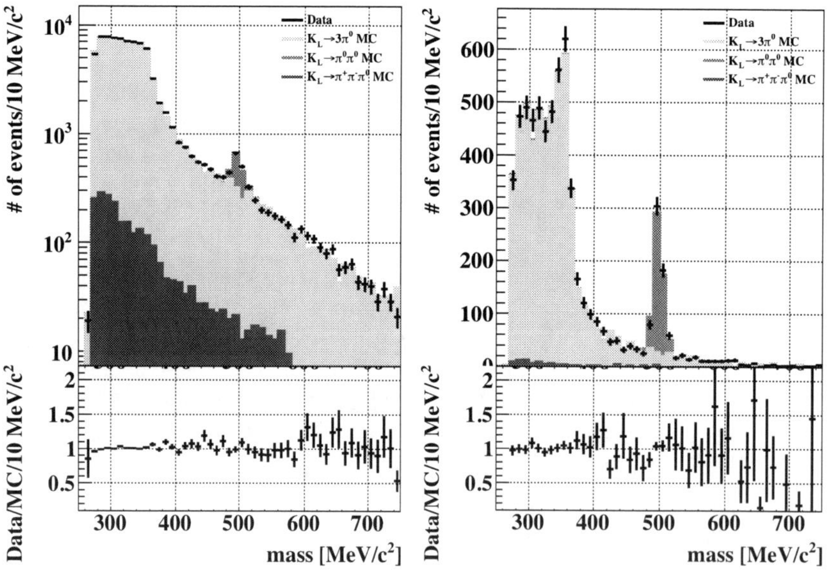

Figure 3. K_L invariant-mass spectrum using the event of 4 photons in the CsI calorimeter. Left) Reconstructed π^0 mass with the information from the CsI calorimeter only. Right) After imposing extra-particle veto for all the hermetic detectors. The ratio of the data and the Monte Carlo is shown in the bottom.

4. Detector upgrades

To reach the sensitivity of the Standard Model prediction, it is necessary to increase the photon detection efficiency in part of the veto detectors. Now, we are planning to :

- increase the number of modules of photon veto in beam (BHPV in Figure 1),
- replace the charge veto in beam to a new one (BHCV in Figure 1),
- add photon veto counters around the decay region (inside MB in Figure 1).

The current BHCV is composed of segmented scintillators, and the count rate is high due to the material. A new BHCV is designed as a gas chamber to reduce the amount of material. We prepared a prototype and checked the performance with the K_L beam during the engineering run; it was confirmed the gas chamber system will be fit to the highest beam intensity at J-PARC.

A new Inner Barrel detector (IB) will be installed to the Main Barrel (MB) to increase the photon detection efficiency; the radiation length increases from 13.5 X_0 to 18.5 X_0 in total. Then the punch-through events will be suppressed by a factor of 50. It is a sandwich calorimeter and consists of 25 layers of lead plates and scintillators read by wavelength shifting fiber.

The new detectors are being constructed, and will be installed during summer in 2015.

5. Summary

The KOTO experiment at J-PARC studies $K_L^0 \to \pi^0\nu\bar{\nu}$. The first 100 hours of physics data in May 2013 is being validated and the analysis for $K_L^0 \to \pi^0\nu\bar{\nu}$ is on going. The KOTO experiment will resume data taking in early 2015; and it is expected to reach the Grossman-Nir bound. We plan to install new detectors in the summer of 2015. These detectors are essential to go further in the $K_L^0 \to \pi^0\nu\bar{\nu}$ sensitivity.

References

[1] J. Brod, M. Gorbahn and E Stamou, *Phys. Rev. D* **83**, 034030 (2011).
[2] Y. Grossman and Y. Nir, *Phys. Lett. B* **398**, 163 (1997).
[3] A. V. Artamonov *et al.* [BNL 949 Collab.], *Phys. Rev.* **79**, 092004 (2009).
[4] J. K. Ahn *et al.*, *Phys. Rev. D* **81**, 072004 (2010).
[5] J. Comfort *et al. J-PARC E14 Collaboration, Proposal for* $K_L^0 \to \pi^0\nu\bar{\nu}$ *Experiment at J-PARC.* (Available at: http://j-parc.jp/researcher/Hadron/en/pac_0606/pdf/p14-Yamanaka.pdf) (2006).
[6] T. Shimogawa, *Nucl. Instrum. Meth.* **A623**, 585 (2010).
[7] T. Masuda, *"KOTO Detector Status", Proceedings for KAON2013.*

Penguin and rare decays in *BABAR*

Simon Akar, on behalf of the *BABAR* Collaboration
LPNHE, IN2P3/CNRS, Université Pierre et Marie Curie-Paris6, Université Denis Diderot-Paris7, F-75252 Paris, France[1]

E-mail: simon.akar@lpnhe.in2p3.fr

Abstract. We present recent results from the *BABAR* Collaboration on radiative decays. These include searches for new physics via measurements of several observables such as the time-dependent *CP* asymmetry in $B^0 \to K^0_S\pi^-\pi^+\gamma$ exclusive decays, as well as direct *CP* asymmetries and branching fractions in $B \to X_s\gamma$ and $B \to X_s\ell^+\ell^-$ inclusive decays.

1. Introduction

Until the end of 2007, the *BABAR* experiment recorded e^+e^- collisions at the $\Upsilon(4S)$ resonance with 471×10^6 $B\overline{B}$ pairs produced, corresponding to an integrated luminosity of $429\,\mathrm{fb}^{-1}$. Seven years after the end of the data-taking period, *BABAR* is still producing many physics results. In the LHC era, *BABAR* is still competitive, especially for channels involving neutral particles such as π^0 or K^0_S. We present here a selection of recent results on radiative decays from the *BABAR* experiment. In the Standard Model (SM), both the $b\to s\gamma$ and $b\to s\ell^+\ell^-$ transitions are quark-level flavor-changing neutral current (FCNC) processes. Since all FCNC processes are forbidden at tree level in the SM, the lowest order diagrams representing these transitions must involve loops. In such processes, QCD corrections can typically be described using an effective Hamiltonian defined as $\mathrm{H_{eff}} \propto \sum_{i=1}^{10} C_i\mathcal{O}_i$, where the C_i and $\mathcal{O}_i$ are, respectively, the short-distance Wilson coefficients and local long-distance operators. Contributions from New Physics (NP) in $b\to s$ transitions may modify the SM values of the Wilson coefficients, and change the values, predicted by the SM, of observables such as branching fractions and *CP* asymmetries.

2. Measurements of direct *CP* asymmetries in $B\to X_s\gamma$ decays using a sum of exclusive decays

In this analysis [1], the direct *CP* asymmetry, A_{CP} for the sum of exclusive final states is measured:

$$A_{CP} = \frac{\Gamma_{\overline{B}^0/B^-\to X_s\gamma} - \Gamma_{B^0/B^+\to X_{\bar{s}}\gamma}}{\Gamma_{\overline{B}^0/B^-\to X_s\gamma} + \Gamma_{B^0/B^+\to X_{\bar{s}}\gamma}} \,. \tag{1}$$

The SM prediction for the asymmetry was found in a recent study to be in the range $-0.6\% < A^{SM}_{CP} < 2.8\%$ [2]. From Ref. [2], another test of the SM is proposed, via the measurement of the difference in A_{CP} in charged and neutral B mesons:

$$\Delta A_{CP} = A_{B^\pm\to X_s\gamma} - A_{B^0/\overline{B}^0\to X_s\gamma} \simeq 0.12\times\frac{\tilde{\Lambda}_{78}}{100\,\mathrm{MeV}}\mathrm{Im}\left(\frac{C_{8g}}{C_{7\gamma}}\right) \,, \tag{2}$$

[1] Now at CPPM, Aix-Marseille Université, CNRS/IN2P3, Marseille, France

where $C_{7\gamma}$ and C_{8g} are the Wilson coefficients corresponding to the electromagnetic dipole and the chromo-magnetic dipole transitions, respectively, and $\tilde{\Lambda}_{78}$ is the interference amplitude. Since in the SM, both $C_{7\gamma}$ and C_{8g} are real, ΔA_{CP} is expected to be zero.

The B meson decays are fully reconstructed in 16 self-tagging final states, which are listed in Ref. [1]. The raw asymmetry is extracted from a simultaneous fit to the *energy-substitued B* meson mass, $m_{\rm ES}$, distributions of B and $\overline{B}$ tagged samples; $m_{\rm ES} \equiv \sqrt{(\sqrt{s}/2)^2 - (p_B^*)^2}$, where p_B^* is the momentum vector of the B in the e^+e^- center-of-mass (CM) frame and $\sqrt{s}$ is the total energy of the e^+e^- system. The direct *CP* asymmetry is obtained after correcting the raw asymmetry for detector effects. Possible dilutions from the presence of peaking backgrounds in the $m_{\rm ES}$ distributions are taken into account as systematic uncertainties. The measured value is $A_{CP} = +(1.73 \pm 1.93 \pm 1.02)\%$, where the first quoted error is statistical and the second is systematic. This corresponds to the most precise measurement to date and is compatible with the SM. The measurement of ΔA_{CP} is obtained by a simultaneous fit to the separate charge and neutral B samples, such as $\Delta A_{CP} = +(5.0 \pm 3.9 \pm 1.5)\%$, where the first quoted error is statistical and the second is systematic.

The interference amplitude, $\tilde{\Lambda}_{78}$, in Eq. 2 is only known as a range of possible values: $17\,\text{MeV} < \tilde{\Lambda}_{78} < 190\,\text{MeV}$. Using the measured values of ΔA_{CP}, a χ^2 minimization is performed for given ${\rm Im}(C_{8g}/C_{7\gamma})$ from all possibles values of $\tilde{\Lambda}_{78}$, as shown in Fig. 1. The plateau of minimum $\chi^2 = 0$ corresponds to the region where a value of $\tilde{\Lambda}_{78}$ always exists in the allowed range such as the theoretical value of ΔA_{CP} matches exactly the experimental one. The 68% and 90% confidence limits are then obtained from the ranges of ${\rm Im}(C_{8g}/C_{7\gamma})$, which yield the minimum χ^2 less than 1 and 4, respectively, such as: $0.07 \le {\rm Im}(C_{8g}/C_{7\gamma}) \le 4.48$, at 68% CL and $-1.64 \le {\rm Im}(C_{8g}/C_{7\gamma}) \le 6.52$, at 90% CL.

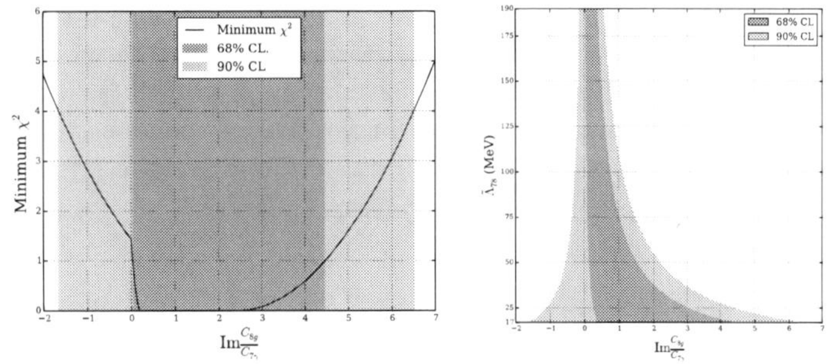

Figure 1. (left) The minimum χ^2 for given ${\rm Im}(C_{8g}/C_{7\gamma})$ from all possible values of $\tilde{\Lambda}_{78}$. 68% and 90% confidence intervals are shown in dark gray and light gray, respectively. (right) The 68% and 90% confidence intervals for ${\rm Im}(C_{8g}/C_{7\gamma})$ as a function of $\tilde{\Lambda}_{78}$.

3. Measurement of the $B \to X_s\,\ell^+\ell^-$ branching fraction and search for direct *CP* violation from a sum of exclusive final states

In the present analysis, the inclusive decay $B \to X_s\,\ell^+\ell^-$ is studied, where X_s is a hadronic system containing exactly one kaon, using a sum over 20 exclusive final states, where $\ell^+\ell^-$ is either e^+e^- or $\mu^+\mu^-$, as listed in Ref. [3].

The total branching fraction (BF), as well as partial BFs in five disjoint dilepton mass-squared $q^2 \equiv m^2_{\ell^+\ell^-}$ bins and four hadronic mass m_{X_s} bins, which are defined in Ref. [3]. Events with q^2 corresponding to signal-like charmonium backgrounds J/ψ and $\psi(2S)$ from B decays are rejected. After requiring the invariant mass of the hadronic system $m_{X_s} < 1.8\,\text{GeV}/c^2$, the entire selection represents $\sim 70\%$ of the inclusive $B \to X_s\,\ell^+\ell^-$ rate. The missing hadronic final states, as well as states with $m_{X_s} > 1.8\,\text{GeV}/c^2$ are accounted for using JETSET fragmentation [4] and theory predictions.

The signal yields are extracted, in each q^2 and m_{X_s} bins, from a simultaneous fit to $m_{\rm ES}$ and a likelihood ratio L_R, which is defined as $L_R \equiv \mathcal{P}_S/(\mathcal{P}_S + \mathcal{P}_B)$ with $\mathcal{P}_S$ ($\mathcal{P}_B$) the probability for a correctly-reconstructed signal ($B\overline{B}$ background) event calculated based on the response of boosted decision trees.

The total BF for $q^2 > 0.1\,{\rm GeV}^2/c^4$ is measured to be $\mathcal{B}(B \to X_s\,\ell^+\ell^-) = 6.73^{+0.70+0.34}_{-0.64-0.25} \pm 0.50 \times 10^{-6}$, where the first uncertainties are statistical, the second experimental systematics and the third model-dependent systematics. The same convention for the quoted uncertainties is used hereafter. This result is less than 2σ above the SM prediction of $\mathcal{B}_{SM} = 4.6 \pm 0.8 \times 10^{-6}$ [5].

In the low mass range, for $1 < q^2 < 6\,{\rm GeV}^2/c^4$, the BF is measured to be $\mathcal{B}^{\rm low}(B \to X_s\,\ell^+\ell^-) = 1.60^{+0.41+0.17}_{-0.39-0.13} \pm 0.18 \times 10^{-6}$, which is in good agreement with the SM predictions of $\mathcal{B}^{\rm low}_{SM}(B \to X_s\,\mu^+\mu^-) = (1.59 \pm 0.11) \times 10^{-6}$ and $\mathcal{B}^{\rm low}_{SM}(B \to X_s\,e^+e^-) = (1.64 \pm 0.11) \times 10^{-6}$ [6]. In the high mass range, for $q^2 > 14.2\,{\rm GeV}^2/c^4$, the BF is measured to be $\mathcal{B}^{\rm high}(B \to X_s\,\ell^+\ell^-) = 0.57^{+0.16+0.03}_{-0.15-0.02} \pm 0.00 \times 10^{-6}$, which is about 2σ higher than the SM predictions for $q^2 > 14.4\,{\rm GeV}^2/c^4$ of $\mathcal{B}^{\rm high}(B \to X_s\,\mu^+\mu^-) = (0.24 \pm 0.07) \times 10^{-6}$ and $\mathcal{B}^{\rm high}(B \to X_s\,e^+e^-) = (0.21 \pm 0.07) \times 10^{-6}$ [6]. The measured partial BFs results in bins of q^2 and m_{X_s} are detailed in Ref [3].

A search for the direct *CP* asymmetry in $B \to X_s\,\ell^+\ell^-$ decays, using the 14 self-tagging final states as listed in Ref. [3], is also performed. For $q^2 > 0.1\,{\rm GeV}^2/c^4$, the measured *CP* asymmetry, $A_{CP}(B \to X_s\,\ell^+\ell^-) = 0.04 \pm 0.11 \pm 0.01$, is consistent with the SM prediction, where it is expected to be suppressed well below the 1%.

4. Time-dependent analysis of $B^0 \to K^0_S\pi^+\pi^-\gamma$ and studies of the $K^+\pi^-\pi^+$ system in $B^+ \to K^+\pi^-\pi^+\gamma$ decays

In $b \to s\gamma$ transitions, the SM predicts that B^0 ($\overline{B}^0$) decays are related predominantly to the presence of right- (left-) handed photons in the final state. Therefore, the mixing-induced *CP* asymmetry in $B \to f_{CP}\gamma$ decays, where f_{CP} is a *CP* eigenstate, is expected to be small.

One of the goals of the present study is to perform a time-dependent analysis of $B^0 \to K^0_S\pi^+\pi^-\gamma$ decays to extract the direct and mixing-induced *CP* asymmetry parameters, $\mathcal{C}_{K^0_S\rho\gamma}$ and $\mathcal{S}_{K^0_S\rho\gamma}$, in the $B^0 \to K^0_S\rho^0\gamma$ mode. However, due to the large natural width of the $\rho^0(770)$, a non negligible amount of $B^0 \to K^{*\pm}(K^0_S\pi^\pm)\pi^\mp\gamma$ events, which do not contribute to $\mathcal{S}_{K^0_S\rho\gamma}$, are expected to lie under the $\rho^0(770)$ resonance and dilute $\mathcal{S}_{K^0_S\rho\gamma}$. We can define a dilution factor $\mathcal{D}_{K^0_S\rho\gamma}$ such as $\mathcal{D}_{K^0_S\rho\gamma} = \mathcal{S}_{K^0_S\pi^+\pi^-\gamma}/\mathcal{S}_{K^0_S\rho\gamma}$, where $\mathcal{S}_{K^0_S\pi^+\pi^-\gamma}$ is the effective value of the mixing-induced *CP* asymmetry measured for the whole $B^0 \to K^0_S\pi^+\pi^-\gamma$ dataset. Since a small number of signal events is expected in this sample, it is difficult to discriminate $B^0 \to K^{*\pm}(K^0_S\pi^\pm)\pi^\mp\gamma$ from $B^0 \to K^0_S\rho^0(\pi^\mp\pi^\pm)\gamma$ decays. Hence the dilution factor needs to be obtained by other means. To do that, the amplitudes of the different resonant modes are extracted in the charged decay channel[2] $B^+ \to K^+\pi^-\pi^+\gamma$, which has more signal events and is related to $B^0 \to K^0_S\pi^+\pi^-\gamma$ by isospin symmetry. Assuming that the resonant amplitudes are the same in both modes, the dilution factor is calculated from those of $B^+ \to K^+\pi^-\pi^+\gamma$. Moreover, the branching fractions of the different $B \to K_{\rm res}\gamma$ intermediate decay modes (where $K_{\rm res}$ designates a kaonic resonance decaying to $K\pi\pi$) are in general not well known. We also use the amplitude analysis of the charged decay mode $B^+ \to K^+\pi^-\pi^+\gamma$ to extract them.

In the charged B-meson decay mode, we first perform a fit to data to extract the yield of $B^+ \to K^+\pi^-\pi^+\gamma$ signal events. The fit uses the knowledge of three discriminating variables, $m_{\rm ES}$, ΔE and a Fisher discriminant output, to discriminate signal events from backgrounds. ΔE is defined as the difference between the expected and reconstructed B meson energy, $\Delta E \equiv E^*_B - \sqrt{s}/2$, where E^*_B is the reconstructed energy of the B in the e^+e^- CM frame. Using

[2] Charge conjugation is implicit throughout the document.

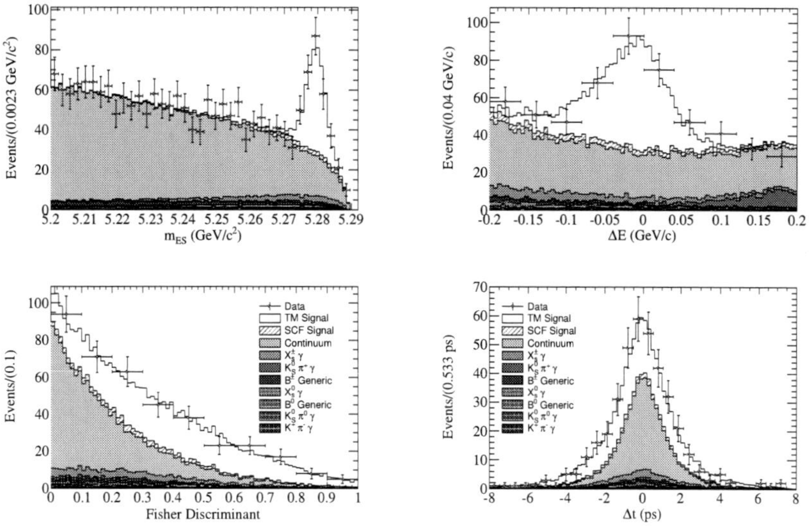

Figure 2. Distributions of $m_{\rm ES}$ (top left), ΔE (top right), the Fisher-discriminant output (bottom left), and Δt (bottom right), showing the fit results on the $B^0 \to K^0_S\pi^+\pi^-\gamma$ data sample. The distributions have their signal/background ratio enhanced by means of the following requirements: $-0.15 \le \Delta E \le 0.10\,{\rm GeV}/c$ ($m_{\rm ES}$), $m_{\rm ES} > 5.27\,{\rm GeV}/c^2$ (ΔE) and $m_{\rm ES} > 5.27\,{\rm GeV}/c^2$; $-0.15 \le \Delta E \le 0.10\,{\rm GeV}/c$ (Fisher and Δt).

information from the maximum likelihood fit, the $K\pi\pi$, $K\pi$ and $\pi\pi$ invariant mass spectra in signal events are extracted using the ${}_s\mathcal{P}lot$ technique [7]. Then fitting the $m_{K\pi\pi}$ and $m_{K\pi}$ spectra, the amplitudes and BFs of the kaonic resonances and the intermediate state resonances, respectively, are extracted, allowing to compute the dilution factor, such as $\mathcal{D}_{K^0_S\rho\gamma} = 0.549^{+0.096}_{-0.094}$, where the quoted uncertainties are sums in quadrature of statistical and systematic uncertainties.

In the neutral B-meson decay mode, we perform a fit to data to extract the effective *CP* asymmetry parameters, using four discriminating variables, $m_{\rm ES}$, ΔE and a Fisher discriminant output and the proper-time difference Δt. Fig. 2 shows the fit projections. We obtain the *CP*-violating parameters $\mathcal{S}_{K^0_S\pi^+\pi^-\gamma} = 0.137 \pm 0.249^{+0.042}_{-0.033}$ and $\mathcal{C}_{K^0_S\pi^+\pi^-\gamma} = -0.390 \pm 0.204^{+0.045}_{-0.050}$, where the first quoted errors are statistical and the second are systematic. Using the dilution factor, extracted from the charged mode analysis, we extract the time-dependent *CP* asymmetry related to the hadronic *CP* eigenstate $\rho^0 K^0_S$ and obtain $\mathcal{S}_{K^0_S\rho\gamma} = 0.249 \pm 0.455^{+0.076}_{-0.060}$, which is compatible with the SM expectation of ~ 0.03 [8].

5. Conclusions

The *BABAR* Collaboration, seven years after the shutdown of the experiment, is still producing competitive results. Three recent analyses have been presented, for which all the results are in agreement with the SM predictions.

References

[1] Lees J *et al.* (BaBar Collaboration) 2014 (*Preprint* 1406.0534)
[2] Benzke M, Lee S J, Neubert M and Paz G 2011 *Phys.Rev.Lett.* **106** 141801 (*Preprint* 1012.3167)
[3] Lees J *et al.* (BaBar Collaboration) 2014 *Phys.Rev.Lett.* **112** 211802 (*Preprint* 1312.5364)
[4] Sjostrand T 1994 *Comput.Phys.Commun.* **82** 74–90
[5] Oliveira A S and Steiner J 2004 *Mon.Not.Roy.Astron.Soc.* **351** 685 (*Preprint* astro-ph/0403281)
[6] Huber T, Hurth T and Lunghi E 2008 *Nucl.Phys.* **B802** 40–62 (*Preprint* 0712.3009)
[7] Pivk M and Le Diberder F R 2005 *Nucl.Instrum.Meth.* **A555** 356–369 (*Preprint* physics/0402083)
[8] Atwood D, Gershon T, Hazumi M and Soni A 2005 *Phys.Rev.* **D71** 076003 (*Preprint* hep-ph/0410036)

XI International Conference on Hyperons, Charm and Beauty Hadrons (BEACH 2014) IOP Publishing
Journal of Physics: Conference Series **556** (2014) 012048 doi:10.1088/1742-6596/556/1/012048

New physics searches with heavy flavour with the ATLAS experiment

Ina Chalupkova on behalf of the ATLAS Collaboration

Institute of Particle and Nuclear Physics, Faculty of Mathematics and Physics, Charles University, V Holesovickach 2, Prague 8, Czech Republic

E-mail: Ina.Chalupkova@cern.ch

Abstract. Flavour changing neutral currents and precision measurements of CP violation are investigated by the ATLAS experiment at the Large Hadron Collider at CERN as probes to physics beyond the Standard Model. This paper presents recent update of flavour tagged time-dependent analysis of $B_s \to J/\psi\ \phi$, angular analysis of $B_d \to K^* \mu^+ \mu^-$ and a search for the rare decay $B_s^0 \to \mu^+ \mu^-$ with measurement of upper limit on its branching fraction. All analyses use 4.9 fb^{-1} of integrated luminosity collected in 2011 at centre-of-mass energy of 7 TeV and the results are in agreement with Standard Model predictions.

1. Introduction

The production of b-hadrons at the Large Hadron Collider (LHC) provides excellent opportunities for indirect searches for new physics. Various rare decay modes occur in Standard Model (SM) as loop diagrams and are sensitive to contributions of physics beyond SM, mainly as a result of the interference with SM diagrams.

The ATLAS experiment [1] is a general purpose detector consisting of an inner tracker, calorimeters and a muon spectrometer. The Inner Detector (ID) detects tracks of charged particles and consists of pixel, silicon strip and transition radiation subdetectors embedded in a 2 T solenoidal magnetic field. Surrounding the ID there are electromagnetic and hadronic calorimeters and the Muon Spectrometer (MS) with toroidal magnetic field.

The ATLAS trigger consists of a hardware-based Level-1 trigger and a 2-staged High Level Trigger (HLT). The muon trigger at Level-1 is based on pattern search for different p_{T} thresholds in the MS. The regions of interest around these patterns are then used as seeds for the HLT reconstruction, matching information from both the MS and ID.

2. Flavour tagged time dependent angular analysis of $B_s^0 \to J/\psi\ \phi$

Events for this measurement [2] are triggered by di-muon triggers searching for $J/\psi \to \mu^+\mu^-$ candidates with p_{T} thresholds of 4 GeV for both muons or 4 GeV and 2 GeV for the first and second muon, respectively. The candidates for $\phi \to K^+K^-$ are reconstructed from tracks with $p_{\mathrm{T}} > 1$ GeV and $|\eta| < 2.5$ not identified as muons and each combination of J/ψ and ϕ is fitted to a common vertex (fixing the di-muon invariant mass to the J/ψ mass). Candidates are accepted if the invariant mass of hadronic track pairs falls within interval (1.0085 GeV $< m_{K^+K^-} <$1.0305) around ϕ mass and if the criteria on vertex fit quality and number of ID hits for tracks are met.

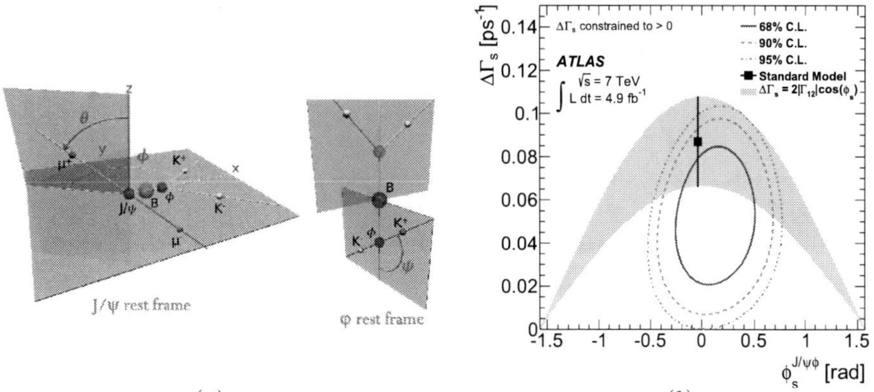

(a) (b)

Figure 1. $B_s^0 \to J/\psi\ \phi$ analysis: (a) definition of transversity angles, (b) the 68% and 95% likelihood contours in $\phi_s - \Delta\Gamma_s$ plane (statistical errors only) with the theoretical prediction of mixing-induced CP violation [2].

This updated analysis includes Opposite-Side Tagging to determine the initial flavour of B-mesons. To calibrate the method, decays $B^\pm \to J/\psi K^\pm$ are used: the flavour of $B^\pm$ at production is known from the kaon charge.

An unbinned maximum likelihood fit is performed on selected events to extract parameters of the signal decay. The likelihood function is defined as a combination of probability density functions $\mathcal{F}_s$ for signal, $\mathcal{F}_{B^0}$ for B^0 background contribution and $\mathcal{F}_{\rm bkg}$ for other backgrounds as

$$\ln\mathcal{L} = \sum_{i=1}^{N}\{\ln(w_i f_s \mathcal{F}_s(m_i,t_i,\Omega_i) + f_s f_{B^0}\mathcal{F}_{B^0}(m_i,t_i,\Omega_i) + (1-f_s(1+f_{B^0}))\mathcal{F}_{\rm bkg}(m_i,t_i,\Omega_i))\}\ ,$$

where N is the number of candidates, w_i is weighting factor to account for trigger efficiency, f_s is fraction of signal candidates, and f_{B^0} is fraction of peaking B^0 events. Values of reconstructed mass m, proper decay time t, their uncertainties σ_m and σ_t, tag probability and transversity angles $\Omega = (\theta_{\rm T}, \psi_{\rm T}, \phi_{\rm T})$, as defined in Fig. 1 (a), are measured from data for each event.

The full fit has 25 free parameters and it extracted 22670 ± 150 signal candidates. The fitted values of weak phase ϕ_s, $B_s - \bar{B}_s$ width difference $\Delta\Gamma_s$ and other physical parameters [2] are

$$\begin{aligned}\phi_s &= 0.12 \pm 0.25\ (\text{stat.}) \pm 0.05\ (\text{syst.})\ \text{rad}\\ \Delta\Gamma_s &= 0.053 \pm 0.021\ (\text{stat.}) \pm 0.010\ (\text{syst.})\ \text{ps}^{-1}\\ \Gamma_s &= 0.677 \pm 0.007\ (\text{stat.}) \pm 0.004\ (\text{syst.})\ \text{ps}^{-1}\\ |A_\parallel(0)|^2 &= 0.220 \pm 0.008(\text{stat.}) \pm 0.009\ (\text{syst.})\\ |A_0(0)|^2 &= 0.529 \pm 0.006(\text{stat.}) \pm 0.012\ (\text{syst.})\\ \delta_\perp &= 3.89 \pm 0.47\ (\text{stat.}) \pm\ 0.11\ (\text{syst.})\ \text{rad}\ .\end{aligned}$$

Figure 1 (b) shows the likelihood contours in $\phi_s - \Delta\Gamma_s$ plane, together with the SM expectations. The values are consistent with those obtained in the previous untagged analysis [3] and the overall uncertainty on ϕ_s is reduced by 40% with respect to the untagged analysis.

3. Angular analysis of $B_d^0 \to K^{*0}\mu^+\mu^-$

The decay $B_d^0 \to K^{*0}\mu^+\mu^-$ with subsequent $K^{*0} \to K^+\pi^-$ decay can be described by four kinematic variables: the invariant mass q^2 of the di-muon system and three angles defining configuration of the final state, θ_L is the angle between μ^+ and direction opposite to B^0 in di-muon rest frame, θ_K is the angle between K^+ and direction opposite to B^0 in K^{*0} rest frame, and ϕ is the angle between di-muon and K^{*0} planes [4].

Signal candidates are selected using single and di-muon triggers with different p_{T} thresholds (mainly $p_{\mathrm{T}} > 4$ GeV or 6 GeV). Decay candidates are reconstructed from two muons and a pair of oppositely charged tracks with $p_{\mathrm{T}} > 0.5$ GeV. Based on Monte Carlo (MC) optimisation, cuts on lifetime significance τ/σ_τ, pointing angle (angle between the B_d^0 momentum and the vector between the primary vertex and the reconstructed B_d^0 vertex) and $p_{\mathrm{T}}(K^{*0})$ are applied to suppress the combinatorial background. In order to remove the resonant background, events with di-muon mass within 3σ from the J/ψ and $\psi(2S)$ peaks are excluded.

Because statistics is insufficient to study the fully differential decay rate, the K^{*0} longitudinal polarisation fraction F_L and the muon forward-backward asymmetry A_{FB} are extracted from projections into 2-dimensional distributions (integrating over ϕ and θ_L or θ_K) as

$$\frac{1}{\Gamma}\frac{\mathrm{d}^2\Gamma}{\mathrm{d}q^2\mathrm{d}\cos\theta_K} = \frac{3}{2}F_L(q^2)\cos^2\theta_K + \frac{3}{4}(1 - F_L(q^2))(1 - \cos^2\theta_K)$$

$$\frac{1}{\Gamma}\frac{\mathrm{d}^2\Gamma}{\mathrm{d}q^2\mathrm{d}\cos\theta_L} = \frac{3}{4}F_L(q^2)(1 - \cos^2\theta_L) + \frac{3}{8}(1 - F_L(q^2))(1 + \cos^2\theta_L) + A_{FB}(q^2)\cos\theta_L.$$

Results of an unbinned maximum likelihood fit in five q^2 bins are shown in Fig. 2, together with comparison with the SM predictions.

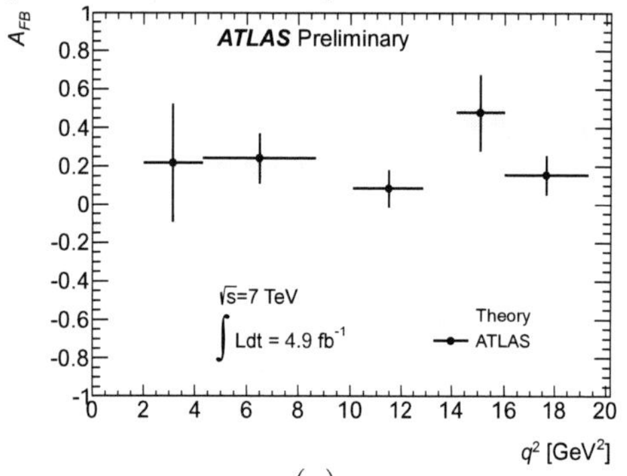

(a)

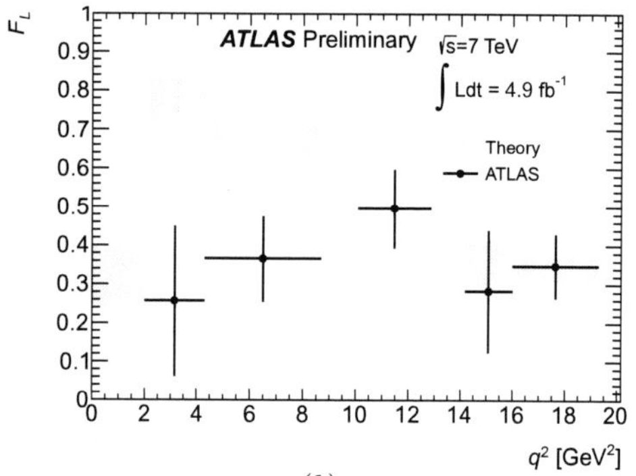

(b)

Figure 2. Results of $B_d^0 \to K^{*0}\mu^+\mu^-$ analysis: (a) forward-backward asymmetry A_{FB} and (b) fraction of longitudinally polarised K^{*0} mesons F_L, with statistical and systematic uncertainties added in quadrature [4]. The SM predictions [5] including theoretical uncertainties are also shown.

4. Search for $B_s^0 \to \mu^+\mu^-$

The $B_s^0 \to \mu^+\mu^-$ branching fraction $\mathcal{B}$ is measured as a relative branching ratio with a reference decay $B^\pm \to J/\psi K^\pm$ in order to minimize systematic uncertainties [6].

To select signal decay candidates, two muons with $p_{\mathrm{T}} > 4$ GeV are required at the Level-1 trigger and a loose cut on their invariant mass are applied at the HLT to select B_s^0 or J/ψ. A preselection is made requiring a good quality of tracks and reconstructed vertices and minimal numbers of hits in the ID subdetectors. Similar cuts are applied to reference channel candidates, leaving approximately 3.9×10^5 signal and 2.5×10^5 reference channel candidates in the signal region (5066 MeV$< m_{\mu^+\mu^-} <$5666 MeV).

Two categories of background are considered: the resonant background due to B decays to hadrons misidentified or decaying to muons is simulated and results are included in the computation of expected background yield for the extraction of upper limit of branching ratio.

The combinatorial background is studied with the $b\bar{b} \rightarrow \mu^+\mu^- X$ MC sample. To improve the agreement of MC simulation with data, a generator level re-weighting procedure is applied (to correct generator bias) and a data-driven re-weighting is performed using half of the reference channel data (avoiding bias to extraction of the reference channel and background yield).

The selection criteria are chosen by performing a 2D optimization of the signal window width and a Boosted Decision Tree (BDT) discriminant, using signal MC and half of the side-band data. Thirteen variables based on the kinematic properties of decay products, vertices and tracks in underlying event are used as inputs to the BDT. The residual combinatorial background in the signal region is determined from interpolation of side-bands using the other half of data.

The estimated number of background events in the signal region is 6.75, the total number of events observed after unblinding is 6. Figure 3 shows the invariant-mass distribution of selected candidates and the behaviour of the observed confidence level ($\mathrm{CL_s}$ [7]). The upper limit of branching fraction is extracted using the $\mathrm{CL_s}$ method and the result is

$$\mathcal{B}(B_s^0 \rightarrow \mu^+\mu^-) < 1.5 \times 10^{-8} \text{ (at 95\% CL)}.$$

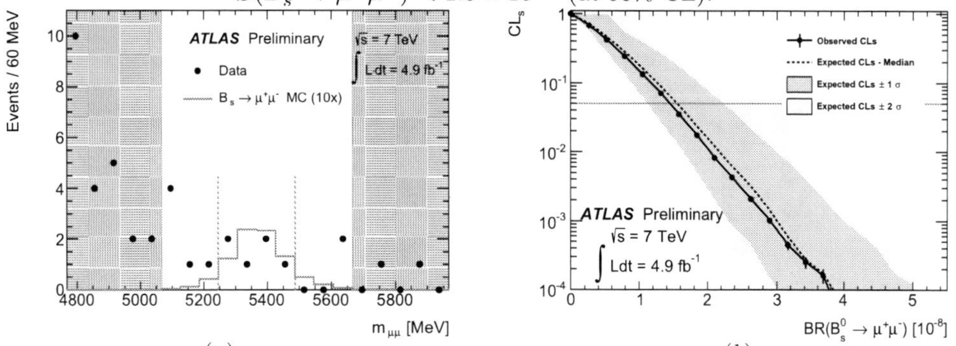

(*a*) (*b*)

Figure 3. (*a*) Invariant-mass distribution of $B_s^0 \rightarrow \mu^+\mu^-$ candidates for data and MC (scaled by a factor of 10) in signal region (dashed vertical lines) and side-bands (grey areas). (*b*) Observed confidence level $\mathrm{CL_s}$ as a function of branching ratio of $B_s^0 \rightarrow \mu^+\mu^-$. The 95% limit is indicated by the red line, green and yellow bands correspond to the $\pm1\sigma$ and $\pm2\sigma$ range of background-only pseudo-experiments with median of expected $\mathrm{CL_s}$ given by the dashed line [6].

5. Conclusions

Presented analyses of rare *b*-hadron decays show no deviations from the SM. They are, however, limited by small statistics of 2011 dataset. Measurements will be updated using the 2012 dataset (21.7 fb^{-1} at centre-of-mass energy of 8 TeV) and with data which will be collected in the future LHC runs.

Acknowledgments

The author acknowledges support of grant LG13009 of Ministry of Education, Youth and Sports of the Czech Republic.

References

[1] ATLAS Collaboration, *JINST* **3**, S08003 (2008)
[2] ATLAS Collaboration, *Phys.Rev.* D **90** 052007, arXiv:1407.1796 [hep-ex]
[3] ATLAS Collaboration, *JHEP* 12 (2012) 072, arXiv:1208.0572 [hep-ex]
[4] ATLAS Collaboration, ATLAS-CONF-2013-038, http://cdsweb.cern.ch/record/1537961
[5] Bobeth C, Hiller G and van Dyk D, *J. Phys.: Conf. Series* **335** 012038, arXiv:1105.2659 [hep-ph]
[6] ATLAS Collaboration, ATLAS-CONF-2013-076, http://cdsweb.cern.ch/record/1562934
[7] Read A L, *J. Phys.* **G28** (2002) 2693

XI International Conference on Hyperons, Charm and Beauty Hadrons (BEACH 2014) IOP Publishing
Journal of Physics: Conference Series **556** (2014) 012049 doi:10.1088/1742-6596/556/1/012049

Rare and forbidden B and tau decays in the Belle

HyoJung Hyun[1] on behalf of the Belle Collaboration
Dept. of Physics, Kyungpook National University, Daegu 702-701, Korea
E-mail: hjhyun@gmail.com

Abstract. A search for the purely leptonic decays $B^+ \to e^+\nu_e$ and $B^+ \to \mu^+\nu_\mu$ was performed using the 772 million $B\overline{B}$ pairs collected by the Belle detector at the KEKB asymmetric-energy e^+e^- collider. There was no evidence of a signal in either decay modes. A heavy neutral lepton-like new particle and heavy neutrino searches were also performed, but no significant signals were observed. Lepton flavor violation in τ decays was studied with an almost full data sample and the upper limits obtained were in the order of 10^{-8}. In this paper, recent results that can give any constraints to new physics are briefly described.

1. Introduction

A 710 fb^{-1} data sample containing 772 million B meson pairs was collected at a resonance of Υ(4S) with the Belle detector [1] at the KEKB accelerator. The KEKB is an energy asymmetric e^+e^- (3.5 on 8.0 GeV) collider [2] operating at the energy of Υ(4S). The Υ(4S) decays exclusively to $B^0\overline{B^0}$ or B^+B^-.

The Belle detector is a large-solid-angle magnetic spectrometer consisting of a silicon vertex detector (SVD), a 50-layer central drift chamber (CDC), an array of silica aerogel threshold cherenkov counters (ACC), time-of-flight scintillation counters (ToF), and a CsI(Tl) crystal electromagnetic calorimeter (ECL). These subdetectors are located inside a superconducting solenoid coil that provides a 1.5 T magnetic field. The Belle detector covers 4π, and the initial state is well known; therefore, this is perfect environment for studies of B decays with neutrinos in their final state.

Here, the study results of the leptonic decays, and new particle searches such as lepton-like neutral particle and heavy neutrino which are based on a 710 fb^{-1} data sample will be described. The lepton flavor violation (LFV) of τ decays highly suppressed in the standard model (SM) is studied and the result is also briefly presented.

2. Leptonic decays: $B^+ \to e^+\nu_e$ and $B^+ \to \mu^+\nu_\mu$

Purely leptonic decays proceed via annihilation of the charged B meson's constituent quarks into a lepton and a neutrino of the same generation. In the SM, this process is mediated by a W^+ boson leading to a branching fraction [3] as follows:

$$\mathcal{B}(B^+ \to l^+ \nu_l) = \frac{G_F^2 m_B m_l^2}{8\pi}\left(1 - \frac{m_l^2}{m_B^2}\right)^2 f_B^2 |V_{ub}|^2 \tau_B,$$

[1] Present address: Pohang Accelerator Laboratory, Jigokro-127-beongil, Nam-gu, Pohang 790-834, Korea

where G_F is the Fermi coupling constant, m_l is the mass of the charged lepton, m_B is the mass of the B^+ meson, τ_B is lifetime of the B^+ meson, V_{ub} is the element of the Cabibbo-Kobayashi-Maskawa (CKM) matrix [4], and f_B is the B decay constant. New physics (NP) contributions might interfere with the SM and modify the expectations. Most prominent is the charged Higgs from the two-Higgs-doublet model [5]; the branching fraction of the B to $l\nu_l$ can be greatly enhanced by as much as $(1\text{-}\tan^2\beta\frac{m_B^2}{m_H^2})^2$. Therefore, this decay can either test the SM or measure the product of the least-well-known parameters, $f_B^2|V_{ub}|^2$. The expected branching fractions in the SM using the achieved value and world average for each parameter [6][7] are in the order of 10^{-11}, 10^{-7}, and 10^{-4} for $B^+ \to e^+\nu_e$ (e mode), $B^+ \to \mu^+\nu_\mu$ (μ mode), and $B^+ \to \tau^+\nu_\tau$ (τ mode), respectively. Due to helicity suppression in the SM, e mode and μ mode have relatively small branching fractions. On the other hand, τ mode has a larger branching fraction and has been measured previously by the Belle [8] and BABAR [9] experiments, resulting in the order of 10^{-4}.

To study the leptonic decays of the e and μ modes [10], the hadronic tagging method based on a hierarchical neural network [11] is used. The signal events are extracted from the signal lepton's momentum (p_l^B) in the rest frame of the B_{sig}. Due to the two-body decay of the signal, we expected the signal events to peak sharply around 2.64 GeV/c, and anticipated a very clean signal with low background. The background in the signal region (2.6 GeV/c $< p_l^B <$ 2.7 GeV/c) was estimated by fitting the sideband of p_l^B and extrapolating it into the signal region. For the peaking background, Monte Carlo (MC) modeling was employed. After all backgrounds were subtracted, the signal event was investigated. In the p_l^B signal region, there were no observed signal events for either search, as shown in Fig. 1. The upper limits at 90% C.L. were set to be 3.4 $\times$ 10^{-6} and 2.7 $\times$ 10^{-6} for e mode and μ mode, respectively. These are by far the most stringent limits.

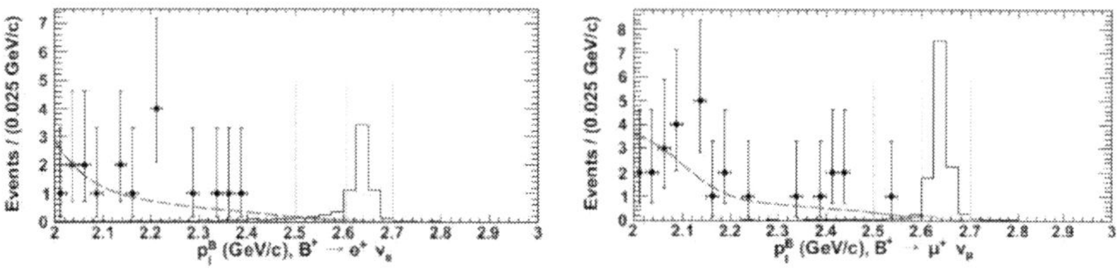

Figure 1. The unbinned maximum likelihood fits of the total background PDF to data for $B^+ \to e^+\nu_e$ (left) and $B^+ \to \mu^+\nu_\mu$ (right). The points with error bars are the experimental data. The dashed and the dotted lines show the background PDF in the sideband and signal regions, respectively. The distributions of the signal MC, displayed as the histogram, are scaled by 10^6 and 40 times the SM expectation for $B^+ \to e^+\nu_e$ and $B^+ \to \mu^+\nu_\mu$, respectively.

3. Searching for massive invisible particle

Figure 2 shows a Feynman diagram for $B^+ \to l^+X$. There is no well-matched theoretical model for X particles; we assumed that massive particles would replace neutrinos in the pure leptonic B decay. The X particle may be a candidate for dark matter or the light neutralino in the supersymmetry (SUSY). We expected the X particle to have no electric charge and to be a spin 1/2 particle. We assumed a range of possible mass for the X particle from 0.1 GeV/c^2 to 1.8 GeV/c^2 in steps of 100 MeV/c^2.

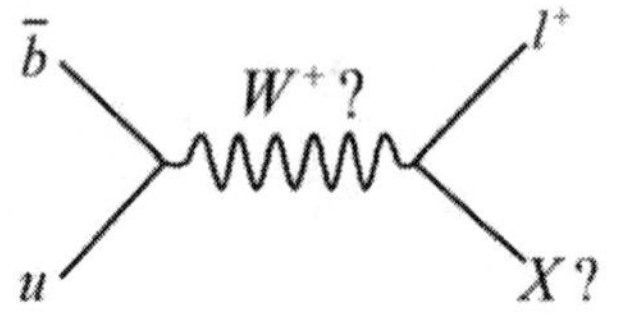

Figure 2. Feynman diagram of $B^+ \to l^+ X$.

The method of analysis was similar to $B^+ \to l^+\nu_l$ as described in Sect. 2; the systematics were also similar. However, a loose momentum cut and loose impact parameter selection were applied. Depending on the mass of the X particle, the peak of p_l^B was changed; therefore, cuts needed to be optimized for each X mass and different cuts were applied for each X. After all cuts were applied and the background was subtracted, the signal region of p_l^B is investigated for each X mass and preliminary upper limits in the branching fractions at 90% C.L. were obtained to be in the order of 10^{-6}.

4. Search for heavy neutrinos

The masses of particles in the SM are generated by the coupling of the Higgs field to the left- and right- handed components of a given particle. There are no right-handed neutrinos in the SM; rather, neutrinos in the SM are strictly massless. However, experimental data on neutrino oscillations have shown that neutrinos are not massless [6]. Many models beyond the SM suggest various explanations for this, and some models assume a heavy neutrino existence [12].

The direct search for heavy neutrinos in $B \to Xl\nu_h$ ($\nu_h \to l^\pm\pi^\mp$) is described in Ref. [13]. The mass range to search for a heavy neutrino is 0.5 GeV/c^2 $\leq M(\nu_h) \leq$ 5.0 GeV/c^2 and approach is divided depending on the mass as follows.

- $M(\nu_h) < 2.0$ GeV/c^2: look for $X = D$ or D^* using recoil mass,
- $M(\nu_h) \geq 2.0$ GeV/c^2: look for $X = D^{(*)}$, light meson and nothing.

Various backgrounds were suppressed by using the number of tracks requirement, strict lepton identification, vertex quality, distance of the lepton-pion vertex, and the distance between the closest associated hit in the SVD/CDC to the vertex of ν_h. By using the number of neutrinos detected, the upper limit of the branching fraction for ν_h-ν_l mixing was obtained and the results are shown along with other experimental results in Fig. 3. The corresponding upper limit for the product branching fraction for heavy neutrino mass (2.0 GeV/c^2) was in the order of 10^{-7}.

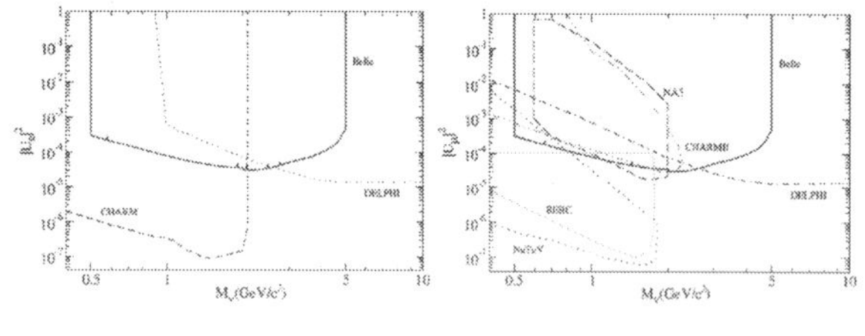

Figure 3. Comparison of the obtained upper limits for $|U_e|^2$ (left) and $|U_\mu|^2$ (right) with the existing experimental results.

5. Lepton flavor violation in tau decays

LFV in charged lepton decays is forbidden in the SM, or is highly suppressed even if neutrino mixing is taken into account. On the other hand, the extension of the SM and many other new models predict LFV with branching fractions as large as 10^{-8} [14]. Moreover, τ lepton is a good tool when it comes to searching for the LFV decays due to enhanced coupling to the NP, as well as the large number of LFV decay modes. In the study result of the LFV in $\tau \to lhh$, l is the electron or muon and h is the charged pion or kaon, as described in Ref. [15].

An 854 fb^{-1} data sample collected at or below resonances of Υ(4S) and Υ(5S) was used and a total of 14 modes were investigated. Blind analysis was performed by searching the signal events on the τ mass-ΔE plane, and the tagging method was used. τ_{tag} and τ_{sig} were required to be 1 prong decay and LFV final state, respectively. For background suppression, various observables were used, for example, an opening angle between the missing momentum and charged track on the tag side, thrust T, and missing mass (m^2_{miss}). To evaluate the branching fraction, elliptical signal regions that contain $\pm 3\sigma$ of the signal MC satisfying all selection criteria were used for each decay mode. As a result, one event was found for $\tau \to \mu\pi h$. For the other 12 modes, no events were observed in the signal regions. The number of observed signal events for all modes agreed with number of expected background events. The upper limits at 90% C.L. were obtained as $(2.0 \sim 8.6) \times 10^{-8}$, representing an improved on previously published upper limits [16] by a factor of about 1.8 on average.

6. Summary

Most stringent upper limits at 90% C.L. on $B^+ \to e^+(\mu^+)\nu$ have been obtained as $3.4(2.7) \times 10^{-6}$. The search for a heavy neutral lepton-like new particle was performed with $B^+ \to l^+X$, where X can be any invisible fermion particle, and preliminary results in the upper limits were in the order of 10^{-6} for 0.1 GeV/c^2 $< M_X <$ 1.8 GeV/c^2. A heavy neutrino search was also performed and the upper limit on mixing of ν_h-ν_l was set in the mass range of 0.5 GeV/c^2 $\sim$ 5.0 GeV/c^2. Up to now, there have been no clues concerning NP contribution from leptonic B decays. We studied 48 different LFV modes in τ decays, 46 of which were analyzed with an almost full data sample and the upper limits obtained on the branching fractions were in the order of 10^{-8}. In particular, the upper limits that were improved by a factor of about 1.8 on average were obtained for $\tau \to l^+hh'$.

References

[1] Abashian A *et al* (Belle Collaboration) 2002 *Nucl. Instrum. Methods Phys. Res., Sect.* A **479** 117
[2] Kurokawa S and Kikutani E 2003 *Nucl. Instrum. Methods Phys. Res., Sect.* A **499** 1
[3] Silverman D and Yao H 1988 *Phys. Rev.* D **38** 214
[4] Cabibbo N 1963 *Phys. Rev. Lett.* **10** 531; Kobayashi M and Maskawa T 1973 *Prog. Theor. Phys.* **49** 652
[5] Hou W S 1993 *Phys. Rev.* D **48** 2342
[6] Beringer J *et al* (Particle Data Group) 2012 *Phys. Rev.* D **86** 010001
[7] Dowdall R J *et al* (HPQCD Collaboration) 2013 *Phys. Rev. Lett.* **110** 222003
[8] Hara K *et al* (Belle Collaboration) 2010 *Phys. Rev.* D **82** 072007(R); Hara K *et al* (Belle Collaboration) 2013 *Phys. Rev. Lett.* **110** 131801
[9] Aubert B *et al* (BABAR Collaboration) 2007 *Phys. Rev.* D **76** 052002; Aubert B *et al* (BABAR Collaboration) 2008 *Phys. Rev.* D **77** 011107(R)
[10] Yook Y *et al* (Belle Collaboration) 2014 arXiv:1406.6356v1[hep-ex]
[11] Feindt M *et al* (Belle Collaboration) 2011 *Nucl. Instrum. Methods Phys. Res., Sect.* A **654** 432
[12] Asaka T, Blanchet S and Shaposhnikov M 2005 *Phys. Lett.* B **631** 151; Asaka T and Shaposhnikov M 2005 *Phys. Lett.* B **620** 17; Harvey J A, Ramond P and Reiss D B 1982 *Nucl. Phys.* B **199** 223; Dimopoulos S, Hall L J and Raby S 1992 *Phys. Rev. Lett.* **68** 1984; Hall L J and Raby S 1995 *Phys. Rev.* D **51** 6524; Dorsner I and Perez P F 2005 *Nucl. Phys.* B **723** 53; Zee A 1980 *Phys. Lett.* **93B** 389; **95B** 290; Zee A 1985 *Phys. Lett.* **161B** 141; Ma E and Sarkar U 1998 *Phys. Rev. Lett.* **80** 5716
[13] Liventsev D *et al* (Belle Collaboration) 2013 *Phys. Rev.* D **87** 071102
[14] Ilakovac A 2000 *Phys. Rev.* D **62** 036010; Black D *et al* 2002 *Phys. Rev.* D **66** 053002; Chen C H and Geng C Q 2006 *Phys. Rev.* D **74** 035010; Arganda E, Herrero M J and Portoles J 2008 *JHEP* **0806** 079; Benbrik R and Chen C H 2009 *Phys. Lett.* B **672** 172; Li Z H, Li Y and Xu H X 2009 *Phys. Lett.* B **677** 150; Li W J P *et al* 2010 *Int. J. Mod. Phys.* A **25** 4827; Liu W, Yue C X and Zhang J 2010 *Eur. Phys. J.* C **68** 197
[15] Miyazaki Y *et al* (Belle Collaboration) 2013 *Phys. Lett.* B **719** 346
[16] Miyazaki Y *et al* (Belle Collaboration) 2010 *Phys. Lett.* B **682** 355

Beauty baryon decays: a theoretical overview

Yu-Ming Wang

Physik Department T31, Technische Universität München, James-Franck-Straße 1, D-85748 Garching, Germany

Institut für Theoretische Teilchenphysik und Kosmologie, RWTH Aachen University, D-52056 Aachen, Germany

E-mail: yuming.wang@tum.de

Abstract. I overview the theoretical status and recent progress on the calculations of beauty baryon decays focusing on the QCD aspects of the exclusive semi-leptonic $\Lambda_b \to p\,\ell\nu$ decay at large recoil and theoretical challenges of radiative and electro-weak penguin decays $\Lambda_b \to \Lambda\,\gamma, \Lambda\,\ell^+\ell^-$.

1. Introduction

Investigating beauty baryon decays has been initiated in the early days of the heavy-quark effective theory (HQET) for the purpose of understanding the general properties of heavy-quark expansion, determining the CKM matrix elements and shedding light on physics beyond the Standard Model (BSM). In contrast to the B-meson decays, the helicity structure of the weak effective Hamiltonian can be extracted uniquely in bottom baryon decays with the aid of the baryon polarization asymmetry. On the theoretical side, QCD dynamics of a certain beauty baryon (i.e., Λ_b) is even simpler than that of B-mesons in the heavy quark limit. It is therefore natural to anticipate that beauty baryon decays will be helpful to sharpen our knowledge of the origin of CP violation and of the strong interaction dynamics, in particular taking into account the fact that no evident BSM signals have been detected yet in the intensive studies of B-meson decays.

The aim of this brief review is to report the current status of theoretical calculations for weak form factors of the semi-leptonic $\Lambda_b \to p\ell\nu$ decay, and to elaborate the primary challenges of factorizing the QCD dynamics for radiative and electro-weak penguin $\Lambda_b \to \Lambda\,\gamma, \Lambda\,\ell^+\ell^-$ decays.

2. Heavy-to-light $\Lambda_b \to p$ form factors at large recoil

A precision determination of the CKM matrix element $|V_{ub}|$ is among the central topics of heavy quark physics. The longstanding tension (approximately $3\,\sigma$) between exclusive and inclusive extractions has triggered enormous attentions for both new-physics (NP) believers and QCD oriented theorists. A measurement of the semi-leptonic $\Lambda_b \to p\ell\nu$ decay spectrum can be used to determine $|V_{ub}|$ complementary to the B-meson decays, provided that $\Lambda_b \to p$ form factors can be computed reliably on the theoretical side.

In general, one can parameterize the (axial)-vector current induced $\Lambda_b \to p$ matrix element in the form

$$\langle N(P')|\bar{u}\,\gamma_\mu\, b|\Lambda_b(P)\rangle = \bar{N}(P')\Big\{ f_1(q^2)\,\gamma_\mu + i\frac{f_2(q^2)}{m_{\Lambda_b}}\,\sigma_{\mu\nu}q^\nu + \frac{f_3(q^2)}{m_{\Lambda_b}}\,q_\mu \Big\}\Lambda_b(P)\,,$$

$$\langle N(P')|u\,\gamma_\mu\gamma_5\,b|\Lambda_b(P)\rangle \;=\; N(P')\Big\{g_1(q^2)\,\gamma_\mu + i\frac{g_2(q^2)}{m_{\Lambda_b}}\,\sigma_{\mu\nu}q^\nu + \frac{g_3(q^2)}{m_{\Lambda_b}}\,q_\mu\Big\}\gamma_5\Lambda_b(P)\,. \quad (1)$$

In the HQET limit, two form factors are sufficient to describe the QCD dynamics of hadronic matrix elements [1]

$$\langle N(P')|u\,\Gamma\,b|\Lambda_b(P)\rangle = N(P')\left[F_1(q^2) + F_2(q^2)\,\not{v}\right]\Gamma\Lambda_b(P)\,, \quad (2)$$

due to the Eikonal interaction of soft gluons with the heavy quark at leading power. At large hadronic recoil, the fast-moving active u-quark can be well approximated by a collinear field

$$\xi_c(x) = \frac{\not{n}_+\,\not{n}_-}{4}\,u(x)\,, \quad (3)$$

and then one can easily derive [2, 3]

$$\begin{aligned}\langle N(P')|u\,\Gamma\,b|\Lambda_b(P)\rangle &= N(P')\,(F_1 + F_2\,\not{v})\,\frac{\not{n}_+\,\not{n}_-}{4}\,\Gamma\,\Lambda_b(P)\,,\\ &= F_1\,N(P')\,\Gamma\,\Lambda_b(P)\,, \end{aligned}\quad (4)$$

where only the soft gluon interaction with the collinear u-quark is taken into account. A remarkable property of $\Lambda_b \to p$ transition matrix elements is that the QCD dynamics can be parameterized by a universal form factor in the combined heavy-quark and large-recoil limits. It needs to be pointed out that the soft-gluon exchange effect, or the so-called "Feynman mechanism", only contributes to $\Lambda_b \to p$ form factors $f_1(q^2)$ and $g_1(q^2)$ at sub-leading power, and the spectator interaction induced by two-collinear-gluon exchange is of leading power parametrically [4]. Such observations are quite distinct from that of $B \to \pi$ transition form factors at large recoil where both soft-gluon exchange and hard-scattering effects contribute at leading power. The formally leading-power spectator interaction for $\Lambda_b \to p$ form factors is however much suppressed numerically [5, 6, 7], due to the large energy of the nucleon (around 2.8 GeV) to be distributed among three valence quarks.

In addition to the fascinating progresses of understanding the factorization of heavy-to-light baryonic form factors, non-perturbative calculations of $\Lambda_b \to p$ form factors are also advanced by a systemic improvement of applying QCD light-cone sum rules (LCSRs) to study the baryon properties [7]. A formidable problem of constructing baryonic sum rule lies in the background pollution of dispersion relations from the unwanted baryons with opposite parity, which cannot be argued to be suppressed after the Borel transformation due to the small mass spitting between a parity doublet. The innovations of [7] consist in the elimination of background contaminations by combining the sum rules from different kinematical structures in the Lorenz decomposition of the correlation function, and the insensitivity of resulting sum rules to the choices of interpolating currents for a given baryon.

Following the strategy of determining the CKM matrix element $|V_{ub}|$ in semi-leptonic $B \to \pi\ell\nu$ decay [8], one can define a similar quantity

$$\Delta\zeta(0, q^2_{max}) \;=\; \frac{1}{|V_{ub}|^2}\int_0^{q^2_{max}} dq^2\,\frac{d\Gamma}{dq^2}(\Lambda_b \to p l \nu_l) \quad (5)$$

with the upper bound constrained by the applicability of LCSRs for $\Lambda_b \to p$ form factors. The prediction of $\Delta\zeta(0, 11\mathrm{GeV}^2)$ from the leading-order sum rule analysis with the axial-vector interpolating current of Λ_b baryon is given by

$$\Delta\zeta(0, 11\mathrm{GeV}^2) \;=\; 5.5^{+2.5}_{-2.0}\ \mathrm{ps}^{-1}\,, \quad (6)$$

the accuracy of which is not comparable with that for $B \to \pi\ell\nu$ including the next-to-leading-order (NLO) QCD corrections. Nevertheless, it is expected that a much higher accuracy can be achieved for QCD improved LCSRs of $\Lambda_b \to p$ form factors, promoting $\Lambda_b \to p\ell\nu$ as another golden channel to extract the matrix element $|V_{ub}|$ exclusively.

3. The FCNC $\Lambda_b \to \Lambda\,\gamma$ and $\Lambda_b \to \Lambda\,\ell^+\ell^-$ decays in QCD

Flavor-changing-neutral-current (FCNC) $b \to s\ell\ell$ induced processes are believed to be among the most sensitive probes of BSM physics. Thanks to a wealth of data accumulated at the LHC, measuring the full angular distributions of $B \to K^*\,\ell^+\ell^-$ is already accessible under the push of NP believers. Some deviations from the SM "expectations" revealed by LHCb measurements, in particular the so-called P_5' anomaly, have trigged hot discussions and debates in the community, stimulating various NP-based explanations. It is certainly of great fun to claim more and more puzzles in precision b-quark physics, however, we need to understand in the first place whether we miss some important pieces in the SM calculations for such sophisticated FCNC processes. Apart from many other conceptual issues, the breakdown of the operator-product-expansion (OPE) above the charm-quark threshold, typically $6 \sim 7\,\mathrm{GeV}^2$, will contaminate the SM calculations within the QCD factorization framework (see [9, 10] for more details).

As discussed in section 2, the QCD dynamics of $\Lambda_b \to \Lambda$ matrix elements can be greatly reduced in the heavy-quark and large-recoil limits. Non-perturbative calculations of $\Lambda_b \to \Lambda$ form factors have been performed extensively at large recoil using LCSRs with Λ-baryon distribution amplitudes (DAs) [11, 12, 13] and Λ_b-baryon DAs [3, 6], and carried out at low recoil with lattice QCD simulations [14]. Perturbative QCD calculation of $\Lambda_b \to \Lambda$ form factors [15] based on the transverse-momentum-dependent (TMD) factorization theorem implies that the parametrically leading-power hard-scattering contribution is strongly suppressed at finite b-quark mass. In the factorization limit, the calculated di-lepton invariant mass (q^2) spectrum of $\Lambda_b \to \Lambda\,\ell^+\ell^-$ at small q^2 [11] can accommodate the binned data from LHCb [16] within sizable uncertainties.

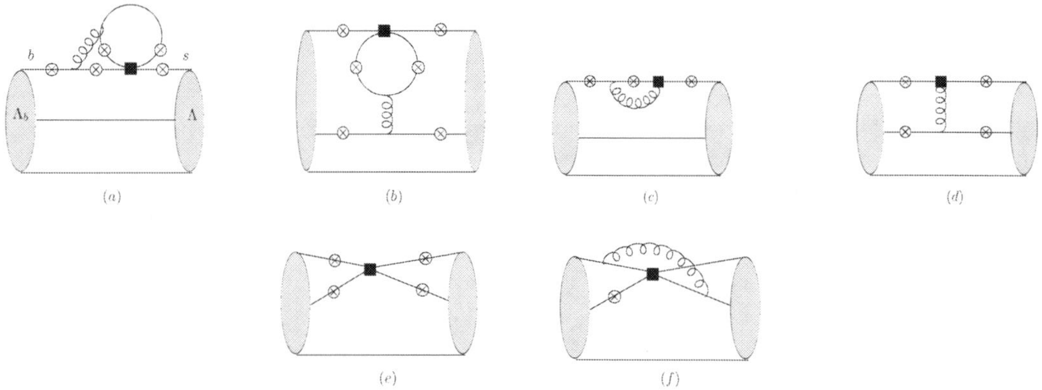

Figure 1. The NLO QCD corrections to the $\Lambda_b \to \Lambda\,\ell^+\ell^-$ decay matrix elements. Black squares indicate proper insertions of four-quark operators and gluonic penguin operator. Crossed circles denote possible virtual-photon emissions.

Anticipating the higher statistics of the LHC RunII data, theoretical calculations of $\Lambda_b \to \Lambda\,\ell^+\ell^-$ decays should be improved with the aid of QCD-based approaches to meet the experimental challenge. Some typical NLO diagrams contributing to the $\Lambda_b \to \Lambda\,\ell^+\ell^-$ transition matrix elements are displayed in figure 1. While the vertex corrections shown in the diagrams (a) and (c) can still be expressed by $\Lambda_b \to \Lambda$ form factors and perturbatively calculable hard coefficients as that of $B \to K^*\,\ell^+\ell^-$, the factorization properties of spectator-interaction and weak-annihilation induced contributions displayed in the diagrams (b) and (d) are drastically distinct from the mesonic counterpart. This can be understood from the fact that one of the two light valence quarks inside the Λ_b-baryon does not participate in the hard scattering process, and the not-light-like separation of this spectator quark and the heavy b-quark defines a general

Bethe-Salpeter wave function of the Λ_b-baryon. We are therefore not allowed to construct a QCD factorization formula to compute one-gluon-exchange spectator interaction and weak annihilation. In this respect, one can resort to the LCSRs approach by constructing proper correlation functions encoding the information of non-local $\Lambda_b \to \Lambda\,\ell^+\ell^-$ matrix elements that we are interested in.

Exploring the physics embedded in the $\Lambda_b \to \Lambda\,\ell^+\ell^-$ decay is also pursued intensively making use of the full angular distributions and baryon polarization information in the SM [17, 18] and beyond [19, 20, 21]. As a by-product, the radiative penguin decay $\Lambda_b \to \Lambda\,\gamma$ can be computed straightforwardly by setting the invariant mass of the lepton pair to zero, provided that the long-distance photon effect is subdominant numerically. This channel is however of great value on the phenomenological side, since the Λ-baryon polarization asymmetry can be used to determine the chirality structure of the electro-magnetic penguin operator uniquely at leading power [22].

4. Conclusions

To summarize, exclusive Λ_b-baryon decays can provide valuable information on the precision determination of CKM matrix elements, competitive with that from B-meson decays potentially, and serve as powerful probes in pursuit of the BSM physics on account of the reduced QCD dynamics in the joint heavy-quark and large-recoil limits. On the experimental side, there has been considerable progress on the measurements of beauty baryon decays at the LHC and the Tevatron. We have good reasonings for a very bright future of beauty-baryon physics which opens a new window to a more fundamental theory attributed to the additional spin information.

Acknowledgements

I would like to thank the organizers for inviting me to this enjoyable workshop and for the finical support.

References

[1] A. V. Manohar and M. B. Wise 2000 *Camb. Monogr. Part. Phys. Nucl. Phys. Cosmol.* **10** 1
[2] T. Mannel and Y. M. Wang 2011 JHEP **1112** 067 [arXiv:1111.1849 [hep-ph]]
[3] T. Feldmann and M. W. Y. Yip 2012 Phys. Rev. D **85** 014035 [Erratum-ibid. D **86** 079901] [arXiv:1111.1844 [hep-ph]]
[4] Wei Wang 2012 Phys. Lett. B **708** 119 [arXiv:1112.0237 [hep-ph]]
[5] C. D. Lu, Y. M. Wang, H. Zou, A. Ali and G. Kramer 2009 Phys. Rev. D **80** 034011 [arXiv:0906.1479 [hep-ph]]
[6] Y. M. Wang, Y. L. Shen and C. D. Lu 2009 Phys. Rev. D **80** 074012 [arXiv:0907.4008 [hep-ph]]
[7] A. Khodjamirian, C. Klein, T. Mannel and Y.-M. Wang 2011 JHEP **1109** 106 [arXiv:1108.2971 [hep-ph]]
[8] A. Khodjamirian, T. Mannel, N. ffen and Y.-M. Wang 2011 Phys. Rev. D **83** 094031 [arXiv:1103.2655 [hep-ph]]
[9] A. Khodjamirian, T. Mannel, A. A. Pivovarov and Y.-M. Wang 2010 JHEP **1009** 089 [arXiv:1006.4945 [hep-ph]]
[10] A. Khodjamirian, T. Mannel and Y. M. Wang 2013 JHEP **1302** 010 [arXiv:1211.0234 [hep-ph]]
[11] Y. M. Wang, Y. Li and C. D. Lu 2009 Eur. Phys. J. C **59** 861 [arXiv:0804.0648 [hep-ph]]
[12] T. M. Aliev, K. Azizi and M. Savci 2010 Phys. Rev. D **81** 056006 [arXiv:1001.0227 [hep-ph]]
[13] L. F. Gan, Y. L. Liu, W. B. Chen and M. Q. Huang 2012 Commun. Theor. Phys. **58** 872 [arXiv:1212.4671 [hep-ph]]
[14] W. Detmold, C.-J. D. Lin, S. Meinel and M. Wingate 2013 Phys. Rev. D **87** 074502 [arXiv:1212.4827 [hep-lat]]
[15] X. G. He, T. Li, X. Q. Li and Y. M. Wang 2006 Phys. Rev. D **74** 034026 [hep-ph/0606025]
[16] R. Aaij *et al.* [LHCb Collaboration] 2013 Phys. Lett. B **725** 25 [arXiv:1306.2577 [hep-ex]]
[17] C. H. Chen and C. Q. Geng 2001 Phys. Rev. D **64** 074001 [hep-ph/0106193]
[18] C. H. Chen and C. Q. Geng 2001 Phys. Rev. D **63** 114024 [hep-ph/0101171]
[19] T. M. Aliev and M. Savci 2006 JHEP **0605** 001 [hep-ph/0507324]
[20] M. J. Aslam, Y. M. Wang and C. D. Lu 2008 Phys. Rev. D **78** 114032 [arXiv:0808.2113 [hep-ph]]
[21] Y. M. Wang, M. J. Aslam and C. D. Lu 2009 Eur. Phys. J. C **59** 847 [arXiv:0810.0609 [hep-ph]]
[22] C. S. Huang and H. G. Yan 1999 Phys. Rev. D **59** 114022 [Erratum-ibid. D **61** 039901] [hep-ph/9811303]

Recent results from b-baryon decays at LHCb

James McCarthy

E-mail: james.mccarthy@cern.ch

Abstract. Recent measurements of b-baryon decays using $3\,\mathrm{fb}^{-1}$ proton-proton collision data collected by the LHCb experiment are presented. Measurements of the branching fractions of $\Lambda_b^0 \to J/\psi\, p\pi^-$, $\Lambda_b^0 \to \Lambda_c^+ D_s^-$, $\Lambda_b^0 \to \Lambda_c^+ D^-$ and $\Lambda_b^0 \to \Lambda_c^+ \pi^-$ decays have been made, as well as precise measurements of the mass and lifetime of the Ξ_b^0 baryon. In addition, the p_T and η dependence of the ratio of Λ_b^0 to B^0 production fractions has been measured.

1. Introduction

Physics involving b-baryons is still a relatively unexplored field in particle physics, but is of particular interest since baryons have a non-zero spin, and so would allow us to probe the helicity structure of the Heavy Quark Expansion Hamiltonian. In order to achieve this, precise measurements of branching fractions, lifetimes and masses of b-baryons are needed. In addition to this, one can look for *CP* asymmetries in b-baryon decays, which can arise due to interference between tree and loop diagrams in these decays. There studies would be relatively easy to carry out, since conservation of baryon number provides a natural method to tag decays; however, no *CP* asymmetries have yet been observed in any baryon decays.

This note presents a selection of recent results involving b-baryon decays, and precise measurements of the Ξ_b^0 baryon. Unless otherwise stated, results are obtained using a sample of $3\,\mathrm{fb}^{-1}$ proton-proton collision data collected in 2011 and 2012 by the LHCb experiment.

2. Observation of the $\Lambda_b^0 \to J/\psi\, p\pi^-$ decay.

This section presents the search for the $\Lambda_b^0 \to J/\psi\, p\pi^-$ decay using the $\Lambda_b^0 \to J/\psi\, pK^-$ decay as a control channel (except for *CP* measurements, charge conjugation is implied throughout). Full details of the results can be found in Ref. [1]. The selection consists of a loose kinematic constraint, followed by particle identification (PID) requirements, which make use of the LHCb detector to separate protons, pions and kaons. Finally, to achieve the optimal separation between signal and background, a multivariate selection is applied, namely a Neural Net (NN). After the selection is applied, an unbinned maximum likelihood fit is performed to the $J/\psi\, ph$ invariant mass distribution, and 2102 ± 61 $\Lambda_b^0 \to J/\psi\, p\pi^-$ candidates are found. This is the first observation of this decay, with a significance of $> 30\sigma$. Using the ratio of yields of signal to control channel, corrected by the ratio of selection efficiencies, η, the ratio of branching fractions is calculated to be

$$\frac{\mathcal{B}(\Lambda_b^0 \to J/\psi\, p\pi^-)}{\mathcal{B}(\Lambda_b^0 \to J/\psi\, pK^-)} = \frac{N(\Lambda_b^0 \to J/\psi\, p\pi^-)}{N(\Lambda_b^0 \to J/\psi\, pK^-)} \times \frac{\eta(\Lambda_b^0 \to J/\psi\, pK^-)}{\eta(\Lambda_b^0 \to J/\psi\, p\pi^-)} = 0.0824 \pm 0.0025\,(\mathrm{stat}) \pm 0.0042\,(\mathrm{syst})$$

where the first uncertainty is statistical, and the second is systematic.

From these extracted yields, we can measure the raw asymmetry for each decay,

$$\mathcal{A}^{\rm raw}(\Lambda_b^0 \to J/\psi\, p\pi^-) = (+7.9 \pm 2.2)\%$$
$$\mathcal{A}^{\rm raw}(\Lambda_b^0 \to J/\psi\, pK^-) = (+1.1 \pm 0.7)\%,$$

where the uncertainties are entirely statistical and

$$\mathcal{A}^{\rm raw} = \frac{N(\Lambda_b^0) - N(\overline{\Lambda}_b^0)}{N(\Lambda_b^0) + N(\overline{\Lambda}_b^0)}. \tag{1}$$

The raw asymmetry is related to the *CP* asymmetry, $\mathcal{A}^{CP}$, via the equation

$$\mathcal{A}^{\rm raw} = \mathcal{A}^{CP} + \mathcal{A}^{\rm prod}(\Lambda_b^0) - \mathcal{A}^{\rm det}(\pi/K) + \mathcal{A}^{\rm det}(p), \tag{2}$$

where $\mathcal{A}^{\rm prod}$ is the asymmetry in production of Λ_b^0 and $\overline{\Lambda}_b^0$, and $\mathcal{A}^{\rm det}$ is the detection asymmetry. Since $\mathcal{A}^{\rm prod}$ (Λ_b^0) and $\mathcal{A}^{\rm det}$ (p) have not been well measured, we measure the difference in $\mathcal{A}^{CP}$ between the$J/\psi\, p\pi^-$ and $J/\psi\, pK^-$ decays, defined as

$$\Delta\mathcal{A}^{CP} = \mathcal{A}^{CP}(J/\psi\, p\pi^-) - \mathcal{A}^{CP}(J/\psi\, pK^-) = \mathcal{A}^{\rm raw}(J/\psi\, p\pi^-) - \mathcal{A}^{\rm raw}(J/\psi\, pK^-) + \mathcal{A}^{\rm det}(\pi) - \mathcal{A}^{\rm det}(K).$$

The difference in detection asymmetries, $\mathcal{A}^{\rm det}$ (π) - $\mathcal{A}^{\rm det}$ (K), is measured using $\overline{B}^0 \to J/\psi\, \overline{K}^{*0}$ decays and is found to be -0.0110 ± 0.0033 [2]. The difference in *CP* asymmetries is found to be $\Delta\mathcal{A}^{CP} = (+5.7 \pm 2.3\,({\rm stat}) \pm 1.2\,({\rm syst}))\%$, which is consistent with no *CP* violation at the level of 2.2σ.

3. $\Lambda_b^0 \to \Lambda_c^+ D^-$ and $\Lambda_b^0 \to \Lambda_c^+ D_s^-$ decays.

This section presents for search for Λ_b^0 baryons decaying into pairs of charm hadrons with Λ_c^+ D^- and Λ_c^+ D_s^- final states, as described in Ref. [3]. The charm hadron resonances are reconstructed through the $D^+ \to K^-\pi^+\pi^+$, $D_s^+ \to K^-K^+\pi^+$ and $\Lambda_c^+ \to pK^-\pi^+$ decays, and a Boosted Decision Tree (BDT) is used to reduce the combinatoric background. Both decays are observed for the first time, with 4633 ± 69 $\Lambda_b^0 \to \Lambda_c^+ D_s^-$ candidates and 262 ± 19 $\Lambda_b^0 \to \Lambda_c^+ D^-$ candidates. The efficiency corrected yields are used to calculate the ratio of branching fractions:

$$\frac{\mathcal{B}(\Lambda_b^0 \to \Lambda_c^+ D^-)}{\mathcal{B}(\Lambda_b^0 \to \Lambda_c^+ D_s^-)} = 0.042 \pm 0.003\,({\rm stat}) \pm 0.003\,({\rm syst}),$$

where the first uncertainty is statistical, and the second is systematic. In order to obtain the absolute branching fractions for these decays, the high statistics $\overline{B}^0 \to D^+ D_s^-$ decay is used, along with the ratio of $\mathcal{B}(\Lambda_b^0 \to \Lambda_c^+ \pi)$ to $\mathcal{B}(\overline{B}^0 \to D^+\pi^-)$. By taking the double ratio of branching fractions, all dependence on the ratio of production fractions and charm hadron branching fractions is removed. The double ratio has been measured to be

$$\left[\frac{\mathcal{B}(\Lambda_b^0 \to \Lambda_c^+ D_s^-)}{\mathcal{B}(\overline{B}^0 \to D^+ D_s^-)}\right] \Big/ \left[\frac{\mathcal{B}(\Lambda_b^0 \to \Lambda_c^+ \pi^-)}{\mathcal{B}(\overline{B}^0 \to D^+ \pi^-)}\right] = 0.96 \pm 0.02\,({\rm stat}) \pm 0.06\,({\rm syst}).$$

From this the absolute branching fractions are found to be $\mathcal{B}(\Lambda_b^0 \to \Lambda_c^+ D_s^-) = (1.1 \pm 0.1) \times 10^{-2}$ and $\mathcal{B}(\Lambda_b^0 \to \Lambda_c^+ D^-) = (4.7 \pm 0.6) \times 10^{-4}$.

From the unbinned maximum likelihood fit to the invariant mass distributions of the Λ_c^+ D_s^+ and D^+ D_s^-, the difference between the mass of the Λ_b^0 baryon and B^0 meson is measured

$$m(\Lambda_b^0) - m(B^0) = 339.72 \pm 0.24\,({\rm stat}) \pm 0.18\,({\rm syst})\,{\rm MeV}/c^2.$$

Using the known mass of the B^0 meson, and taking an average with the Λ_b^0 baryon mass measured previously by LHCb [4], the mass of the Λ_b^0 baryon is measured to be $m(\Lambda_b^0) = 5619.36 \pm 0.26\,{\rm MeV}/c^2$.

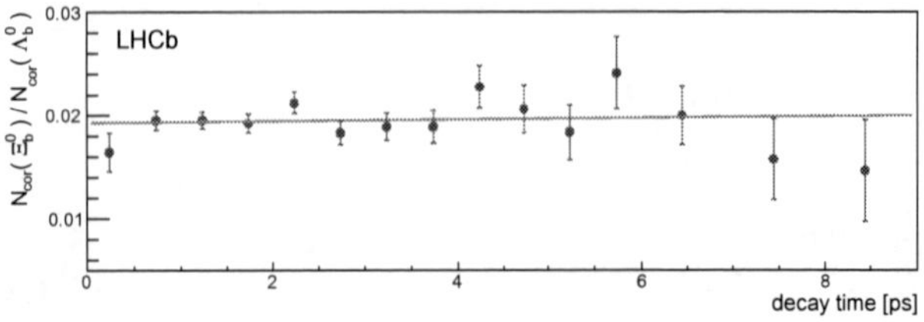

Figure 1. Efficiency corrected yield as a function of decay time, fit with an exponential.

4. Mass and lifetime measurements of the Ξ_b^0 baryon

This section presents precision measurements of the Ξ_b^0 baryon. Full details can be found in Ref. [5]. The Ξ_b^0 baryon is measured using the $\Xi_b^0 \to \Xi_c^+\pi^-$ decay, with the $\Lambda_b^0 \to \Lambda_c^+\pi^-$ decay used as a control channel. Both charm baryons are reconstructed through decays to $pK^-\pi^+$ final state, and a BDT is used to obtain a clean sample of 3775 ± 71 Ξ_b^0 candidates and $(180.8 \pm 0.5) \times 10^3$ Λ_b^0 candidates. An unbinned maximum likelihood fit to the invariant mass distributions is used to extract the difference in mass of the Ξ_b^0 and Λ_b^0 baryons, which is measured to be

$$m(\Xi_b^0) - m(\Lambda_b^0) = 172.44 \pm 0.39\,(\text{stat}) \pm 0.17\,(\text{syst})\,\text{MeV}/c^2.$$

Using the mass of the Λ_b^0 baryon measured in Section 3, the mass of the Ξ_b^0 baryon is measured to be $m(\Xi_b^0) = 5791.80 \pm 0.39\,(\text{stat}) \pm 0.17\,(\text{syst}) \pm 0.26(\Lambda_b^0)\,\text{MeV}/c^2$, which is a vast improvement on previous measurements. In addition, the first measurement of the lifetime of the Ξ_b^0 is made by plotting the efficiency corrected ratio of Ξ_b^0 to Λ_b^0 yields as a function of decay time, shown in Figure 1. This is then fit with with an exponential of the the form $e^{-\beta t}$ where

$$\beta = \frac{1}{\tau(\Xi_b^0)} - \frac{1}{\tau(\Lambda_b^0)}. \tag{3}$$

From the fit we find $\beta = (0.40 \pm 1.21) \times 10^{-2}\,\text{ps}^{-1}$, corresponding to the ratio of Ξ_b^0 to Λ_b^0 lifetimes:

$$\frac{\tau(\Xi_b^0)}{\tau(\Lambda_b^0)} = \frac{1}{1 - \beta\tau(\Lambda_b^0)} = 1.006 \pm 0.018\,(\text{stat}) \pm 0.010\,(\text{syst}).$$

Using the value of $\tau(\Lambda_b^0)$ measured in Ref. [6], the lifetime of the Ξ_b^0 baryon is measured for the first time to be $\tau(\Xi_b^0) = 1.477 \pm 0.026\,(\text{stat}) \pm 0.014\,(\text{syst}) \pm 0.013(\Lambda_b^0)\,\text{ps}$.

5. Λ_b^0 production and $\Lambda_b^0 \to \Lambda_c^+\pi^-$ branching fraction measurements.

This section presents the measurement of Λ_b^0 production fractions, $f_{\Lambda_b^0}/f_d$, using a sample of proton-proton collision data at $\sqrt{s} = 7\,\text{TeV}$ corresponding to an integrated luminosity of $1\,\text{fb}^{-1}$. Full details can be found in Ref. [7]. It has been known for some time that the production of Λ_b^0 baryons depends on the transverse momentum, p_T, of the system. Measurements made by LEP [8] were inconsistent with those made by CDF [9], and the p_T dependence was explicitly measured by LHCb using semileptonic Λ_b^0 decays [10]. This analysis is an update of this measurement using the hadronic $\Lambda_b^0 \to \Lambda_c^+\pi^-$ and $\overline{B}^0 \to D^+\pi^-$ decays, and the production fractions are measured as a function of p_T and pseudorapidity, η. The ratio of Λ_b^0 to B^0 production fractions is then given by

$$\frac{f_{\Lambda_b^0}}{f_d}(X) = \frac{\mathcal{B}(B^0 \to D^+\pi^-)}{\mathcal{B}(\Lambda_b^0 \to \Lambda_c^+\pi^-)} \frac{\mathcal{B}(D^+ \to K^-\pi^+\pi^+)}{\mathcal{B}(\Lambda_c^+ \to pK^-\pi^+)} \times \frac{N_{\Lambda_b^0 \to \Lambda_c^+\pi^-}(X)\epsilon_{B^0 \to D^+\pi^-}(X)}{N_{B^0 \to D^+\pi^-}(X)\epsilon_{\Lambda_b^0 \to \Lambda_c^+\pi^-}(X)} \tag{4}$$

$$\equiv S \times \frac{N_{\Lambda_b^0 \to \Lambda_c^+\pi^-}(X)\epsilon_{B^0 \to D^+\pi^-}(X)}{N_{B^0 \to D^+\pi^-}(X)\epsilon_{\Lambda_b^0 \to \Lambda_c^+\pi^-}(X)} \tag{5}$$

$$\equiv S \times R(X).$$

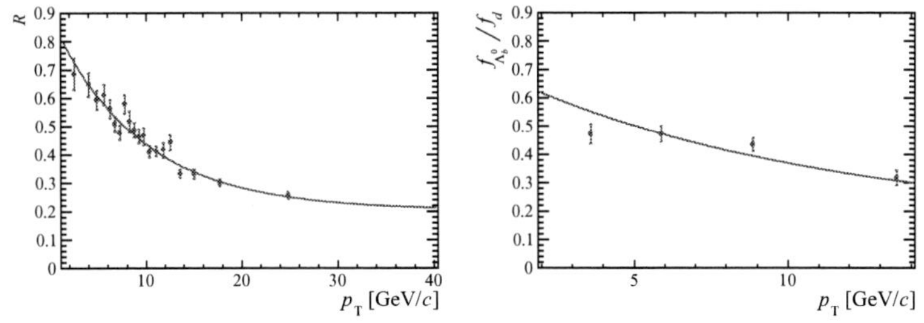

Figure 2. a) Efficiency corrected ratio of yields as a function of p_T, fit with an exponential function. b) Ratio of production fractions as a function of p_T from [10], fit with the parameters from the fit in a)

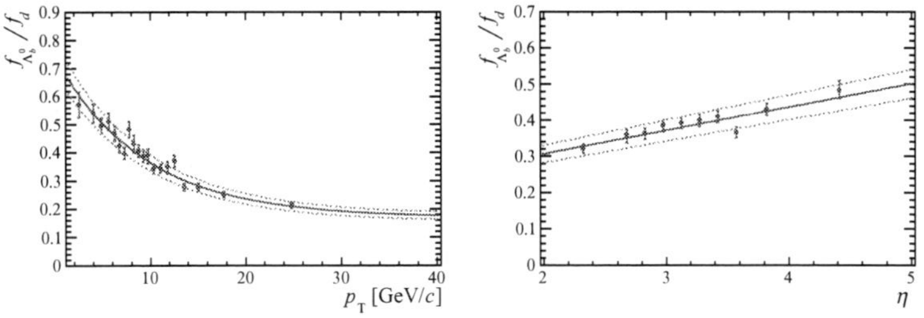

Figure 3. Ratio of production fractions as a function of a) p_T, fit with an exponential function, and b) as a function of η, fit with a linear function. The bands indicate the uncertainties from the semileptonic measurement.

The efficiency corrected ratio of yields, $R(p_T)$, is plotted as a function of p_T in Figure 2a), and fit with an exponential function. The parameters of the exponential are then fixed, and the same fit is performed to the ratio $f_{\Lambda_b^0}/f_d$ measured in the semileptonic decay, allowing the normalisation to float. This is shown in Figure 2b). From S, the branching fraction of the $\Lambda_b^0 \to \Lambda_c^+\pi^-$ decay is extracted, and found to be $\mathcal{B}(\Lambda_b^0 \to \Lambda_c^+\pi^-) = (4.46 \pm 0.36) \times 10^{-3}$. This is the most precise measurement of a branching fraction of a Λ_b^0 decay measured to date.

Figure 3a) shows the ratio of efficiency corrected yields multiplied by the scale factor, S. This corresponds to a measurement of the p_T dependence of the ratio of production fractions, and the dependence is fit with an exponential function. Figure 3b) shows $f_{\Lambda_b^0}/f_d$ as a function of η, which is fit with a linear function.

6. Summary

In summary, many new measurements of b-baryon decays performed by the LHCb experiment have been presented, including the branching fractions of $\Lambda_b^0 \to J/\psi p\pi^-$, $\Lambda_b^0 \to \Lambda_c^+ D_s^-$, $\Lambda_b^0 \to \Lambda_c^+ D^-$ and $\Lambda_b^0 \to \Lambda_c^+\pi^-$ decays. Precise measurements of the mass and lifetime of the Ξ_b^0 baryon have been made, and the p_T and η dependence of the ratio of Λ_b^0 to B^0 production fractions has been measured. The programme of analyses at LHCb involving baryons is extensive, and more results can be expected in the near future.

References

[1] Aaij R *et al.* (LHCb collaboration) 2014 *JHEP* **07** 103 (*Preprint* 1406.0755)
[2] Aaij R *et al.* (LHCb collaboration) 2013 *Phys. Rev. Lett.* **110** 031801 (*Preprint* 1210.4492)
[3] Aaij R *et al.* (LHCb collaboration) 2014 *Phys. Rev. Lett.* **112** 202001 (*Preprint* 1403.3606)
[4] Aaij R *et al.* (LHCb collaboration) 2013 *Phys. Rev. Lett.* **110** 182001 (*Preprint* 1302.1072)
[5] Aaij R *et al.* (LHCb collaboration) 2014 *Phys. Rev. Lett.* **113** 032001 (*Preprint* 1405.7223)
[6] Aaij R *et al.* (LHCb collaboration) 2014 *Phys. Lett.* **B734** 122 (*Preprint* 1402.6242)
[7] Aaij R *et al.* (LHCb collaboration) 2014 to appear in JHEP (*Preprint* 1405.6842)
[8] Amhis Y *et al.* (Heavy Flavor Averaging Group) 2012 (*Preprint* 1207.1158)
[9] Aaltonen T *et al.* (CDF Collaboration) 2009 *Phys.Rev.* **D79** 032001 (*Preprint* 0810.3213)
[10] Aaij R *et al.* (LHCb collaboration) 2012 *Phys. Rev.* **D85** 032008 (*Preprint* 1111.2357)

Charmonium with an effective Morse molecular potential

Mario Everaldo de Souza

Departamento de Fsica, Universidade Federal de Sergipe, So Cristovo, 49100-000, Brazil

E-mail: mariodesouza.ufs@gmail.com

Abstract. The Morse molecular potential is used for the first time as an effective potential for the overall interaction in charmonium. This procedure allows the calculation of the rotational contributions of P states, the radii of five S states, and an absolute threshold for bound states. The calculation of the latter provides important information on the character of the recently found levels $X(3915)$, $X(3940)$, $\Psi(4040)$, $X(4050)$, $X(4140)$, $\Psi(4160)$, $X(4160)$, $X(4250)$, $X(4260)$, $X(4350)$, $\Psi(4415)$, $X(4430)$, and $X(4660)$.

1. Introduction

The pioneering work of Eichten *et al.* [1], and many other subsequent works, have shown that the static potential plays an important role in the description of heavy mesons. Despite efforts lasting more than three decades, some important features of the static interaction are not yet fully understood. That is why several approximate methods and effective potentials have been proposed for the description of the overall interaction in quarkonia. A partial list of the articles on these two subjects is found in reference [2].

It has been established in the literature that for heavy mesons non-relativistic quantum mechanics with constituent masses for the quark and antiquark can be applied. In this work we use the Morse molecular potential for the first time to describe some low energy states of charmonium and to find its effective potential parameters. This unique description allows us to calculate the contribution of the rotational energy of P states, and the radii of five S states. It also permits the calculation of the absolute threshold for the bound states of charmonium that sheds light on the character of the recently found levels [3] $X(3915)$, $X(3940)$, $\Psi(4040)$, $X(4050)$, $X(4140)$, $\Psi(4160)$, $X(4160)$, $X(4250)$, $X(4260)$, $X(4350)$, $\Psi(4415)$, $X(4430)$, and $X(4660)$.

It is not necessary to consider confinement because we only deal with low energy levels. Confinement is not well understood and there are models of hadrons that do not take it into account. For example, the original MIT bag model treats confinement only at the wall by making the vector current null at it [4]. In the case of the chiral bag model [5], confinement is treated by means of the continuity of the axial vector current at the wall. All the experimental data used below for the energies of charmonium states and the quark c mass were taken from the Particle Data Group [3]. This work is based on the recently published article by de Souza [6] on bottomonium.

2. The Morse Molecular Potential

The Morse molecular potential [7]

$$V(x) = D\left(e^{-2\alpha x} - 2e^{-\alpha x}\right) \tag{1}$$

is widely used in the description of diatomic molecules where $x = (r - a)/a$, and α is a parameter to be found from the fitting. For $r = a$ the potential has its minimum value, $-D$. The first term in the above expression takes care of the well-known repulsion of the strong force for very small distances [8] and the second term takes care of the overall attraction due to QCD forces. For $|x| < 1$ this potential can be expanded about the minimum up to order 3 in x and the expression

$$V(x) = -D + \frac{1}{2}ka^2x^2 - \lambda ka^3x^3 \tag{2}$$

is produced, where $\lambda = \alpha/2a$, and $k = m\omega^2$, in which m is the reduced mass of the constituent quark and antiquark. That is, $m = (1/2)M_c$ in which M_c is the mass of the c quark. The constant ω is defined below and is found from the fitting. For this potential the solution of the Schrödinger equation yields the expression [7, 9]

$$\begin{aligned} E_{vL} &= h\omega\left(v+\frac{1}{2}\right) - A\left(v+\frac{1}{2}\right)^2 + B_L L(L+1) \\ &\quad -D_L L^2(L+1)^2 - C_{vL}\left(v+\frac{1}{2}\right)L(L+1) + \ldots \end{aligned} \tag{3}$$

for the vibrational and rotational levels above the minimum of the potential, where the quantum numbers v and L are equal to $0, 1, 2, 3, \ldots$ and A, B_L, C_L, D_L and C_{vL} are constants to be found from the fitting. In this last equation the first term describes harmonic vibrations, the second term takes into account the anharmonicity of the potential, the third term describes rotations with constant moment of inertia, the fourth term represents the centrifugal distortion and the fifth term represents the coupling between vibration and rotation. This expression can also be written as [9]

$$\begin{aligned} E_{vL} &= h\omega\left(v+\frac{1}{2}\right) - \frac{(h\omega)^2}{4D}\left(v+\frac{1}{2}\right)^2 + B_L L(L+1) \\ &\quad -D_L L^2(L+1)^2 - C_{vL}\left(v+\frac{1}{2}\right)L(L+1) + \ldots \end{aligned} \tag{4}$$

where ω and D are related to α, a and m by

$$\omega^2 = \frac{2\alpha^2 D}{ma^2}. \tag{5}$$

The constant B_L is given by $B_L = h^2/2ma^2$. In the fitting below we disregarded the last term of Eqs. (3) and (4).

3. The Fitting

The first two levels, $\eta_c(1S)$ and $J/\Psi(1S)$, are a hyperfine doublet due to the spin-spin interaction, but as our Hamiltonian does not depend on spin, we can use the S states $J/\Psi(1S)$ and $\Psi(2S) \equiv \Psi(3686)$ for the fitting. We consider now the P states $\chi_{c0}(1P)$, $\chi_{c1}(1P)$ and $\chi_{c2}(1P)$ which is a triplet due to the spin-orbit interaction whose contribution is given by

$$\Delta E_{SL} = \Delta\left[J(J+1) - L(L+1) - S(S+1)\right], \tag{6}$$

Table 1. The levels considered in the fitting of charmonium.

(v, L)	Particle	Mass (MeV/c^2)
$(0, 0)$	$J/\Psi(1S)$	3096.916 ± 0.011
$(0, 1)$	$\chi_c(1P)$	3549.7 ± 37.9
$(1, 0)$	$\Psi(3686)$	3686.09 ± 0.04
$(0, 2)$	$\Psi(3770)$	3778.1 ± 1.2

where $J = \left|\vec{J}\right| = \left|\vec{L} + \vec{S}\right|$. For the states $\chi_{c0}(1P)$, $\chi_{c1}(1P)$ and $\chi_{c2}(1P)$ we have $S = 1$ and $L = 1$, and thus, $J = 0, 1$ and 2, respectively. Using these values in Eq. (6) we obtain the average values $\Delta_{1P} = (27.7 \pm 13.6)$ Mev/c^2, and $E_{1P} = (3549.7 \pm 37.9)$ Mev/c^2 for the degenerate level, $\chi_c(1P)$. The state $\Psi(3770)$, which is classified as 1^3D_1, is a very well established state, experimentally found with a mass of (3778.1 ± 1.2) Mev/c^2. Table 1 presents a summary of the levels used in the fitting with the corresponding values of v and L.

4. Results and discussion

4.1. Results for the potential parameters

Fitting the levels of table 1 to Eq. (3) gives the following values for its parameters: $h\omega = (8062.0 \pm 0.1)$ Mev/c^2, $A = (3736.4 \pm 0.1)$ Mev/c^2, $B_L = (282.8 \pm 58.3)$ Mev/c^2, $D_L = (28.3 \pm 5.1)$ Mev/c^2. From Eqs. (3) and (4) we obtain that $A = (h\omega)^2/4D$, and thus $D = (4348 \pm 0.5)$ Mev/c^2.

Using the values of D and $h\omega$ above in Eq. (5) we obtain $a/\alpha = (13.15 \pm 0.05) \times 10^{-2}$ fm, and from the value of B_L we find that $a = (0.28 \pm 0.05)$ fm which is a very reasonable figure since the Compton wavelength of charmonium is about 0.69 fm if we use a constituent mass of 1.7 GeV/c^2 . Using the above values of a and a/α we obtain $\alpha = 2.15 \pm 0.39$.

As we showed above, a molecular potential is harmonic about its minimum, and thus we can calculate the value for the constant $k = m\omega^2$ which can be written as $k = mc^2(h\omega)^2/(hc)^2$. Using the above values we obtain $k \approx 1.44 \times 10^3$ GeV/fm$^2 \approx 2.31 \times 10^{23}$N/m which is a quite fair number. For a distance of 0.3 fm it produces a force $F \approx 10^8$N. The following simple calculation shows that this is a reasonable number. If we calculate the average force necessary for producing a work of 4 GeV in 0.3 fm we obtain a force of about 10^6N.

4.2. Number of S and P states

As we see in table 1 the quantum number n of QCD models corresponds to $v + 1$ in the present model. The maximum number of v can be calculated for a molecular potential by using the simple following reasoning. In Eq.(3) the second term is always smaller than the first term, and thus we obtain $v < \frac{\hbar\omega}{A} - \frac{1}{2} = 1.65792 \pm 0.00004$ and thus the possible values of v are only 0 and 1. Therefore, the possible values for n are only 1 and 2.

We do a similar reasoning for calculating the maximum number of L. In Eq.(3) the fourth term is always smaller than the third term, and so we obtain $L(L+1) < \frac{B_L}{D_L} = 9.99 \pm 2.06$ which is satisfied for $L = 0, 1, 2, 3$. This means, respectively, S, P, D and F states.

4.3. Discussion on the recently found levels

We found above that $D = (4348 \pm 0.5)$ Mev/c^2 which is a very important result and states that there is no charmonium bound state above (4348.8 ± 0.5) Mev/c^2. Therefore, the states $\Psi(4415)$, $X(4430)$, and $X(4660)$ are not bound states of charmonium. On the other hand this

Table 2. Radii of five states of charmonium calculated with the use of Eq. (7).

(v, L)	Particle	Radius (fm)
$(0,0)$	$J/\Psi(1S)$, $\eta_c(1S)$	0.35 ± 0.06
$(1,0)$	$\Psi(3686)$, $\eta_c(2S)$	0.49 ± 0.09
$(2,0)$	$\Psi(3S)$	0.63 ± 0.11

result shows that the recently found states $X(3915)$, $X(3940)$, $\Psi(4040)$, $X(4050)$, $X(4140)$, $\Psi(4160)$, $X(4160)$, $X(4250)$, $X(4260)$, $X(4350)$ may be bound states of charmonium. But as we saw in section *4.2* above the maxium number of n is two and, thus, if they are charmonium states, none of them is a $3S$ state. They have to be $2P$, $2D$, and $2F$ states.

4.4. Radii of some S states

As it was shown above the Morse potential, when expanded about its minimum, yields Eq. (2). For such a potential Robinett [10] obtained the following equation for the average value of position for S states

$$\langle r\rangle_v \quad = \quad a + \frac{3\alpha h\omega}{2m\omega^2 a}\left(v + \frac{1}{2}\right) = a + \frac{3ah\omega}{4\alpha D}\left(v + \frac{1}{2}\right), \tag{7}$$

where we have taken into account Eq. (5). We can identify these average values with the radii of charmonium states. We only calculate the radii of the states $J/\Psi(1S)$, $\eta_c(1S)$, $\Psi(3686)$, $\eta_c(2S)$, and $\Psi(3S)$ because the other upper states are far from the bottom of the potential. Using the above values for the constants we obtain the results shown in table 2 for the radii of these five states of charmonium.

5. Conclusion

The fitting of some energy levels of charmonium to the Morse molecular potential makes possible the calculation of parameters of the effective molecular potential, prediction of the radii of five states and sheds some light on the character of the states $X(3915)$, $X(3940)$, $\Psi(4040)$, $X(4050)$, $X(4140)$, $\Psi(4160)$, $X(4160)$, $X(4250)$, $X(4260)$, $X(4350)$, $\Psi(4415)$, $X(4430)$, and $X(4660)$. Therefore, the above results add important information for the understanding of charmonium and complement information obtained from QCD models.

6. References

[1] Eichten E, Gottfried K, Kinoshita T, Kogut J, Lane K D and Yan T -M 1975 *Phys. Rev. Lett.* **34** 369
[2] http://inspirehep.net/record/129333/references
[3] Beringer J *et al.* (Particle Data Group) 2012 (updated to 2014) *Phys. Rev. D* **86** 010001
[4] Chodos A, Jaffe R L, Johnson K, Thorn C B and Weiskopf V F 1974 *Phys. Rev. D* **9** 3471
[5] Hosaka A and Toki H 1996 *Phys. Reports* **277**(2-3) 65
[6] de Souza M E 2012 *Nucl. Phys. B Proc. Suppl.* **00** 1
[7] Flugge S 1974 *Practical Quantum Mechanics* vol 1 (New York: Springer-Verlag, New York, 1974) pp 180-181
[8] Aoki S *et al.* (HAL QCD collaboration) 2010 *Chinese Physics C* **34** 1229
[9] MIT penCourseWare, Lecture # 3 Supplement Fall 2008 *Small-Molecule Spectroscopy and Dynamics*
[10] Robinett R W 1997 *Am. J. Phys.* **65**(3) 190

Journal of Physics: Conference Series **556** (2014) 012053 doi:10.1088/1742-6596/556/1/012053

B_c^+ physics at LHCb

Xuhao Yuan
On behalf of the LHCb collaboration
Center for High Energy Physics, Tsinghua University, Beijing, China
E-mail: xuhao.yuan@cern.ch

Abstract. We present recent LHCb results in B_c^+ physics with four topics. Using $3\,\text{fb}^{-1}$ of data collected in 2011 and 2012 produced in pp collisions, a baryonic decay mode, $B_c^+ \to J/\psi\, p\bar{p}\pi^+$, is observed for the first time. With $1\,\text{fb}^{-1}$ of data collected in 2011, the first measurement of the relative branching ratio $\mathcal{B}(B_c^+ \to J/\psi\,\pi^+)/\mathcal{B}(B_c^+ \to J/\psi\,\mu^+\nu_\mu)$ is performed. With $2\,\text{fb}^{-1}$ of data collected in 2012, the lifetime of the B_c^+ meson is measured with the semileptonic decay $B_c^+ \to J/\psi\,\mu^+\nu_\mu$. Finally, evidence is presented for the decay $B_c^+ \to J/\psi\, 3\pi^+2\pi^-$ using pp collision data, corresponding to an integrated luminosity of $3\,\text{fb}^{-1}$.

1. Introduction

In the Standard Model, the B_c^+ meson is the only meson family composed of two different heavy flavor quarks. Similar to quarkonia, B_c^+ has a rich spectroscopy. At LHCb, the most precise mass measurement of the ground state of B_c^+ comes from the decay of $B_c^+ \to J/\psi\, D_s^+$ in Ref. [1]: $6276.3 \pm 1.4 \pm 0.4\,\text{MeV}/c^2$. Recently ATLAS reported a search for excited state of the B_c^+ meson and new state is observed through its hadronic transition to the ground state, $B_c^+(2S) \to B_c^+\pi^+\pi^-$. The mass of that state is measured as $m(B_c^+(2S)) = 6842 \pm 4 \pm 5\,\text{MeV}/c^2$ [2]. The B_c^+ meson provides a wide range of decay modes, including the b quark decay, c quark decay and the annihilation modes [3]. At Tevatron, only two decay modes are observed, $B_c^+ \to J/\psi\,\pi^+$ [4] and $B_c^+ \to J/\psi\,\mu^+\nu_\mu$ [5]. Due to the large production rate at the LHC, more and more decays are observed at LHCb, such as $B_c^+ \to J/\psi\,\pi^+\pi^-\pi^+$ [6], $B_c^+ \to \psi(2S)\pi^+$ [7], $B_c^+ \to J/\psi\, D_s^{(*)+}$ [1], $B_c^+ \to J/\psi\, K^+$ [8], $B_c^+ \to J/\psi\, K^+K^-\pi^+$ [9] and the first c quark decay mode, $B_c^+ \to B_s^0\pi^+$ [10]. However, there are still many unclear issues about B_c^+ physics. The B_c^+ lifetime is measured with large uncertainty, many other decays are not observed jet, especially for the annihilation modes and so far there is no absolute branching ratio measurement for B_c^+ mesons. Therefore, more experimental studies on B_c^+ physics are still needed.

2. First observation of $B_c^+ \to J/\psi\, p\bar{p}\pi^+$

The baryonic decay modes of b hadrons are very important to understand baryon production mechanism and to search for new baryons and tetra-quarks. Besides, in baryonic decays of B^+ and B^0, some special behaviors are observed, but they are not yet well understood [11][12].

In this analysis, the pre-selection and multivariate selection based on boosted decision tree are used for the event selection. In order to select the proton, the proton identification is also optimised. Figure 1 shows the fitting curves to the invariant mass distributions of the $B_c^+ \to J/\psi\, p\bar{p}\pi^+$ decay and the $B_c^+ \to J/\psi\,\pi^+$ decay, which is taken as the controlled channel.

With the $3\,\text{fb}^{-1}$ of data collected by LHCb in 2011 and 2012, 23.9±5.3 signals of $B_c^+ \to J/\psi\, p\bar{p}\pi^+$ are observed with a significance of 7.3 σ. The branching fraction of the $B_c^+ \to J/\psi\, p\bar{p}\pi^+$ decay with respect to the $B_c^+ \to J/\psi\,\pi^+$ decay is measured as [13]:

$$\frac{\mathcal{B}(B_c^+ \to J/\psi\, p\bar{p}\pi^+)}{\mathcal{B}(B_c^+ \to J/\psi\,\pi^+)} = 0.143^{+0.039}_{-0.034}(\text{stat}) \pm 0.013(\text{syst}). \tag{1}$$

With the factorization assumption, this result is consistent with the B^0 baryonic decay, $\mathcal{B}(B^0 \to D^{\star -} p\bar{p}\pi^+)/\mathcal{B}(B^0 \to D^{\star -}\pi^+) = 0.17 \pm 0.02$ [14]. Because the low Q-value, the decay $B_c^+ \to J/\psi\, p\bar{p}\pi^+$ provides a precise measurement of the B_c^+ mass. The B_c^+ mass is measured, with corrections for the momentum scale and the final state radiation. Combining with LHCb previous measurements [1][15], the result is $m(B_c^+) = 6274.7 \pm 0.9(\text{stat}) \pm 0.8(\text{syst})\,\text{MeV}/c^2$ [13], which is the most precise B_c^+ mass measurement up to date.

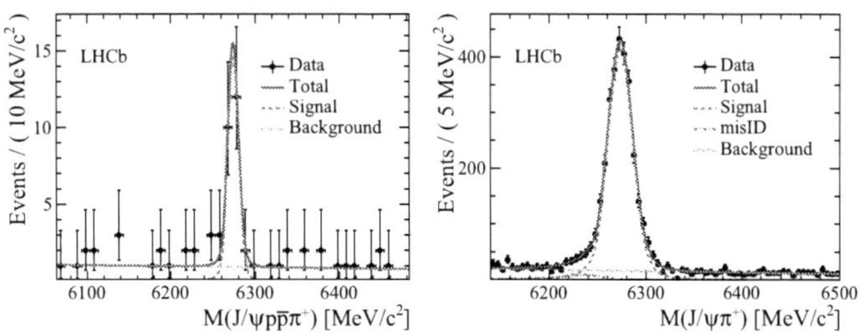

Figure 1. Fitting curves for invariant mass distributions of $B_c^+ \to J/\psi\, p\bar{p}\pi^+$ (left) and the controlled channel $B_c^+ \to J/\psi\,\pi^+$ (right), where the black dots represent the data, the blue lines stand for the fitting functions, with the components of signals (red dashed line), combinatorial background (green dotted line) and the misidentified background from $B_c^+ \to J/\psi\, K^+$ (pink dashed line)

3. Measurement of $\mathcal{B}(B_c^+ \to J/\psi\,\pi^+)/\mathcal{B}(B_c^+ \to J/\psi\,\mu^+\nu_\mu)$

The decays of $B_c^+ \to J/\psi\,\pi^+$ and $B_c^+ \to J/\psi\,\mu^+\nu_\mu$ are used in the studies of B_c^+properties, such as the measurements of B_c^+ lifetime and mass. However, there is no experimental determination of the relative size of these two decays, and the theoretical predictions have large spread among different models [16][17][18][19][20][21]. Additionally, the decays of $B_c^+ \to J/\psi\,\pi^+$ and $B_c^+ \to J/\psi\,\mu^+\nu_\mu$ are used respectively as references for the measurement of the production cross section of B_c^+ by Tevatron [5] and LHCb [15] respectively. Therefore, the measurement of the relative branching ratio $\mathcal{B}(B_c^+ \to J/\psi\,\pi^+)/\mathcal{B}(B_c^+ \to J/\psi\,\mu^+\nu_\mu)$ will provide a comparison between the B_c^+ production cross sections measured at Tevatron and LHCb.

In the $B_c^+ \to J/\psi\,\mu^+\nu_\mu$ decay, the main challenge is caused by the undetected neutrino. According to simulation study, most of the background events from $B \to J/\psi X$ will be suppressed by the requirement $m(J/\psi\,\mu^+) > 5.3\,\text{GeV}/c^2$. The signal-to-background likelihood-ratio method is used for further selection. The signal yields are determined from the fits to invariant mass distributions of the $B_c^+ \to J/\psi\,\mu^+\nu_\mu$ and $B_c^+ \to J/\psi\,\pi^+$ decays with $1\,\text{fb}^{-1}$ of data sample collected by LHCb in 2011, as shown in Figure 2. The relative branching ratio in the selected mass region $m(J/\psi\,\mu^+) > 5.3\,\text{GeV}/c^2$ is measured [22]

$$R(m(J/\psi\,\mu^+) > 5.3\,\text{GeV}/c^2) \equiv \frac{\mathcal{B}(B_c^+ \to J/\psi\,\pi^+)}{\mathcal{B}(B_c^+ \to J/\psi\,\mu^+\nu_\mu)} = 0.271 \pm 0.016(\text{stat}) \pm 0.016(\text{syst}),$$

Journal of Physics: Conference Series **556** (2014) 012053 doi:10.1088/1742-6596/556/1/012053

and the ratio is extrapolated to the full mass region [22]

$$R \equiv f \times R(m(J/\psi\,\mu^+) > 5.3\,\mathrm{GeV}/c^2) = 0.0469 \pm 0.0028(\mathrm{stat}) \pm 0.0046(\mathrm{syst}),$$

where the ratio f is determined from theoretical predictions. This result is consistent with the predictions of the potential model [19], Bethe-Salpter equation [20] and light-front quark model [21], but lower than other predictions [16][17][18].

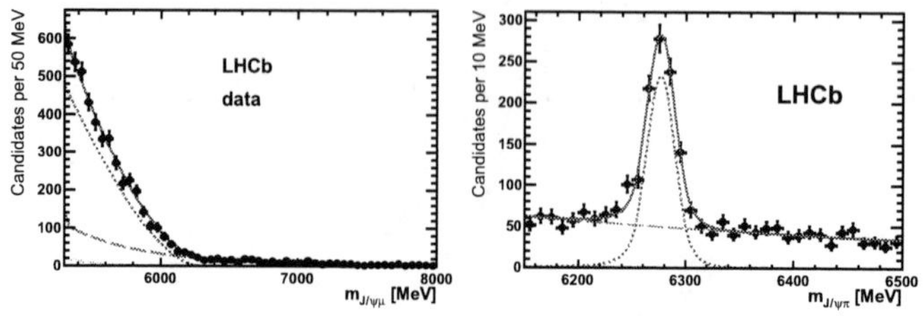

Figure 2. Fitting curves for invariant mass distributions of $B_c^+ \to J/\psi\,\mu^+\nu_\mu$ (left) and $B_c^+ \to J/\psi\,\pi^+$ (right), where the black dots represent the data, the blue lines stand for the fitting functions, with the components of signals (red dashed line), feed-down background or combinatorial background (green dotted line) and $B_{u,d,s} \to J/\psi X$ background or the misidentified background from $B_c^+ \to J/\psi K^+$ (pink dashed line) respectively

4. Measurement of the B_c^+ lifetime with $B_c^+ \to J/\psi\,\mu^+\nu_\mu$

It is predicted that the decay mechanism of B_c^+ mesons is dominated by c-quark decay modes [3]. Therefore, the lifetime of B_c^+ mesons is less than other B mesons, but very close to that of charm mesons. The theoretical predictions of the B_c^+ lifetime have large spread and the world average value was $\tau(B_c^+) = 0.452 \pm 0.033$ ps [14], leading to the large uncertainty which dominates the systematic uncertainty in most of B_c^+ analyses.

In this analysis, the semileptonic decay $B_c^+ \to J/\psi\,\mu^+\nu_\mu$ is used. Comparing with the B_c^+ hadronic decay $B_c^+ \to J/\psi\,\pi^+$, the semileptonic decay has higher statistics and more clear signature with three muons. However, due to the undetected neutrino, the decay $B_c^+ \to J/\psi\,\mu^+\nu_\mu$ can only be partially reconstructed at LHCb and the kinematics of this decay is model dependent. Therefore, the B_c^+ lifetime has to be determined by the 2D fit to the distributions of invariant mass $m(J/\psi\,\mu^+)$ and the pseudo-decay time t_{ps}, which is defined as $t_{ps} = m(J/\psi\,\mu^+) \times (\vec{p}\cdot\vec{L})/|\vec{p}|^2$, where, $\vec{p}$ represents the sum of the J/ψ and μ^+momenta, and $\vec{L}$ means the distance between the primary vertex and the B_c^+ decay vertex. With $2\,\mathrm{fb}^{-1}$ of data sample collected by LHCb in 2012, the B_c^+ lifetime is measured as $\tau(B_c^+) = 509 \pm 8(\mathrm{stat}) \pm 12(\mathrm{syst})$ fs [23]. Comparing with the previous measurements by CDF and D0, the result of LHCb is the most precise measurement of $\tau(B_c^+)$ to date.

5. Evidence of $B_c^+ \to J/\psi\, 3\pi^+ 2\pi^-$ decay

The $B_c^+ \to J/\psi\, 3\pi^+ 2\pi^-$ decay is a new b-quark decay mode of B_c^+ meson and the study of this decay can provide powerful tests of the form factors of $B_c^+ \to J/\psi$ transition and the spectral functions for virtual W^+ boson into light mesons. In this analysis, the $3\,\mathrm{fb}^{-1}$ of data are used for selecting the signals. With a fit to the invariant mass distributions of B_c^+ mesons, 32 ± 8 signal events are observed with a significance of 4.5 σ [24], which is the first evidence of this decay. Taking the $B_c^+ \to J/\psi\,\pi^+$ decay as the controlled channel, the branching fraction of

$B_c^+ \to J/\psi 3\pi^+ 2\pi^-$ with respect to that of $B_c^+ \to J/\psi\pi^+$ is measured [24]

$$\frac{\mathcal{B}(B_c^+ \to J/\psi 3\pi^+ 2\pi^-)}{\mathcal{B}(B_c^+ \to J/\psi\pi^+)} = 1.74 \pm 0.44(\text{stat}) \pm 0.24(\text{syst}). \tag{2}$$

The result is consistent with the theoretical predictions in Ref. [25]; with the factorization assumption, it is also in agreement with $B^0$and B^+meson decays [14]

$$\frac{\mathcal{B}(B_c^+ \to J/\psi 3\pi^+ 2\pi^-)}{\mathcal{B}(B_c^+ \to J/\psi\pi^+)} \approx \begin{cases} \dfrac{\mathcal{B}(B^0 \to D^{*-} 3\pi^+ 2\pi^-)}{\mathcal{B}(B^0 \to D^{*-}\pi^+)} = 1.70 \pm 0.34, \\ \dfrac{\mathcal{B}(B^+ \to \bar{D}^{*0} 3\pi^+ 2\pi^-)}{\mathcal{B}(B^0 \to \bar{D}^{*0}\pi^+)} = 1.10 \pm 0.24. \end{cases} \tag{3}$$

6. Summary

LHCb provides good opportunities for B_c^+ physics, as many new B_c^+ decays are observed and their properties are measured. The first baryonic decay mode of B_c^+ is observed and, combined with LHCb previous measurements, the B_c^+ mass measurement is the most precise measurement to date. For the first time, the relative ratio $\mathcal{B}(B_c^+ \to J/\psi\pi^+)/\mathcal{B}(B_c^+ \to J/\psi\mu^+\nu_\mu)$ is measured and is consistent with theoretical predictions. The most precise measurement of the B_c^+ lifetime is performed with the decay of $B_c^+ \to J/\psi\mu^+\nu_\mu X$. The first evidence of a new decay mode $B_c^+ \to J/\psi 3\pi^+ 2\pi^-$ is reported.

References

[1] RAaij *et al.* [LHCb Collaboration] 2013 *Phys. Rev.* D **87** no. 11 112012 arXiv:1304.4530 [hep-ex]
[2] G. Aad *et al.* [ATLAS Collaboration] 2014 arXiv:1407.1032 [hep-ex]
[3] I. P. Gouz, V. V. Kiselev, A. K. Likhoded, V. I. Romanovsky and O. P. Yushchenko 2004 *Phys. Atom. Nucl.* **67** 1559-1570 hep-ph/0211432
[4] A. Abulencia *et al.* [CDF Collaboration] 2006 *Phys. Rev. Lett.* **96** 082002 hep-ex/0505076
[5] F. Abe *et al.* [CDF Collaboration] 1998 *Phys. Rev. Lett.* **81** 2432-2437 hep-ex/9805034
[6] R. Aaij *et al.* [LHCb Collaboration] 2012 *Phys. Rev. Lett.* **108** 251802 arXiv:1204.0079 [hep-ex]
[7] R. Aaij *et al.* [LHCb Collaboration] 2013 *Phys. Rev.* D **87** 071103 arXiv:1303.1737 [hep-ex]
[8] R. Aaij *et al.* [LHCb Collaboration] 2013 *JHEP* **1309** 075 arXiv:1306.6723 [hep-ex]
[9] R. Aaij *et al.* [LHCb Collaboration] 2013 *JHEP* **1311** 094 arXiv:1309.0587 [hep-ex]
[10] R. Aaij *et al.* [LHCb Collaboration] 2013 *Phys. Rev. Lett.* **111** no. 18 181801 arXiv:1308.4544 [hep-ex]
[11] K. Abe *et al.* [Belle Collaboration] 2002 *Phys. Rev. Lett.* **88** 181803 hep-ex/0202017
[12] Y. -J. Lee *et al.* [Belle Collaboration] 2004 *Phys. Rev. Lett.* **93** 211801 hep-ex/0406068
[13] R. Aaij *et al.* [LHCb Collaboration] 2014 *Phys. Rev. Lett.* **113** 152003 arXiv:1408.0971 [hep-ex]
[14] J. Beringer *et al.* [Particle Data Group Collaboration] 2012 *Phys. Rev.* D **86** 010001
[15] RAaij *et al.* [LHCb Collaboration] 2012 *Phys. Rev. Lett.* **109** 232001 arXiv:1209.5634 [hep-ex]
[16] C. -H. Chang and Y. -Q. Chen 1994 *Phys. Rev.* D **49** 3399-3411
[17] P. Colangelo and F. De Fazio 2000 *Phys. Rev.* D **61** 034012 hep-ph/9909423
[18] M. A. Ivanov, J. G. Korner and P. Santorelli 2006 *Phys. Rev.* D **73** 054024 hep-ph/0602050
[19] D. Ebert, R. N. Faustov and V. O. Galkin 2003 *Phys. Rev.* D **68** 094020 hep-ph/0306306
[20] A. Abd El-Hady, J. H. Munoz and J. P. Vary 2000 *Phys. Rev.* D **62** 014019 hep-ph/9909406
[21] H. -W. Ke, T. Liu and X. -Q. Li 2014 *Phys. Rev.* D **89** 017501 arXiv:1307.5925 [hep-ph]
[22] R. Aaij *et al.* [LHCb Collaboration], 2014 *Phys. Rev.* D **90** 032009 arXiv:1407.2126 [hep-ex]
[23] R. Aaij *et al.* [LHCb Collaboration] 2014 *Eur. Phys. J.* C **74** 2839 arXiv:1401.6932 [hep-ex]
[24] R. Aaij *et al.* [LHCb Collaboration] 2014 *JHEP* **1405** 148 arXiv:1404.0287 [hep-ex]
[25] A. V. Luchinsky 2012 *Phys. Rev.* D **86** 074024 arXiv:1208.1398 [hep-ph]

Electromagnetic structure of charmed hadrons

G. Erkol[1], K. U. Can[2], B. Isildak[1], M. Oka[2], A. Ozpineci[3], T. T. Takahashi[4]

[1] Department of Natural and Mathematical Sciences, Faculty of Engineering, Ozyegin University, Nisantepe Mah. Orman Sok. No:13, Alemdag 34794 Cekmekoy, Istanbul Turkey
[2] Department of Physics, H-27, Tokyo Institute of Technology, Meguro, Tokyo 152-8551 Japan
[3] Physics Department, Middle East Technical University, 06531 Ankara, Turkey
[4] Gunma National College of Technology, Maebashi, Gunma 371âĂŘ8530, Japan

Abstract. We compute the electromagnetic structures of D and D* mesons, the singly charmed Σ_c, Ω_c and the doubly charmed Ξ_{cc}, Ω_{cc} baryons in 2+1 flavor Lattice QCD. We extract the charge radii and the magnetic moments of these hadrons. In general, charmed hadrons are found to be more compact as compared to light hadrons. The magnetic moments of the singly charmed baryons are found to be dominantly determined by the light quark and the role of the charm quark is significantly enhanced when it is doubly represented.

1. Introduction

Electromagnetic form factors reveal valuable information about the size and the shape of the hadrons. Determining these form factors is an important step in our understanding of the hadron properties in terms of quark-gluon degrees of freedom. There has been an increasing activity in determining the electromagnetic structure of octet mesons and baryons where the challenge is to understand these quantities directly from QCD, the theory of quarks and gluons. Currently, only the spectrum of the charmed baryons are accessible by experiments and future charm factories like BES-III and PANDA at GSI are expected to probe the charm sector.

In this work, we report on our recent results on the electromagnetic structure of D and D* mesons and the singly charmed Σ_c, Ω_c and the doubly charmed Ξ_{cc}, Ω_{cc} baryons as obtained in 2+1 flavor Lattice QCD [1, 2, 3]. In particular we compute the charge radii and the magnetic moments of these hadrons.

2. Lattice formulation and setup

Here we give the details of our calculations for charmed baryons and refer the reader to Ref. [1] for the charmed mesons. Electromagnetic form factors of baryons can be calculated by considering the baryon matrix elements of the electromagnetic vector current $V_\mu = \sum_q e_q \bar{q}(x)\gamma_\mu q(x)$, where q runs over the quark content of the given baryon:

$$\langle \mathcal{B}(p)|V_\mu|\mathcal{B}(p')\rangle = \bar{u}(p)\left[\gamma_\mu F_{1,\mathcal{B}}(q^2) + i\frac{\sigma_{\mu\nu}q^\nu}{2m_\mathcal{B}}F_{2,\mathcal{B}}(q^2)\right]u(p). \quad (1)$$

Here $q_\mu = p'_\mu - p_\mu$ is the transferred four-momentum; $u(p)$ denotes the Dirac spinor for the baryon with four-momentum p^μ and mass $m_\mathcal{B}$. The Sachs form factors $F_{1,\mathcal{B}}(q^2)$ and $F_{2,\mathcal{B}}(q^2)$

are related to the electric and magnetic form factors by

$$G_{E,\mathcal{B}}(q^2) = F_{1,\mathcal{B}}(q^2) + \frac{q^2}{4m_{\mathcal{B}}^2} F_{2,\mathcal{B}}(q^2), \tag{2}$$

$$G_{M,\mathcal{B}}(q^2) = F_{1,\mathcal{B}}(q^2) + F_{2,\mathcal{B}}(q^2). \tag{3}$$

We use the following ratio

$$R(t_2, t_1; \mathbf{p}', \mathbf{p}; \mathbf{\Gamma}; \mu) = \frac{\langle F^{\mathcal{B}V_\mu \mathcal{B}'}(t_2,t_1;\mathbf{p}',\mathbf{p};\mathbf{\Gamma})\rangle}{\langle F^{\mathcal{B}\mathcal{B}}(t_2;\mathbf{p}';\Gamma_4)\rangle} \times \left[\frac{\langle F^{\mathcal{B}\mathcal{B}}(t_2-t_1;\mathbf{p};\Gamma_4)\rangle \langle F^{\mathcal{B}\mathcal{B}}(t_1;\mathbf{p}';\Gamma_4)\rangle \langle F^{\mathcal{B}\mathcal{B}}(t_2;\mathbf{p}';\Gamma_4)\rangle}{\langle F^{\mathcal{B}\mathcal{B}}(t_2-t_1;\mathbf{p}';\Gamma_4)\rangle \langle F^{\mathcal{B}\mathcal{B}}(t_1;\mathbf{p};\Gamma_4)\rangle \langle F^{\mathcal{B}\mathcal{B}}(t_2;\mathbf{p};\Gamma_4)\rangle}\right]^{1/2}, \tag{4}$$

where the baryonic two-point and three-point correlation functions are respectively defined as:

$$\langle F^{\mathcal{B}\mathcal{B}}(t; \mathbf{p}; \Gamma_4)\rangle = \sum_{\mathbf{x}} e^{-i\mathbf{p}\cdot\mathbf{x}} \Gamma_4^{\alpha\alpha'} \times \langle 0|T[\eta_{\mathcal{B}}^{\alpha}(x)\bar{\eta}_{\mathcal{B}}^{\alpha'}(0)]|0\rangle, \tag{5}$$

$$\langle F^{\mathcal{B}V_\mu \mathcal{B}'}(t_2, t_1; \mathbf{p}', \mathbf{p}; \mathbf{\Gamma})\rangle = -i \sum_{\mathbf{x_2},\mathbf{x_1}} e^{-i\mathbf{p}\cdot\mathbf{x_2}} e^{i\mathbf{q}\cdot\mathbf{x_1}} \Gamma^{\alpha\alpha'} \langle 0|T[\eta_{\mathcal{B}}^{\alpha}(x_2) V_\mu(x_1) \bar{\eta}_{\mathcal{B}'}^{\alpha'}(0)]|0\rangle, \tag{6}$$

with $\Gamma_i = \gamma_i\gamma_5\Gamma_4$ and $\Gamma_4 \equiv (1+\gamma_4)/2$.

The baryon interpolating fields are chosen as

$$\begin{aligned}
\eta_{\Xi_{cc}}(x) &= \epsilon^{ijk}[c^{Ti}(x)C\gamma_5\ell^j(x)]c^k(x), \\
\eta_{\Sigma_c}(x) &= \epsilon^{ijk}[\ell^{Ti}(x)C\gamma_5 c^j(x)]\ell^k(x), \\
\eta_{\Omega_{cc}}(x) &= \epsilon^{ijk}[c^{Ti}(x)C\gamma_5 s^j(x)]c^k(x), \\
\eta_{\Omega_c}(x) &= \epsilon^{ijk}[s^{Ti}(x)C\gamma_5 c^j(x)]s^k(x),
\end{aligned} \tag{7}$$

where $\ell = u$ for the doubly charged $\Xi_{cc}^{++}(ccu)/\Sigma_c^{++}(cuu)$ and $\ell = d$ for the singly charged $\Xi_{cc}^{+}(ccd)/\Sigma_c^{+}(cdd)$ baryons. Here i, j, k denote the color indices and $C = \gamma_4\gamma_2$. t_1 is the time when the external electromagnetic field interacts with a quark and t_2 is the time when the final baryon state is annihilated. When $t_2 - t_1$ and $t_1 \gg a$, the ratio in Eq. (4) reduces to the desired form [3]

$$R(t_2, t_1; \mathbf{p}', \mathbf{p}; \Gamma; \mu) \xrightarrow[t_2-t_1\gg a]{t_1\gg a} \Pi(\mathbf{p}', \mathbf{p}; \Gamma; \mu). \tag{8}$$

We extract the form factors $G_{E,\mathcal{B}}(q^2)$ and $G_{M,\mathcal{B}}(q^2)$ by choosing appropriate combinations of Lorentz direction μ and projection matrices Γ:

$$\Pi(\mathbf{0}, -\mathbf{q}; \Gamma_4; \mu = 4) = \left[\frac{(E_{\mathcal{B}}+m_{\mathcal{B}})}{2E_{\mathcal{B}}}\right]^{1/2} G_{E,\mathcal{B}}(q^2), \tag{9}$$

$$\Pi(\mathbf{0}, -\mathbf{q}; \Gamma_j; \mu = i) = \left[\frac{1}{2E_{\mathcal{B}}(E_{\mathcal{B}}+m_{\mathcal{B}})}\right]^{1/2} \epsilon_{ijk}\, q_k\, G_{M,\mathcal{B}}(q^2). \tag{10}$$

Here, $G_{E,\mathcal{B}}(0)$ gives the electric charge of the baryon. Similarly, the magnetic moment can be obtained from the magnetic form factor $G_{M,\mathcal{B}}$ at zero momentum transfer.

We have run our simulations on $32^3 \times 64$ lattices with 2+1 flavors of dynamical quarks and we use the gauge configurations that have been generated by the PACS-CS collaboration [4] with the nonperturbatively $\mathcal{O}(a)$-improved Wilson quark action and the Iwasaki gauge action. We use the gauge configurations at $\beta = 1.90$ with the clover coefficient $c_{SW} = 1.715$ and they have a lattice spacing of $a = 0.0907(13)$ fm ($a^{-1} = 2.176(31)$ GeV). We consider four different hopping parameters for the sea and the u,d valence quarks, $\kappa_{sea}, \kappa_{val}^{u,d} = 0.13700, 0.13727, 0.13754$ and 0.13770, which correspond to pion masses of $\sim$ 700, 570, 410, and 300 MeV. The hopping parameter for the s sea quark is fixed to $\kappa_{sea}^{s} = 0.1364$.

Journal of Physics: Conference Series **556** (2014) 012054 doi:10.1088/1742-6596/556/1/012054

3. Results and discussion

Our results for the charmed mesons indicate that they are more compact as compared to light mesons [1]. As for the charmed baryons we find that the electric charge radii of Ω_{cc}^{+} and Ξ_{cc}^{+} [2] are about the same size but as compared to that of the proton they are smaller. Replacing the light d-quark by an s-quark does not seem to have any effect on the charge radii. Σ_c^{++} baryon has the largest electric charge radius amongst the baryons that we considered. Magnetic charge radii also show a similar pattern as the electric charge radii. The magnetic charge radii of Σ_c^{++} and Σ_c^0 are close to that of the proton's, which is $\langle r^2_{M,p}\rangle = 0.604$ fm^2 [5]. In the case of Ω_c and Ω_{cc}, their electric and magnetic charge radii are somewhat dependent on the sea quarks. However, their magnetic moments are almost independent of the sea-quark effects. A more detailed discussion can be found in Ref. [3].

An analysis of the individual quark contributions [3] shows that the c quark core shifts the centre of mass towards itself thus shrinking the baryon. The charm quark contributions are independent of the quark content of the baryons and the contributions from the u/d- and s-quark are roughly the same. As a result, we find Σ_c^{++} and Ξ_{cc}^{++}, as well as, Ω_{cc}^{+} and Ξ_{cc}^{+} to have almost the same sizes.

Doubly represented light quarks give the dominant contribution to the magnetic moments and the opposite signs of the light and heavy quark contributions suggest that their spins are generally anti-aligned within the baryon. Σ_c^{++} has the largest magnetic moment of all and the strange baryons Ω_c and Ω_{cc} have somewhat smaller moments. It is interesting to compare these values with the experimental magnetic moment of the proton, which is $\mu_p = 2.793\ \mu_N$ [5]. Comparing our magnetic moment results with several other models, although the signs match we see a quantitative disagreement [3].

In conclusion, we have extracted the electromagnetic structures of D and D* mesons, the singly charmed Σ_c, Ω_c and the doubly charmed Ξ_{cc}, Ω_{cc} baryons from 2+1-flavor simulations of QCD on a $32^3 \times 64$ lattice. Our results imply that the charmed baryons are more compact with respect to baryons that are composed of only light quarks, *e.g.*, the proton. The existence of the heavy quark shrinks the baryons and doubly charmed baryons are more compact than the singly charmed baryons of the same charge. The size of the baryon is increased when the u/d is replaced by the s quark in a qQQ system. Ω_{cc} has the smallest radius and Σ_c^{++} and Σ_c^0 baryons have larger and roughly the same magnetic radii. The magnetic moments are dominantly determined by the doubly represented quarks. The signs of the magnetic moments are correctly reproduced on the lattice. However, in general we see an underestimation of the magnetic moments as compared to what has been found with other theoretical methods.

Table 1. The electric and magnetic charge radii in fm^2, the values of magnetic form factors at $Q^2 = 0$ ($G_{M,B}(0)$), the magnetic moments in nuclear magnetons, for $B \equiv \Sigma_c^{++}$, Σ_c^0, Ω_{cc}, Ω_c at the chiral point. Charge radii and magnetic moments results are from quadratic and linear fits, respectively.

$\langle r^2_{E,\Sigma_c^{++}}\rangle$ [fm^2]	$\mu_{\Sigma_c^0}$ [μ_N]	$\mu_{\Sigma_c^{++}}$ [μ_N]	$\langle r^2_{M,\Sigma_c^0}\rangle$ [fm^2]	$\langle r^2_{M,\Sigma_c^{++}}\rangle$ [fm^2]
0.240(44)	-0.875(103)	1.499(202)	0.568(130)	0.656(142)
$\langle r^2_{E,\Omega_{cc}}\rangle$ [fm^2]	μ_{Ω_c} [μ_N]	$\mu_{\Omega_{cc}}$ [μ_N]	$\langle r^2_{M,\Omega_c}\rangle$ [fm^2]	$\langle r^2_{M,\Omega_{cc}}\rangle$ [fm^2]
0.064(15)	-0.627(43)	0.402(10)	0.343(52)	0.176(24)

Acknowledgments
All the numerical calculations in this work were performed on National Center for High Performance Computing of Turkey (Istanbul Technical University) under project number 10462009. The unquenched gauge configurations employed in our analysis were generated by PACS-CS collaboration [4]. We used a modified version of Chroma software system [6]. This work is supported in part by The Scientific and Technological Research Council of Turkey (TUBITAK) under project numbers 110T245, 114F261 and in part by KAKENHI under Contract Nos. 22105503, 24540294 and 22105508.

References
[1] K. U. Can, G. Erkol, M. Oka, A. Ozpineci and T. T Takahashi 2013 Phys.Lett. B **719** 103
[2] K. U. Can, G. Erkol, B. Isildak, M. Oka and T. T Takahashi 2013 Phys. Lett. B **726** 703
[3] K. U. Can, G. Erkol, B. Isildak, M. Oka and T. T Takahashi *J. High Energy Phys.* JHEP05(2014)125
[4] S. Aoki et al. (PACS-CS Collaboration) 2009 Phys. Rev. D **79** 034503
[5] J. Beringer et al. (Particle Data Group) 2012 Phys. Rev. D **86** 010001
[6] R. G. Edwards and B. Joo (SciDAC Collaboration, LHPC Collaboration, UKQCD Collaboration), *Nucl.Phys.Proc.Suppl.* 2005 **140** 832

Journal of Physics: Conference Series **556** (2014) 012055 doi:10.1088/1742-6596/556/1/012055

Rare charm decays

Svjetlana Fajfer

Faculty of Mathematics and Physics, University of Ljubljana, Jadranska 19, 1000 Ljubljana, Slovenia and J. Stefan Institute, Jamova 39, P. O. Box 3000, 1001 Ljubljana, Slovenia

E-mail: svjetlana.fajfer@ijs.si

Abstract. Rare charm decays offer the possibility to search for signals beyond Standard Model in the up-quark sector. CP conserving and CP violating contributions within the Standard Model and beyond are reviewed for inclusive $c \to u\gamma$ and $c \to ul^+l^-$ and exclusive $D \to V\gamma$, $D \to P(V)l^+l^-$and $D \to l^+l^-$ decays.

1. Introduction

For many years flavor changing neutral current (FCNC) processes were considered as an important tool in searches of new physics. Many experimental results for K and B physics, as well as a number theoretical approaches, enable to study down-quark sector very precisely. Accordingly, new physics (NP) was expected to be observed first in the down-quark sector. The up-quark sector was not considered to be easily accessible for search of new physics, due to difficulties in the treatment of charm meson dynamics. In D decays long distance contributions dominate over short distance effects. The main task is how to separate information on short distance dynamics, within Standard Model (SM) or in its extensions. This is a longstanding problem in rare charm decays. In addition, in FCNC charm processes the GIM mechanism results in the interplay of CKM parameters and masses of down-like quarks leading to a strong amplitudes suppression. In the last few years additional motivation for charm physics appeared after LHCb collaboration found unexpectedly large CP violating asymmetry in charm decays [1, 2, 3]. That triggered many studies and additional checks of the observed anomaly. Although discrepancy has decreased after new measurements [3, 4], new studies of rare charm decays appeared in the meantime questioning possibilities to search for CP violating signals in rare charm decays.

This review is organized as follows: in Sec. 2 SM contributions to $c \to u\gamma$ and $c \to ul^+l^-$ decay modes and exclusive decay channels are reviewed. In Sec. 3 NP contributions in inclusive and exclusive decay channels are discussed. Sec. 4 contains a discussion on CP violation in rare charm decays. The last section contains a summary.

2. Standard Model contributions in rare charm inclusive and exclusive decay modes

The most common way to approach these processes is to use the effective low-energy Lagrangian and the operator product expansion. Within the Standard Model (SM) short distance contributions in $c \to u\gamma$ and $c \to ul^+l^-$ transitions are described by:

$$\mathcal{L}_{eff}^{SD} = \frac{G_F}{\sqrt{2}} V_{cb}^* V_{ub} \sum_{i=7,9,10} C_i Q_i, \tag{1}$$

The operators are then:

$$Q_7 = \frac{e}{8\pi^2} m_c F_{\mu\nu} \bar{u}\sigma_{\mu\nu}(1+\gamma_5)c, \ Q_9 = \frac{e^2}{16\pi^2} \bar{u}_L \gamma_\mu c_L \bar{l}\gamma^\mu l, \ Q_{10} = \frac{e^2}{16\pi^2} \bar{u}_L \gamma_\mu c_L \bar{l}\gamma^\mu \gamma_5 l. \tag{2}$$

In (1) C_i denote, as usual, effective Wilson coefficients (they are determined at the scale $\mu = m_c$), $F_{\mu\nu}$ is the electromagnetic field strenght and $q_L = \frac{1}{2}(1-\gamma_5)q$. In the case of the $c \to u\gamma$ decay only C_7 contributes, while in the case of $c \to ul^+l^-$ all three Wilson coefficents are present. At the one-loop level contributions coming from penguin diagrams are strongly GIM suppressed giving a branching ratio $\sim 10^{-18}$ [5, 6, 7, 8, 9]. The QCD corrections enhance this rate to $BR(c \to u\gamma)_{SM} = 2.5 \times 10^{-8}$ [10, 11]. Within SM the short distance contribution coming from $Q_{7,9}$ leads to the branching ratio [8, 12, 13, 14]

$$BR(D \to X_u e^+ e^-)_{SM}^{SD} \simeq 3.7 \times 10^{-9}. \tag{3}$$

However, this short distance contribution is overshadowed by long distance contributions, which are the result of the nonleptonic D decays [8, 12]. The branching ratio for the inclusive decay is:

$$BR(D \to X_u e^+ e^-)_{SM}^{LD} \sim \mathcal{O}(10^{-6}). \tag{4}$$

The amplitude for the $D \to V\gamma$ decay can be most generally written as:

$$\begin{aligned} \mathcal{A}[D(p) \to V(p', \epsilon')\gamma(q,\epsilon)] &= -iA_{CP}\epsilon_{\mu\nu\alpha\beta} q^\mu \epsilon^{*\nu} p^\alpha \epsilon^{*\prime,\beta} \\ &+ A_{PV}[(\epsilon^{*\prime,\beta} \cdot q)(\epsilon^{*\nu} \cdot q) - (p \cdot q)(\epsilon^{*\nu} \epsilon^{*\nu})]. \end{aligned} \tag{5}$$

The authors of [15] have reinvestigated long distance dynamics. Using the QCD sum rules result for the tensor form factors ($T^\rho \simeq T^\omega \simeq 0.7 \pm 0.2$) they found that parity conserving (violating) amplitudes are $(A_{PC,PV}^{\rho,\omega})^{SD} \simeq 0.6(2) \times 10^{-9}/m_D |C_7(m_c)/0.4 \cdot 10^{-2}|$ where superscripts ρ, ω denote the appropriate vector meson state V. For the determination of short distance contribution one has to know the matrix element of the Q_7 operator. In the calculations of it the tensor form-factors are present [15]. The long distance contribution was estimated by knowing that the relation $BR(D^0 \to K^{*0}\gamma)/BR(D^0 \to K^{*0}\rho^0) = BR(D^0 \to \phi\gamma)/BR(D^0 \to \phi\rho^0)$ is a consequence of vector meson dominance [15] $|(A_{PC,PV}^V)^{LD}| = [32\pi/2m_D^3(1-\frac{m_V^2}{m_D^2})^{-3}\Gamma(D \to V\gamma)]^{1/2}$, what gives, for $V = \phi$, $|(A_{PC,PV}^\phi)^{LD}| = 5.9(4) \times 10^{-8}/m_D$. These estimations are close to the previously determined ones in [5, 6].

The SM short distance contributions to $D^0 \to \gamma\gamma$ and $D^0 \to \mu^+\mu^-$ can be determined using the effective Lagrangian (1), while in both decay modes the dominant contribution comes from long distance effects [12, 9]. Recently $D^0 \to \gamma\gamma$ and $D^0 \to l^+l^-$ were reconsidered in [13]. The branching ratio coming from long and short distance contributions are $BR_{LD}^{SM}(D^0 \to \gamma\gamma) \simeq (1-3) \times 10^{-8}$, $BR_{SD}^{2-loops}(D^0 \to \gamma\gamma) \simeq (3.6-8.1) \times 10^{-12}$. In the $D^0 \to l^+l^-$ decay also the SM long distance contribution dominates over the short distance ones. The authors of [13] considered contributions coming from $\gamma\gamma$ intermediate states due to long distance dynamics in $D^0 \to \gamma\gamma$ arriving at the value $BR(D^0 \to \mu^+\mu^-) \sim (2.7-8.0) \times 10^{-13}$. Recently LHCb improved the bound on the branching ratio $BR(D^0 \to \mu^+\mu^-) \leq 6.2 \times 10^{-9}$ [17].

Among all exclusive decay modes containing lepton pair in the final state, the simplest one for experimental searches are $D^+ \to \pi^+\ell^+\ell^-$ and $D_s^+ \to K^+\ell^+\ell^-$. Close to the ϕ resonant peak

Decay mode	Branching ratio	Reference
$D \to \rho(\omega)\gamma$	0.6×10^{-5}	Isidori & Kamenik 2012
$D \to K^+K^-\gamma$	1.35×10^{-5} (ϕ)	Isidori & Kamenik 2012
$D \to X_u l^+ l^-$	$\mathcal{O}(10^{-6})$	Paul et al, 2011
$D^+ \to \pi^+ l^+ l^-$	2×10^{-6}	Fajfer et al, 2007
$D_s^+ \to K^+ l^+ l^-$	6×10^{-6}	Fajfer et al, 2007
$D \to \pi^+ K^- l^+ l^-$	$\mathcal{O}(10^{-5})$	Cappiello et al, 2013
$D \to \pi^+ \pi^- l^+ l^-$	$\mathcal{O}(10^{-6})$	Cappiello et al, 2013
$D \to K^+ K^- l^+ l^-$	$\mathcal{O}(10^{-7})$	Cappiello et al, 2013
$D \to \pi^- K^+ l^+ l^-$	$\mathcal{O}(10^{-8})$	Cappiello et al, 2013
$D \to \gamma\gamma$	$(1-3) \times 10^{-8}$	Paul et al, 2010
$D \to \mu^+\mu^-$	$(7-8) \times 10^{-13}$	Paul et al, 2010

Table 1. Branching ratios for charm meson decays. The second column contains the SM theoretical predictions in which the long-distance contribution is dominant. The last column contains the most recent references.

the long distance amplitude for $D^+ \to \pi^+\mu^+\mu^-$ decay is, to a good approximation, determined by non-factorizable contributions of four-quark operators in $\mathcal{H}^s$. The width of the ϕ resonance is very narrow ($\Gamma_\phi/m_\phi \approx 4 \times 10^{-3}$) and well separated from other vector resonances in the $\bar{q}q$ spectrum of $D \to P\ell^+\ell^-$. Relying on the vector meson dominance hypothesis the q^2-dependence of the decay spectrum close to the resonant peak follows the Breit-Wigner shape [12, 22, 21]

$$\mathcal{A}^{\phi}_{\rm LD}\left[D \to \pi\phi \to \pi\ell^-\ell^+\right] = \frac{iG_F}{\sqrt{2}}\lambda_s \frac{8\pi\alpha}{3}\, a_\phi e^{i\delta_\phi} \frac{m_\phi\Gamma_\phi}{q^2 - m_\phi^2 + im_\phi\Gamma_\phi}\, \bar{u}(k_-)\,\gamma_\mu p^\mu\, v(k_+)\,. \qquad (6)$$

The result for branching ratios for $D \to \pi\ell^-\ell^+$ are given in Table 1. A very nice study of LHCb [18] gave for the first time non-resonant branching fractions $BR(D^+ \to \pi^+\mu^+\mu^-) \leq 7.3(8.3) \times 10^{-8}$.

The detailed analysis of the semileptonic four body $D \to hhl^+l^-$ decays was done in the work of Ref. [23]. The dominant long-distance contributions (bremsstrahlung and hadronic effects) are calculated and total branching ratios and the (m_{ll}^2, m_{hh}^2) Dalitz plots are presented. Branching ratios for these decay modes are also presented in Table 1.

3. New physics searches in rare charm decays

The long distance contribution dominates for two or three order of magnitude over the short distance contribution in the above discussed decays. In inclusive $D \to X_u\gamma$ or exclusive $D \to V\gamma$ decays only a huge increase of the C_7^{eff} Wilson coefficient would lead to observable effects in the branching ratio. However, such contributions within a particular model of new physics are strongly constrained by other low-energy phenomenological constraints. One of the most popular extension of the SM is the MSSM. Following the discussion in [21] the model with non-universal soft-breaking terms, knowing that the gluino (according to LHC bounds) cannot be lighter than 1.3 TeV, would give $Br(c \to u\gamma)_{gluino} \sim 5 \times 10^{-8}$. A rather high mass of gluino would also give rise to SM $BR(c \to ul^+l^-)$ by about factor 2. As noticed in [14, 21, 22] other SM extensions might give a larger increase of both inclusive branching ratios. However, even if the short distance contribution is enlarged by NP, long distance effects screen it in the branching ratios for all exclusive rare charm decays.

The authors of [19] have analyzed warped extra dimensional models contributions to the chromo-magnetic dipole operator. They explained why the calculation of the coefficient of the

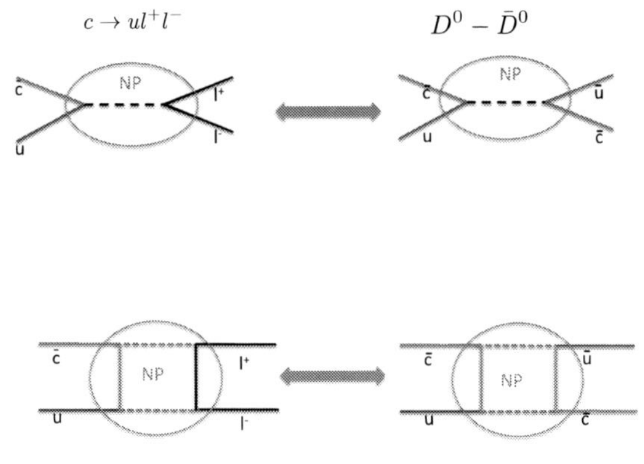

Figure 1. NP contributions to $c \to ul^+l^-$ and $D^0 - \bar{D}^0$ at tree and loop level.

lowest dimension (5D) operator has to be finite. They found that the branching ratio for the inclusive decay is $BR(D^0 \to X_u\gamma) \simeq 1 \times 10^{-8}$.

If one considers contributions of a new scalar or pseudoscalar particle mediating the decay $c \to u\ell^+\ell^-$, then the same particle would contribute to $D^0 - \bar{D}^0$ oscillations and the physical observable from this process immediatelly strongly constrain couplings of this operator. The same holds for the flavor changing Z or new Z' boson. If new physics is generated at the loop level in $c \to u\ell^+\ell^-$ then it contributes to $D^0 - \bar{D}^0$ at the loop level too, as presented in Fig. 1. A number of NP models have been explored in literature [14, 21, 22, 26]. Even in the cases when relatively large couplings are allowed in the certain model, the effects are screened by long distance dynamics. The branching ratios for the inclusive $D \to X_ul^+l^-$ and $D \to P/Vl^+l^-$ decays, when calculated within existing theoretical approaches cannot receive any measurable impact of NP. The only relevant question is, can we find such observables that the short distance dynamics is uncovered? The differential branching ratio might get a rather small enhancement at high lepton invariant mass [21, 26]. The new physics detection in these decay modes was also discussed. It was found that two angular asymmetries, namely the T-odd di-plane asymmetry and the forward-backward di-lepton asymmetry offer direct tests of new physics due to tiny SM backgrounds [14].

The effects of the extra heavy up vector-like quark on the decay spectrum of $D^+ \to \pi^+l^+l^-$ and $D_s^+ \to K^+l^+l^-$ decays were also considered in Ref. [22, 21]. It was found that there is a tiny increase of the differential decay rate in the region of large di-lepton mass. The R-parity violating supersymmetric model can also modify short distance dynamics in $c \to ul^+l^-$decays. The relevant parameters were constrained using current upper bound on the $D^+ \to \pi^+l^+l^-$ decay rate. Present bounds still allow small modification of the SM differential decay rate distribution [21].

The authors of [20] analyzed effects of models with a warped extra dimension on rare charm decays. New degrees of freedom in these models are bounded above a few TeV. In rare charm decays they can leave an order of magnitude signatures larger than what the SM can generate, due to contributions to C_9 and C_{10} Wilson coefficients. This can be achieved even without giving

Decay mode	size	Reference
$D \to \rho(\omega)\gamma$	$\leq 3\%$	Zwicky et al, 2012
$D \to K^+K^-\gamma$	$\leq 0.7\%$	Isidori & Kamenik 2012
$D \to X_u l^+ l^-$	$\leq 3\%$	Paul et al, 2012
$D^+ \to \pi^+\mu^+\mu^-$	$\leq 1\%$	Fajfer & Košnik, 2013
$D^+ \to h h \mu^+\mu^-$	$\leq 1\%$	Cappiello et al, 2013

Table 2. CP violating asymmetries for charm rare decays, size and the original reference.The four last decay modes have the CP asymmetry in the vicinity ϕ resonance.

the up-quark sector a special dynamical advantage. The effects are at the tree level and can be larger than the loop-level enhancement found in the Little Higgs models [14].

4. Rare D decays and direct CP violation

Since the end of 2011 the direct CP violating asymmetry in charm decays has been the subject of many studies. The experimental results of LHCb [1, 2, 3, 4] indicated unexpected large CP violation in the difference of CP violating asymmetries for $D \to \pi^+\pi^-(K^+K^-)$. Using the updated result of the LHCb collaboration, HFAG [3] produced the world average CP asymmetry, $\Delta A_{CP} = A_{CP}(D \to K^+K^-) - A_{CP}(D \to \pi^+\pi^-) = (-0.253 \pm 0.104)\%$. The size of the CP violating asymmetry in $D \to \pi^+\pi^-(K^+K^-)$ is still being questioned. However, all these measurements and studies stimulated a number of new analyses of possible CP violating effects in rare charm decays. The CP violating symmetry is introduced as:

$$A_{CP} = \frac{\Gamma(D^0 \to f) - \Gamma(\bar{D}^0 \to f)}{\Gamma(D^0 \to f) + \Gamma(\bar{D}^0 \to f)}. \quad (7)$$

The authors [15] investigated CP violating observables in $D \to V\gamma \to P^+P^-\gamma$ decays. If CP violation is due to NP effects [15, 16], then it is most likely a result of the chromomagnetic operator Q_8 contribution [19]. In the case of rare D decays CP violation results from mixing of Q_8 into Q_7 under QCD renormalization.

$$\mathrm{Im}[C_7^{NP}(m_c)]| \simeq \mathrm{Im}[C_8^{NP}(m_c)]| \simeq 0.02 \times 10^{-2}. \quad (8)$$

The imaginary part of C_7^{SM} is two orders of magnitude smaller. The NP contribution is comparable in size with the real part of the SM $|C_7^{SM-eff}(m_c)| = (0.5 \pm 0.1) \times 10^{-2}$. This means that if the phase of the long distance contribution can be neglected and the relative strong phase is maximal, the CP asymmetry can reach the $\mathcal{O}(1\%)$ level. The current world average of the CP violating asymmetry in ΔA_{CP}, following the work of [15] leads to a CP asymmetry in $D \to K^+K^-\gamma$ of the order 1%.

In rare decays $D \to P\ell^+\ell^-$ the CP violating effects might arise due to the interference of resonant part of the long distance contribution and the new physics affected short distance contribution. The appropriate observables, the differential direct CP asymmetry and partial decay width CP asymmetry are introduced in a model independent way [25]. If supersymmetric and Z'-enhanced scenarios are assumed and if the size of Wilson coefficients C_9 and C_{10} is compatible with the observed CP asymmetry in nonleptonic charm decays and flavor constraints, it was found in [23] that new physics effects in $D^0 \to h_1 h_2 l^+ l^-$ might reach the $\sim 1\%$ level. In Table 2 predictions for the size of CP violating asymmetries in rare charm decays are presented.

It is interesting that the direct CP-violating asymmetry in neutral D-meson decays leads to stronger model parameter constraints than the current experimental bound on the neutron electric dipole moment for certain class of NP models, as noticed in [28].

5. Summary

The SM contribution to rare charm decays are rather well known. For all decay modes amplitudes are fully dominated by long distance dynamics. Branching ratios itself cannot reveal information on the size of short distance contribution. One has to construct and measure additional observables, as differential rate, forward-backward asymmetries, T-odd asymmetries. New physics might be hidden only in the short distance contribution and therefore a measurement of these new observables might help in distinguishing new physics contributions from Standard Model ones. Interesting signals of NP might arise in $D \to \rho(\omega)\gamma$ and $D \to K^+K^-\gamma$, as well as in decays with the leptonic pair in the final state $D \to X_u l^+ l^-$, $D^+ \to \pi^+\mu^+\mu^-$, $D^+ \to hh\mu^+\mu^-$. Search for CP violating signals in rare charm decays became important with the experimental hint for a presence of a new CP violating source in nonleptonic charm decays. The study of rare charm decays were revived and a number of studies of CP violation in rare charm decays were done. If the CP violation in D nonleptonic decay will remain on the present level, then one expects that in all rare charm decays that CP violating asymmetries might be of of the order 1%.

Acknowledgments

This work is supported in part by the Slovenian Research Agency.

References

[1] Aaij R *et al.* [LHCb Collaboration] 2012 *Phys. Rev. Lett.* **108** 111602
[2] Aaltonen T *et al.* [CDF Collaboration] 2012 *Phys. Rev. D* **85** 012009
[3] HFAG http://www.slac.stanford.edu/xorg/hfag/
[4] Aaij R *et al.* [LHCb Collaboration] 2014 *JHEP* **1407** 041
[5] Burdman E, Golowich E, Hewett J. L. and Pakvasa S 1995 *Phys. Rev. D* **52** 6383
[6] Fajfer S, Prelovsek S and Singer P 1999 *Eur. Phys. J. C* **6** 471
[7] Fajfer S, Prelovsek S and Singer P 1998 *Phys. Rev. D* **58** 094038
[8] Fajfer S, Prelovsek S and Singer P 2001 *Phys. Rev. D* **64** 114009
[9] Fajfer S, Singer P and Zupan J 2003 *Eur. Phys. J. C* **27** 201
[10] Greub C, Hurth T, Misiak M and Wyler D 1996 *Phys. Lett. B* **382** 415
[11] Kim Q Ho and Pham X Y 2000 *Phys. Rev. D* **61** 013008
[12] Burdman E, Golowich E, Hewett J L and Pakvasa S 2002 *Phys. Rev. D* **66** 014009
[13] Paul A , Bigi I I and Recksiegel S 2010 *Phys. Rev. D* **82** 094006 [Erratum-ibid. D **83** 019901]
[14] Paul A, Bigi I I and Recksiegel S 2011 *Phys. Rev. D* **83** 114006
[15] Isidori G and Kamenik J F, 2012 *Phys. Rev. Lett.* **109** 171801
[16] Altmannshofer W, PrimulandoR, Yu C T and Yu F 2012 *JHEP* **1204** 049
[17] Aaij R*et al.* [LHCb Collaboration] 2013 *Phys. Lett. B* **725** 15
[18] Aaij R *et al.* [LHCb Collaboration] 2013 *Phys. Lett. B* **724** 203
[19] Delaunay C, Kamenik J F, Perez G and Randall L 2012 *JHEP* **1301** 027
[20] Paul A, de La Puente A and Bigi I I 2014 *Phys. Rev. D* **90** 014035
[21] Fajfer S, Kosnik N and Prelovsek S 2007 *Phys. Rev. D* **76** 074010
[22] Fajfer S and Prelovsek S 2006 *Phys. Rev. D* **73** 054026
[23] Cappiello L,Cata O and D'Ambrosio G 2013 *JHEP* **1304** 135
[24] Fajfer S, Greljo A, Kamenik J F and Mustac I, 2013 *JHEP* **1307** 155
[25] Fajfer S and Kosnik N 2013 *Phys. Rev. D***87** 054026
[26] Golowich E, Hewett J L and Pakvasa S and Petrov A A 2009 *Phys. Rev. D* **79** 114030
[27] Lyon J and R. Zwicky 2012 (Preprint 1210.6546 [hep-ph])
[28] Fajfer S and Eeg J O 2014 *Phys. Rev. D* **89** 095030.

Charm mixing and CP violation at LHCb

Artur Ukleja
National Centre for Nuclear Research, Warsaw, Poland
E-mail: artur.ukleja@fuw.edu.pl

Abstract. The LHCb experiment has collected the world's largest sample of charmed hadrons. Using data corresponding to an integrated luminosity of 3.0 fb^{-1} recorded in 2011 and 2012, measurements of direct and indirect CP violation in the charm sector and of D^0 mixing parameters were performed. Results from several decay modes are presented with complementary time-dependent and time-integrated analyses.

1. Introduction

The charm sector is a promising place to probe for the presence of physics beyond the Standard Model (SM) because CP violation (CPV) in the charm sector is expected to be very small in the SM [1]. Since evidence of $D^0 - \bar{D}^0$ oscillations was first reported [2],[3],[4] there is growing interest in this subject. Details of the measurement of $D^0 - \bar{D}^0$ oscillations achieved recently at the LHCb are discussed in Section 2. CPV arises when two or more amplitudes with different weak and strong phases contribute to the same final state. This is possible in the singly Cabibbo-suppressed (SCS) D decays, where significant tree and penguin contributions can be expected. The time-integrated CP asymmetry measurements in two-body SCS D^0 decays are discussed in Section 3. The searches for CP asymmetry in three-body SCS D^+ decays are discussed in Section 4. Conclusions are presented in Section 5.

2. Search for CP in $\mathbf{D^0 - \bar{D}^0}$ oscillations

To measure $D^0 - \bar{D}^0$ oscillations, the D^0 flavour has to be identified at both the production and decay. The flavour of the D^0 meson at the production time is identified by the charge of the slow pion (π_s) from $D^{*+} \to D^0\pi_s^+$ and $D^{*-} \to \bar{D}^0\pi_s^-$ decays. The flavour at the time of decay is identified using the almost flavour-specific $K^\pm\pi^\mp$ final state. The $D^0 \to K^+\pi^-$ decays are called wrong-sign (WS) decays and the $D^0 \to K^-\pi^+$ decays are called right-sign (RS) decays. The RS decays are dominated by the Cabibbo-favoured (CF) decay amplitude, whereas the WS amplitude includes contributions from both the doubly-Cabibbo-suppressed (DCS) $D^0 \to K^+\pi^-$ decay and $D^0 - \bar{D}^0$ mixing followed by the favoured $\bar{D}^0 \to K^+\pi^-$ decay. Assuming negligible CPV, the time-dependent ratio, R, of WS to RS decay rates is approximated by [1]

$$R(t) = R_D + \sqrt{R_D}y'\frac{t}{\tau} + \frac{x'^2 + y'^2}{4}(\frac{t}{\tau})^2, \tag{1}$$

where t/τ is the decay time expressed in units of the average D^0 lifetime (τ), R_D is the ratio of DCS to CF decay rates, $x' = x\cos\delta + y\sin\delta$, $y' = y\cos\delta - x\sin\delta$, where δ is the strong phase difference between DCS and CF amplitudes. The parameters $x = \Delta m/\Gamma$ and $y = \Delta\Gamma/2\Gamma$, where Δm and $\Delta\Gamma$ are the mass and the decay width differences between the two mass eigenstates and Γ is the average D^0 decay width.

The measurement of the time-dependent WS/RS ratio is performed with 3 fb^{-1} in thirteen bins of D^0 decay time, chosen to have a similar number of candidates in each bin [5]. The number of RS and WS decays are determined using fits to the $M(D^0\pi_s^+)$ distributions in each bin. The measured WS/RS ratios as a function of t/τ are fit separately for D^{*+} and D^{*-} decays to the time-dependence of Equation 1. Three fits are performed to the data; the first allows direct and indirect CPV; the second allows only indirect CPV; and the third is a CP-conserving fit. Figure 1 shows the central values and confidence regions in the (x'^2, y') plane. Assuming no CPV, the resulting of the probabilities of observing CP asymmetry as large or larger than those in data, *p*-value, for the fit with direct and indirect (indirect only) CPV allowed, is 91% (81%), showing that the data are compatible with CP symmetry. This allows to limit the amount of CPV in $D^0 - \bar{D}^0$ mixing, using the relations $x^{\pm} = |q/p|^1(x\cos\phi \pm y\sin\phi)$ and $y = |q/p|^1(y\cos\phi \mp x\sin\phi)$, to $0.75 < |q/p| < 1.24$ at 68.3% CL and the amount of direct CPV in the DCS decay to $A_D = (R_D^+ - R_D^-)/(R_D^+ + R_D^-) = (-0.7 \pm 1.9)\%$, where R_D^+ (R_D^-) is the DCS/CF rate ration for D^0 ($\bar{D}^0$) decays. Assuming CP conservation, the mixing parameters are measured to be $x'^2 = (5.5 \pm 4.9) \times 10^{-5}$, $y' = (4.8 \pm 1.0) \times 10^{-3}$ and $R_D = (3.568 \pm 0.066) \times 10^{-3}$.

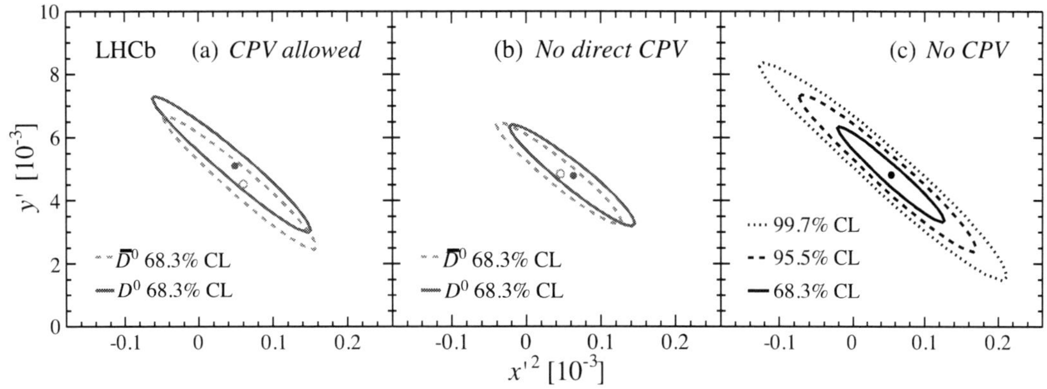

Figure 1. Two-dimensional confidence regions in the (x'^2, y') plane obtained (a) without any restriction on CPV, (b) assuming no direct CPV, and (c) assuming CP conservation. The dashed (solid) curves in (a) and (b) indicate the contours of the mixing parameters associated with $\bar{D}^0$ (D^0) decays. The best-fit values are shown with open or filled points.

3. A search for time-integrated CPV in $D^0 \to h^-h^+$ decays

The CP asymmetry of a decay of a D^0 meson to a CP eigenstate, $A_{CP}(f)$, can be expressed in terms of two contributions, a direct component associated with CPV in the decay amplitudes and an indirect component associated with CPV in the mixing or in the interference between mixing and decay. It can be written to the first order as [6]

$$A_{CP}(f) = a_{CP}^{dir}(f) + \frac{\langle t \rangle}{\tau} a_{CP}^{ind}, \tag{2}$$

where $a_{CP}^{dir}(f)$ is a direct CPV for the decay, $\langle t \rangle$ is the average proper time of the reconstructed sample, and a_{CP}^{ind} is the indirect CPV. The indirect component is universal for CP eigenstates in the SM whilst the direct component depends in general on the final state.

The time-integrated CPV asymmetry difference between $D^0 \to K^-K^+$ and $D^0 \to \pi^-\pi^+$, ΔA_{CP}, is performed with 3 fb^{-1} in which the D^0 mesons are produced in inclusive semileptonic B meson decays to the $D^0\mu^-\bar{\nu}_\mu X$ final state, where X means other possible particles [7]. The charge of the accompanying muon is used to identify the flavour of the D^0 meson ($\bar{B} \to D^0\mu^-\bar{\nu}_\mu X$

or $B \to \bar{D}^0\mu^+\nu_\mu X$). The measured raw time-integrated asymmetry is a sum of physics and detector asymmetries,

$$A_{RAW}(f) = A_{CP}(f) + A_D(\mu^+) + A_P(B), \tag{3}$$

where $A_D(\mu^+)$ is the detection asymmetry for selecting a muon from semileptonic B decay chain and $A_P(B)$ is the production asymmetry of the B mesons. The detector and production asymmetries are removed by the difference of $A_{RAW}(K^-K^+)$ and $A_{RAW}(\pi^-\pi^+)$.

A binned maximum likelihood fit to the mass distributions is performed to determine the numbers of signal candidates as 2.1×10^3 for $D^0 \to K^-K^+$ decays and 0.77×10^3 for $D^0 \to \pi^-\pi^+$ decays. The fractional difference in an average decay time of the D^0 candidates for the two samples is $\Delta\langle t\rangle/\tau = 0.014 \pm 0.004$. The small value of $\Delta\langle t\rangle/\tau$ implies that the measured ΔA_{CP} is equal to the difference in direct CPV with negligible indirect CPV correction. For the first time in LHCb, also the individual CP asymmetry in $D^0 \to K^-K^+$ decays is measured. In this case the determination of the detector and production asymmetries uses control samples of CF $\bar{B} \to D^0(\to K\pi)\mu\nu X$, $D^+ \to K\pi\pi$ and $D^+ \to \bar{K}^0(\to \pi\pi)\pi$ decays where CPV effects are negligible.

The measured value of ΔA_{CP} and the CP asymmetry in the $D^0 \to K^-K^+$ channel are found to be $\Delta A_{CP} = [+0.14 \pm 0.16(stat) \pm 0.08(syst)]\%$ and $A_{CP}(K^-K^+) = [-0.06 \pm 0.15(stat) \pm 0.10(syst)]\%$. By combining the above results, the CP asymmetry in the $D^0 \to \pi^-\pi^+$ decays is extracted as $A_{CP}(\pi^-\pi^+) = [-0.20 \pm 0.19(stat) \pm 0.10(syst)]\%$. The results show that there is no significant CPV in SCS $D^0 \to K^-K^+, \pi^-\pi^+$ decays at the level of 10^{-2}.

4. Search for CPV in the decay $D^+ \to \pi^-\pi^+\pi^+$

Many-body decays benefit from rich resonance structures with interfering amplitudes modulated by strong-phase variations across the phase space. Searches for localised asymmetries can bring complementary information on the nature of the CPV to the time-integrated decay rate measurements. The investigation of the Cabibbo-suppressed decay $D^+ \to \pi^-\pi^+\pi^+$ with 1 fb^{-1} is performed across the Dalitz plot using two model-independent techniques, one binned and the other unbinned searches [8].

The binned method, as employed in previous LHCb analyses [9],[10], is based on a bin-by-bin comparison between the D^+ and D^- Dalitz plots. For each bin of the Dalitz plot, the significance of the difference between the number of D^+ and D^- candidates, S^i_{CP}, is computed as

$$S^i_{CP} \equiv \frac{N_i^+ - \alpha N_i^-}{\sqrt{\alpha(N_i^+ + N_i^-)}}, \qquad \alpha \equiv \frac{N^+}{N^-}, \tag{4}$$

where N_i^+ (N_i^-) is the number of D^+ (D^-) candidates in the ith bin and N^+ (N^-) is the sum of N_i^+ (N_i^-) over all bins. The parameter α removes the contribution of global asymmetries which may arise due to production and detection asymmetries, as well as from CPV. In the absence of localised asymmetries, the S^i_{CP} values follow a standard normal distribution. Therefore, CPV can be detected as a deviation from this behaviour.

Two binning schemes are used, a uniform grid with bins of equal size and an adaptive binning schemes where the bins have the same population (Figure 2). No single bin in any of the binning schemes presents an absolute S^i_{CP} value larger than 3. Assuming no CPV, the probabilities of observing local asymmetries across the phase-space of the D^+ meson decay as large or larger than those in data are above 50% in all the tested binned schemes. All results are consistent with no CPV .

The unbinned method is based on the concept of nearest-neighbour events (kNN) in a combined D^+ and D^- sample to test whether they share the same parent distribution function [11],[12],[13]. A test statistic, T, is defined as the average fraction of like-charged neighbour pairs. For the null hypothesis of no CPV, this follows a Gaussian distribution

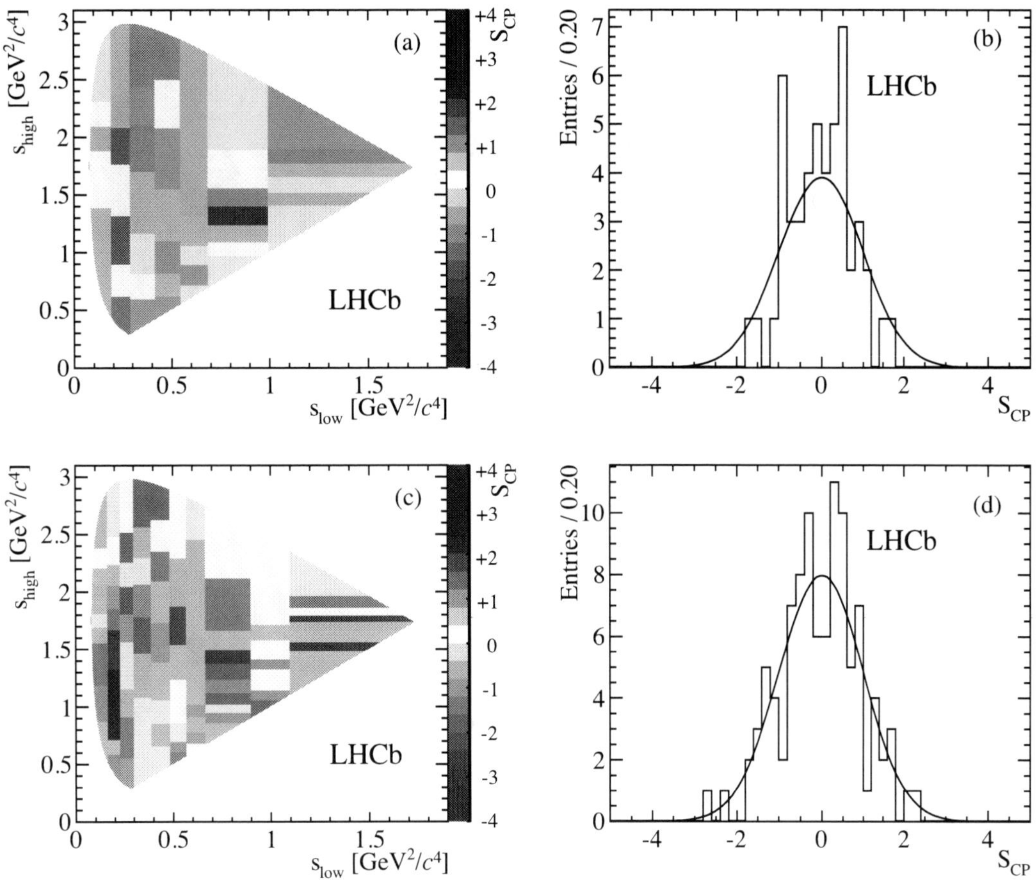

Figure 2. Distributions of S^i_{CP} across the D^+ Dalitz plot, with the adaptive binning scheme of uniform population for the total $D^+ \to \pi^-\pi^+\pi^+$ decays with (a) 49 and (c) 100 bins. The corresponding one-dimensional S^i_{CP} distributions (b) and (d) are shown with a standard normal function superimposed (solid line).

with predictable mean μ_T and variance σ^2_T [8]. CPV can be detected as a deviation from this behaviour.

The kNN method is applied with the two region definitions (Figure 3). The pull values of T and the corresponding p-values for the hypothesis of no CPV are shown in Figure 4. All p-values are above 20%, consistent with no CP asymmetry.

5. Conclusions

Using data collected with the LHCb experiment in 2011 and 2012 $D^0 - \bar{D}^0$ oscillations are observed by evaluating the time-dependence of the ratio of $D^0 \to K^+\pi^-$ and $D^0 \to K^-\pi^+$ decays. Assuming CP conservation, the mixing parameters are measured to be $x'^2 = (5.5 \pm 4.9) \times 10^{-5}$, $y' = (4.8 \pm 1.0) \times 10^{-3}$ and $R_D = (3.568 \pm 0.066) \times 10^{-3}$. Study of D^0 and $\bar{D}^0$ decays separately shows no evidence for CPV. The measurement of time-integrated difference in CP asymmetry between $D^0 \to K^-K^+$ and $D^0 \to \pi^-\pi^+$ decays also does not show evidence for direct CP violation. No evidence for CP violation is found in searches for CP asymmetries in three-body SCS decays $D^+ \to \pi^-\pi^+\pi^+$.

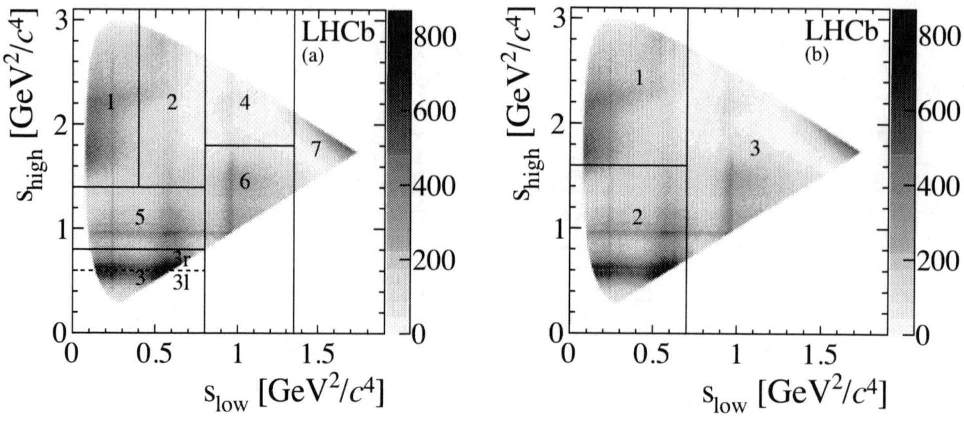

Figure 3. Dalitz plot for $D^+ \to \pi^-\pi^+\pi^+$ candidates divided into (a) seven regions R1-R7 and (b) three regions P1-P3.

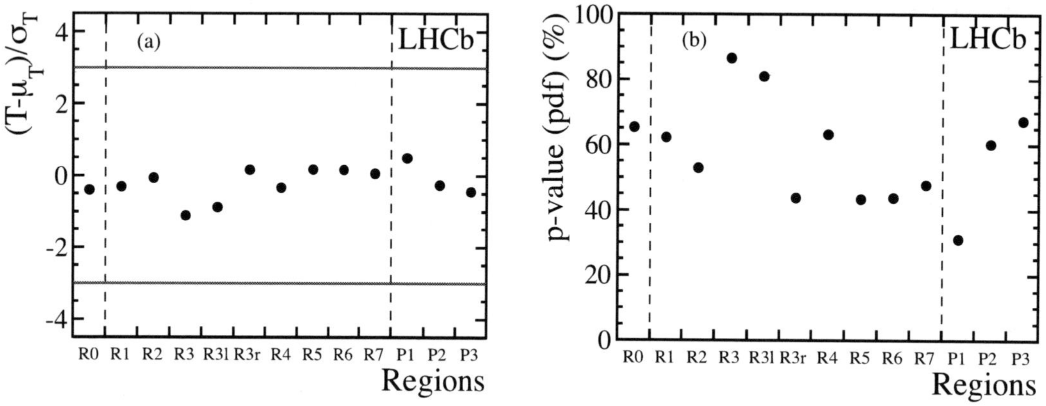

Figure 4. (a) Pull values of T and (b) the corresponding p-values for $D^+ \to \pi^-\pi^+\pi^+$ candidates restricted to each region obtained the kNN method. The horizontal blue lines in (a) represents pull values -3 and $+3$. The region R0 corresponds to the full Dalitz plot.

References

[1] Bianco S, Fabbri F L, Benson D and Bigi I 2003 *Riv. Nuovo Cim.* **26 N 7** 1-200
[2] Aubert B *et al.* [BABAR Collaboration] 2007 *Phys. Rev. Lett.* **98** 211802
[3] Staric M *et al.* [Belle Collaboration] 2007 *Phys. Rev. Lett.* **98** 211803
[4] Aaltonen T *et al.* [CDF Collaboration] 2008 *Phys. Rev. Lett.* **100** 121802
[5] Aaij R *et al.* [LHCb Collaboration] 2013 *Phys. Rev. Lett.* **111** 251801
[6] Aaltonen T *et al.* [CDF Collaboration] 2012 *Phys. Rev.* D **85** 012009
[7] Aaij R *et al.* [LHCb Collaboration] 2014 *JHEP* **1407** 041
[8] Aaij R *et al.* [LHCb Collaboration] 2014 *Phys. Lett.* B **728** 585
[9] Aaij R *et al.* [LHCb Collaboration] 2011 *Phys. Rev.* D **84** 112008
[10] Aaij R *et al.* [LHCb Collaboration] 2013 *Phys. Lett.* B **726** 623
[11] Henze N 1988 *The Annals of Statistics* **16 No 2** 772
[12] Schilling M F 1986 *J.Am.Stat.Assoc* **81** 799
[13] Williams M 2010 *JINST* **5** P09004

Implications of CP-violation in charmed hadrons

Alexey A. Petrov

Department of Physics and Astronomy, Wayne State University, Detroit, MI 48201, USA
Michigan Center for Theoretical Physics, University of Michigan, Ann Arbor, MI 48109, USA

E-mail: apetrov@wayne.edu

Abstract. I discuss theoretical implications of recent experimental progress in understanding CP-violation in charmed mesons. I review recent standard model predictions and attempts to constrain beyond the standard model scenarios using observations of charm hadron transitions.

1. Introduction

CP-violation has never been observed in the up-quark sector. This is in stark contrast to the bottom-type quarks, where CP-violating effects have been seen in both s and b-quark transitions. While current experimental constraints do not support a possibility of large new physics (NP) enhancement of the expected signals in the up-quark sector, a modest NP contribution is still possible. It is therefore important to constrain the size of possible NP interactions, or, at least determine the size of the standard model (SM) signal.

Among the studies of flavor physics in the up quark sector, charm transitions play an utmost important role. The charm quark is the only up-type quark that can undergo flavor oscillations, which makes it the only currently viable probe of $\Delta Q = 2$ interactions there. Besides that, the availability of large statistical samples of charm data makes charm system one of the most convenient places to put constraints on NP interactions [1]. Depending on the source of CP-violating interactions, it often convenient to classify the CP-violating contributions in charm according to the number of units of charm quantum number the corresponding operators change. In other words,

(I) **CP-violation in $\Delta C = 2$ transitions.** Both SM and NP can generate local operators at the charm quark mass scale that change charm quantum number by two units. Yet, NP CP-violating interactions can have a different origin from the SM ones, where they came from the same Yukawa couplings – or complex-valued matrix elements of the Cabibbo-Kobayashi-Maskawa (CKM) quark mixing matrix [2] – and contribute to both $\Delta C = 2$ and $\Delta C = 1$ effective operators. Phenomenologically, one can distinguish CP violation that originates in the $D^0\overline{D}^0$ mixing matrix (aka "indirect" CP-violation) from other sources of CP-violation. The non-diagonal entries in the $D^0 - \overline{D}^0$ mass matrix,

$$\left[M - i\frac{\Gamma}{2}\right]_{ij} = \begin{pmatrix} A & p^2 \\ q^2 & A \end{pmatrix} \tag{1}$$

can lead to manifest effects of CP-violation when $R_m^2 = |p/q|^2 = (2M_{12} - i\Gamma_{12})/(2M_{12}^* - i\Gamma_{12}^*) \neq 1$. Also, CP violation in the interference of decays with and without mixing can be

Journal of Physics: Conference Series **556** (2014) 012057 doi:10.1088/1742-6596/556/1/012057

detected. This type of CP violation is possible for a subset of final states to which both D^0 and $\overline{D}^0$ can decay.

(II) CP violation in the $\Delta C = 1$ decay amplitudes (or "direct" CP-violation). This type of CP violation occurs when the absolute value of the decay amplitude for D to decay to a final state f (A_f) is different from the one of the corresponding CP-conjugated amplitude ("direct CP-violation"). This can happen if the final state in the decay can be reached via at least two different "pathways" whose amplitudes are associated with different weak and strong phases,

$$A_f = |A_1|\, e^{i\delta_1} e^{i\phi_1} + |A_2|\, e^{i\delta_2} e^{i\phi_2}, \tag{2}$$

where ϕ_i represent weak phases (odd under CP), and δ_i represents strong phases (even under CP). This ensures that the CP-conjugated amplitude, $\overline{A}_{\overline{f}}$ is different from A_f.

Contrary to the case of bottom quarks, standard model interactions do not produce a large CP-violating signal in the charmed system. This argument stems from the fact that all quarks that build up initial and final hadronic states in weak decays of charm mesons or baryons belong to the first two generations. This implies that, at tree level, those transitions are governed by a 2×2 Cabibbo quark mixing matrix. This matrix is real, so no CP-violation is possible in the dominant tree-level diagrams which describe the decay amplitudes. In the standard model, CP-violating amplitudes can be introduced by including penguin or box operators induced by virtual b-quarks, which gives the necessary access to the third generation of quarks. However, their contributions are strongly suppressed by the small combination of CKM matrix elements $V_{cb}V_{ub}^*$. Explicit evaluations of b-quark contributions to mixing asymmetries yield results of the order of $0.1 - 1\%$ [3] with similar predictions for decay amplitudes [4]. How would those effects manifest themselves in experimental observables?

While CPT-symmetry requires the total widths of D and $\overline{D}$ to be the same, the partial decay widths $\Gamma(D \to f)$ and $\Gamma(\overline{D} \to \overline{f})$ could be different in the presence of CP-violation, which would be signaled by a non-zero value of the asymmetry

$$a_f = \frac{\Gamma(D \to f) - \Gamma(\overline{D} \to \overline{f})}{\Gamma(D \to f) + \Gamma(\overline{D} \to \overline{f})}, \tag{3}$$

which, depending on the final and initial state, can be generated by both $\Delta C = 1$ and $\Delta C = 2$ interactions.

Asymmetries of Eq. (3) can be introduced for both charged and neutral D-mesons. In the latter case a much richer structure becomes available due to the interplay of CP-violating contributions to decay and mixing amplitudes [5, 6], which can play the role of a "second pathway,"

$$\begin{aligned} a_f &= a_f^d + a_f^m + a_f^i, \\ a_f^m &= -R_f \frac{y'}{2}\left(R_m - R_m^{-1}\right)\cos\phi, \\ a_f^i &= R_f \frac{x'}{2}\left(R_m + R_m^{-1}\right)\sin\phi, \end{aligned}$$

where a_f^d, a_f^m, and a_f^i represent CP-violating contributions from decay, mixing and interference between decay and mixing amplitudes respectively (we shall introduce a_f^d momentarily). Note that for the final states that are also CP-eigenstates $f = \overline{f}$ and $y' = y$.

Early experimental hints [7] at larger, about a few percent, values of the combined asymmetry,

$$\Delta A_{CP} = a_{KK} - a_{\pi\pi}, \tag{4}$$

led to active theoretical reevaluations of the size of the SM contribution to this asymmetry, as observation of larger CP violation in charm decays or mixing was assumed to be an unambiguous sign for new physics. We shall discuss those models in Section 3.

Finally, it is convenient to define, for a given final state f, a measure that parameterizes CP violating contributions,

$$\lambda_f = \frac{q}{p}\frac{\overline{A}_f}{A_f} = R_m e^{i(\phi+\delta)}\left|\frac{\overline{A}_f}{A_f}\right|, \tag{5}$$

where A_f and $\overline{A}_f$ are the amplitudes for $D^0 \to f$ and $\overline{D}^0 \to f$ transitions respectively and δ is the CP-conserving strong phase difference between A_f and $\overline{A}_f$. In Eq. (5) ϕ represents the convention-independent CP-violating phase difference between the ratio of decay amplitudes and the mixing matrix.

2. CP-violation in $\Delta C = 2$ transitions.

The non-diagonal entries in the mixing matrix of Eq. (1) lead to mass eigenstates of neutral D-mesons that are different from the weak eigenstates. They, however, are related by a linear transformation,

$$|D_{\frac{1}{2}}\rangle = p|D^0\rangle \pm q|\overline{D}^0\rangle, \tag{6}$$

where the complex parameters p and q are obtained from diagonalizing the $D^0\overline{D}^0$ mass matrix of Eq. (1). Note that if CP-violation is neglected, $p = q = 1/\sqrt{2}$. The mass and width splittings between mass eigenstates are

$$x_D = \frac{m_1 - m_2}{\Gamma_D}, \qquad y_D = \frac{\Gamma_1 - \Gamma_2}{2\Gamma_D}, \tag{7}$$

where Γ_D is the average width of the two neutral D meson mass eigenstates. The mixing parameters above can be related to real and imaginary parts of the correlation function,

$$\begin{aligned} x_D &= \frac{1}{2m_D\Gamma_D}\mathrm{Re}\left[\langle\overline{D}^0|H^{\Delta C=2}|D^0\rangle + \langle\overline{D}^0|i\int d^4x \mathrm{T}\left[H^{\Delta C=1}(x)H^{\Delta C=1}(0)\right]|D^0\rangle\right], \\ y_D &= \frac{1}{2m_D\Gamma_D}\mathrm{Im}\left[\langle\overline{D}^0|i\int d^4x \mathrm{T}\left[H^{\Delta C=1}(x)H^{\Delta C=1}(0)\right]|D^0\rangle\right]. \end{aligned} \tag{8}$$

Because of the absence of superheavy down-type quarks destroying Glashow-Iliopoulos-Maiani (GIM) cancellation, it is expected that x_D and y_D should be rather small in the standard model. Moreover, since the integral in Eq. (8) is not necessarily dominated by short distances, the expectation is that the theoretical calculation of x_D and y_D is rather uncertain. A fit to the current database of experimental analyses by the Heavy Flavor Averaging Group (HFAG) gives, allowing for CP-violation [8, 9]

$$\begin{aligned} &x_D = 0.41^{+0.14}_{-0.15}\%, \quad y_D = 0.63^{+0.07}_{-0.08}\% \ , \\ &|q/p| = 0.93^{+0.09}_{-0.08} \ , \quad \phi = -8.7^{+8.7}_{-9.1} \text{ degrees}. \end{aligned} \tag{9}$$

CP-violation induced by $\Delta C = 2$ operators can be best studied in the "super weak" limit[1], that is, neglecting direct CP-violation in the decay amplitudes.

It is easy to see that in this limit the number of "experimental" parameters (x_D, y_D, $|q/p|$, and ϕ of Eq. (9)), exceeds the number of "theoretical" parameters ($|M_{12}|$, $|\Gamma_{12}|$, and

[1] We borrow this terminology from kaon physics, where this phrase was coined by L. Wolfenstein to understand origins of CP-violation in that system.

$\phi_{12} = \arg(M_{12}/\Gamma_{12})$, so one can write a "theory-independent" constraint relation among $D^0\overline{D}^0$ mixing amplitudes [10, 11],

$$\frac{x_D}{y_D} = \frac{1-|q/p|}{\tan\phi}. \tag{10}$$

Current experimental results $x/y \approx 0.8 \pm 0.3$ imply that amount of CP-violation in the $D^0\overline{D}^0$ mixing matrix is comparable to CP-violation in the interference of decays and mixing amplitudes.

Now, even though theoretical calculations of x_D and y_D are quite uncertain, the values $x_D \sim y_D \sim 1\%$ are natural in the standard model [12], which is approximately what observed experimentally. Also, as was argued earlier, CP-violation asymmetries in charm mixing are quite small. Can we still use charm data to constrain beyond the SM scenarios?

This question can be answered using an effective field theory approach. Heavy NP degrees of freedom cannot be directly produced in charm meson decays, but can nevertheless affect the effective $|\Delta C| = 2$ Hamiltonian by changing Wilson coefficients and/or introducing new operator structures[2]. By integrating out those new degrees of freedom associated with new interactions at a high scale M, we are left with an effective hamiltonian written in the form of a series of operators of increasing dimension. It turns out that a model-independent study of NP $|\Delta C| = 2$ contributions is possible, as any NP model will only modify Wilson coefficients of those operators [14, 15],

$$\mathcal{H}_{NP}^{|\Delta C|=2} = \frac{1}{M^2}\left[\sum_{i=1}^{8} \mathrm{C}_i(\mu)\, Q_i\right], \tag{11}$$

where C_i are dimensionless Wilson coefficients, and the Q_i are the effective operators:

$$\begin{aligned}
Q_1 &= (\overline{u}_L^\alpha \gamma_\mu c_L^\alpha)\,(\overline{u}_L^\beta \gamma^\mu c_L^\beta)\,, & Q_5 &= (\overline{u}_R^\alpha c_L^\beta)\,(\overline{u}_L^\beta c_R^\alpha)\,,\\
Q_2 &= (\overline{u}_R^\alpha c_L^\alpha)\,(\overline{u}_R^\beta c_L^\beta)\,, & Q_6 &= (\overline{u}_R^\alpha \gamma_\mu c_R^\alpha)\,(\overline{u}_R^\beta \gamma^\mu c_R^\beta)\,,\\
Q_3 &= (\overline{u}_R^\alpha c_L^\beta)\,(\overline{u}_R^\beta c_L^\alpha)\,, & Q_7 &= (\overline{u}_L^\alpha c_R^\alpha)\,(\overline{u}_L^\beta c_R^\beta)\,,\\
Q_4 &= (\overline{u}_R^\alpha c_L^\alpha)\,(\overline{u}_L^\beta c_R^\beta)\,, & Q_8 &= (\overline{u}_L^\alpha c_R^\beta)\,(\overline{u}_L^\beta c_R^\alpha)\,,
\end{aligned} \tag{12}$$

where α and β are color indices. In total, there are eight possible operator structures that exhaust the list of possible independent contributions to $|\Delta C| = 2$ transitions.

No CP-violation has been observed in charm transitions yet. However, available experimental constraints of Eq. (9) can provide some tests of CP-violating NP models. For example, a set of constraints on the imaginary parts of Wilson coefficients of Eq. (11) can be placed,

$$\begin{aligned}
&\mathrm{Im}\,[C_1] \le 1.1 \times 10^{-7}\left[\frac{M}{1\ \mathrm{TeV}}\right]^2, & &\mathrm{Im}\,[C_4] \le 1.1 \times 10^{-8}\left[\frac{M}{1\ \mathrm{TeV}}\right]^2,\\
&\mathrm{Im}\,[C_2] \le 2.9 \times 10^{-8}\left[\frac{M}{1\ \mathrm{TeV}}\right]^2, & &\mathrm{Im}\,[C_5] \le 3.0 \times 10^{-8}\left[\frac{M}{1\ \mathrm{TeV}}\right]^2.\\
&\mathrm{Im}\,[C_3] \le 1.1 \times 10^{-7}\left[\frac{M}{1\ \mathrm{TeV}}\right]^2, & &
\end{aligned} \tag{13}$$

The constraints on $C_6 - C_8$ are identical to those on $C_1 - C_3$ [14]. Note that Eq. (13) implies that new physics particles, for some unknown reason, have highly suppressed couplings to charmed quarks. Alternatively, the tight constraints of Eq. (13) probe NP at very high scales: $M \ge (4-10) \times 10^3$ TeV for tree-level NP-mediated charm mixing and $M \ge (1-3) \times 10^2$ TeV for loop-dominated mixing via new physics particles.

Other observables could also be interesting in the "superweak" limit. For example, an "untagged" CP-violating asymmetry where one compares decays of a D to a state f to which

[2] NP can also affect $|\Delta C| = 1$ transitions and thus contribute to y_D. For more details, see [13].

both D^0 and $\overline{D}^0$ states can decay (but not a CP-eigenstate) without identifying flavor of the initial meson [16, 17] can be considered,

$$A_{CP}^U(f) = \frac{\Sigma_f - \Sigma_{\overline{f}}}{\Sigma_f + \Sigma_{\overline{f}}}, \quad \text{where} \quad \Sigma_f = \Gamma(D^0 \to f) + \Gamma(\overline{D}^0 \to f), \tag{14}$$

Examples of those states are $f = \pi\rho$, KK^*, $K^+\pi^-$, etc.

3. CP-violation in $\Delta C = 1$ transitions.

In principle, $D^0\overline{D}^0$ mixing is not required for the observation of CP-violation[3]. For charged D-decays the only contribution to the asymmetry of Eq. (3) comes from the multi-component structure of the $\Delta C = 1$ decay amplitude of Eq. (2). In this case,

$$a_f = \frac{2Im\,(A_1 A_2^*) \sin\delta}{|A_1|^2 + |A_2|^2 + 2Re A_1 A_2^* \cos\delta} = 2r_f \sin\phi \sin\delta, \tag{15}$$

where $\delta = \delta_1 - \delta_2$ is the CP-conserving phase difference and ϕ is the CP-violating one. $r_f = |A_2/A_1|$ is the ratio of amplitudes. Both r_f and δ are extremely difficult to compute reliably in D-decays. However, the task can be significantly simplified if one only concentrates on detection of new physics in CP-violating asymmetries in the current round of experiments [6], i.e. at the $\mathcal{O}(1\%)$ level. This is the level at which a_f is currently probed experimentally, see, e.g. [19]. As follows from Eq. (15), in this case one should expect $r_f \sim 0.01$.

Naively, the standard model asymmetries are safely below this estimate. First, Cabibbo-favored ($A_f \sim \lambda^0$) and doubly Cabibbo-suppressed ($A_f \sim \lambda^2$) decay modes proceed via amplitudes that share the same weak phase, so no CP-asymmetry is generated[4]. On the other hand, singly-Cabibbo-suppressed decays ($A_f \sim \lambda^1$) readily have a two-component structure, receiving contributions from both tree and penguin amplitudes. In this case the same conclusion follows from the consideration of the charm CKM unitarity, $V_{ud}V_{cd}^* + V_{us}V_{cs}^* + V_{ub}V_{cb}^* = 0$.

In the Wolfenstein parameterization of CKM, the first two terms in this equation are of the order $\mathcal{O}(\lambda)$ (where $\lambda \simeq 0.22$), while the last one is $\mathcal{O}(\lambda^5)$. Thus, the CP-violating asymmetry is naively expected to be at most $a_f \sim 10^{-3}$ in the SM. This naive estimate was called in question when a somewhat large value of A_{CP} of Eq. (4) was reported [7] to be

$$\Delta A_{CP} = (-0.678 \pm 0.147)\%. \tag{16}$$

Naively writing decay amplitudes as

$$\begin{aligned} A_{KK} &= \frac{G_F}{\sqrt{2}}\lambda\left[(T + E + P_{sd}) + \lambda^4 e^{-i\gamma} P_{bd}\right], \\ A_{\pi\pi} &= \frac{G_F}{\sqrt{2}}\lambda\left[(-(T + E) + P_{sd}) + \lambda^4 e^{-i\gamma} P_{bd}\right], \end{aligned} \tag{17}$$

where T, E, $P_{q_1q_2}$ represent tree, exchange, and penguin amplitudes with q_1q_2 intermediate quarks respectively, we can see that in the flavor $SU(3)_F$ limit $\Delta A_{CP} \approx 2a_{KK}$, i.e. somewhat above the naive prediction for the direct CP-violating asymmetry in the standard model.

Re-evaluations of the SM contributions proposed enhancement of penguin amplitudes via various mechanisms, such as variants of $\Delta I = 1/2$ rule for D-decays, large $1/m_c$ corrections,

[3] But it is quite important to produce equal amounts of decaying particles and antiparticles, which is a consideration at LHC [18].

[4] Technically, there is a small, $\mathcal{O}(\lambda^4)$ phase difference between the dominant tree T amplitude and exchange E amplitudes.

new fits of flavor-flow diagrams, etc. [20]. Alternatively, a number of new physics scenarios were proposed [21] to explain the result of Eq. (16), as well as means to test them in other charm transitions [22] (see also [23]). The discussions of ΔA_{CP} somewhat muted when LHCb released new analysis of the same quantity, now reporting [9]

$$\Delta A_{CP} = (-0.329 \pm 0.121)\%. \tag{18}$$

While the central value and statistical significance of of this results are much smaller, the question of how to reliably compute ΔA_{CP} is by no means settled [24], so the discussion must continue.

References

[1] For recent reviews see M. Artuso, B. Meadows and A. A. Petrov, Ann. Rev. Nucl. Part. Sci. **58**, 249 (2008). S. Bianco, F. L. Fabbri, D. Benson and I. Bigi, Riv. Nuovo Cim. **26N7**, 1 (2003); G. Burdman and I. Shipsey, Ann. Rev. Nucl. Part. Sci. **53**, 431 (2003); A. Ryd and A. A. Petrov, Rev. Mod. Phys. **84**, 65 (2012); X. Q. Li, X. Liu and Z. T. Wei, Front. Phys. China **4**, 49 (2009).
[2] I. I. Bigi and A. I. Sanda, *CP violation* (Cambridge University Press, 2000).
[3] M. Bobrowski, A. Lenz, J. Riedl and J. Rohrwild, JHEP **1003**, 009 (2010); I. I. Y. Bigi and A. I. Sanda, arXiv:hep-ph/9909479.
[4] F. Buccella, M. Lusignoli, G. Mangano, G. Miele, A. Pugliese and P. Santorelli, Phys. Lett. B **302**, 319 (1993).
[5] S. Bergmann, *et. al*, Phys. Lett. B **486**, 418 (2000);
[6] Y. Grossman, A. L. Kagan and Y. Nir, Phys. Rev. D **75**, 036008 (2007).
[7] R. Aaij *et al.* [LHCb Collaboration], Phys. Rev. Lett. **108**, 111602 (2012); T. Aaltonen *et al.* [CDF Collaboration], Phys. Rev. Lett. **109**, 111801 (2012); B. R. Ko [Belle Collaboration], PoS ICHEP **2012**, 353 (2013).
[8] B. Aubert *et al.* [BABAR Collaboration], Phys. Rev. Lett. **98**, 211802 (2007); B. R. Ko *et al.* [Belle Collaboration], Phys. Rev. Lett. **112**, 111801 (2014); T. Peng *et al.* [Belle Collaboration], Phys. Rev. D **89**, 091103 (2014).
[9] Heavy Flavor Averaging Group (HFAG), FPCP14 average, http://www.slac.stanford.edu/xorg/hfag/charm/
[10] Y. Grossman, Y. Nir and G. Perez, Phys. Rev. Lett. **103**, 071602 (2009).
[11] A. L. Kagan and M. D. Sokoloff, Phys. Rev. D **80**, 076008 (2009).
[12] A. F. Falk, Y. Grossman, Z. Ligeti and A. A. Petrov, Phys. Rev. D **65**, 054034 (2002); A. F. Falk, *et. al*, Phys. Rev. D **69**, 114021 (2004); J. Donoghue, E. Golowich, B. Holstein and J. Trampetic, Phys. Rev. D33, 179 (1986); L. Wolfenstein, Phys. Lett. B164, 170 (1985); P. Colangelo, G. Nardulli and N. Paver, Phys. Lett. B242, 71 (1990); T.A. Kaeding, Phys. Lett. B357, 151 (1995); for other approaches, see H. Georgi, Phys. Lett. B297, 353 (1992); T. Ohl, G. Ricciardi and E. Simmons, Nucl. Phys. B403, 605 (1993); I. Bigi and N. Uraltsev, Nucl. Phys. B **592**, 92 (2001).
[13] E. Golowich, S. Pakvasa and A. A. Petrov, Phys. Rev. Lett. **98**, 181801 (2007).
[14] O. Gedalia, Y. Grossman, Y. Nir and G. Perez, Phys. Rev. D **80**, 055024 (2009).
[15] E. Golowich,et. al, Phys. Rev. D **76**, 095009 (2007).
[16] A. A. Petrov, Phys. Rev. D **69**, 111901 (2004).
[17] E. Won, B. R. Ko and B. Golob, arXiv:1409.7997 [hep-ex].
[18] W. K. Lai, A. K. Leibovich and A. A. Petrov, Phys. Rev. D **90**, 054022 (2014)
[19] CDF Collaboration, CDF note 10296.
[20] M. Golden and B. Grinstein, Phys. Lett. B **222**, 501 (1989); D. Pirtskhalava and P. Uttayarat, Phys. Lett. B **712**, 81 (2012); B. Bhattacharya, M. Gronau and J. L. Rosner, Phys. Rev. D **85**, 054014 (2012); J. Brod, A. L. Kagan and J. Zupan, Phys. Rev. D **86**, 014023 (2012); J. Brod, Y. Grossman, A. L. Kagan and J. Zupan, JHEP **1210**, 161 (2012); E. Franco, S. Mishima and L. Silvestrini, JHEP **1205**, 140 (2012), and others.
[21] G. Isidori, J. F. Kamenik, Z. Ligeti and G. Perez, Phys. Lett. B **711**, 46 (2012); G. F. Giudice, G. Isidori and P. Paradisi, JHEP **1204**, 060 (2012); W. Altmannshofer, R. Primulando, C. T. Yu and F. Yu, JHEP **1204**, 049 (2012); G. Hiller, Y. Hochberg and Y. Nir, Phys. Rev. D **85**, 116008 (2012).
[22] G. Isidori and J. F. Kamenik, Phys. Rev. Lett. **109**, 171801 (2012); J. Lyon and R. Zwicky, arXiv:1210.6546 [hep-ph].
[23] A. A. Petrov, Phys. Lett. B **399**, 172 (1997) J. F. Donoghue and A. A. Petrov, Phys. Rev. D **53**, 3664 (1996)
[24] A. Lenz, arXiv:1311.6447 [hep-ph].

Charm CP violation and mixing at Belle

Byeong Rok Ko (on behalf of the Belle Collaboration)
Korea University, SEOUL 136-713, Republic of Korea
E-mail: brko@hep.korea.ac.kr

Abstract. We present charm CP violation and mixing measurements at Belle. They are the first observation of $D^0 - \bar{D}^0$ mixing in e^+e^- collisions from $D^0 \to K^+\pi^-$ decays, the most precise mixing and indirect CP violation parameters from $D^0 \to K^0_S\pi^+\pi^-$ decays, and the time-integrated CP asymmetries in $D^0 \to \pi^0\pi^0$ and $D^0 \to K^0_S\pi^0$ decays. Our mixing measurement in $D^0 \to K^+\pi^-$ decays excludes the no-mixing hypothesis at the 5.1 standard deviation level. The mixing parameters $x = (0.56 \pm 0.19^{+0.03+0.06}_{-0.09-0.09})\%$, $y = (0.30 \pm 0.15^{+0.04+0.03}_{-0.05-0.06})\%$ and indirect CP violation parameters $|q/p| = (0.90^{+0.16+0.05+0.06}_{-0.15-0.04-0.05})$, $\arg(q/p) = (-6 \pm 11 \pm 3^{+3}_{-4})^\circ$ measured from $D^0 \to K^0_S\pi^+\pi^-$ decays, and the time-integrated CP asymmetries $A_{CP}^{D^0\to\pi^0\pi^0} = (-0.03 \pm 0.64 \pm 0.10)\%$ and $A_{CP}^{D^0\to K^0_S\pi^0} = (-0.21 \pm 0.16 \pm 0.07)\%$ are the most precise measurements to date. Our measurements here are consistent with predictions of the standard model.

The magnitudes of CP violation (CPV) and mixing rate in the charm system are very small in the standard model (SM) [1, 2]. Therefore, D^0-$\bar{D}^0$ mixing and CPV measurements provide a unique probe to search for physics beyond the SM [3, 4]. $D^0 - \bar{D}^0$ mixing occurs since the mass eigenstates D_1 and D_2 are different from the flavor eigenstates D^0 and $\bar{D}^0$. The mass eigenstates can be written in terms of the flavor eigenstates, namely, $|D_{1,2}\rangle = p|D^0\rangle \pm q|\bar{D}^0\rangle$. The phenomenology of meson mixing is described by two parameters, $x = \Delta m/\Gamma$ and $y = \Delta\Gamma/2\Gamma$, where Δm and $\Delta\Gamma$ are the mass and width differences between the two mass eigenstates, and Γ is the average decay width of the mass eigenstates. Indirect CPV parameters are $|q/p|$ and $\arg(q/p)$, where the former and the latter are responsible for CPV in mixing and that in interference between the decays with and without mixing, respectively.

In this proceedings, we present charm CPV and mixing measurements from the data recorded with the Belle detector [5] at the e^+e^- asymmetric-energy collider KEKB [6].

1. Mixing in $D^0 \to K^+\pi^-$ decays

We present the first observation of $D^0 - \bar{D}^0$ mixing from an e^+e^- collision experiment by measuring the time-dependent ratio of the $D^0 \to K^+\pi^-$ (**W**rong **S**ign) to $D^0 \to K^-\pi^+$ (**R**ight **S**ign) decay rates [7]. Assuming CP conservation and that the mixing parameters are small ($|x| \ll 1$ and $|y| \ll 1$), the time-dependent RS and WS decay rates are $\Gamma_{\rm RS}(\tilde{t}/\tau) \approx |\mathcal{A}_{\rm CF}|^2 e^{-\frac{\tilde{t}}{\tau}}$ and $\Gamma_{\rm WS}(\tilde{t}/\tau) \approx |\mathcal{A}_{\rm CF}|^2 e^{-\frac{\tilde{t}}{\tau}}\left(R_D + \sqrt{R_D}y'\frac{\tilde{t}}{\tau} + \frac{x'^2+y'^2}{4}\left(\frac{\tilde{t}}{\tau}\right)^2\right)$, respectively, to second order in the mixing parameters, where $\tilde{t}$ is the true proper decay time, $\mathcal{A}_{\rm CF}$ is the **C**abibbo-**F**avored decay amplitude, τ is the D^0 lifetime, R_D is the ratio of DCS (**D**oubly **C**abibbo-**S**uppressed) to CF decay rates, $x' = x\cos\delta + y\sin\delta$, and $y' = y\cos\delta - x\sin\delta$, where δ is the strong phase difference between the DCS and CF decay amplitudes. The time-dependent ratio of WS to RS decay rates then

$$R(t/\tau) = \frac{\int_{-\infty}^{+\infty} \Gamma_{\rm WS}(\tilde{t}/\tau)\mathcal{R}(t/\tau - \tilde{t}/\tau)d(\tilde{t}/\tau)}{\int_{-\infty}^{+\infty} \Gamma_{\rm RS}(\tilde{t}/\tau)\mathcal{R}(t/\tau - \tilde{t}/\tau)d(\tilde{t}/\tau)}, \qquad (1)$$

where t is the reconstructed proper decay time and $\mathcal{R}(t/\tau - \tilde{t}/\tau)$ is the resolution function of the real decay time, $\tilde{t}$. Figure 1 shows the time-dependent ratios of WS to RS decays together with the two hypothesis tests, with (line) and without mixing (dots). The χ^2 difference between the "no-mixing" and "mixing" hypotheses, $\Delta\chi^2 = \chi^2_{\rm no-mixing} - \chi^2_{\rm mixing}$, is 29.3 for two degrees of freedom, corresponding to a probability of 4.3×10^{-7}; this implies the no-mixing hypothesis is excluded at the 5.1 standard deviation level. Thus, we observe $D^0 - \bar{D}^0$ mixing for the first time in an e^+e^- collision experiment. We also show this in Figure 2 with the 1σ, 3σ, and 5σ contours around the best fit point in the (x'^2, y') plane.

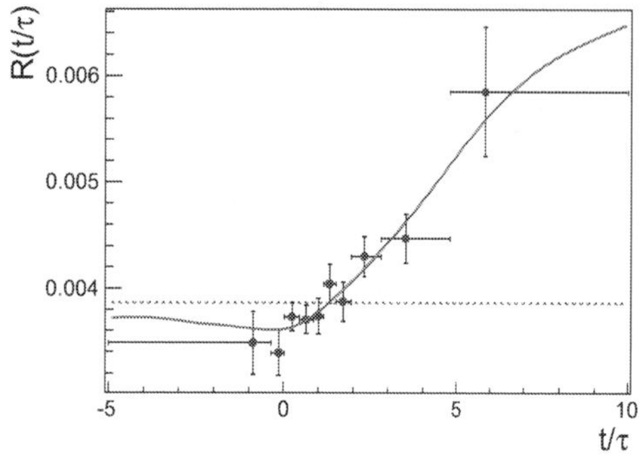

Figure 1. The time-dependent ratios of WS to RS decay rates. Points with error bars reflect the data and their total uncertainties. The lines show the fit with (solid) and without (dashed) the mixing hypothesis.

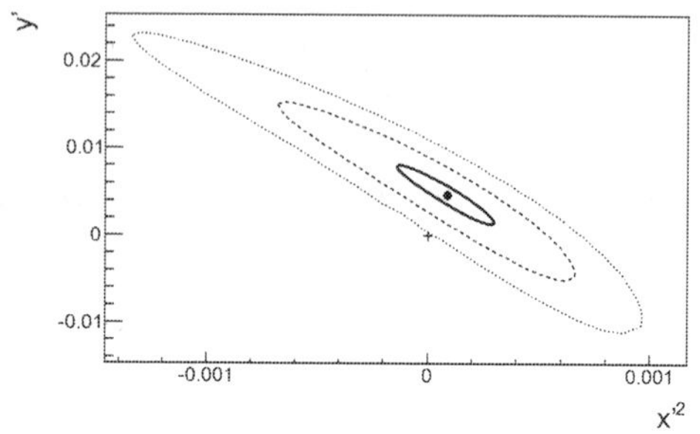

Figure 2. Best-fit point and contours in the (x'^2, y') plane. The solid, dashed, and dotted lines, respectively, correspond to 1, 3, and 5 standard Gaussian deviations from the best fit. The cross is the no-mixing point.

2. Mixing and indirect CPV in $D^0 \to K^0_S\pi^+\pi^-$ decays

We present the most precise mixing and indirect CP violation parameters from $D^0 \to K^0_S\pi^+\pi^-$ decays [8] using the time-dependent Dalitz fit analysis [9]. The time-dependent decay matrix elements of $D^0 \to K^0_S\pi^+\pi^-$ and $\bar{D}^0 \to K^0_S\pi^+\pi^-$ are

$$\begin{aligned}\mathcal{M}(m^2_+, m^2_-, t) &= g_+(t)\mathcal{A}(m^2_+, m^2_-) + \frac{q}{p}g_-(t)\bar{\mathcal{A}}(m^2_+, m^2_-) \text{ and} \\ \bar{\mathcal{M}}(m^2_+, m^2_-, t) &= g_+(t)\bar{\mathcal{A}}(m^2_+, m^2_-) + \frac{p}{q}g_-(t)\mathcal{A}(m^2_+, m^2_-),\end{aligned} \qquad (2)$$

where $m^2_\pm = m^2_{K^0_S\pi^\pm}$, $\mathcal{A}(m^2_+, m^2_-) = \sum a_j e^{i\delta_i} A_j(m^2_+, m^2_-)$, $\bar{\mathcal{A}}(m^2_+, m^2_-) = \sum \bar{a}_j e^{i\bar{\delta}_i} A_j(m^2_+, m^2_-)$, $g_\pm(t) = (e^{-i\lambda_1 t} \pm e^{-i\lambda_2 t})/2$, and $\lambda_i = m_i - i\Gamma_i/2$, where a_j, δ_j, and $A_j(m^2_+, m^2_-)$ are amplitude, phase, and resonance matrix element, respectively. The decay matrix element squared $\mathcal{M}^2$ then contains the mixing parameters x and y as well as q/p of which magnitude and argument are indirect CPV parameters. Two separate time-integrated Dalitz fits to D^0 and $\bar{D}^0$ samples show no direct CPV resulting in $a_j \approx \bar{a}_j$ and $\delta \approx \bar{\delta}$. Hence, we search for indirect CPV assuming no direct CPV, namely, $\mathcal{A}(m_+^2, m_-^2) = \bar{\mathcal{A}}(m_+^2, m_-^2)$. The best fit model for $\mathcal{A}(m_+^2, m_-^2)$ is found to be a sum of twelve Breit-Wigner for P- and D-wave resonances, K-matrix [10] and LASS [11] models for $\pi\pi$ and $K\pi$ S-wave states, respectively, without non-resonant decay. Figure 3 shows the proper-time distribution superimposing the fit under both direct and indirect CP conservation and the fit results are $x = (0.56 \pm 0.19^{+0.03+0.06}_{-0.09-0.09})\%$ and $y = (0.30 \pm 0.15^{+0.04+0.03}_{-0.05-0.06})\%$. With allowing indirect CPV, the fit returns $x = (0.56 \pm 0.19^{+0.04+0.06}_{-0.08-0.08})\%$, $y = (0.30 \pm 0.15^{+0.04+0.03}_{-0.05-0.07})\%$, $|q/p| = (0.90^{+0.16+0.05+0.06}_{-0.15-0.04-0.05})$, and $\arg(q/p) = (-6 \pm 11 \pm 3^{+3}_{-4})^\circ$. Our measurements of indirect CPV parameters are the most accurate to date, but consistent with no CPV. Figure 4 show the two-dimensional (x, y) confidence-level (C.L.) contours for the CP-conserved and CPV-allowed fits, where the no-mixing point $(x = 0, y = 0)$ is excluded with 2.5 standard deviations.

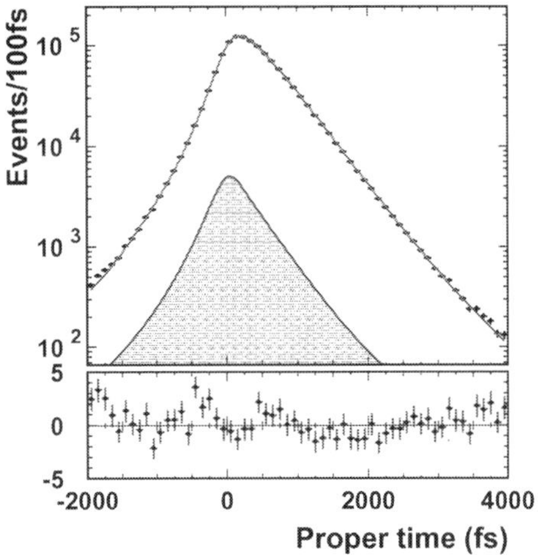

Figure 3. The proper-time distribution for events in the signal region (points) and fit projection for the CP conserved fit (curve). The shaded region shows the combinatorial components. The residuals are shown below the plot.

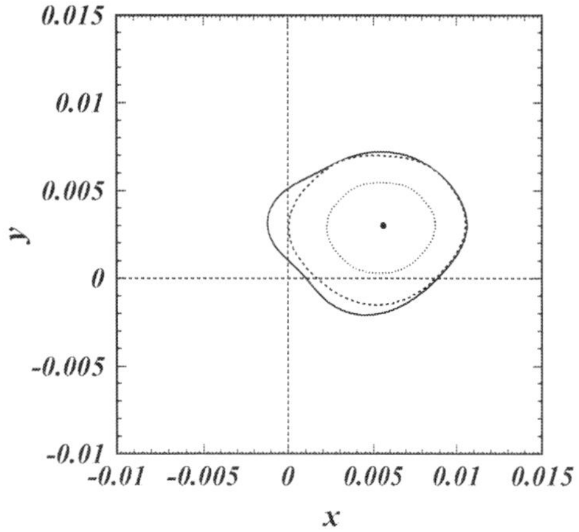

Figure 4. Central value (point) and C.L. contours for (x, y) : dotted (dashed) corresponds to 68.3% (95%) C.L. contour for CP-conserved Dalitz fit, and solid corresponds to 95% C.L. contour for CPV-allowed fit with statistical, experimental and model uncertainties included.

3. Time-integrated CPV in $D^0 \to \pi^0\pi^0$ and $D^0 \to K_S^0\pi^0$ decays

We present the time-integrated CPV asymmetry in $D^0 \to f$, $A_{CP}^{D^0 \to f}$, where f is the final state $\pi^0\pi^0$ or $K_S^0\pi^0$ [12]. It is measured through the reconstruction asymmetry

$$A_{\rm rec} = \frac{N_{\rm rec}^{D^{*+}\to D^0\pi_s^+} - N_{\rm rec}^{D^{*-}\to \bar{D}^0\pi_s^-}}{N_{\rm rec}^{D^{*+}\to D^0\pi_s^+} + N_{\rm rec}^{D^{*-}\to \bar{D}^0\pi_s^-}} \approx A_{CP}^{D^0\to f} + A_{FB}^{D^{*+}} + A_\epsilon^{\pi_s^+} (+A_\epsilon^{\bar{K}^0}), \qquad (3)$$

where $N_{\rm rec}$ is the number of reconstructed signal events, A_{FB} is the forward-backward asymmetry, $A_\epsilon^{\pi_s^+}$ is the detection asymmetry between positively and negatively charged soft pions, and $A_\epsilon^{\bar{K}^0}$ is the asymmetry due to different nuclear interactions between $\bar{K}^0$ and K^0 [13], thus included to the decay $D^0 \to K_S^0\pi^0$. Once we remove $A_\epsilon^{\pi_s^+}$ with the correction in Ref. [14]

and $A_\epsilon^{\bar{K}^0}$ with the value in Ref. [15], then $A_{\rm rec}$ becomes to have $A_{CP}^{D^0\to f}$ and $A_{FB}^{D^{*+}}$ only, where the latter is an odd function of the cosine of the polar angle of the D^{*+} momentum in the center-of-mass system ($\cos\theta^*_{D^{*+}}$). Using the antisymmetry of $A_{FB}^{D^{*+}}$ in $\cos\theta^*_{D^{*+}}$, we obtain A_{CP} as a function of $|\cos\theta^*_{D^{*+}}|$ as shown in Figure 5. The central A_{CP} values obtained from a least-square minimization are $A_{CP}^{D^0\to\pi^0\pi^0} = (-0.03\pm0.64\pm0.10)\%$ and $A_{CP}^{D^0\to K_S^0\pi^0} = (-0.21\pm0.16\pm0.07)\%$ which are the most precise measurements to date revealing no CPV, thus consistent with the SM.

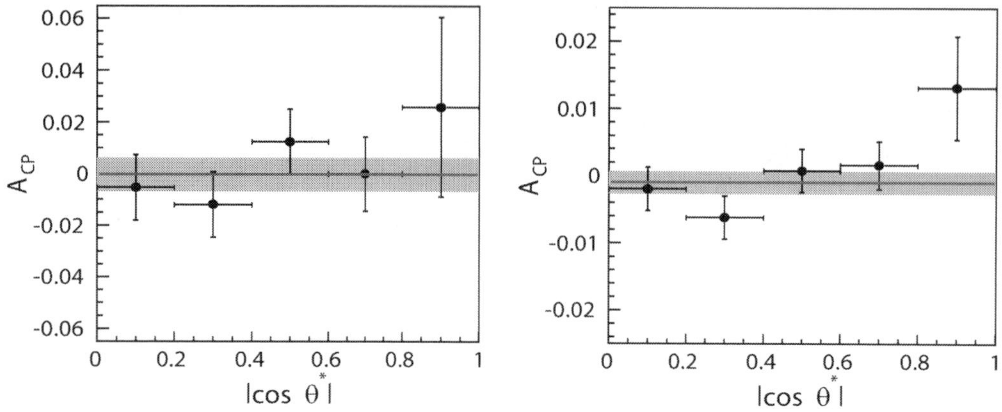

Figure 5. The A_{CP} values in the decays $D^0 \to \pi^0\pi^0$ (left) and $D^0 \to K_S^0\pi^0$ (right) as a function of $|\cos\theta^*|$. The solid lines and shaded regions represent the central value and 1σ interval of the A_{CP}.

Acknowledgments

B. R. Ko acknowledges support by NRF Grant No. NRF-2014R1A2A2A01005286.

References

[1] A. A. Petrov, Int. J. Mod. Phys. A **21**, 5686 (2006).
[2] A. F. Falk, Y. Grossman, Z. Ligeti, and A. A. Petrov, Phys. Rev. D **65**, 054034 (2002).
[3] E. Golowich, J. Hewett, S. Pakvasa, and A. A. Petrov, Phys. Rev. D **79**, 114030 (2009).
[4] Y. Grossman, A. L. Kagan, and Y. Nir, Phys. Rev. D **75**, 036008 (2007).
[5] A. Abashian *et al.* (Belle Collaboration), Nucl. Instrum. Meth. Phys. Res., Sect. A **479**, 117 (2002); also see the detector section in J. Brodzicka *et al.*, Prog. Theor. Exp. Phys. **2012**, 04D001 (2012).
[6] S. Kurokawa and E. Kikutani, Nucl. Instrum. Meth. Phys. Res., Sect. A **499**, 1 (2003), and other papers included in this volume; T. Abe *et al.*, Prog. Theor. Exp. Phys. **2013**, 03A001 (2013) and following articles up to 03A011.
[7] B. R. Ko *et al.* (Belle Collaboration), Phys. Rev. Lett. **112**, 111801 (2014).
[8] T. Peng *et al.* (Belle Collaboration), Phys. Rev. D **89**, 091103(R) (2014).
[9] D. M. Asner *et al.* (CLEO Collaboration), Phys. Rev. D **72**, 012001 (2005).
[10] V. V. Anisovich and A. V. Sarantsev, Eur. Phys. J. A **16**, 229 (2003).
[11] D. Aston *et al.* (LASS), Nucl. Phys. B **296**, 493 (1988).
[12] N. K. Nisar *et al.* (Belle Collaboration), Phys. Rev. Lett. **112**, 211601 (2014).
[13] B. R. Ko, E. Won, B. Golob, and P. Pakhlov, Phys. Rev. D **84**, 111501 (2011).
[14] B. R. Ko *et al.* (Belle Collaboration), Phys. Rev. Lett. **106**, 211801 (2011).
[15] B. R. Ko *et al.* (Belle Collaboration), Phys. Rev. Lett. **109**, 021601 (2012), [Erratum-ibid. **109**, 119903 (2012)].

Review of Neutrino Mixing and Results from Daya Bay

L Whitehead
University of Houston, Houston, TX USA
For the Daya Bay Collaboration
E-mail: lawhitehead@uh.edu

Abstract. Experimental observations have established that neutrinos undergo flavor oscillations as they propagate due to quantum mechanical mixing between the mass states and flavor states. The Daya Bay reactor neutrino experiment has observed the disappearance of electron-type antineutrinos from nuclear reactor cores at the Daya Bay nuclear power complex located in China. This observation allowed Daya Bay to make a measurement of the last neutrino mixing angle, which was previously only known to be small in comparison to the other neutrino mixing angles. An overview of the current status of neutrino oscillation measurements will be presented, followed by the most recent results from Daya Bay and prospects for JUNO, a future neutrino experiment in China.

1. Neutrino Oscillations

Neutrino mixing between the three flavor states (ν_e, ν_μ, ν_τ) and the mass states (ν_1, ν_2, ν_3) can be described by three mixing angles (θ_{12}, θ_{13}, θ_{23}) and one phase parameter, δ. Due to this mixing, neutrinos undergo flavor oscillations in which a neutrino created in one flavor state can be observed at a later time in a different flavor state. The probability for flavor oscillations depends on the mixing parameters above and the differences in the squared masses of the neutrinos, Δm^2_{21} and Δm^2_{31}, where $\Delta m^2_{ij} = m^2_i - m^2_j$ and $\Delta m^2_{31} = \Delta m^2_{32} + \Delta m^2_{21}$. All three mixing angles have been measured, as well as the magnitudes of both the mass splittings. The measured values of these parameters are summarized in Table 1. Both the phase parameter δ and the mass hierarchy (the sign of the larger mass splitting Δm^2_{31}) are unknown at this time. The octant of θ_{23} is also an open question, because the current measured values allow for both $\sin^2\theta_{23} < 0.5$ and $\sin^2\theta_{23} > 0.5$ within the experimental uncertainties.

Prior to 2012, neutrino experiments had only been able to set limits on the mixing angle θ_{13}. Long-baseline accelerator neutrino experiments can measure θ_{13} by observing muon neutrino to electron neutrino transitions. The T2K experiment has recently confirmed electron neutrino appearance in a muon neutrino beam with a significance of 7.3σ [1]. However, the muon neutrino to electron neutrino transition probability also depends on δ and the sign of Δm^2_{31}, both of which are unknown. Another way to measure θ_{13} is to search for the disappearance of electron antineutrinos from a reactor. The electron antineutrino survival probability in this channel is

$$P_{ee} = 1 - \cos^4\theta_{13}\sin^2(2\theta_{12})\sin^2\Delta_{21} - \sin^2(2\theta_{13})(\cos^2\theta_{12}\sin^2\Delta_{31} + \sin^2\theta_{12}\sin^2\Delta_{32}), \quad (1)$$

where $\Delta_{ij} = 1.267\Delta m^2_{ij}\frac{L(\mathrm{m})}{E(\mathrm{MeV})}$. At short baselines of around 1 km, the $\sin^2(2\theta_{13})$ term is dominant. Due to the ability to accumulate large statistics in a relatively short amount of time and the lack of dependence on δ or the mass hierarchy in this channel, a short-baseline reactor experiment is ideal for measuring θ_{13}.

Table 1. Summary of measured values

Measured Value	Description
$\tan^2\theta_{12} = 0.44 \pm 7\%$	Combined solar and KamLAND[2]
$\sin^2(2\theta_{13}) = 0.084 \pm 6\%$	Daya Bay
$\sin^2\theta_{23} = 0.51 \pm 11\%$	T2K[3]
$\Delta m^2_{21} = 7.5 \times 10^{-5}\ \mathrm{eV}^2 \pm 3\%$	Combined solar and KamLAND[2]
$\lvert\Delta m^2_{31}\rvert \approx \lvert\Delta m^2_{32}\rvert = 2.4 \times 10^{-3}\ \mathrm{eV}^2 \pm 4\%$	MINOS/MINOS+[4], T2K[3], Daya Bay

2. The Daya Bay Experiment

Daya Bay measures the antineutrino spectrum from six reactor cores located at the Daya Bay nuclear power complex in China. The Daya Bay detectors are located underground in experimental halls (EHs) built beneath mountains that provide shielding from cosmic ray background. There are two near sites (EH1 and EH2) close to the reactor cores and one far site (EH3) at a distance of $\sim$ 2 km from the cores. Multiple functionally-identical antineutrino detectors (ADs) are placed at each site. The relative near-far measurement between nearly identical ADs allows for the cancellation of detector-related systematic uncertainties. Figure 1 shows a cross-sectional diagram of an AD. The ADs use $\sim$20 tons of gadolinium-loaded liquid scintillator as a target. Surrounding the target region is $\sim$20 tons of undoped liquid scintillator to detect gammas that escape the target volume. The outer region of each AD contains $\sim$40 tons of mineral oil to shield the inner volumes. The three regions are separated by acrylic vessels. The ADs are instrumented with 8-inch photomultiplier tubes (PMTs) mounted on the wall of the containment vessel in the mineral oil region. Reflective panels on the top and bottom increase the light collection and improve the uniformity of the energy response. Three automated calibration units (ACUs) are located on top of the containment vessel. Each ACU contains an LED and radioactive sources that can be lowered into the detector. Antineutrinos are detected via the inverse beta decay (IBD) interaction, $\overline{\nu}_e p \to e^+ n$. The signal is a coincidence between the prompt energy due to the positron and the delayed energy ($\sim$8 MeV) due to the neutron capture on gadolinium. The neutron capture time on gadolinium is approximately 30 μs. A muon system is employed to shield against natural radioactivity and fast neutrons and to tag cosmic muons that produce background in the ADs. The muon veto system consists of an active water shield and resistive plate chambers (RPCs). The ADs are placed inside the water pool such that there is at least 2.5 m of water shielding in every direction. The water pool is instrumented with 8-inch PMTs to detect the Cherenkov light emitted by muons passing through the water. The pool is divided into inner and outer regions separated by reflective Tyvek sheets. The RPC modules are installed on top of the pool.

Two ADs were installed in EH1 and took data for 90 days beginning in August 2011. This data set was used for a side-by-side comparison of ADs that established the residual uncorrelated detector uncertainty [5]. During this period, one AD was installed in EH2 and two ADs were installed in EH3. Beginning in December 2011, data was taken in this 6-AD configuration for a total of 217 days. The first observation of electron antineutrino disappearance at Daya

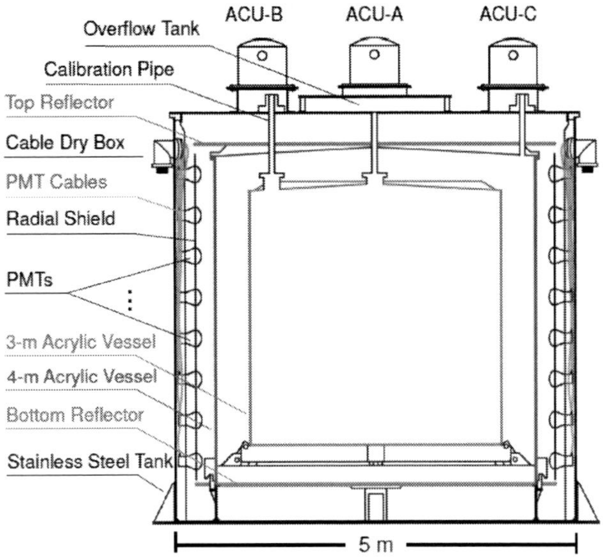

Figure 1. Diagram of a Daya Bay antineutrino detector.

Bay was based on 55 days of data and measured a non-zero value of θ_{13} with a significance of 5.2σ [6]. An updated measurement was published based on 139 days of data [7]. The first spectral measurement by Daya Bay was based on the entire 217 days of data in the 6-AD configuration [8]. In the summer of 2012, the two final ADs were installed and a series of special calibrations were performed. Daya Bay has been taking data with all 8 ADs (two at each near site and four at the far site) since October 2012. The results presented in the next section are based on 621 days of data-taking, including the entire 6-AD period.

3. Recent Results from Daya Bay

IBD interactions are selected by first rejecting events caused by PMT light emission. Candidate events are selected by requiring the coincidence of a prompt signal with energy between 0.7 and 12 MeV and a delayed signal with energy between 6 and 12 MeV within a time window of 1-200 μs. Candidate coincidences occurring within 0.6 ms of a water pool muon, 1 ms of an AD muon, and 1 s of an AD showering muon are rejected from the sample. A multiplicity cut is performed to ensure that only isolated candidate pairs are selected. In total, Daya Bay has observed more than one million IBD interactions.

The top panel of Figure 2 shows the measured spectrum at the far site compared to the expected spectrum based on the near site data both without oscillations and with the best fit oscillations included. The bottom panel shows the ratio of the measured to expected spectrum. The observed relative rate deficit and relative spectral distortion are highly consistent with the oscillation interpretation.

Because oscillations of reactor antineutrinos at a baseline near 1 km are mostly due to the Δ_{3i} terms in Equation 1 and $\Delta m^2_{21} << |\Delta m^2_{31}| \approx |\Delta m^2_{32}|$, we define the effective mass-squared difference $\Delta_{ee} = \cos^2\theta_{12}\sin^2\Delta_{31} + \sin^2\theta_{12}\sin^2\Delta_{32}$ for the analysis. Figure 3 shows the 68.3%, 95.5%, and 99.7% confidence level allowed regions for $\sin^2(2\theta_{13})$ and $|\Delta m^2_{ee}|$ based on the comparison of the rates and and prompt energy spectra measured by the near site and far

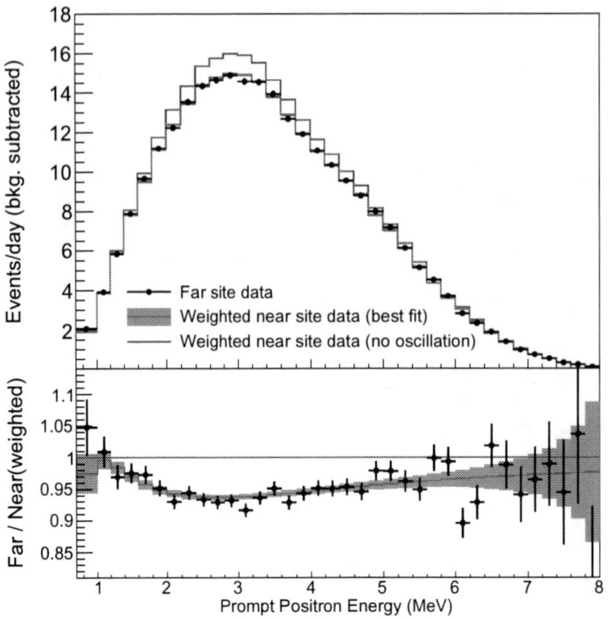

Figure 2. The top panel shows the measured spectrum at the far site compared to the expected spectrum based on the near site data both without oscillations and with the best fit oscillations included. The bottom panel shows the ratio of the measured to expected spectrum. The uncertainties are statistical only.

site ADs. The best fit values of $\sin^2(2\theta_{13}) = 0.084 \pm 0.005$ and $|\Delta m^2_{ee}| = 2.44^{+0.10}_{-0.11} \times 10^{-3}$ eV^2 are also indicated in the figure.

Daya Bay has also recently announced an independent measurement of $\sin^2(2\theta_{13}) = 0.083 \pm 0.018$ by performing a rate-only analysis using neutron captures on hydrogen (nH) during the 217 days of the 6-AD data period. Combining this result with the result based on neutron captures on gadolinium (nGd) yields improved precision on $\sin^2(2\theta_{13})$. Because the nH sample has different systematic uncertainties than the nGd sample, results derived from the nH sample will be important in the future when the systematic uncertainty in the nGd sample becomes the dominant uncertainty in $\sin^2(2\theta_{13})$ [9].

A search for a light sterile neutrino has been performed using the 217 days of data from the 6-AD period. A minimal extension of the standard model, the 3 (active) + 1 (sterile) neutrino mixing model, was used in the analysis. The data are consistent with the standard neutrino model and place stringent limits on $\sin^2(2\theta_{14})$ in the region of 10^{-3} eV^2 $< \Delta m^2_{41} <$ 0.1 eV^2, a region which was largely unexplored. The precision of this result is dominated by statistics, and the sensitivity to $\sin^2(2\theta_{14})$ is expected to approximately double with three additional years data [10].

4. JUNO

Because the reactor antineutrino disappearance probability depends on both Δm^2_{31} and Δm^2_{32} as shown in Equation 1, a Fourier transform of the L/E spectrum of reactor antineutrinos could reveal the mass hierarchy [11]. JUNO is a planned next-generation neutrino experiment that will seek to determine the mass hierarchy by a precision measurement of the reactor antineutrino

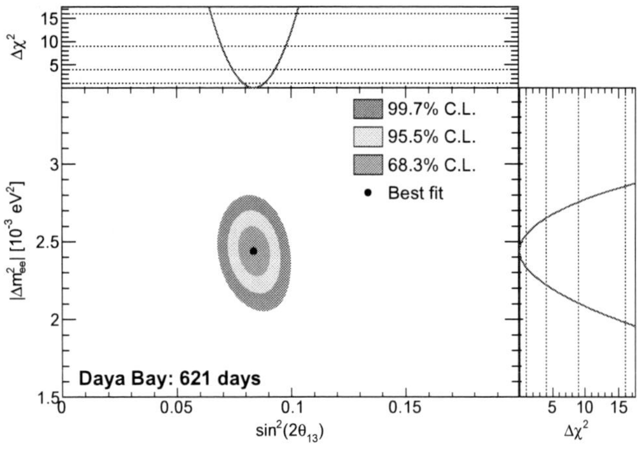

Figure 3. The 68.3%, 95.5%, and 99.7% confidence level allowed regions for $\sin^2(2\theta_{13})$ and $|\Delta m^2_{ee}|$ based on the comparison of the rates and and prompt energy spectra measured by the near site and far site ADs. The best fit values are indicated by the point. Also shown is the dependence of $\Delta\chi^2$ on each of the parameters.

spectrum at a baseline of 60 km. Unlike long-baseline accelerator neutrino experiments, JUNO's approach has the advantage of not depending on δ or θ_{23}. However, very good energy resolution is necessary to achieve the mass hierarchy determination. Figure 4 shows a sketch of the JUNO detector, a 20-kt liquid scintillator detector with nearly 80% photocathode coverage to achieve an energy resolution of $3\%/\sqrt{E}$. With six years of data, JUNO's sensitivity to the mass hierarchy is expected to be at least 3σ. JUNO will have a rich physics program in addition to the mass hierarchy measurement, including precision oscillation parameter measurements, supernova neutrinos, geoneutrinos, solar neutrinos, sterile neutrinos, atmospheric neutrinos, or other exotic searches. Civil construction is expected to begin in 2014, with data-taking scheduled for 2020 [12].

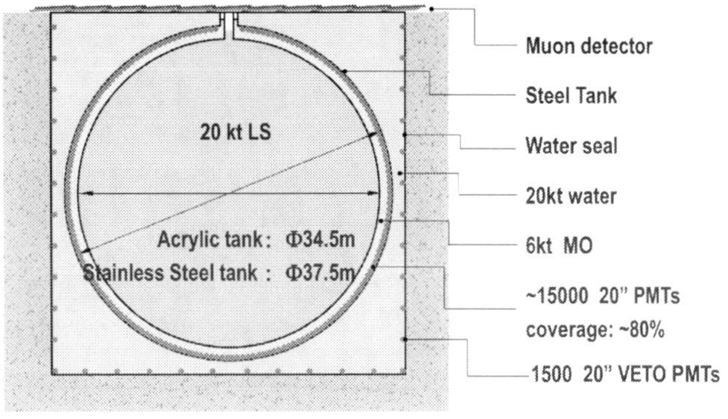

Figure 4. Planned design of the JUNO detector

5. Summary

In summary, Daya Bay has measured $\sin^2(2\theta_{13}) = 0.084 \pm 0.005$ and $|\Delta m^2_{ee}| = 2.44^{+0.10}_{-0.11} \times 10^{-3}$ eV^2 with 621 days of data. This is the most precise measurement of $\sin^2(2\theta_{13})$ to date, and the most precise measurement of $|\Delta m^2|$ in this channel, comparable in precision and consistent with the measurements from muon neutrino disappearance experiments. The precision of both parameters is expected to reach 3% with approximately three more years of data.

6. References

[1] K. Abe *et al.* [T2K Collaboration], *Phys. Rev. Lett.* **112**, 061802 (2014) [arXiv:1311.4750 [hep-ex]].
[2] B. Aharmim *et al.* [SNO Collaboration], *Phys. Rev.* C **88**, 025501 (2013) [arXiv:1109.0763 [nucl-ex]].
[3] K. Abe *et al.* [T2K Collaboration], *Phys. Rev. Lett.* **112**, 181801 (2014) [arXiv:1403.1532 [hep-ex]].
[4] A. Sousa, presented at the XXVI International Conference on Neutrino Physics and Astrophysics, Boston, MA USA, 2014.
[5] F. P. An *et al.* [Daya Bay Collaboration], *Nucl. Instrum. Meth.* A **685**, 78 (2012) [arXiv:1202.6181 [physics.ins-det]].
[6] F. P. An *et al.* [DAYA-BAY Collaboration], *Phys. Rev. Lett.* **108**, 171803 (2012) [arXiv:1203.1669 [hep-ex]].
[7] F. P. An *et al.* [Daya Bay Collaboration], *Chin. Phys.* C **37**, 011001 (2013) [arXiv:1210.6327 [hep-ex]].
[8] F. P. An *et al.* [Daya Bay Collaboration], *Phys. Rev. Lett.* **112**, 061801 (2014) [arXiv:1310.6732 [hep-ex]].
[9] F. P. An *et al.* [Daya Bay Collaboration], *Phys. Rev.* D **90**, 071101 (2014) [arXiv:1406.6468 [hep-ex]].
[10] F. P. An *et al.* [DAYA-BAY Collaboration], *Phys. Rev. Lett.* **113**, 141802 (2014) [arXiv:1407.7259 [hep-ex]].
[11] L. Zhan, Y. Wang, J. Cao and L. Wen, *Phys. Rev.* D **78**, 111103 (2008) [arXiv:0807.3203 [hep-ex]].
[12] L. Wen, presented at the XXVI International Conference on Neutrino Physics and Astrophysics, Boston, MA USA, 2014.

CP Violation in the Neutrino Sector

Sandip Pakvasa
University of Hawaii, Honolulu, Hawaii 96822

Abstract. I will summarise the current status of CP violation in the neutrino sector, with a somewhat historical perspective.

1. Some Prehistory:

As everyone knows, the need for a particle like the neutrino was suggested by Pauli [1] in his famous letter in 1930 to account for the continuous energy spectrum of the electrons emitted in nuclear beta-decay. To be fair, a lot of the credit goes to the experimenters, especially, Ellis and Wooster [2] who labored prodigiously during the period between 1921 and 1929 to establish conclusively that the beta decay spectrum was indeed continuous and answered all the criticisms of Meitner. Without their work, Pauli would not have been driven to " the desperate remedy". The next big step was the 1933 paper by Fermi [3] which laid out a detailed theory of beta decay incorporating Pauli's neutrino and containing the first fresh application of the quantum field theory (developed by Dirac, Heisenberg and Pauli as recently as 1929) to a new phenomenon. Amazingly, what Fermi wrote down is still almost valid, except needing a factor of $(1 + \gamma_5)/2$ to allow for maximal parity violation! In 1935, Yukawa [4] introduced a meson field with a mass scale of about 100 MeV and coupling strongly to nucleons to account for the strength and range of the nuclear force. Anderson and Neddermeyer[5],while trying to understand the "hard" component in cosmic rays found that it was due to charged particles of intermediate mass. Street and Stevenson[6] independently reached the same conclusion and so did the Nishina group[7] in Tokyo. Its mass seemed to be approximately 100 MeV and so in the range expected of the Yukawa's meson. But this particle did not behave as a Yukawa meson should! It had the wrong coupling strength to nucleons and incorrect lifetime. The final clincher was the fact that negative mesons were not absorbed by nuclei at the expected rate(Conversi et al.[8]). Sakata and Inoue proposed in 1943(not published till 1946 due to wartime problems[9]) to solve this puzzle by suggesting that the pion of Yukawa is indeed produced in cosmic rays but decays into new particles, now called muon and μ-neutrino and also proposed the the correct decay scheme $\pi \rightarrow \mu \rightarrow e$, they also proposed the correct spins spin 1/2 for both particles (in contrast to the Marshak-Bethe[10] proposal). This scheme was completely confirmed in cosmic ray data as observed by Lattes, Occhialini and Powell(1947)[11]. After more data on the muon capture by nucleons and muon decay became available and the similarity of the various coupling strengths became clear, this observation that the couplings in beta decay, muon decay and muon capture were comparable was elevated to the principle of Universality of Weak Interactions. This observation was made by Klein, by Puppi, by Tiomno and Wheeler and by Lee, Rosenbluth and Yang in the period 1948-9[12]. In the period 1947 onwards the field of particle physics exploded with the discovery of "strange" particles (Kaons, Λ hyperons etc....). In 1953, Gell-Mann and Nishijima(with Nakano) provided a simple classification scheme [13] with the introduction of strangeness quantum number. In 1954-5 Gell-Mann and Pais[14] analysed the behaviour of $K^0 - \bar{K}^0$ oscillations and the mass-lifetime eigenstates K_1 and K_2. The decays of the kaons became very puzzling when the analysis by Dalitz(1953)[15] was applied to the three pion decay mode of the K^0 particle. It seemed to imply that either parity is not conserved or there is something very strange going on. In 1956 Lee and Yang[16] presented the analysis of several possible experiments to test parity conservation in weak interactions. Very shortly, the experimental results of the Co-60 experiment of Ambler-Wu et al[17], and the $\pi - \mu$ decay experiment[18] of Lederman et al and Friedman-Telegdi appeared proving conclusively maximal parity violation in weak interactions. Within a few months(1957), the Universal V-A interaction was proposed by Sudarshan and Marshak[19] and by Feynman and Gell-Mann[20]:

$$H_W = J^+ J \tag{1}$$

$$\text{with } J(V - A) = \bar{p}n + \bar{e}\nu \ + \bar{\mu}\nu + \bar{p}\Lambda \tag{2}$$

This was very successful except for a small problem. According to the data on the beta decay of Λ hyperons, the coupling strength for the strangeness changing current was much smaller than required by universality. In 1960, Gell-Mann and Levy[21] proposed a modified version of "Universality" to account for this. They suggested that the form of the hadronic current be modified as:

$$\begin{aligned} J &= (\bar{p}n \ + \epsilon\bar{p}\Lambda) \, / \left(1 + \epsilon^2\right)^{1/2} \\ &= (\cos\theta\bar{p}n + \sin\theta p\bar{\Lambda}) \end{aligned} \tag{3}$$

2. History

In 1962, an important step was taken by Maki, Nakagawa and Sakata[22] in which they proposed that there is similar mixing in the leptonic sector, and further that the Neutrino flavor states are mixtures of mass eigenstates:

$$\begin{aligned} \nu_e &= \cos\phi\nu_1 \ + \sin\phi\nu_2 \\ \nu_\mu &= \text{-} \sin\phi\nu_1 \ + \cos\phi\nu_2 \end{aligned} \tag{4}$$

They also discussed neutrino mixing and oscillations, estimating the oscillation time in terms of δm^2 and L/E. While the general notion of neutrino oscillations had been raised earlier(1957) by Pontecorvo[23], this was the first time that neutrino flavor oscillations were discussed. A similar analysis was carried out by Gribov and Pontecorvo in 1969[24]. In 1963, Cabibbo[25] analysed with great success semileptonic decays of all hyperons, using the mixing angle θ, and with the baryons as well as the weak current transforming as 8 under Gell-Mann's flavor SU(3)[26]. 1964 marked the discovery of CP violation in pionic decays of K_L by Christenson, Cronin, Fitch and Turlay[27]. In 1973 there appeared the seminal paper of Kobayashi and Maskawa[28]. They observed that with two families of quarks and leptons it is not possible to have CP non-conservation in the unified gauge theory of electroweak interactions of Glashow, Salam and Weinberg[29]; and proposed several ways to make it possible. One possibility was to add a third family of quarks (t,b). In 1975, this possibility was analysed and shown to be viable and to be able to account for the observed CP violation in Kaon decays by several groups by Pakvasa and Sugawara[30]; by Maiani[31]; and by Ellis, Gaillard and Nanopoulos[32]. In 1975, the tau lepton was discovered by Perl et al at SPEAR[33]. Its decays showed that it was accompanied by its own neutrino. In 1977, there appeared two papers with analyses of a weak current with three lepton families, including a mixing between three neutrinos with a complex (hence CP non-conserving) unitary 3X3 matrix: Lee, Pakvasa, Shrock and Sugawara; Fritzsch[34]. The 3X3 unitary matrix U is:

$$U = \begin{pmatrix} U_{e1} & U_{e2} & U_{e3} \\ U_{\mu 1} & U_{\mu 2} & U_{\mu 3} \\ U_{\tau 1} & U_{\tau 2} & U_{\tau 3} \end{pmatrix} \cdot V_M \tag{5}$$

where the matrix V_M is given by

$$V_M = \begin{pmatrix} 1 & 0 & 0 \\ 0 & e^{i\alpha 1/2} & 0 \\ 0 & 0 & e^{i\alpha 2/2} \end{pmatrix} \tag{6}$$

and defines the Majorana phases[35] α_i which are not observable in oscillations. With the PDG[36] parametrisation of the unitary matrix U, the Jarlskog invariant[37] for the neutrino sector which is a measure of CP Non-conservation is given by

$$\begin{aligned} J &= J_{Max} \sin\delta, \text{with} \\ J_{max} &= c_{12} s12 c_{23} s_{23} c_{13}^2 s_{13}^2 \end{aligned} \tag{7}$$

and is approximately equal to $3.10^{-2}\sin(\delta)$ [38] to be compared to the corresponding J in the quark sector: $J_{CKM} \sim 3.10^{-5}$[36].

There are very simple tests of CP conservation that can be easily stated[39]:

$$P_{\alpha_\beta}(t) \neq P_{\alpha_\beta}(t) \quad \text{etc.} \tag{8}$$

More quantatively, it can be shown that[40]:

$$\Delta_{\mu e} = P_{\beta\alpha}(L/E) - P_{\bar{\beta}\bar{\alpha}}(L/E) \tag{9}$$

$$= -16 J_{\mu e} \sin(\Delta_{12}) \sin(\Delta_{23}) \sin(\Delta_{31}) \tag{10}$$

where

$$\Delta_{ij} = \delta m_{ij}^2 \; L/4E \tag{11}$$

$$J_{\mu e} = Im(U_{\mu i} U_{\mu j}^* U_{ei}^* U_{ej}) \tag{12}$$

$$= c_{12} s_{12} c_{23} s_{23} c_{13}^2 s_{13}^2 \tag{13}$$

One can also look for CP violating effects even without access to antineutrino beams. For example, flavor conversion probability over a long time (or large L/E) may contain terms both odd and even in t, to wit

$$P = A + B\cos(at) + C\sin(at) \tag{14}$$

and if the coefficient C is not zero, then CP is violated[38].

The current knowledge of the neutrino mass/mixing matrix can be summarised (roughly) as follows:

$$\begin{array}{ll} \delta m_{21}^2 & \sim 7.5.10^{-5} \; eV^2 \\ \delta m_{31}^2 & \sim 2.4.10^{-3} \; eV^2 \\ s_{12}^2 & \sim 0.3 \\ s_{23}^2 & \sim 0.42 \\ s_{13}^2 & \sim 0.22 \end{array} \tag{15}$$

In addition the current fits show a slight preference for $\cos(\delta) < 0$[38]. The mass hierarchy is unknown, leaving a twofold ambiguity. The normal hierarchy(NH) refers to $m_3 > m_2 > m_1$, and the inverted hierarchy(IH) refers to $m_2 > m_1 > m_3$. That $m_2 > m_1$ is deduced from the fit of the solar neutrino data to the matter effects in the sun.

Everything so far has been in vacuum. However, we know that most neutrino beams in real life travel thru matter for at least some fraction(sometimes quite large) of their path. Since matter effects are not CP invariant, this induces additional "fake" CP violating effects into the oscillation probabilities. For example, the expression for $\Delta\mu e$ now becomes[41]:

$$\Delta_{\mu e}(m) = 2xyg\left[-\sin\delta\sin\Delta(f+\bar{f}) + \cos\delta\cos\Delta(f-\bar{f})\right] \tag{16}$$

where $\Delta = \Delta_{31}$; and x, y, g and $f, \bar{f}$ are known functions of the parameters and the electron density in the matter.

3. Now

This means that to extract information about the CP phase δ and to detect "true" CP violation in neutrino oscillation one has to extract them using this expression and study the sensitivity with respect to δ of various proposed experimental configurations. It is crucial to have knowledge of the mass hierarchy to do this successfully. A number of proposed future facilities will be attempting to do this. The proposed facilities at LBNE, LBNO, T2HK and also ESSnuSB are some examples. The aim is to determine that the phase δ is not 0 or π and find its value to at least 3 σ. It appears that at LBNE, with a near detector, and a 35 kT LAr far detector, a dedicated run for 4-5 years may be able to yield a measurement to confirm a non-zero δ over a large range and thus confirm CP non-conservation[42].

As mentioned earlier, these experiments based on neutrino oscillations can only measure the phase δ but not the Majorana phases, should the neutrino be a Majorana particle. Can those phases be observed and measured? Now if neutrinoless double beta decay is observed, and neutrino mass is directly measured in tritium beta decay, then one can combine these measurements

$$\begin{aligned} M_{ee} &= \left|| U_{e1} \right|^2 m_1 + | U_{e2} |^2 m_2 e^{i\alpha_1} + | U_{e3} |^2 m_3 e^{i\alpha_2} | \\ m_e &= \sum_i | V_{ei} |^2 m_i \end{aligned} \tag{17}$$

In the reasonable approximation that $\delta m_{12}^2 << \delta m_{13}^2$, one can show that only one Majorana phase is left, and can be written as:

$$\cos\alpha_1 \cong 1 - \frac{2}{s_{12}^2}\left(1 - M_{ee}^2/m_e^2\right) \tag{18}$$

This shows that, in principle, it is possible to extract a value for this Majorana phase[43]. However, the uncertainties in the nuclear matrix elements of the neutrinoless double beta decay render this impossible[43]. Some day that these uncertainties will be under better control to make this possible.

My final topic is that of CP violation and the baryon asymmetry in the universe (BAU). The evidence for this is very obvious in the world around us and out to distant galaxies and is quantified as:

$$\eta_B = (n_B - n_{\bar{B}})/n_\gamma \sim 6.10^{-10} \tag{19}$$

We have known since Sakharov's ground breaking paper in 1967[44] that to account for this asymetry there are three essential ingredients needed:

(i) Baryon number violation
(ii) CP violation and C violation
(iii) Departure from thermal equilibrium.

Now, we already have C and CP violation in weak interactions and the departure from thermal equilibrium is supplied by the expansion of the universe. In the attractive scenarios using leptogenesis[45], the B violation is to be supplied by the sphalerons of the Standard Model which act at high temperature and conserve B-L but not B and L separately. The basic idea is that the Right Handed neutrinos needed in the see-saw mechanism violate Lepton number and CP in their decays to lighter particles, and generate a lepton asymmetry thereby, which also carries CP violation. After the lepton asymmetry and a net lepton number L has been created at some high scale, as the universe cools and the temperature drops to the electroweak scale, the sphaleron converts the lepton asymmetry into a baryon asymmetry[46]. As B-L is conserved, a net non-zero B is left. In general, the CP violation that contributes to create this asymmetry is in the heavy neutral lepton sector at the high scale. But in some specific flavor models, it may be actually related to the phase δ in the PMNS (Pontecorvo, Maki, Nakagawa, Sakata) matrix[47]. In one example, provided θ_{13} is large enough, and the hierarchy is Normal, a reasonable value of δ can give rise to the observed BAU[48]. But testing these ideas is a long shot. I remark in passing that predicting the phase δ, or for that matter, any of the mass mixing parameters is very far from realisation,to put it mildly.

4. Conclusions/Outlook:

There are some hints from the current best fits that δ is not equal to zero. Future Long Baseline experiments will eventually be able to detect CP violation in neutrino oscillations, and measure a non-zero value for δ to a reasonable degree of accuracy. Ongoing searches for neutrinoless double beta decay may find a non-zero rate, and if the error bars on the nuclear matrix elements are reduced sufficiently, we may be able to measure one of the Majorana phases. To test the connection between the baryon asymmetry and the measured value of δ is a rather long shot. In any case, it should be obvious that the whole question of observable CP violation in neutrino sector is a long term program.

5. Acknowledgments:

I thank the organisers, and especially Nick Solomey for encouragement. I thank Danny Marfatia in help with preparing this talk, and with the selection of the material. This work was supported in part by the U.S. D.O.E. under grant #DE-FG02-04ER-41291 and in part by the Alexander von Humboldt Foundation. I also thank Professor Jose Valle for hospitality during my stay in Valencia where some of this work was done.

6. References

[1] Paul W *Letter to a Physicists' gathering at Tubingem* Dec. 4, 1930. Repr in Pauli W. collected scientific papers, Ed. Kroving R and Woisskopf V 1969 Vol. **2** p. 1313, Interscience, NY.
[2] Ellis C D and Wooster W A 1927 *Proc. RoySoc* A**117** 109.
[3] Fermi E 1933 *La Riciera Scientifica* , 12.
[4] Yukawa H *Proc. Phys. Math. Soc. Japan* 1935 17 **48**.
[5] Anderson C D and Neddlermeyer S H *Phys. Rev.* 1937 **51**, 884.
[6] Street J C and Stevenson E C 1937 *Phys. Rev.* **51**, 1005.
[7] Nishina Y, Takeuchi M and Ichimaya T 1937 *Phys. Rev.* **52**, 1198.
[8] Conversi M, Pancini F and Piccioni O 1945 *Phys. Rev.* **71** 209.
[9] Sakata S and Inoue T 1946 *Prog. Theoret. Phys.*, **1**, 143.
[10] Marshak R and Bethe H 1947 *Phys. Rev.* **72**, 506.
[11] Lattes C M G, Occhialini G P S and Powell C F 1947 *Nature* 160, 453.

[12] Klein O, Nature, 161, 897 (1948) Puppi G Nuov. Cim. 5, 587 (1948); Tiomno J and Wheeler J 1949 *Rev. Mod. Phys.*, **21**, 144; Lee T D, Rosenbluth M N, and Yang C N 1949 *Phys. Rev.*, **75**,905.
[13] M. Gell-Mann M 1952*Nuov. Cim.*, **4**, 52; Nakano T and Nishijima K 1953 *Prog. Theoret. Phys.*, **10**, 581.
[14] Gell-Mann M and Pais A 1955 *Phys. Rev.* **97** 1387.
[15] Dalitz R H *Phil. Mag* 1953 **44** 1068.
[16] Lee T D and Yang C N 1956 *Phys. Rev.* **104**, 254.
[17] Wu C S, Ambler E et al. 1957 *Phys. Rev.* **105**, 1413.
[18] Garwin R, Lederman L and Weinrich L 1957 *Phys. Rev.* **105**, 1415; Friedman J L and Telegdi V L *Phys. Rev.* **105**, 1681.
[19] Sudarshan E C G and Marshak R E 1957 *Proc. Conf. on Mesons and Newly Discovered Particles*, Padua-Venice (Conf. held Sept, 1957), ed. Zonichelli N, p. v-14.; *Phys. Rev.* **109**, 1860.
[20] R P Feynman and Gell-Mann 1957 *Phys. Rev.* **109**, 193.
[21] Gell-Mann M and Levy M 1960 *Nuov. Cim.* **16**, 705.
[22] Maki Z, Nakagawa M and Sakata S 1962 *Prog. Theoret. Phys.* **28**, 870.
[23] Pontecorvo B M, Zh. Eksp. 1957 *Teor.Fiz.* **34**, 247.
[24] Gribov V N and Pontecorvo B M 1963 *Phys. Lett.* B**26**, 443.
[25] Cabibbo N 1963 *Phys. Rev. Lett.* **10**, 531.
[26] Gell-Mann M Caltech Report CTSL-20 (1961), Ne'eman Y 1961 *Nucl. Phys.* B**26**, 222.
[27] Christenson J H et al., 1964 *Phys. Rev. Lett.* **13**, 138.
[28] Kobayashi M and Maskawa T 1973 *Prog. Theoret. Phys.* **49**, 652.
[29] Glashow S L 1980 *Rev. Mod.Phys.* **52**, 539; Salam A 1980 ibid., 52, 525: Weinberg S, ibid., 52, 515.
[30] Pakvasa S and Sugawara H, 1976 *Phys. Rev.* D**14**, 305.
[31] Maiani L 1976 *Phys. Lett.* B**62**, 183.
[32] Ellis J, Gaillard M and Nanopoulos D 1976 *Nucl. Phys. B109* 213.
[33] Perl M L 1975 et al., *Phys. Rev. Lett.* **35**, 22.
[34] Lee B W, Pakvasa S, Shrock R and Sugawara H 1977 *Phys. Rev. Lett.*, **38**, 937; Fritzsch H *Phys. Lett.* B**69**, 451.
[35] Bilenky S M, Hosek J and Petcov S 1980 *Phys. Lett.* **B94**, 495; J. Schechter J and Valle J W F 1981 *Phys. Rev.* D**22**, 2227; M. Doi M, Kotani T, Nishiura H and Takasugi E 1981 *Phys. Lett.* B**102**, 323; J. Bernabeu J and Pascual P 1982 *Nucl. Phys.* B**228**, 21.
[36] Particle Data Group Collaboration (Olive K et al.), Chin J 2014 *Phys.* C**38**, 090001.
[37] Jarlskog C 1985 *Phys. Rev. Lett.* **55**, 1039.
[38] Gonzalez-Garcia M C, Maltoni M and Schwetz T, arXiv:1409.5439.
[39] Pakvasa S, *In High Energy Physics-1980*, AIP Conf. Proc. No. 68, ed. Durand L and Pondrom L G, (AIP, New York, 1981). Cabibbo N 1978 *Phys. Lett.* B**72** 333; Barger V and Whisnant K 1980 *Phys. Rev. Lett.* **45**, 2084.
[40] Barger V et al., Ref. 39.
[41] Freund M 2001 *Phys. Rev.* D**64**, 053003; Barger V, Marfatia D and Whisnant K 2001 *Phys. Rev.* **D65**, 073023.
[42] Barger V et al., arXiv:1405.1054.
[43] Barger V et al. 2002 *Phys. Lett.* B**540**, 247, hep-ph/0205290; for a more optimistic point of view, see Dodelson S and Lykken J, arXiv:1405.6310.
[44] Sakharov A 1967 *JETP Lett.* **5** 24.
[45] Fukugita M and Yanagida T 1986 *Phys. Lett.* B**174**, 45.
[46] Kuzmin V, Rubakov V and Shaposnikov M 1985 *Phys. Lett.* B**155**, 36.
[47] Frampton P et al. 2002 *Phys. Lett.* B**548**, 118.
[48] Pascoli S, Petcov S and Riotto A 2007 *Phys. Rev.* D**75**, 083511, hep-ph/0609125; *Nucl. Phys.* B**774**, 1, hep-ph/9611338.

Journal of Physics: Conference Series **556** (2014) 012061 doi:10.1088/1742-6596/556/1/012061

Latest Results from T2K and Hyper-Kamiokande Perspectives

Francesca Di Lodovico [1]
Queen Mary, University of London
E-mail: f.di.lodovico@qmul.ac.uk

Abstract. We present a review of the latest results on the oscillation mixing parameters obtained by the T2K collaboration and foreseen for the Hyper-Kamiokande experiment by the Hyper-Kamiokande Working Group.

1. Introduction

Neutrino oscillations are governed by the 3×3 Pontecorvo-Maki-Nakagawa-Sakata [1], [2] mixing matrix and parameterized by two mass-squared differences, Δm^2_{21} and Δm^2_{32}; three mixing angles, θ_{12}, θ_{23}, and θ_{13}; and a complex CP-violating phase, δ_{CP}. In recent years, the mass-squared differences and the mixing angles have all been measured to be non-zero [3]. All the parameters except for δ_{CP}, the θ_{23} octant (maximal or non-maximal) and the mass hierarchy have been previously measured as summarized in Reference [3]. The mass hierarchy, MH, is defined normal mass hierarchy (NH) if $\Delta m^2_{31} > 0$ and inverted mass hierarchy (IH) if $\Delta m^2_{31} < 0$.

Both the T2K [4] and Hyper-Kamiokande (Hyper-K) [5] experiments aim to measure θ_{23} and $|\Delta m^2_{31}| \approx |\Delta m^2_{32}|$ precisely via ν_μ disappearance. The T2K and Hyper-K experiments can also measure ν_μ to ν_e appearance that is sensitive to θ_{13} and can explore δ_{CP}, especially through the CP-odd term, in the oscillation probability equation. Hyper-K is specifically designed to measure the CP violation in the neutrino sector, thanks to the increased beam power with respect to T2K and a much larger far detector.

In proceeding, the latest results from T2K and the foreseen sensitivity of Hyper-K for the measurement of the oscillation parameters is presented.

2. The T2K Experiment

An intense and high purity ν_μ beam is produced at J-PARC by colliding a 30 GeV proton beam with a graphite target, then focusing the resulting charged hadrons by magnetic horns prior to decay into neutrinos. The far detector, Super-Kamiokande (SK), is situated 2.5° off-axis from the neutrino beam resulting in a narrow-band energy spectrum peaked at 0.6 GeV, which maximizes the ν_e appearance probability at a baseline of $L = 295$ km and minimizes high energy backgrounds. This baseline corresponds to a matter effect correction of $|x| \approx 5\%$ [6]. The near detector complex, 280 m from the average neutrino production point, consists of an on-axis (INGRID) detector to constrain the beam direction and an off-axis (ND280) detector to constrain the neutrino flux and cross sections.

[1] Partially supported by the grant ERC-207282.

The flux prediction is based on simulations tuned and constrained by hadron production data from the NA61/SHINE experiment and in-situ proton beam monitoring. The NEUT simulation package is used for the neutrino interaction model, with prior constraints based on external neutrino, pion and nucleon scattering cross section measurements.

The SK analysis uses a single-ring sample, which enhances charged current (CC) quasi-elastic (QE) events, separated into μ-like (ν_μ) and e-like (ν_e) sub-samples. The ND280 analysis selects charged current (CC) ν_μ interactions and separates the sample based on the number of reconstructed pions and decay electrons: CC0π, CC1π and CC Other. These topologies provide a strong constraint on the flux and interaction model governing CCQE scattering and resonant pion production, the signal and main background to the SK analysis, respectively. The reduction in the uncertainty on the SK predicted event rates due to the ND280 data is shown in Table 1. The SK and ND280 detector errors are constrained by calibration data and control samples such as cosmic rays and atmospheric neutrinos. More details of the SK and ND280 analyses can be found in previous T2K publications e.g. [7], [8] and the references therein.

Table 1. Summary of the effect of the systematic errors on the SK ν_e and ν_μ candidate total event rates. The prior uncertainties in () brackets do not include the ND280 data.

Systematic Error Source	Relative Uncertainty (%)	
	ν_μ Candidates	ν_e Candidates
Flux & Xsec. ND280 Constrained (Prior)	2.7 (21.7)	3.1 (26.0)
Xsec. ND280 Independent	5.0	4.7
Pion Hadronic Interactions	3.5	2.3
SK Detector	3.6	2.9
Total ND280 Constrained (Prior)	7.6 (23.4)	6.8 (26.8)

We estimate oscillation parameters using an unbinned maximum likelihood fit to the SK spectrum for the parameters $\sin^2(\theta_{23})$ and either Δm^2_{32} or Δm^2_{13} for the normal and inverted mass hierarchies respectively, and all 45 systematic parameters. The fit uses 73 unequal-width energy bins, and interpolates the spectrum between bins. Oscillation probabilities are calculated using the full three-flavor oscillation framework. Matter effects are included with an Earth density of $\rho = 2.6\,g/cm^3$ [6], δ_{CP} is unconstrained in the range $[-\pi, \pi]$, and other oscillation parameters are fit with constraints $\sin^2(\theta_{13}) = 0.0251 \pm 0.0035$, $\sin^2(\theta_{12}) = 0.312 \pm 0.016$, and $\Delta m^2_{21} = (7.50 \pm 0.20) \times 10^{-5}$ eV$^2/c^4$ [3]. Two-dimensional confidence regions in the oscillation parameters are constructed using the Feldman-Cousins method [9], with systematics incorporated using the Cousins-Highland method [10]. Figure 1 shows 68% and 90% confidence regions for the oscillation parameters for both normal and inverted hierarchies.

We calculate one-dimensional (1D) limits using a new method inspired by Feldman-Cousins [9] and Cousins-Highland [10] that marginalizes over the second oscillation parameter. The 1D 68% confidence intervals are $\sin^2(\theta_{23}) = 0.514^{+0.055}_{-0.056}$ (0.511 ± 0.055) and $\Delta m^2_{32} = 2.51 \pm 0.10$ ($\Delta m^2_{13} = 2.48 \pm 0.10$) $\times 10^{-3}$eV$^2/c^4$ for the NH (IH). The best fit corresponds to the maximal possible disappearance probability for the three-flavor formula.

To measure δ_{CP}, we use a frequentist-based analysis. The best fits for the oscillation parameters after minimizing over all parameters are shown in Table 2 for the NH and IH assumptions. The errors are based on the 1D constant-$\Delta\chi^2$ profile for each parameter. Allowed intervals for δ_{CP} with the reactor constraint are shown in Figure 2 for the frequentist-based analysis including a Feldman-Cousins (FC) critical $\Delta\chi^2$ correction ($\Delta\chi^2_c$) and the Bayesian analysis using the posterior probability and Confidence Intervals. $\delta_{CP} \approx -\pi/2$ is preferred since the T2K data alone prefers a larger θ_{13} compared to the reactor data.

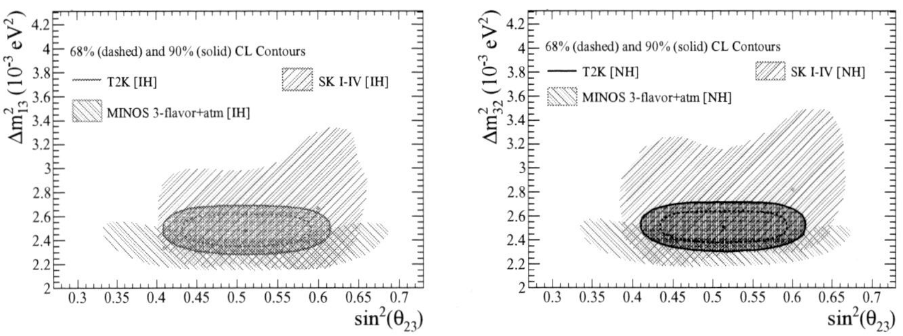

Figure 1. Contours comparing T2K Run 1-4 result with Super-K [11] and MINOS [12] for the Normal (Inverted) Hierarchy in the left (right) plot. The Super-K is a 3-flavor analysis of the atmospheric neutrino data using SK Runs I-IV. The MINOS result is a 3-flavor analysis using both of their ν_μ-disappearance and ν_e-appearance beam neutrino samples along with their atmospheric neutrino sample. For the MINOS contour, both the normal and inverted hierarchy contours are made with respect to a common minimum that is located in the inverted hierarchy parameter space.

Table 2. Best-fits and 1D constant-$\Delta\chi^2$ 68% confidence intervals (errors) for the oscillation parameters assuming each MH with and without the reactor constraint. Δm^2_{32} (Δm^2_{13}) is used for the NH (IH) assumption. The errors are not shown for δ_{CP} in the T2K-only case since there is no strong constraint. The errors for the other parameters in the reactor-constrained case are not yet calculated and will be shown in a future publication, while the exclusion region for δ_{CP} is shown in Figure 2.

	MH	$\lvert\Delta m^2_{32,13}\rvert$ [$\times 10^{-3}$ eV^2]	$\sin^2\theta_{23}$	$\sin^2\theta_{13}$	δ_{CP} (rad)
T2K-only	NH	$2.51^{+0.11}_{-0.12}$	$0.524^{+0.057}_{-0.059}$	$0.0422^{+0.0128}_{-0.0212}$	1.9
	IH	2.49 ± 0.12	$0.523^{+0.073}_{-0.065}$	$0.0491^{+0.0149}_{-0.0211}$	1.0
Reactor-constrained	NH	2.51	0.527	0.0248	-1.55
	IH	2.48	0.533	0.0252	-1.56

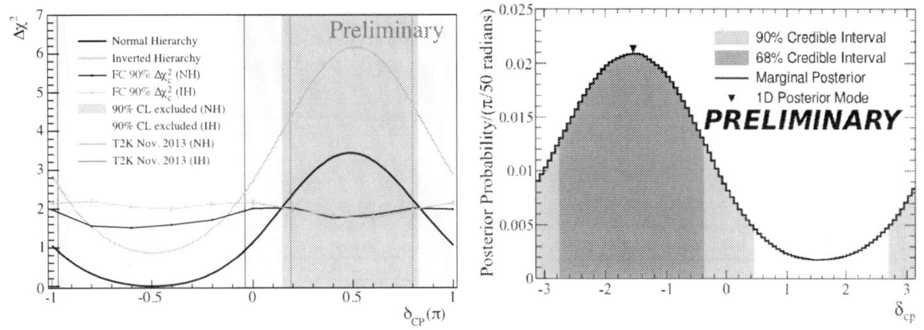

Figure 2. Left: The $\Delta\chi^2$ profile for δ_{cp}, showing the 90% CL regions based on the FC $\Delta\chi^2_c$ for NH and IH. Right: The posterior probability and 68% and 90% Confidence Intervals for δ_{CP}, marginalized over all the other parameters including the MH with priors π(NH) = π(IH) = 0.5.

Journal of Physics: Conference Series **556** (2014) 012061 doi:10.1088/1742-6596/556/1/012061

Table 3. The expected number of ν_e and ν_μ candidate events for the appearance and disappearance final states, respectively. NH, $\sin^2 2\theta_{13} = 0.1$ and $\delta_{CP} = 0$ are assumed. Background is categorized by the flavor before oscillation.

Appearance

	signal		BG					total
	$\nu_\mu \to \nu_e$	$\overline{\nu}_\mu \to \overline{\nu}_e$	ν_μ	$\overline{\nu}_\mu$	ν_e	$\overline{\nu}_e$	NC	
ν mode	3016	28	11	0	503	20	172	3750
$\overline{\nu}$ mode	396	2110	4	5	222	265	265	3397

Disappearance

	signal		BG				total
	$\nu_\mu \to \nu_\mu$	$\overline{\nu}_\mu \to \overline{\nu}_\mu$	ν_e	$\overline{\nu}_e$	NC	$\nu_\mu \to \nu_e$	
ν mode	17225	1088	11	1	999	49	19372
$\overline{\nu}$ mode	10066	15597	7	7	1281	6	26964

3. The Hyper-Kamiokande Experiment

The Hyper-Kamiokande experiment [5] is the next generation flagship experiment for the study of neutrino oscillations, nucleon decays, and astrophysical neutrinos. The detector is a third generation underground water Cherenkov (WC) detector situated in Kamioka, Japan. It consists of a 1 million tonne water target which is about 20 times larger than that of the existing Super-Kamiokande detector. It will serve as the far detector for a long baseline neutrino oscillation experiment planned for the upgraded J-PARC beam. It will also serve as a detector capable of observing proton decays, atmospheric neutrinos, and neutrinos from astronomical origins enabling measurements that far exceed the current world best measurements. Hyper-K has a sensitivity to the mass hierarchy through the atmospheric neutrino measurements and will be able to make a definitive measurement. A recent update of the previous sensitivity studies presented in 2011 [5] is based on the latest value of θ_{13} (it was not known at that time) and more generally the latest knowledge of the oscillation parameters. It uses a framework for the oscillation analysis developed by T2K and reported in [13], and the latest systematic errors based on the experience and prospects of the T2K experiment. The document, that contains an update on the R&D of the experiment as well, was submitted in April 2014 to the J-PARC PAC. An integrated beam power of 7.5 MW$\times 10^7$ sec is assumed in this study. It corresponds to 1.56×10^{22} protons on target (POT) with 30 GeV J-PARC beam. The ratio of neutrino and anti-neutrino running time is assumed to be 1:3. The oscillation parameters used for the sensitivity analysis are: $\sin^2 2\theta_{13}$ (0.1, fitted), δ_{CP} (0, fitted), $\sin^2\theta_{23}$ (0.5, fitted), Δm^2_{32} (2.4×10^{-3} eV^2, fitted), mass hierarchy (normal, fitted), $\sin^2 2\theta_{12}$ (0.8704, fixed) and Δm^2_{12} (7.6×10^{-5} eV^2, fixed), where in parenthesis the nominal values used in the fits and the treatment used in the fits are indicated. The criteria to select ν_e and ν_μ candidate events are based on those developed for and established with the SK and T2K experiments, and the corresponding number of expected events is shown is Table 3. Figure 3 shows the expected significance to exclude $\sin\delta_{CP} = 0$ (the CP conserved case). The significance is calculated as $\sqrt{\Delta\chi^2}$, where $\Delta\chi^2$ is the difference of χ^2 for the *trial* value of δ_{CP} and for $\delta_{CP} = 0°$ or $180°$ (the smaller value of difference is taken). We have also studied the case with a reactor constraint but the result changes only slightly. Figure 3 (left) shows the fraction of δ_{CP} for which $\sin\delta_{CP} = 0$ is excluded with $3\,\sigma$ and $5\,\sigma$ of significance as a function of the integrated beam power. The normal mass hierarchy is assumed. The results for the inverted hierarchy is almost the same. *CP* violation in the lepton sector can be observed with $3(5)\,\sigma$ significance for 76(58)% of the possible values of δ_{CP}.

Figure 3 (right) shows the $1\,\sigma$ uncertainty of δ_{CP} as a function of the integrated beam power.

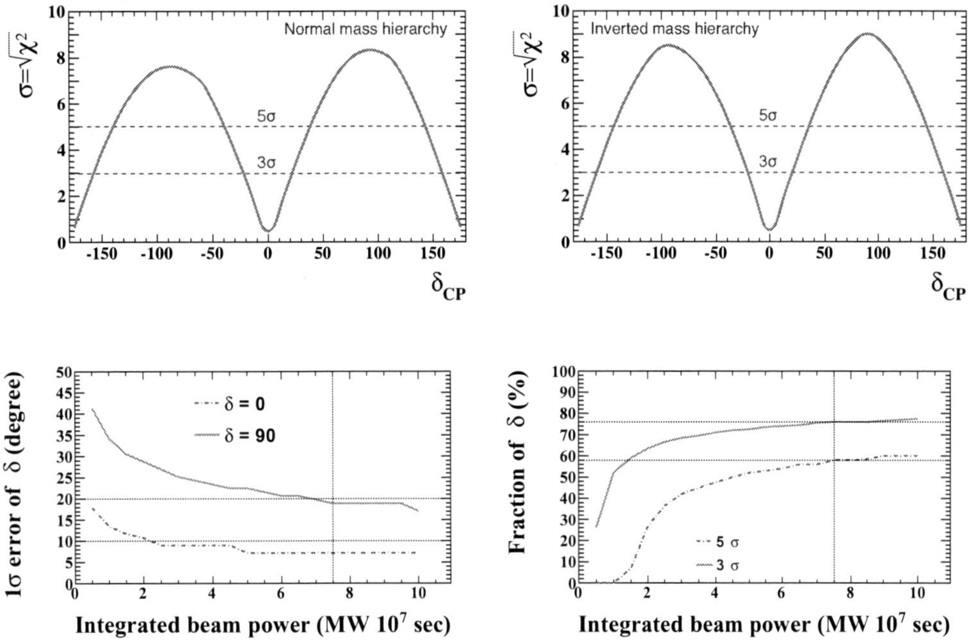

Figure 3. Upper row: expected significance to exclude $\sin\delta_{CP} = 0$. Left: normal hierarchy case. Right: inverted hierarchy case. Bottom row: fraction of δ_{CP} for which $\sin\delta_{CP} = 0$ can be excluded with $3\,\sigma$ (red solid line) and $5\,\sigma$ (blue dashed line) significance as a function of the integrated beam power (NH). Right plot: expected $1\,\sigma$ uncertainty of δ_{CP} as a function of integrated beam power.

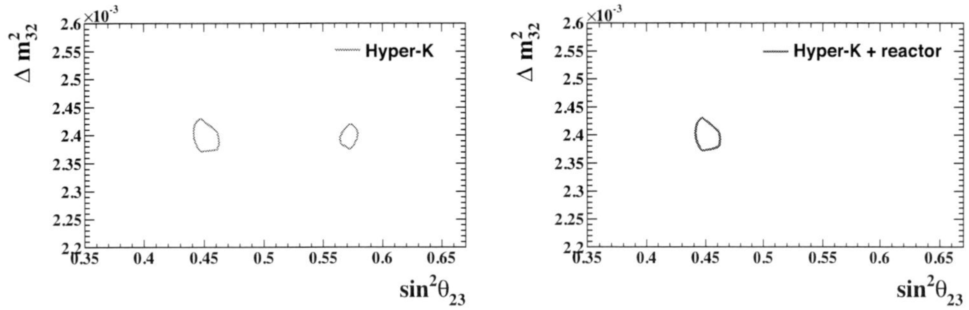

Figure 4. 90% CL allowed regions in the $\sin^2\theta_{23}$–Δm^2_{32} plane. The true values are $\sin^2\theta_{23} = 0.45$ and $\Delta m^2_{32} = 2.4 \times 10^{-3}$ eV^2. Effect of systematic uncertainties is included. Left plot: Hyper-K only. Right plot: With reactor constraint.

With 7.5 MW$\times 10^7$sec of exposure (1.56×10^{22} POT), the value of δ_{CP} can be determined to better than 19° for all values of δ_{CP}.

The use of ν_μ sample in addition to ν_e enables us to also measure $\sin^2\theta_{23}$ and Δm^2_{32} and Hyper-K will be able to provide a precise measurement of $\sin^2\theta_{23}$ and Δm^2_{32}. Figure 4 shows the 90% CL allowed regions on the $\sin^2\theta_{23}$-Δm^2_{32} plane, for the true values of $\sin^2\theta_{23} = 0.45$ and $\Delta m^2_{32} = 2.4 \times 10^{-3}$ eV^2. With a constraint on $\sin^2 2\theta_{13}$ from the reactor experiments, the

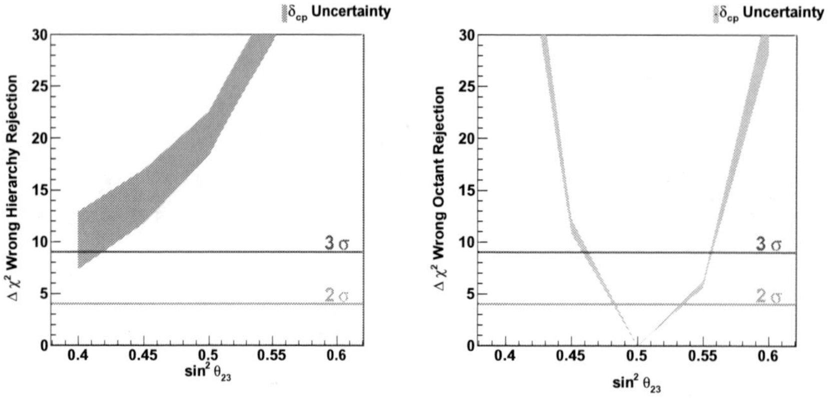

Figure 5. Atmospheric neutrino sensitivities for a ten year exposure of Hyper-K assuming the mass hierarchy is normal. Left: the $\Delta\chi^2$ discrimination of the wrong hierarchy hypothesis as a function of the assumed true value of $\sin^2\theta_{23}$. Right: the discrimination between the wrong octant for each value of $\sin^2\theta_{23}$. The uncertainty from δ_{cp} is given by the thickness of the band.

octant degeneracy is resolved and θ_{23} can be precisely measured.

Atmospheric neutrinos can provide an independent and complementary information to the accelerator beam program on the study of neutrino oscillation. Assuming a 10 year exposure, Hyper-K's sensitivity to the mass hierarchy and the octant of θ_{23} by atmospheric neutrino data are shown in Figure 5. Depending upon the true value of θ_{23} the sensitivity changes considerably, but for all currently allowed values of this parameter the mass hierarchy sensitivity exceeds $3\,\sigma$ independent of the assumed hierarchy. If θ_{23} is non-maximal, the atmospheric neutrino data can be used to discriminate the octant at $3\,\sigma$ if $\sin^2 2\theta_{23} < 0.99$.

4. Conclusion and Outlook

The first T2K combined ν_μ disappearance and ν_e appearance analysis based on 0.657×10^{21} POT is presented. T2K is producing the leading measurement on θ_{23} and, combined with reactor neutrino data, non-trivial exclusion intervals in δ_{CP}. Hyper-Kamiokande is the next generation long baseline neutrino experiments. It will be able to measure δ_{CP} with $3(5)\,\sigma$ significance for 76(58)% of the possible values of δ_{CP} to better than 19° for all values of δ_{CP}.

References

[1] Maki Z, Nakagawa M and S. Sakata 1962 *Prog. Theor. Phys.* **28** 870.
[2] Pontecorvo B 1968 *Sov. Phys. JETP* **26** 984.
[3] Beringer J *et al.* (Particle Data Group) 2012 *Phys. Rev.* D **86** 010001.
[4] Abe K *et al.* (T2K Collaboration) 2011 *Nucl. Instrum. Methods* A **659** 106.
[5] Abe K *et al.* (Hyper-Kamiokande Working Group) 2011 arXiv:1109.3262 [hep-ex].
[6] Hagiwara K, Okamura N and Senda K 2011 JHEP **1109** 082.
[7] Abe K *et al.* (T2K Collaboration) 2014 *Phys. Rev. Lett.* **112**.
[8] Abe K *et al.* T2K Collaboration) 2014 *Phys. Rev. Lett.* **112** 181801.
[9] Feldman G and Cousins R D 1998 *Phys. Rev.* D **57** 3873.
[10] R. D. Cousins and V. Highland, Nucl. Instrum. Methods A **320**, 331 (1992).
[11] Himmel A 2014 *AIP Conf.Proc.* 1604 345.
[12] Adamson P *et al.* (MINOS Collaboration) 2014 arXiv:1403.0867 [hep-ex].
[13] Abe K *et al.* (T2K Collaboration) 2014 arXiv:1409.7469 [hep-ex].

Journal of Physics: Conference Series **556** (2014) 012062 doi:10.1088/1742-6596/556/1/012062

Theory of oscillations and sterile neutrinos

Antonio Palazzo

Max-Planck-Institut für Physik (Werner Heisenberg Institut), Föhringer Ring 6, 80805 München, Germany

E-mail: palazzo@mpp.mpg.de

Abstract. We present a concise review of the theoretical status of neutrino oscillations within the (standard) 3-flavor framework and the (non-standard) 4-flavor scheme endowed with one additional sterile species (the so-called 3+1 scheme). We emphasize the slight overall preference that recently emerged for CP-violation in the 3-flavor analysis and highlight the unique role of the global data combination in the near future. After a brief introduction of the motivations for light (eV) sterile neutrinos, we discuss the bounds on their mixing with the electron neutrinos, attainable from the solar sector. The upper limit so obtained is independent of the reactor neutrino fluxes, whose calculations are affected by systematic uncertainties not completely under control. Finally, we highlight the possibility to explore sub-eV "super-light" sterile neutrinos exploiting the θ_{13}-dedicated reactor experiments also commenting on the robustness of the 3-flavor results within the enlarged 3+1 scheme.

1. Introduction

Neutrino physics is living one of its most exciting epochs due to the recent observation of a relatively large value of the third mixing angle θ_{13}. This discovery provides concrete hopes of completing the description of the 3-flavor framework with the determination of the leptonic CP-violation (CPV) and the neutrino-mass-hierarchy (NMH). But the field of neutrino physics may even open the way to unexpected avenues. The neutrino sector may be richer than commonly believed and not confined to the 3-flavor framework. Several anomalies have recently emerged both in laboratory and cosmological data analyses, which already indicate some cracks in the standard 3-flavor picture, pointing towards new physics. New light sterile neutrinos may provide the solution of these puzzling results. In what follows, we present a concise review of the theoretical status of neutrino oscillations within the (standard) 3-flavor framework and the (non-standard) 4-flavor scheme endowed with one new additional sterile species (the so-called 3+1 scheme).

2. The 3-flavor paradigm

After many years of observations performed with atmospheric, solar, accelerator and reactor experiments, neutrino oscillations have been firmly established and the measurement of the mass-mixing properties is now entering a precision era. Except for a few results obtained in short-baseline (SBL) setups — on which we will return later — all the existing phenomenology is consistent with a simple 3-flavor scheme. In such a scenario, the three known flavor fields (ν_e, ν_μ, ν_τ) are mixed with the three mass ones (ν_1, ν_2, ν_3), through a unitary matrix U (the so-called PMNS matrix). A common parameterization (adopted by the Particle Data Group [1])

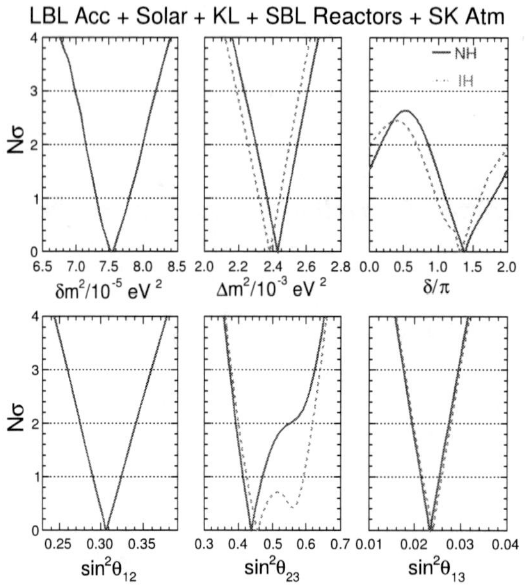

Figure 1. Bounds on the oscillation parameters in terms of standard deviations Nσ from the best fit. Solid (dashed) lines refer to NH (IH). The horizontal dotted lines mark the 1σ, 2σ and 3σ levels for each parameter.

for such a matrix is $U = R_{23}\Gamma_\delta R_{13}\Gamma^\dagger_\delta R_{12}$, where the R_{ij}'s are real Euler rotations with angles $\theta_{ij} \in [0, \pi/2]$, while $\Gamma_\delta = \text{diag}(1, 1, e^{i\delta})$ embeds a Dirac CP-violating phase $\delta \in [0, 2\pi]$.

2.1. Status of the 3-flavor parameters

The current neutrino phenomenology implies that the three-neutrino mass spectrum $\{m_i\}_{i=1,2,3}$ is formed by a "doublet" of relatively close states and by a third "lone" neutrino state. In the most frequently adopted labeling convention, the lightest (heaviest) neutrino in the doublet is called ν_1 (ν_2) and the corresponding squared-mass difference is defined as $\delta m^2 = m_2^2 - m_1^2 > 0$. Thanks to solar neutrinos, and more precisely to the matter (MSW) effects involved in the flavor conversion inside the sun, we know that the lightest eigenstate ν_1 is the one having the larger superposition with the electron neutrino ($|U_{e1}|^2 \simeq 0.674, |U_{e2}|^2 \simeq 0.303$). The lone state is labeled as ν_3 and the sign of $m_3^2 - m_{1,2}^2$ distinguishes two physically different cases. In one case, the mass spectrum is said to obey a "normal" ordering or hierarchy (NH) since, in this configuration, the lone state ν_3, which has a very small component of the electron neutrino flavor ($|U_{e3}|^2 \simeq 0.023$) is the heaviest mass eigenstate. Conversely, in the other case, the lone state ν_3 is the lightest mass eigenstate and the mass hierarchy is said to be inverted (IH). While the first case (NH) may naively seem more natural, both orderings are perfectly admissible and the one effectively realized in Nature must be determined through experiments. It is convenient to define the second independent larger squared-mass splitting as $\Delta m^2 = m_3^2 - (m_1^2 + m_2^2)/2$, so that the two hierarchies (NH and IH) are simply related by the transformation $+\Delta m^2 \to -\Delta m^2$. For phenomenological and historical reasons it is common practice to dub the δm^2-driven (Δm^2-driven) oscillations as the "solar" ("atmospheric") flavor transitions.

Figure 1 displays the estimates of the mass-mixing parameters obtained by the global data analysis [2]. The plot shows the number of standard deviations Nσ from the best fit values. We have a good determination of the "solar" ($\delta m^2, \theta_{12}$) parameters (at the level of

2.6% and 5.4% respectively) and "atmospheric" ($|\Delta m^2|, \theta_{23}$) parameters (at the level of 2.9% and 11% respectively). It is important to notice that while the (more precise) estimate of the two mass-splittings is currently dominated by the measurement of the frequency of the oscillation pattern(s) performed through "artificial" (reactor and accelerator) neutrino source experiments, the (less precise) estimate of the two large mixing angles (θ_{12}, θ_{23}) — directly related to the amplitude of the oscillations — is determined by an interplay of "natural" (solar and atmospheric) and "artificial" (reactor and accelerator) neutrino source measurements. The third mixing angle θ_{13} (often called "reactor angle") has been determined to be non-zero in 2012 by the three dedicated reactor experiments Daya Bay, Double-Chooz and RENO and by the accelerator experiment T2K, which have confirmed earlier indications coming from global analyses [3] (see also [4, 5, 6]). The current estimate of θ_{13} is dominated by Daya-Bay and the uncertainty on this parameter is $\sim 8\%$. Paradoxically, among the three mixing angles, the atmospheric one (θ_{23}), which was the first to be measured, is currently the least well-known. In particular, it is not known if it is maximal or not and at the moment there is no robust indication in favor of any of two octants. Indeed, in our preliminary analysis (not shown) including the results presented at the Neutrino 2014 conference the slight preference for the first octant evident in Fig. 1 tends to decrease. In contrast, the same preliminary analysis confirms the weak preference (slightly below the 90% C.L.) for a non-zero value of the CP-violating phase $\delta \sim -\pi/2$ apparent in Fig. 1. The current analysis is insensitive to the neutrino mass hierarchy.

2.2. Importance of the leptonic CP-violating phase δ

The determination of the CP-violating phase δ is important *per se* as it would complete our knowledge of the electroweak interactions, complementing the relevant information coming from the quark sector. But the discovery of the leptonic CPV may have other and far reaching implications. In particular, it may have some role in leptogenesis, and therefore it may be linked to the origin of the matter-antimatter asymmetry of our Universe (see below).

In the PMNS matrix, the imaginary parts of all the quartic products $U_{\alpha i}U^*_{\alpha j}U^*_{\beta i}U_{\beta j}$ are equal up to a sign and related by antisymmetric (in both flavor and mass indexes) coefficients

$$\Im[U_{\alpha i}U^*_{\alpha j}U^*_{\beta i}U_{\beta j}] \equiv J \sum_{\gamma=e,\mu,\tau} \sum_{k=1,2,3} \epsilon_{\alpha\beta\gamma}\, \epsilon_{ijk} \,. \tag{1}$$

where J is the so-called Jarlskog invariant, which gives a parameterization-independent measure of CP violation [7]. In the usual parametrization, one finds the expression

$$J = c_{12}s_{12}c_{23}s_{23}c^2_{13}s_{13}\sin\delta \equiv J^{\max}\sin\delta \,, \tag{2}$$

where $c_{ij} \equiv \cos\theta_{ij}$ and $s_{ij} \equiv \sin\theta_{ij}$. From (2) one realizes that CP-violation can occur only if all the three mixing angles are different from zero. Using the current best fit values of the three mixing angles one derives

$$J^{\max} \simeq 3.0 \times 10^{-2} \,, \tag{3}$$

which is not far from the maximum value theoretically attainable $1/6\sqrt{3} \simeq 0.1$ (see [8]). The value of $\delta \sim -\pi/2$ currently preferred by data implies a best fit $J^{\rm best} \simeq -3 \times 10^{-2}$. It is interesting to make a comparison with the quark sector, in which case the Jarlskog invariant has been precisely determined and is $J^{\rm CKM} \simeq 3 \times 10^{-5}$ [1]. Therefore, we see that in the eventuality of confirmation of the current preference for $\delta \simeq -\pi/2$, the lepton sector would be a source of CP-violation three orders of magnitude larger than that provided by the quark sector. Interestingly, the current best estimates of θ_{13} and δ satisfy the inequality

$$|\sin\theta_{13}\sin\delta| \simeq 0.15 > 0.11 \,, \tag{4}$$

which is the condition for a successful leptogenesis having as the sole source of CP-violation the Dirac phase δ [9].

2.3. Role of the global analyses in the near term

In the next years the role of global analyses will become more important than ever. Whilst for the third mixing angle θ_{13} we had at our disposal dedicated experiments capable to transform the first weak indications into a solid observation (θ_{13} is now known to be non-zero at the incredible statistical significance of twelve standard deviations), the situation appears to be completely different for the remaining neutrino properties. From all the recent sensitivity studies it has clearly emerged that, at least for the next five years, it will be extremely difficult (if not impossible) for a single experiment to provide definitive information (say an evidence at the 3 sigma level) for any of the two searched properties CPV and NMH. Rather, each experiment will (in the best case) provide some hint with low statistical significance. In this situation, it is clear that the combination of all the available data will be a powerful tool enabling us to get the most from the neutrino data. Important synergies can manifest only at the level of the global combination, where parameter degeneracies affecting the single datasets may be reduced. Last but not least, the global analyses will test the mutual consistency of different pieces of data possibly evidencing discrepancies, which may be the harbingers of new physics beyond the Standard Model.

3. Light (eV) and super-light (sub-eV) sterile neutrinos

Several anomalies have recently emerged in short baseline (SBL) oscillation experiments, which indicate some "cracks" in the standard picture and point towards the existence of new physics. These are the LSND and MiniBooNE anomalies and the so-called "reactor and gallium anomalies". Furthermore, the latest cosmological data suggests an excess in the radiation content of the Universe (the so-called "dark radiation anomaly").

An intriguing hypothesis is that new light sterile neutrino species — Standard Model gauge singlets with mass of O(eV) — may be responsible for such inconsistencies, although they seem unable to provide a completely convincing explanation (see [10] for a critical discussion). Right-handed neutrinos appear in many extensions of the Standard Model, usually in the context of see-saw mechanisms, where their natural mass scale lies at the GUT energies. However, it is safe to affirm that the mass of these particles cannot be predicted in a robust way, and indeed it is not uncommon to find models accommodating lower masses down to the sub-eV domain.

3.1. An independent test of the low-energy anomalies

As mentioned above, a consistent interpretation of the anomalous results in terms of sterile neutrino oscillations is problematic. The most important difficulty is related to the tension (arising in all schemes with extra sterile neutrinos) among the $\nu_\mu \to \nu_e$ appearance results and the (joint $\nu_e \to \nu_e$ and $\nu_\mu \to \nu_\mu$) disappearance searches (see [10]). In this perspective, it is essential to carefully put under test any single piece of evidence, irrespective of all the others. Here, we focus on the reactor and gallium anomalies, which *taken alone* are fully consistent with a non-zero admixture of the electron neutrino with new sterile species, favoring the range $|U_{e4}|^2 \sim [0.020, 0.060]$ at the 2σ level [11] (see also [12]).

As we have shown in [13], where we have presented the analytical treatment of the solar MSW transitions in a general 3+s scheme, the solar sector data (Solar and KamLAND) offer a sensitive probe of such an admixture. In a subsequent paper [14], we have shown how the first evidence for non-zero θ_{13} (coming a that time only from Double-CHOOZ and T2K) further improved the sensitivity of the solar sector to U_{e4}. In these proceedings we present an updated version of the analysis performed in [14] by incorporating the more recent (strongest) constraints on θ_{13}, which is now known with far better precision.

In the left panel of Fig. 2, the diagonal bands indicate the region allowed by the combined solar and KamLAND data in the plane $[\sin^2\theta_{13}, \sin^2\theta_{14}]$. We recall that in the parametrization adopted in [13, 14] it is $U_{e4}^2 \equiv \sin^2\theta_{14}$. We stress that the KamLAND analysis has been

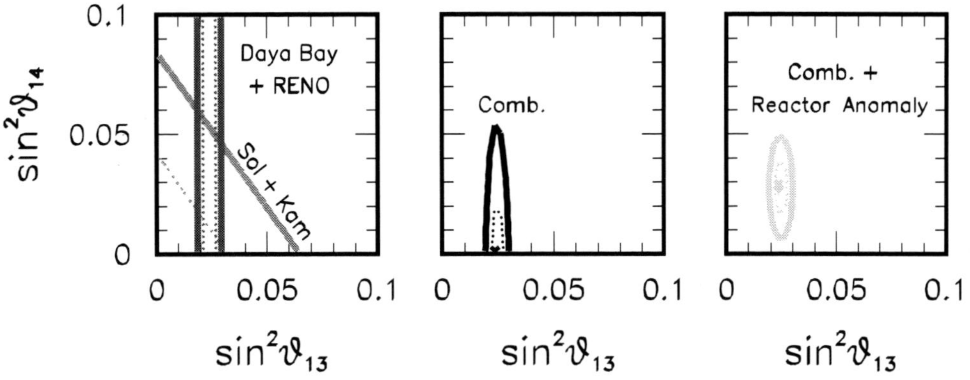

Figure 2. Left panel: regions allowed by the solar sector data (diagonal bands) and by the dual-baseline reactor (Daya Bay and RENO) experiments (vertical bands). Middle panel: regions allowed by their combination. Right panel: combination of the constraints in the middle panel with those coming from the reactor anomaly taken from [11]. The contours refer to $\Delta\chi^2 = 1$ (dotted line) and $\Delta\chi^2 = 4$ (solid line).

performed using only the spectral shape information so as to render its results independent of the reactor antineutrino flux normalization. In the same panel, the vertical bands identify the range allowed for θ_{13} by the combination of the dual-baseline reactor experiments Daya Bay and RENO. The superposition of the regions allowed by the two datasets (solar+KamLAND and Daya Bay+RENO) evidences their complementarity in constraining the two mixing angles. Their combination, shown in the central panel of Fig. 2, leads to the strong upper bound

$$\sin^2\theta_{14} \equiv |U_{e4}^2| < 0.041 \qquad (90\%\,\text{C.L.})\,. \tag{5}$$

As already stressed, this bound does not depend on the normalization of the reactor fluxes and thus represents an independent constraint on $|U_{e4}|^2$. Therefore, it makes sense to combine such a bound with the information coming from the reactor anomaly. The result of such an exercise is shown in the right panel of Fig. 2, where we have taken the reactor anomaly likelihood from [11]. The effect of the bound in Eq. (5) is a downshift of the 2σ range allowed for $|U_{e4}|^2$ from $[0.011, 0.054]$ to $[0.005, 0.050]$, and a sensitive reduction of the overall statistical significance (from $\sim 3\sigma$ to $\sim 2.5\sigma$) of the indication of $\theta_{14} > 0$.

3.2. Probing super-light sterile neutrinos at reactor θ_{13}-dedicated experiments

Three dedicated reactor experiments, Daya Bay, Double Chooz and RENO, have recently performed a precision measurement of the third standard mixing angle θ_{13} exploiting a multiple baseline comparison of $\nu_e \to \nu_e$ disappearance driven by the atmospheric mass-squared splitting. In [15] we have shown how the same technique can be used to put stringent limits on the oscillations of the electron neutrino into a fourth "super-light" sterile species characterized by a mass-squared difference lying in the range $[10^{-3} - 10^{-1}]\,\text{eV}^2$. Here, due to space limitations, we only present the results of the analysis obtained in [15], referring the reader to the original work for more details. Figure 3 displays the constraints on the admixture $|U_{e4}|^2$ obtained by a general 4-flavor analysis, in which θ_{13} is marginalized away.

The comparison of the first three panels evidences the complementarity of the three experiments induced by the different baselines involved in the three setups. Hence, in the more complicated 4-flavor scenario, in contrast to the simpler 3-flavor scheme, the fit is not dominated by the sole experiment Daya Bay. Very recently, the Daya Bay collaboration has presented the

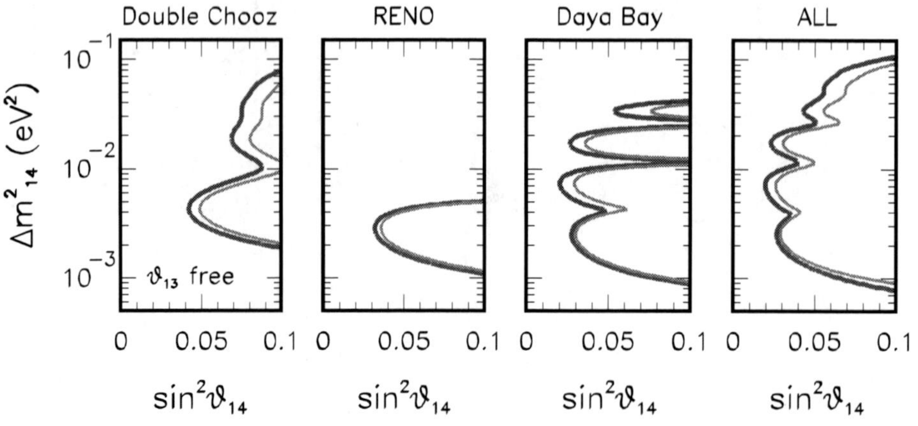

Figure 3. Results obtained from the 4-flavor analysis. The contours refer to 2 d.o.f. 90% C.L. (blue thick line) and 99% (red thin line).

first results of a 4-flavor analysis [16]. Unlike our study, which could make use only of the total rate information, Daya Bay exploits the full spectral information, obtaining more stringent constraints. We close this discussion on light sterile neutrinos with a final important remark. As we have shown in [15], the estimate of θ_{13} obtained by the combination of the three reactor experiments is substantially independent of the new mass-squared splitting provided that this is not too low ($> 6 \times 10^{-3}\,\mathrm{eV}^2$). Therefore, the 3-flavor picture is robust with respect to the perturbations induced by a fourth neutrino.

Acknowledgments

I would like to thank Cristina Lazzeroni and co-organizers for inviting me to this very interesting and enjoyable conference and for their kind hospitality in Birmingham. I acknowledge support from the European Community through a Marie Curie IntraEuropean Fellowship, grant agreement no. PIEF-GA-2011-299582, "On the Trails of New Neutrino Properties". I also acknowledge partial support from the European Union FP7 ITN Invisibles (Marie Curie Actions, PITN-GA-2011-289442).

References

[1] Beringer J *et al* [Particle Data Group Collaboration] 2012 *Phys. Rev.* D **86** 010001
[2] Capozzi F, Fogli G L, Lisi E, Marrone A, Montanino D and Palazzo A 2014 *Phys. Rev.* D **89** 093018
[3] Fogli G L, Lisi E, Marrone A, Palazzo A and Rotunno A M 2008 *Phys. Rev. Lett.* **101** 141801
[4] Balantekin A B and Yilmaz D 2008 *J. Phys.* G **35**, 075007
[5] Schwetz T, Tortola M and Valle J W F 2011 *New J. Phys.* 13 063004
[6] Gonzalez-Garcia M C, Maltoni M and Salvado J 2010 *J. High Energy Phys.* JHEP04 (2010) 056
[7] Jarlskog C 1985 *Phys. Rev. Lett.* **55** 1039
[8] Giunti C and Kim C W 2007 *Fundamentals of Neutrino Physics and Astrophysics (Oxford University Press)*
[9] Pascoli S, Petcov S T and Riotto A 2007 *Nucl. Phys.* B **774** 1
[10] Palazzo A 2013 *Mod. Phys. Lett.* A **28** 1330004
[11] Giunti C, Laveder M, Li Y F, Liu Q Y and Long H W 2012 *Phys. Rev.* D **86** 113014
[12] Mention G *et al* 2011 *Phys. Rev.* D **83**, 073006
[13] Palazzo A 2011 *Phys. Rev.* D **83** 113013
[14] Palazzo A 2012 *Phys. Rev.* D **85** 077301
[15] Palazzo A *J. High Energy Phys.* JHEP10 (2013) 172
[16] An F P *et al* [DAYA-BAY Collaboration] 2014 *Phys. Rev. Lett.* **113** 141802

Journal of Physics: Conference Series **556** (2014) 012063 doi:10.1088/1742-6596/556/1/012063

The GERDA experiment: results and perspectives

Carla Macolino[1] **on behalf of the GERDA Collaboration**
INFN, Laboratori Nazionali del Gran Sasso, L'Aquila, Italy
E-mail: carla.macolino@lngs.infn.it

Abstract. The Germanium Detector Array, GERDA, at Laboratori Nazionali del Gran Sasso (Italy), is designed to search for Majorana neutrinos via neutrinoless double beta ($0\nu\beta\beta$) decay of ^{76}Ge. GERDA completed the Phase I in 2013, after an exposure of 21.6 kg·yr and with a background of about 0.01 cts/(keVkgyr): no signal was found and a limit on the half-life of $T^{0\nu}_{1/2} > 2.1 \cdot 10^{25}$ yr (90% C.L.) was established. The previous claim of $0\nu\beta\beta$ observation for ^{76}Ge is strongly disfavoured in a model independent way. The commission for GERDA Phase II is currently ongoing and about 20 kg of additional enriched Ge diodes will be deployed. Pulse-shape analysis, together with the liquid argon instrumentation will allow to reach a background level one order of magnitude lower than in Phase I. In this paper the measurement of the half-life of $0\nu\beta\beta$ decay from GERDA Phase I and the expected sensitivity for Phase II are discussed.

1. Introduction

The observation of neutrinoless double beta decay ($0\nu\beta\beta$) probes fundamental questions about lepton number violation and neutrino properties, indeed it would give direct information on the possible "Majorana" nature of the neutrino and could determine the absolute neutrino mass complementary to other techniques. $0\nu\beta\beta$ decay could be observed experimentally as a narrow peak at the end-point of the $2\nu\beta\beta$ decay energy spectrum, corresponding to the Q-value ($Q_{\beta\beta}$) of the decay, given by the sum energy of the two emitted electrons. The GERDA experiment [1, 2, 3] searches for neutrinoless double beta decay of ^{76}Ge, in which ^{76}Ge (Z=32) would decay into ^{76}Se (Z=34) and two electrons. GERDA detectors are germanium semiconductors with an enrichment fraction in ^{76}Ge of about 86%. The crystals act as both the decay source and a 4π detector and their very good energy resolution translate in a narrow $0\nu\beta\beta$ peak at $Q_{\beta\beta}$=2039 keV.

2. The GERDA experiment

The GERDA experimental setup consists of an array of bare ^{enr}Ge (Ge detectors enriched in ^{76}Ge) semiconductors, placed in strings into a cryostat filled with liquid argon (LAr) and surrounded by an additional shield of ultra-pure water. Liquid argon is the cooling medium for the ^{enr}Ge detectors and also acts as shield against external gamma radiation from the cryostat walls. The water buffer provides the Cherenkov medium for the muon veto, attenuates the flux of the external γ radiation and absorbs neutrons. Further details about the GERDA experimental setup are reported in ref. [3]. Data acquisition of GERDA Phase I started in November, 2011 with nine p-type ^{enr}Ge semi-coaxial (HPGe) detectors (from the previous HdM

[1] presently also at INFN, Gran Sasso Science Institute, L'Aquila, Italy

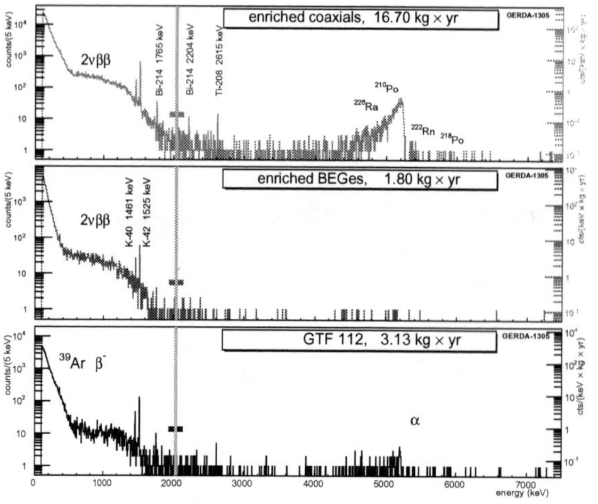

Figure 1. Spectra from enriched semi-coaxial (top), enriched BEGe (middle) and non-enriched (bottom) detectors of GERDA Phase I. The y-axis scale on the right-hand side indicates the number of counts divided per (keV·kg·yr). Plots from ref. [6].

and IGEX experiments) with a total mass of about 20.7 kg (17.7 kg enriched and 3 kg not enriched). Five Broad Energy GErmanium diodes [2] (BEGes), were deployed on July 2012, with total mass of about 3.6 kg. The energy scale is determined by calibrating the detectors with ^{228}Th sources on a weekly basis. The exposure-weighted average energy resolution (FWHM), extrapolated at $Q_{\beta\beta}$, is (4.8±0.2) keV for semi-coaxial detectors and (3.2±0.2) keV for BEGes. Events in the region of interest (in the interval $Q_{\beta\beta}\pm20$ keV) were kept "blinded", *i.e.* not processed, until the calibration was finalized and all the selection cuts and analyses were fixed. The experimental energy spectra for the enriched and natural detectors are shown in fig. 1. The green boxes indicate the blinded window at $Q_{\beta\beta}\pm20$ keV. Visible gamma peaks are from ^{40}K and ^{42}K decays and from the decay chains of ^{226}Ra and ^{232}Th. The low energy part of the spectrum is dominated by the β-decay of ^{39}Ar and events from $2\nu\beta\beta$ decay populate the range from 600 to 1800 keV; the half-life of $2\nu\beta\beta$ decay for ^{76}Ge has been measured after collecting an exposure of 5.04 kg·yr (see ref. [5]). Above 4000 keV the background is dominated by α decays of ^{210}Po and ^{226}Ra. The energy spectrum from semi-coaxial and BEGe detectors is fitted to a background model in the range between 570 and 7500 keV (for further details see ref. [6]): the background is mainly due to sources close to the detectors or on the detector surface and no peak is expected to appear in the region of interest. In this region, with the exclusion of ±5 keV around the expected position of the single escape peak from ^{208}Tl (2104 keV) and of the γ line from ^{214}Bi (2119 keV), the interpolated value for the background index (BI) is $\mathrm{BI}=1.75^{+0.26}_{-0.24}\times10^{-2}$ cts/(keV·kg·yr) for semi-coaxial detectors and $\mathrm{BI}=3.6^{+1.3}_{-1.0}\times10^{-2}$ cts/(keV·kg·yr) for the BEGe detectors. Detector signals are read out independently by a charge sensitive amplifier; the signal is then digitized by 100 MHz Flash ADCs and physical parameters like energy and risetime of the signal are reconstructed by digital filters [7]. The specific pulse shape of $0\nu\beta\beta$ events in GERDA detectors is used to discriminate them from background events. In GERDA Phase I two different methods for Pulse Shape Discrimination (PSD) have been developed, according to the different characteristics of the pulses and electric field distributions of semi-coaxial and BEGe detectors [8]. For a review of the GERDA experiment see also ref. [9].

[2] manufactured by Canberra in Olen, Belgium

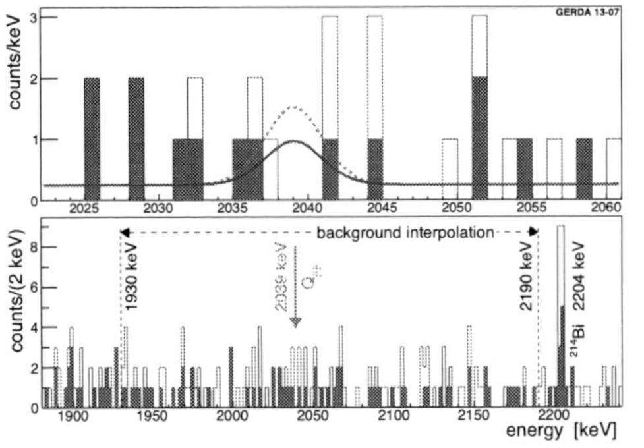

Figure 2. Energy spectrum from all ^{enr}Ge detectors before (filled) and after (open) the PSD selection. In the upper panel the expectation based on the central value of the half-life estimated by ref. [12] is also shown (red), together with the 90% C.L. limit estimated by GERDA Phase I (blue). In the lower panel the energy window used for the backgrund interpolation is indicated. Plot from ref. [4].

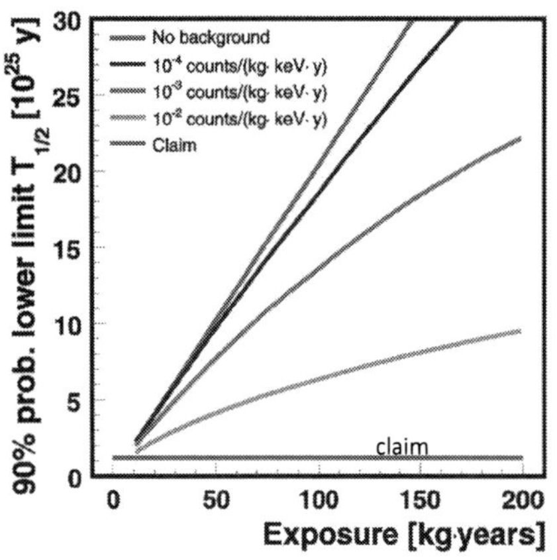

Figure 3. GERDA sensitivity on $0\nu\beta\beta$ half-life as a function of the collected exposure. The different lines correspond to different values of the background level. The expected background for GERDA Phase II is 10^{-3} counts/(keV·kg·yr).

3. Results on $0\nu\beta\beta$ decay of ^{76}Ge

After the collection of an exposure of 21.6 kg·yr from GERDA Phase I, a limit on the half-life of $0\nu\beta\beta$ decay in ^{76}Ge was established [4]. Phase I data were divided into three different sets: data from the BEGe detectors ("BEGe" set), data from semi-coaxial detectors in the time period when the BEGe detectors were deployed ("silver" set) and remaining data from semi-coaxial detectors ("golden" set). "Unblinded" data showed a flat background in the region of interest, with seven events observed while 5.1±0.5 were expected from background interpolation. After the PSD was applied, three events from the semi-coaxial detectors and the one from the BEGe detector were classified as background. No event remains in the energy window $Q_{\beta\beta}\pm\sigma_E$ and the corresponding BI becomes 10^{-2} cts/(keV·kg·yr). The observed spectrum was fitted with a profile likelihood fit; the fitted function contains three constant terms for the background (for the three data sets) and a Gaussian peak, centered at $Q_{\beta\beta}$ and with standard deviation according to the energy resolution. The four corresponding parameters of the function were the three terms for the background and $1/T^{0\nu}_{1/2}$. The limit on the half-life is $T^{0\nu}_{1/2} > 2.1 \cdot 10^{25}$ yr (90% C.L.) and the best fit value for the number of $0\nu\beta\beta$ events is $N^{0\nu}=0$. The corresponding limit on the number of signal events is $N^{0\nu} < 3.5$ counts. The median sensitivity for the 90% C.L. limit, given the background levels and the efficiencies, is $T^{0\nu}_{1/2} > 2.4 \cdot 10^{25}$ yr. The systematic uncertainties due to detector parameters, selection efficiency, energy resolution and energy scale, were folded into the half-life estimation; they weaken the limit by about 1.5%. A Bayesian analysis [10] was also performed (using the BAT toolkit [11]) with the same fit and a flat prior distribution for $1/T^{0\nu}_{1/2}$ between 0 and 10^{-24} yr^{-1}. The corresponding result for the limit is $T^{0\nu}_{1/2} > 1.9 \cdot 10^{25}$ yr, with a median sensitivity of $T^{0\nu}_{1/2} > 2.0 \cdot 10^{25}$ yr. The GERDA result does not support the previous claim of $0\nu\beta\beta$ decay observation in ^{76}Ge [12]. The Bayes factor, *i.e.* the ratio between the probability that the observed data D are produced according to the model H_1 ($0\nu\beta\beta$ with half-life $T^{0\nu}_{1/2}$ from ref. [12]) and the probability that they are produced according to the model

H_0 (only background), is $P(D|H_1)/P(D|H_0) = 0.024$. In fig. 2 the "unblinded" spectrum is shown, together with the likelihood fit and the expectation based on the claim from ref. [12]. A combined profile likelihood fit, when GERDA data are combined with data from the HdM [13] and IGEX [14] experiments, gives again $N^{0\nu} = 0$ as best fit and $T^{0\nu}_{1/2} > 3.0 \cdot 10^{25}$ yr (90% C.L.). Considering this limit, the phase-space factor for the ^{76}Ge [15] and the nuclear matrix element calculations reported in refs. from [16] to [22] (scaling the different g_A and R_A parameters according to ref. [23]), the derived upper limits on the effective electron neutrino mass range between 0.2 and 0.4 eV. A Bayesian analysis gives the same limit and a Bayes factor equal to $P(D|H_1)/P(D|H_0) = 2{\cdot}10^{-4}$.

4. Phase II upgrades

The main goal of GERDA Phase II is to increase the sensitivity with respect to Phase I, by lowering the background level and increasing the total collected exposure. The background level will be reduced thanks to the implementation of a LAr scintillation veto and the procurement of 30 additional enriched BEGe detectors, to achieve a total mass of about 40 kg. The scintillation veto consists of the detection of the 128 nm scintillation light generated in liquid argon by radioactive background decays or cosmic muons, accompanied with the emission of gamma particles which eventually excite the argon. A curtain made of light-guiding fibers surrounding the detector strings will collect the light, eventually read out by Silicon Photo-Multipliers (SiPMs) on the top of the array. In addition, light will be also directly detected by PMTs coated by wavelength shifter and placed on top and bottom of the detector array. Residual background contamination will be rejected by the Pulse Shape Discrimination, as described in ref. [8]. The expected background index for GERDA Phase II, when the combination of LAr veto and Pulse Shape Discrimination is used, is of the order of 10^{-3} cts/(keV·kg·yr). The commissioning of the Phase II upgrade of GERDA is presently ongoing. The sensitivity of GERDA as a function of the total collected exposure is shown in fig. 3. With the increased total mass of enriched germanium an exposure of 100 kg·yr will be reached in about 3 years. The corresponding sensitivity on the half-life of $0\nu\beta\beta$ decay is $T^{0\nu}_{1/2} \simeq 1.4 \cdot 10^{26}$ yr.

References

[1] The GERDA collaboration 2004 *Letter Of Intent*
[2] The GERDA collaboration 2004 *Proposal* http://www.mpi-hd.mpg.de/GERDA
[3] The GERDA collaboration 2013 *Eur. Phys. J. C* **73** 2330
[4] The GERDA collaboration 2013 *Phys. Rev. Lett.* **111** 122503
[5] The GERDA collaboration 2013 *J. Phys. G: Nucl. Part. Phys.* **40** 035110
[6] The GERDA collaboration 2013 *Eur. Phys. J. C* **74** 2764
[7] Agostini M et al. 2011 *J. Instrum.* **6** P08013
[8] The GERDA collaboration 2013 *Eur. Phys. J. C* **73** 2583
[9] Macolino C on behalf of the GERDA collaboration 2014 *Mod. Phys. Lett.* A**29** 1430001
[10] Caldwell A and Kröninger K 2006 *Phys. Rev. D* **74** 092003
[11] Caldwell A, Kollar D and Kröninger K 2009 *Comput. Phys. Commun.* **180** 2197
[12] Klapdor-Kleingrothaus H V et al. 2004 *Phys. Lett. B* **586** 198
[13] Klapdor-Kleingrothaus H V et al. 2001 *Eur. Phys. J. A* **12** 147
[14] Alseth C E et al. 2002 *Phys. Rev. D* **65** 092007
[15] Kotila J and Iachello F 2012 *Phys. Rev. C* **85** 034316
[16] Rodriguez T R and Martinez-Pinedo G 2010 *Phys. Rev. Lett.* **105** 252503
[17] Menendez J et al. 2009 *Nucl. Phys. A* **818** 139
[18] Barea J, Kotila J and Iachello F 2013 *Phys. Rev. C* **87** 014315
[19] Suhonen J and Civitarese O 2010 *Nucl. Phys. A* **847** 207
[20] Meroni A, Petcov S T and Simkovic F 2013 *JHEP* **1302** 25
[21] Simkovic F, Rodin V, Faessler A and Vogel P 2013 *Phys. Rev. C* **87** 045501
[22] Mustonen M T and Engel J 2013 *Phys. Rev. C* **87** 064302
[23] Smolnikov A and Grabmayr P 2010 *Phys. Rev. C* **81** 028502

The Heavy Photon Search experiment at Jefferson Laboratory

Andrea Celentano, for the HPS Collaboration
INFN-Genova, Via Dodecaneso 33, 16147 Genova, IT
E-mail: andrea.celentano@ge.infn.it

Abstract. The Heavy Photon Search experiment (HPS) at Jefferson Laboratory will search for a new $U(1)$ massive gauge boson, or "heavy-photon", mediator of a new fundamental interaction, called "dark-force", that couples to ordinary photons through kinetic mixing. HPS has sensitivity in the mass range 20 MeV - 1 GeV and coupling ε^2 between 10^{-5} and 10^{-10}. The HPS experiment will look for the e^+e^- decay of the heavy photon, by resonance search and detached vertexing, in an electron beam fixed target experiment. HPS will use a compact forward spectrometer, which employs silicon microstrip detectors for vertexing and tracking, and a $PbWO_4$ electromagnetic calorimeter for energy measurement and fast triggering.

1. Introduction

The "heavy-photon", or A', is an hypothetical massive gauge boson associated to an additional $U(1)$ hidden symmetry, as predicted by several "Beyond the Standard Model" (BSM) theories [1, 2]. The A' couples to ordinary photons through a "kinetic mixing" mechanism [3, 4], i.e. through a loop-order effect, mediated by a GUT-scale massive particle that carries both the standard-model weak hypercharge and the dark-force equivalent. This induces an effective parity-conserving interaction $\varepsilon e A'_\mu J^\mu_{\rm EM}$ of the A' to the electromagnetic current $J^\mu_{\rm EM}$, with coupling εe, where the parameter ε can naturally be in the range 10^{-12} - 10^{-2} [5, 6, 7]. Consequently, standard model charged particles acquire a "milli-charge" εe for the interaction with the "heavy-photon", resulting in em-like processes such as A'-strahlung (the emission of an A' by an high-energy electron) and the A' decay to electron-positron pairs.

Recently, a possible role of heavy photons in dark matter physics has been suggested, generating great interest to directly search for the A' [8, 9]. Results from indirect and direct dark matter searches have been re-interpreted as potential signals of dark matter interacting with heavy photons. For example, the positron excess in the cosmic-ray flux recently reported by AMS [10], confirming previous results from PAMELA [11] and FERMI [12], can be explained assuming that dark matter is charged under the "dark-force" interaction, and decays or annihilates to A' particles, that in turn decay to e^+e^- pairs. In this picture, the A' acts as a "mediator" between the Standard Model and the dark matter sector. The non-observed excess in anti-proton fraction suggests that the A' mass is lower than $\simeq 2$ GeV [13].

In the simplest scenarios, the two main parameters that determine the characteristics of the A' and thus the experimental search strategies are: the kinetic mixing parameter ε and the mass M_A. While a huge range of mixing parameters and masses are possible, it is natural that ε be around 10^{-3}, and necessary that masses be around GeV, if the positron excess is to be

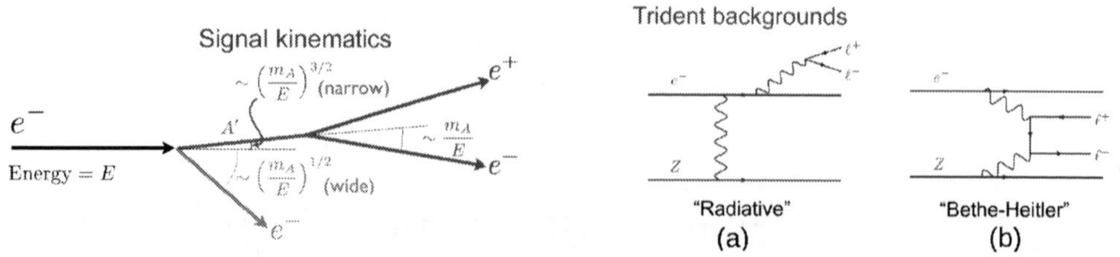

Figure 1. Left: the foreseen kinematics for the A' production by Bremsstrahlung off an incoming electron and the subsequent decay to an e^+e^- pair. Right: sample diagrams of radiative trident (γ^*) and Bethe-Heitler trident reactions that comprise the primary background to the A' search.

explained [14]. Other than explaining astrophysics anomalies, an A' in this mass and coupling range can explain the muon anomalous magnetic moment [15]. This parameter space region has been investigated by a large number of experiments, with so far no positive results (see [16] for a complete review).

High luminosity, fixed target experiments provide an unique way to probe this unknown region [17]. In these experiments, an electron beam impinges on a high Z target, producing heavy photons that subsequently decay to pairs of fermions. The A' is then reconstructed by measuring the decay products momenta.

The kinematics of this process, sketched in Figure 1 (left panel), are characterized by a very-forward emitted A' that carries most of the incident electron beam energy E. This results in a very small opening angle of the decay products, of the order of $\theta \simeq M_A/E$. Therefore, the A' decay products will be emitted at a very low angle with respect to the beam-line, and will carry most of the beam energy. Furthermore, depending on the value of the parameters, long-lived A' are possible in the aforementioned parameter space, with displaced vertexes up to hundreds of meters.

The main background sources associated to this reaction are QED trident processes, i.e. Bethe-Heitler trident reactions and radiative trident production (see Figure 1, right-panel). Although the Bethe-Heitler process has a much larger total cross-section, it can be significantly reduced by exploiting its very different kinematics. Instead, the radiative trident process is an irreducible background, with the same kinematic as the signal. However, it presents a smooth, continuous distribution in the e^+e^- invariant mass distribution, in contrast to the peak at $M_{e^+e^-} = M_A$ for the signal, and occurs promptly at the production target, while the signal can be displaced for small ε values.

The Heavy Photon Search (HPS) experiment at Jefferson Laboratory [18] is a fixed-target experiment specifically designed to search for dark-photons in the mass range between 20 MeV and 1 GeV and coupling ε^2 between 10^{-5} and 10^{-10}. HPS will search for the e^+e^- decay of the heavy photon, through complementary techniques: resonance search (the traditional "bump-hunting") and detached vertexing. In particular, HPS will explore a unique region corresponding to small cross sections, out of reach from collider experiments, where thick absorbers, as used in beam-dump experiments, are not allowed due to the relatively short A' decay length (< 1 m).

2. The HPS experiment at Jefferson Laboratory

The HPS experiment will run in Hall B of Thomas Jefferson National Accelerator Facility (JLab). The high duty-cycle, high-quality electron beam, with energy tunable between 1.1 and 11 GeV and current up to 250 nA, will permit to reach high luminosity while keeping beam-related backgrounds under control. The HPS detector was designed to match the foreseen signal kinematics described above. Mandatory requirements are: a large and uniform acceptance in

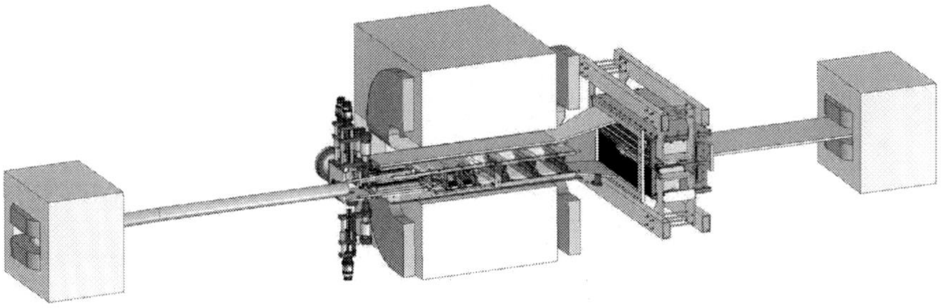

Figure 2. Schematic of the HPS detector setup. From left to right: the first dipole magnet, the target system, the silicon vertex tracker (SVT) inside the analyzing dipole magnet, the electromagnetic calorimeter (ECal), the third dipole magnet.

the forward region close to the beam, down to $\theta \simeq 15$ mrad, excellent track reconstruction efficiency for electrons and positrons, good angular and momentum resolution to reconstruct invariant mass precisely, and excellent vertex resolution to discriminate displaced A' decays from prompt QED backgrounds. Radiation hardness is also a critical issue, given the high radiation environment, particularly near the target position. Finally, the material budget must be as low as possible, to minimize multiple scattering and preserve the angular resolution.

The HPS detector is made of two main sub-components, a $\simeq 1$ m long silicon tracker inside a dipole magnet (SVT), to measure charged particle trajectories and vertexes, and a fast lead-tungstate electromagnetic calorimeter (ECal), to measure particle energies, identify electrons and provide a fast trigger signal. The detector setup is shown in Figure 2. HPS employs a three-magnet system with the second dipole serving as an analyzing magnet. The tungsten target is located in front of the analyzing magnet, at about 10 cm from the first SVT layer. To reduce beam-related backgrounds, the target and the SVT are mounted in a vacuum chamber. The ECal is mounted outside the vacuum chamber.

The SVT is made of 6 layers of Si modules, each with two sensors. The introduction of a small stereo angle in between the two (50 or 100 mrad) provides three-dimensional tracking and vertexing. Each layer in one half is supported on a common support plate with independent cooling and readout. The SVT is split in two modules, mounted on top and on bottom of the beam plane, leaving a central "dead-region" where the degraded beam can pass through the detector unobstructed. Sensors are readout by the APV25 chip, developed for the CMS experiment at the CERN LHC [19], with a custom DAQ system based on the Advanced Telecom Communications Architecture (ATCA) technology.

The electromagnetic calorimeter is also split into two halves. Each half of the ECal consists of 221 $PbWO_4$ crystals arranged in five rows. Each row is made of 46 crystals, except the one closest to the beam that has only 37. The light from each crystal is read out by a Large Area Avalanche Photodiode (APD) glued on the back surface. Signals from the APDs are read using custom-made amplier boards before being sent to the data acquisition electronics. Each signal is acquired by a 250 MHz, 12 bit Flash Analog to Digital Converter (FADC), with 8 μs latency. The on-board FPGA processes the digitized samples to provide information about the energy and the timing associated with each hit.

2.1. HPS reach

The projected HPS reach is shown in Figure 3, together with published exclusion limits from other experiments. The green band $a_{\mu,\pm 2\sigma}$ ($a_{\mu,\pm 5\sigma}$) corresponds to the region where the A' can explain the discrepancy between the experimental value of the muon anomalous magnetic

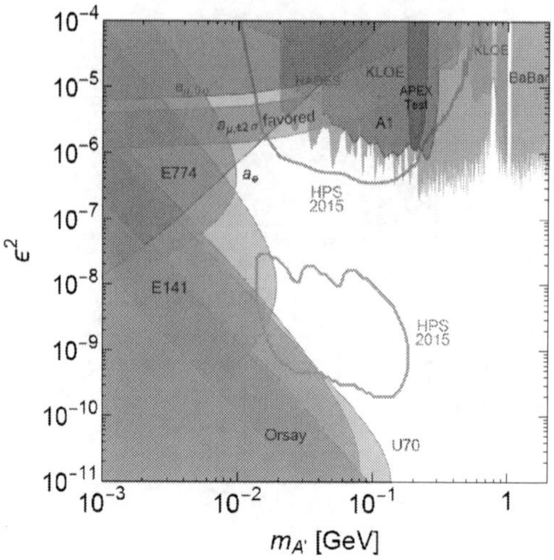

Figure 3. The HPS projected reach. The continuous yellow line corresponds to the 2σ limit. The preferred muon g-2 region is reported, together with published exclusion limits from other experiments. A detailed description of all these results can be found in [16]. The newest results reported by the PHENIX collaboration are not included in this plot [20].

moment, considering a 2σ (5σ) error, and the corresponding Standard Model calculation. The red band a_e, instead, corresponds to the exclusion region from the electron magnetic momentum.

The reach was computed by assuming a one-week run at 1.1 GeV, a one-week run at 2.2 GeV, and a two-weeks run at 4.4 GeV. The two regions in the reach plot correspond to the two analysis techniques employed in the experiment: the resonance search (at larger ε^2) and the detached vertex search (at lower ε^2). They are not connected since a minimum displacement of $\simeq 20$ mm at 2.2 GeV ($\simeq 10$ mm at 6.6 GeV) is required for the vertex-based search.

References

[1] Feldman D, Kors B and Nath P 2007 *Phys. Rev.* D **75** 023503
[2] Andreas S, Goodsell M D and Ringwald A 2013 *Phys. Rev.* D **87** 025007
[3] Holdom B 1986 *Phys. Lett.* B **166** 196
[4] Galison P and Manohar A 1984 *Phys. Lett.* B **136** 279
[5] Goodsell M, Ramos-Sanchez S and Ringwald A 2012 *JHEP* **1201** 021
[6] Cicoli M, Goodsell M, Jaeckel J and Ringwald A 2011 *JHEP* **1107** 114
[7] Essig R, Schuster P and Toro N 2009 *Phys. Rev.* D **80** 015003
[8] Arkani-Hamed N, Finkbeiner D P, Slatyer T R and Weiner N 2009 *Phys. Rev.* D **79** 015014
[9] Pospelov M and Ritz A 2009 *Phys. Lett.* B **671** 391-7
[10] Aguilar M *et al.* 2013 *Phys. Rev. Lett.* **110** 141102
[11] Adriani O *et al.* 2011 *Phys. Rev. Lett.* **106** 201101
[12] Ackermann M *et al.* 2012 *Phys. Rev. Lett.* **108** 011013
[13] Adriani O *et al.* 2009 *Phys. Rev. Lett.* **102** 051101
[14] Hewett J L *et al.* 2012 Fundamental Physics at the Intensity Frontier *Preprint* arXiv:1205.2671 [hep-ex]
[15] Pospelov M 2009 *Phys. Rev.* D **80** 095002
[16] Essig R *et al.* 2013 Dark sectors and new, light, weakly-coupled particles *Preprint* arXiv:1311.0029 [hep-ph]
[17] Bjorken J D, Essig R, Schuster P and Toro N 2009 *Phys. Rev.* D **80** 075018
[18] Adrian P H *et al.* 2013 Heavy Photon Search Experiment at Jefferson Laboratory: proposal for 2014-2015 run
[19] French M, Jones L, Morrissey Q, Neviani A, Turchetta R *et al.* 2001 *Nucl. Instr. Meth.* A **466** 359-65
[20] Adare A *et al.* [PHENIX Collaboration] 2014 Closing the Door for Dark Photons as the Explanation for the Muon g-2 Anomaly *Preprint* arXiv:1409.0851 [nucl-ex]

Muon charged Lepton Flavor Violation search in Europe: the $\mu^+ \to e^+\gamma$ and the $\mu^+ \to e^+e^-e^+$ decays

Angela Papa

Paul Scherrer Institut, 5232 Villigen, OSRA/007, Switzerland

E-mail: angela.papa@psi.ch

Abstract. Lepton flavor violation (LFV) research is currently one of the most exciting branches of particle physics. Flavor violating processes, such as $\mu^+ \to e^+\gamma$ and $\mu^+ \to e^+e^-e^+$, which are strongly suppressed in the Standard Model (SM), are very sensitive to new physics. The MEG experiment and the Mu3e experiment, which search for the $\mu^+ \to e^+\gamma$ and the $\mu^+ \to e^+e^-e^+$ decay respectively, are two precision physics experiments at the forefront of field. They are housed at the Paul Scherrer Institut (PSI), in Switzerland, which provides the most intense continuous muon beam in the world. A summary of the status of the two experiments is given.

1. Introduction

The Standard Model supplemented with massive neutrinos (SM) and enriched by the latest discovery of the Higgs-like particle at the Large Hadron Collider (LHC), summarizes our present best knowledge of particle physics. In spite of its extraordinary success and ability to account for a huge quantity of experimental data, there exist both strong theoretical reasons in particle physics and significant observational hints from astro-particle physics for new physics beyond the SM.

Different experimental approaches can be pursued to address these fundamental physics questions. Amongst others a powerful way is to search for SM forbidden or strongly suppressed (rare) processes which can reveal new physics via indirect production of Beyond Standard Model (BSM) particles, strongly enhancing the probability of these processes to occur [1, 2], exploring physics energy scales up to $\approx 10^4$ TeV.

In the last years, flavor physics became one of the most exciting branches of particle physics due to the high sensitivity to new physics in the so called charged lepton flavor violation (cLFV) processes. Indeed, the simplest and most reliable theoretical SM extensions predict measurable charged lepton flavor violating processes. Furthermore, the observation of neutrino oscillations has clearly demonstrated that neutral lepton flavor is not conserved.

Muonic rare channels such as the $\mu^+ \to e^+\gamma$ decay, the $\mu^+ \to e^+e^+e^-$ decay and $\mu^-N \to e^-N$ conversion are the most promising LFV processes, the so called "golden muonic channels" [3]. The effective lagrangian, which describes in a model independent way the above processes, contains two possible terms contributing to cLFV [4]:

$$\mathcal{L}_{cLFV} = \frac{m_\mu}{(k+1)\Lambda^2}\overline{\mu}_R\sigma_{\mu\nu}e_L F^{\mu\nu} + \frac{k}{(k+1)\Lambda^2}\overline{\mu}_R\gamma_\mu e_L \overline{f}\gamma^\mu f, \tag{1}$$

where the first term describes the photonic interaction, associated with the photon field $F^{\mu\nu}$, while the second term describes the fermionic interaction, associated with the fermion f. In the formula m_μ is the muon mass, Λ represents the effective mass scale and k governs the relative size of the two different types of operators. While $\mu^+ \to e^+\gamma$ proceeds only via the first term, the $\mu^+ \to e^+e^+e^-$ decay and the $\mu^- N \to e^- N$ conversion may occur also through the second one. If nature prefers the $k = 0$ case, the $\mu^+ \to e^+\gamma$ decay is favored and appears as the most sensitive discovery channel. On the other hand for the $k \neq 0$ cases the $\mu^+ \to e^+e^+e^-$ decay and the $\mu^- N \to e^- N$ conversion become more and more prominent as k grows [5]. A complementary search approach is the only way to reveal the nature of the physics which induces cLFV if an evidence of it is found.

Rare decay searches require the use of high beam intensities and demand detectors able to work in critically high background environments. Two of the three golden processes can be studied in Europe, at the Paul Scherrer Institut (PSI) as the unique laboratory in the world delivering the highest continuous (DC) positive muon beam. We will focus on them, starting from the experiment that has just completed its data acquisition and is presently finalizing the data analysis.

2. The MEG experiment and its upgrade

The MEG experiment [6] searches for the $\mu^+ \to e^+\gamma$ decay and has recently set the most stringent upper limit on its branching ratio $\mathcal{B}(\mu^+ \to e^+\gamma) < 5.7 \times 10^{-13}$ at 90 % C.L. [7], using the 2009-2011 data sample. It is a factor 20 better than the previous limit set by the MEGA experiment [8] and also the strongest upper limit among all the other particle decays. The sample statistic has been increased by a factor two adding the 2012-2013 data sample to the previous one and the analysis is ongoing. We reached the planned sensitivity, whose limit comes from the accidental background. The strong scientific motivation of searching for cLFV pushed the collaboration to think about an upgrade of the experiment, aiming at enhancing the sensitivity by a factor 10 [9]. The upgrade preserves the general idea of the previous experiment in terms of the $\mu^+ \to \mathrm{e}^+\gamma$ signature and the detectors used to extract it.

The signature of a $\mu^+ \to \mathrm{e}^+\gamma$ decay at rest is a back-to-back, mono-energetic, time coincident photon and positron pair.

A new positron-tracking chamber (DCH) with significantly improved acceptance, resolutions and efficiency is under construction. The DCH provides a measurement of the positron momentum and the emission direction (which is described by the polar θ and azimuthal ϕ angles), from which the positron energy (E_e) and the decay vertex coordinates on the target plane ($Z_{proj,e}$ and Y_e) are extracted[1]. The positron trajectory is measured up to the point where the positron reaches the new timing counter (TC), with a strong reduced presence of passive material and an increased number of hits per track with respect to the current detector. The new TC, which provides a measurements of the positron time t_e, is made by a large number of small ultra-fast scintillator plates coupled with silicon photomultiplier (SiPM). The high segmentation allows to work at a higher muon beam rate and to reach a better timing resolution. The energy (E_γ), the time (t_γ) and the interaction point (described by the (u, v) coordinates on the LXe front face and the depth w) of the photon are measured with a large (900 liters) homogeneous liquid xenon calorimeter (LXe). The LXe upgrade namely refers to the replacement of the current PMTs (2 inch diameter) of the calorimeter front face with smaller Muti-Pixel Photon Counter (MPPC Hamamatsu) 12 $\times$ 12 mm^2 for better energy and position resolutions. Two completely new detectors are considered to be added to the framework of the MEGII experiment: a) an active target which provides a continuous beam monitoring and a direct measurement of the

[1] We refer to a cylindrical coordinate system (r,ϕ,z) with origin in the centre of the DCH. The z-axis is along the incoming muon beam. The axis defining $\phi = 0$ (the x-axis of the corresponding Cartesian coordinate system) is directed opposite to the centre of the liquid xenon detector.

Table 1. Comparison between the MEG and the MEGII detector.

Detector Resolutions (σ)	MEG	MEGII (Upgrade scenario)
e^+ energy (keV)	305	130
e^+ timing (ps)	70	35
e^+ θ (mrad)	10.6	5.3
e^+ ϕ (mrad)	7.5	5.0
e^+ vertex (mm) Z_{proj}/Y	1.9 / 1.3	1.6 / 0.7
γ energy (%) (w <2 cm)/(w >2 cm)	2.6 / 1.7	1.3 / 1.0
γ timing (ps)	67	similar
γ position (mm) $u/v/w$	5 / 5 / 6	2.6 / 2.2 / 5
γ-e^+ timing (ps)	127	84
Detector Efficiency (%)		
e^+	40	90
γ	63	69
trigger	≈ 99	≈ 99

decay vertex; b) a radiative decay counters, enabling to increase the capability of rejecting the accidental background. The upgraded detectors demand also new calibration and monitoring methods: a dedicated monochromatic positron beam with an energy very close to the MEG signal is under study to fully explore the new DCH and TC. A new DAQ system (WaveDREAM) will provide a larger bandwidth of the waveform digitization and a better channel inter-calibration implying better detector resolutions. A summary of the current detector performances and the expected ones for the MEG II is given in Tab. 1.

3. The Mu3e experiment

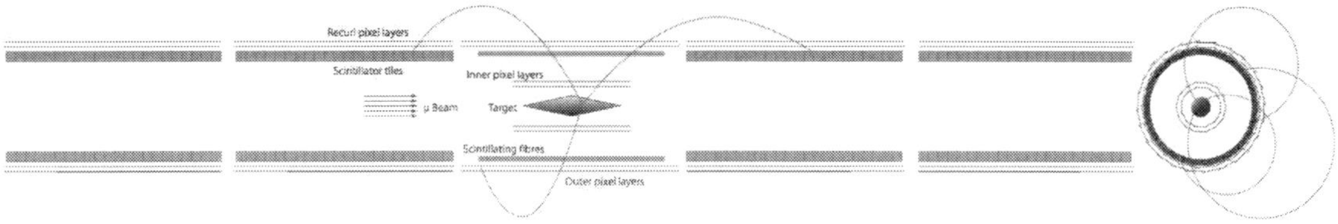

Figure 1. Schematic view of the final Mu3e experimental setup.

The approved Mu3e experiment [10] will search for the $\mu^+ \to e^+e^+e^-$ decay aiming at a sensitivity of a few $\times\ 10^{-16}$, four orders of magnitude better than the previous upper limit on the $\mu^+ \to e^+e^+e^-$ decay set by the SINDRUM experiment $\mathcal{B}(\mu^+ \to e^+e^+e^-) < 1.0 \times 10^{-12}$ [11]. This expected strong improvement is possible because the previous experiment was limited only by statistics. Furthermore the achieved momentum resolution of the SINDRUM tracking system (4.2 % in σ) leaves space for a large improvement. Indeed the Mu3e experiment, based on a silicon tracker, allows to enhance the sensitivity at least by two orders of magnitude using the presently available beam intensity (10^8 muons/s). In order to explore the 10^{-16} sensitivity region a DC positive muon beam delivering 10^9 muons/s is required. Currently there are no such (pulsed or DC) high-intensity muon sources available in the world. A feasibility study for a DC muon beam aiming at 10^{10} muons/s (more than what is required by Mu3e to cover other physics programmes) is undergoing at PSI.

The $\mu^+ \rightarrow e^+e^+e^-$ decay signal is defined by its final state: two positrons and one electron without any additional neutrinos. The muons are stopped in the target. All the tracks originating from the decay share a single common vertex and they are coincident in time. The invariant mass of the three tracks, measured at the vertex position, is identical to the muon mass. Any background to the signal comes from internal conversion and accidental processes that mimic the signal. The main background source is given by the $\mu^+ \rightarrow e^+e^+e^-\nu\overline{\nu}$ decay. In order to reject that a very good energy resolution is requested (below 1 MeV/c on the reconstructed momentum sum resolution). The accidental background is not coincident in time or space and the total momentum does not fulfill the requirements given above. To suppress these kinds of backgrounds a high vertex (200 μm) and time resolution (100 ps) are needed.

The proposed Mu3e detector is based on two double layers of High Voltage Monolithic Active Pixel Sensors (HV-MAPS) [12] around a hollow double cone target. The outer two pixel sensor layers are extended upstream and downstream to provide precise momentum measurements in an extended region with the help of re-curling electrons. The silicon detector layers are supplemented by two timing systems, a scintillating fibre tracker in the central part and scintillating tiles inside the re-curl layers. Precise timing of all tracks is necessary for event building and to suppress accidental combinatorial background. The entire detector is built in a cylindrical shape around a beam pipe, with a total length of approximately 2 m, inside a 1 T solenoid magnet with 1 m inside diameter and 2.5 m total length.

A schematic view of the detector with two sets of re-curl stations for high intensity physics runs is shown in Fig. 1.

Acknowledgments
We are grateful for the support and co-operation provided by PSI as the host laboratory and to the technical and engineering staff of the institutes of both experiments.

References
[1] Hewett J L et al 2012 *arXiv:1205.2671v1 [hep-ex]*
[2] Raidal M et al 2008 *Eur. Phys. J.* C **57** 13
[3] Marciano W J Mori T and Roney J M 2008 *Annu. Rev. Nucl. Part. Sci.* **58** 315
[4] Kuno Y and Okada Y 2001 *Rew. Mod. Phys.* **73** 151
[5] de Gouvea A and Vogel P 2013 *arXiv:1303.4097v2 [heo=ph]*
[6] Adam J et al (MEG Collaboration) 2013 *Eur. Phys. J.* C **73** 2365
[7] Adam J et al (MEG Collaboration) 2013 *Phy. Rev. Lett.* **110** 201801
[8] Brooks M L et al. (MEGA collaboration) 1999 *Phys. Rev. Lett.* **83** 1521
[9] Baldini A M et al (MEG Collaboration) 2013 *arXiv:1301.7225v2 [physics.ins-det]*
[10] Blondel A et al (Mu3e Collaboration) 2013 *arXiv:1301.611v1[physics.ins-det]*
[11] Bellgardt U et al (SINDRUM Collaboration)1988 *Nucl. Phys.* B **299** 1
[12] Peric I 2007 *Nucl. Instr. Meth.* A **582** 876

Theory overview of $(g-2)_\mu$

Liliana Velasco-Sevilla

Department of Physics and Technology, University of Bergen, PO Box 7803, 5020 Bergen, Norway

E-mail: liliana.velascosevilla@gmail.com

Abstract. For more than a decade the SM determination of the anomalous magnetic moment of the muon, $(g-2)_\mu$, has shown a deviation from its experimental value. There is currently a more than 3 standard deviations difference. However, the SM determination could be greatly improved, mainly by the reduction of the uncertainties associated to the calculation of the hadronic contributions to $(g-2)_\mu$. Supersymmetric models are capable to explain the difference between experimental and SM values of $(g-2)_\mu$. In these models, the study of the correlations to other observables, for example BR$(\mu \to e\gamma)$, can help to select models where constraints from the LHC are compatible and can give definite predictions.

1. SM overview

Current status. A lepton has a magnetic moment, g_ℓ, aligned along its spin: $\vec{\mu}_\ell = g_\ell \frac{Q_\ell e}{2m_\ell}\vec{s}$, where Q_ℓ, is the charge of the lepton, m_ℓ and s its spin. Without quantum corrections $g_\ell = 2$, but with quantum corrections the *anomalous moment*, $a_\ell \equiv (g-2)_\ell/2$, has a non zero value. In the Standard Model (SM), the largest contribution to a_μ comes from the one loop, 1ℓ, QED correction, Fig. 1, but EW and Hadronic corrections are crucial to determine the departure of the SM with the experimental value [1]. The up to date value of $a_\mu^{\rm QED}$ has been computed at 5ℓ [2] and the main uncertainty comes from the fine structure constant α. The other uncertainties come, respectively, from the lepton mass ratios, the 4ℓ and the 5ℓ corrections. This value is presented in Tab. 1. The $a_\mu^{\rm EW}$ contribution has been calculated up to 2ℓ, [3], and is presented in the second row of Tab. 1. Its main uncertainty comes from hadronic considerations at 2ℓ, and unknown 3ℓ effects. The quoted value includes a reevaluation using the latest constraints from the LHC Higgs data, based on previous calculations from [4, 5]. The hadronic contributions

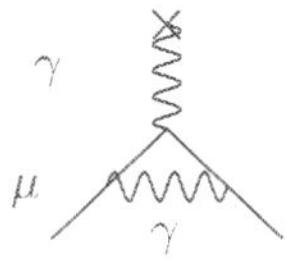

Figure 1. 1ℓ correction to a_μ in QED.

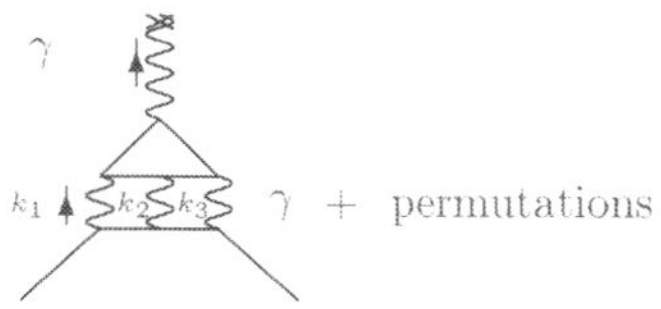

Figure 2. a_μ^{HLbL}: Hadronic Light by Light contributions to a_μ.

are divided in $a_\mu^{HLO,HNLO}$ and Hadronic light by light contributions: a_μ^{HLbL}. The LO Hadronic contribution is calculated from the dispersion relation

$$a_\mu^{HLO} = \left(\frac{\alpha m_\mu}{3\pi}\right)^2 \int_{m_\pi^2}^{\infty} \frac{ds}{s^2} K(s) R(s)\,, \quad R \equiv \frac{\sigma_{\rm tot}(e^+e^- \to \text{hadrons})}{\sigma(e^+e^- \to \mu^+\mu^-)} \tag{1}$$

where the kinematic form factor varies from $K(s = m_\pi^2) = 0.4$ to $K(s \to \infty) = 0$. The dispersion relation connects the bare cross section for e^+e^- annihilation into hadrons to the hadronic vacuum polarization contritubution to a_μ. There are presently different analyses, each with different models to calculate the dispersion relation, Eq. (1), based on existing experimental data sets. However, there has not been a consensus on how these values should be put together. The most recent calculations can be found in [6, 7, 8, 9]. Their values are respectively $a_\mu^{HLO} = (6909.6\pm, 46.5, 6812.3 \pm 45.1, 6923 \pm 42, 6949 \pm 43) \times 10^{-11}$. Taking into account an average of these values, we obtain $(6\,898 \pm 44) \times 10^{-11}$. The value of the NLO Hadronic contribution, a_μ^{HNLO}, has been calculated just by one collaboration, [9], and is presented in Tab. 1.

The contribution a_μ^{HLbL}, Fig. 2, must be determined using hadronic models, but currently there is not a single computation valid for all values of the momenta, k_i of the photons, Fig. 2. There are computations done in two limits: long and short distance, which correspond respectively to small momenta k_i, and large or mixed momenta k_i. Most part of the long-distance evaluations agree basically on QCD chiral & large N_c, (c = color) limit and how to take into account the π^0 contributions modulated by $\pi^0\gamma^*\gamma^*$ form factors (see for example [10]). However, up to now, there has been just an independent evaluation of the short distance limit [11]. Despite of this, here is an agreed value of leading collaborations [12], quoted in Tab. 1.

Table 1. Summary of SM results. The results marked with * has been computed here, taking into account references [6, 7, 8, 9]. There is not an agreed official value of the LO contribution, but taking as an average the four references quoted, we obtain $\Delta a_\mu(E821 - SM) = 311 \pm 81$.

	Contribution ($\times 10^{-11}$) UNITS
QED [2]	$116\,584\,718.951 \pm 0.009 \pm 0.019 \pm 0.007 \pm 0.077_\alpha$
EW [3]	154 ± 1
HVP(LO) [*]	$6\,898 \pm 44$
HVP(NLO) [9]	-98.4 ± 0.7
HLbL [12]	105 ± 26
Total SM [*]	$116\,591\,778 \pm 44_{\rm HLO} \pm 26_{\rm HNLO} \pm 2_{\rm other}\ (\pm 51_{\rm tot})$

If we compare the average value a_μ of Tab. 1 to the experimental value from E821, $a_\mu^{E821} = (116592089 \pm 63) \times 10^{-11}$, and the revised value of $\lambda = \mu_\mu/\mu_p$, we obtain $\Delta a_\mu(E821 - SM) = 311 \pm 81$. Since there are different analyses of a_μ^{HLO}, what we can say in general is that there is currently a more than 3σ difference between SM and experimental results.

Future improvements in the theoretical error estimation. The evaluation of a_μ^{HLO} from $e^+e^- \to$ *hadrons* requires a better combination from different experimental data sets used to determine the dispersion relations, Eq. (1), in particular some of the old sets should be revisited. The expected error reduction from the HLO evaluation of [8] is to bring 42×10^{-11} down to 26×10^{-11} [13]. Current lattice estimates for the error are at the $\sim 5\%$ and the plan is to reduce them to $\sim 2\%$, within the next five years and within a decade to be competitive to the theoretical estimates taking into account the experimental information of $e^+e^- \to$ *hadrons* [14]. Since a_μ^{HLbL} cannot be evaluated by dispersions relations, a big decrease is not expected but some experimental data can help to pin down related amplitudes. The ultimate goal is to reduce the uncertainty from 25% to 10%. A calculation on the lattice even with an error $\sim$ 30% would have an impact since it could pin down important difference in the models used to determine short and long distance limits of a_μ^{HLbL} [13, 14].

2. Supersymmetry overview

In the MSSM, at 1ℓ, the relevant parameters to determine the supersymmetric (SUSY) contribution to a_μ are: $m_{\tilde{p}}$, $\tan\beta = \frac{\langle H_u \rangle}{\langle H_d \rangle}$, μ $(\mu H_u H_d)$, where $m_{\tilde{p}}$ is the scale of the particles involved in the 1ℓ correction to a_μ, Fig. 3. When all the relevant SUSY masses $m_{\tilde{p}}$ are of the same order, the contribution to a_μ has the form $a_\mu^{\rm SUSY} = 120 \times 10^{-11}\ \tan\beta\ {\rm sign}(\mu)\ (100\ {\rm GeV/m_{\tilde{p}}})^2$.

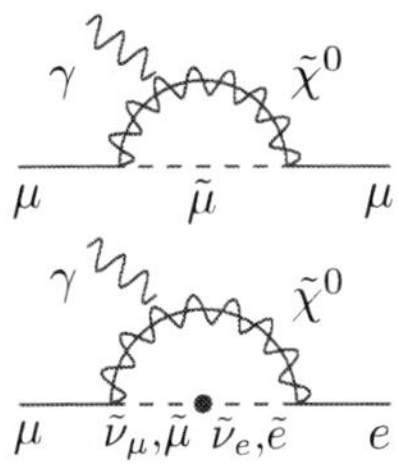

Figure 3. Top: example of a SUSY diagram mediating a_μ. Bottom: The flavor-violating analogous diagram, mediating the transition $\mu \to e\gamma$.

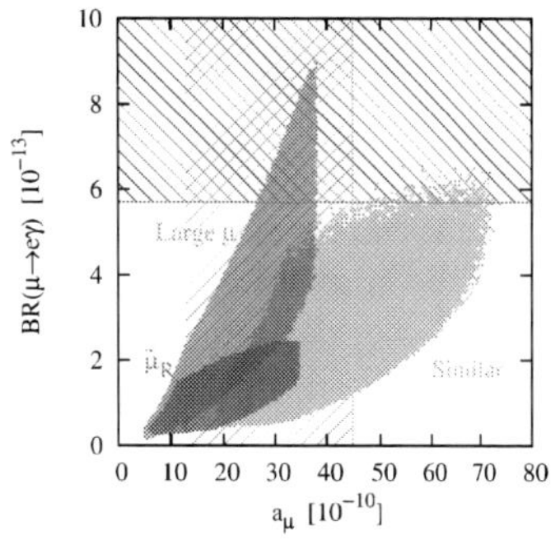

Figure 4. SUSY contribution to a_μ vs BR($\mu \to e\gamma$) for similar SUSY masses (light blue), large μ (orange), and heavy left-handed sleptons ($\tilde{\chi}^0 - \tilde{\mu}_R$, violet). In each case, $\delta_{LL} = \delta_{RR} = 2 \times 10^{-5}$, $\tan\beta = 50$.

The LHC searches for SUSY particles, indicate that $m_{\tilde{q}}$, $m_{\tilde{g}} \gtrsim 1$ TeV (the masses of the squarks -in general- and the gluino, respectively). Since a_μ requires $m_{\tilde{p}} = O(100)$ GeV, we should consider that the supersymmetric particles involved in a_μ should be smaller than those for which the LHC has put constraints above the 1 TeV range, hence we need $m_{\tilde{q}}$, $m_{\tilde{g}} \gg m_{\tilde{p}} = m_{\tilde{\ell}}$, $m_{\tilde{\chi}^\pm}$, $m_{\tilde{\chi}^0}$. These last three masses are respectively the masses of the sleptons, the charginos and the neutralinos. With these considerations, a solution to the discrepancy between experimental and SM determinations to a_μ is still possible. Furthermore, when there are some big hierarchies between some sfermions and the sleptons relevant to a_μ, 2ℓ effects could be important [15]. In [16] we have studied cases where there is a hierarchy among the particles entering into the 1ℓ calculation of $a_\mu^{\rm SUSY}$ and hence, the typical behavior of the general 1ℓ corrections does not longer hold. A nice example is the case where the lightest neutralino contribution dominates and we can see an increase of the value of $a_\mu^{\rm SUSY}$ as the value of μ grows. This result holds when the hierarchy is such that $\mu > M_2 > M_1$, where $M_{i=1,2}$ are the masses of the gauginos coupled to the bino and the $SU(2)_L$ gauginos, respectively. In this case, $a_\mu \approx g_1^2 \frac{m_\mu^2}{48\pi^2} \frac{\mu \tan\beta}{M_1} F\left[m^2_{\tilde{\chi}^0_i}/m^2_{\tilde{\ell}_m}\right]$, where in the range of a_μ^{SUSY} that fits the difference $\Delta a_\mu(E821 - SM)$, the function $F[m^2_{\tilde{\chi}^0_i}/m^2_{\tilde{\ell}_m}]$ is $O(1)$ and the value of μ can be up to 5 TeV. This example is a clear departure from the typical behavior at 1ℓ and is relevant for constraints at the LHC, and as we will see next, for correlations to BR($\mu \to e\gamma$). We call this scenario "Large-μ".

3. Correlation to $\mu \to e\gamma$

The SUSY contributions to a_μ and to the decay $\mu \to e\gamma$ are given by very similar Feynman diagrams. There has been already more than a decade (please see references in [16]) of works reporting correlations in specific scenarios, in particular if a_μ is dominated by a single diagram. In [16] we disentangled at which degree the correlations depended on the specific relations on the masses of the MSSM. We have shown how the correlations are weakened by significant cancellations between diagrams in large parts of the MSSM parameter space. Nevertheless, the order of magnitude of BR($\mu \to e\gamma$) for a fixed flavor-violating parameter can often be predicted. We can define flavor violating parameters $(\delta^l_{12})_{XX} \equiv m^2_{\tilde{X}_{12}}/\sqrt{m^2_{\tilde{X}_{11}} m^2_{\tilde{X}_{22}}}$, for $X = L, R$, left and right contributions. What we found is that by fixing this parameter, we can study the correlations between BR($\mu \to e\gamma$) and a_μ in a more systematic way. In Fig. 4 we show a plot of BR($\mu \to e\gamma$) vs. a_μ showing three different regions. In blue (light) we show an scenario where SUSY masses are varied between 300 and 600 GeV, $\tan\beta = 50$ and $(\delta^l_{12})_{LL} = 2 \times 10^{-5}$. The Large-$\mu$ region, which we have described above, is shown in orange (dark) and finally a region labeled by "$\tilde{\mu}_R$" is shown towards the left-bottom corner. The vertical green-shaded area corresponds to the 2σ difference between experimental and SM determinations, using a_μ^{HLO} from [8], and the rest of the values from Tab. 1, $\Delta a_\mu(E821 - SM) = 287 \pm 80$. The red horizontal line corresponds to the upper limit of BR($\mu \to e\gamma$) [17]. We can see that for similar SUSY masses, there is not a correlation, but when we go to a particular hierarchy we start to see a behavior were we can identify a correlation. The use of this correlation is as follows. We can write BR($\mu \to e\gamma$) $= \frac{3\pi^2 e^2}{G_F^2 m_\mu^4}\left(|a_{\mu e\gamma L}|^2 + |a_{\mu e\gamma R}|^2\right)$, where $a_{\mu e\gamma X}$, $X = L, R$ are the contributions to the amplitude of the process $\mu \to e\gamma$, with Left and Right chiralities, respectively (see. eq. 5 of [16]). By calculating the values of $a_{\mu e\gamma X}$ and a_μ for the different scenarios, we are able to write $|a_{\mu e\gamma X}| = |a_\mu| f\, m^2_{\tilde{p}}$, where $f = O(1)$ number and now $m^2_{\tilde{p}}$ is the value of the SUSY mass involved in the specific scenario, for example in the Large-μ scenario, the only particles involved will be the particles in the loop of the top figure of Fig. 3. Now, the requirement of a_μ to lie within the difference of the experimental and SM values fixes $m^2_{\tilde{p}}$, the coefficient f changes with the scenario, but the point is that then $|a_{\mu e\gamma X}| \propto (\delta^l_{12})_{XX}$ is constrained if we want to satisfy the bound on BR($\mu \to e\gamma$) and this cannot be evaded by simply raising the value of the parameter $m^2_{\tilde{p}}$. The ultimate application of this kind of analyses would be attained if BR($\mu \to e\gamma$) could eventually be measured: that would give definite indications for the required values of all relevant mass parameters.

Acknowledgments

I thank D. Stöckinger for collaboration, discussions and encouragement. I also would like to thank the organizers of BEACH 2014, specially C. Lazzeroni.

References

[1] Bennett G W *et al.* (Muon $(g-2)$ Collaboration) 2006 *Phys. Rev.* **D73** 072003
[2] T Aoyama M Hayakawa T K and Nio M 2012 *Phys. Rev. Lett.* **109** 111808
[3] Gnendiger C, Stöckinger D and Stöckinger-Kim H 2013 *Phys. Rev.* **D88** 053005
[4] Czarnecki A, Krause B and Marciano W J 1996 *Phys.Rev.Lett.* **76** 3267–3270
[5] Knecht M, Peris S, Perrottet M and de Rafael E 2004 *JHEP* **0403** 035
[6] Jegerlehner F and Szafron R 2011 *Eur.Phys.J.* **C71** 1632
[7] Benayoun M, David P, DelBuono L and Jegerlehner F 2013 *Eur.Phys.J.* **C73** 2453
[8] Davier M, Hoecker A, Malaescu B and Zhang Z 2011 *Eur.Phys.J.* **C71** 1515
[9] Hagiwara K, Liao R, Martin A D, Nomura D and Teubner T 2011 *J.Phys.* **G38** 085003
[10] Knecht M, Nyffeler A, Perrottet M and de Rafael E 2002 *Phys.Rev.Lett.* **88** 071802
[11] Melnikov K and Vainshtein A 2004 *Phys.Rev.* **D70** 113006
[12] J. Prades, E. de Rafael and A. Vainshtein, (*Preprint* arXiv:0901.0306)
[13] Blum T, Denig A, Logashenko I, de Rafael E, Lee Roberts B *et al.* 2013, (*Preprint* arXiv: 1311.2198)
[14] Benayoun M, Bijnens J, Blum T, Caprini I, Colangelo G *et al.* 2014 (*Preprint* arXiv: 1407.4021)
[15] von Weitershausen P, Schäfer M, Stöckinger-Kim H and Stöckinger D 2010 *Phys. Rev.* **D81** 093004
[16] Kersten J, Park J-h, Stöckinger D and Velasco-Sevilla L 2014 *JHEP* **1408** 118
[17] Adam J *et al.* (MEG Collaboration) 2013 *Phys. Rev. Lett.* **110** 201801

Journal of Physics: Conference Series **556** (2014) 012067 doi:10.1088/1742-6596/556/1/012067

LFV and exotics at the NA62 experiment

Viacheslav Duk [1]
INFN Perugia, Italy
E-mail: `Viacheslav.Duk@cern.ch`

Abstract. The NA62 experiment at the CERN SPS is aimed at measuring the branching fraction of the ultrarare decay $K^+ \to \pi^+\nu\bar{\nu}$ with $\sim$10% precision by collecting $\sim 10^{13}$ kaon decays in two years of data taking. This amount of data will allow to carry out a wide program of searching for rare and forbidden (within the Standard Model) K^+ and π^0 decays, including sterile neutrinos, lepton flavor violating modes, exotic particles (e.g. "dark photons"). The expected performances of the NA62 setup will allow to improve existing limits for several decay modes.

1. Introduction

Although the lepton flavor violation (LFV) was observed in neutrino mixing, the predictions for the branching ratios (BR) of LFV meson decays within SM are very small. On the other side, many extensions of the SM give rise to the lepton flavor and/or number violation. This leads to a straightforward conclusion that any observation of a LFV process in the meson decays is a clear signal of the New Physics beyond the SM.

The high intensity of kaon beams and clear experimental signature allow to have a good sensitivity to LFV processes. At present the upper limits reach $10^{-11} - 10^{-12}$. Upper limits for the main decay modes are summarized in Table 1.

[1] for the NA62 collaboration: G. Aglieri Rinella, R. Aliberti, F. Ambrosino, B. Angelucci, A. Antonelli, G. Anzivino, R. Arcidiacono, I. Azhinenko, S. Balev, J. Bendotti, A. Biagioni, C. Biino, A. Bizzeti, T. Blazek, A. Blik, B. Bloch-Devaux, V. Bolotov, V. Bonaiuto, M. Bragadireanu, D. Britton, G. Britvich, N. Brook, F. Bucci, V. Buescher, F. Butin, E. Capitolo, C. Capoccia, T. Capussela, V. Carassiti, N. Cartiglia, A. Cassese, A. Catinaccio, A. Cecchetti, A. Ceccucci, P. Cenci, V. Cerny, C. Cerri, O. Chikilev, R. Ciaranfi, G. Collazuol, P. Cooke, P. Cooper, G. Corradi, E. Cortina Gil, F. Costantini, A. Cotta Ramusino, D. Coward, G. D'Agostini, J. Dainton, P. Dalpiaz, H. Danielsson, J. Degrange, N. De Simone, D. Di Filippo, L. Di Lella, N. Dixon, N. Doble, V. Duk, V. Elsha, J. Engelfried, T. Enik, V. Falaleev, R. Fantechi, L. Federici, M. Fiorini, J. Fry, A. Fucci, L. Fulton, S. Gallorini, L. Gatignon, A. Gianoli, S. Giudici, L. Glonti, A. Goncalves Martins, F. Gonnella, E. Goudzovski, R. Guida, E. Gushchin, F. Hahn, B. Hallgren, H. Heath, F. Herman, D. Hutchcroft, E. Iacopini, O. Jamet, P. Jarron, K. Kampf, J. Kaplon, V. Karjavin, V. Kekelidze, S. Kholodenko, G. Khoriauli, A. Khudyakov, Yu. Kiryushin, K. Kleinknecht, A. Kluge, M. Koval, V. Kozhuharov, M. Krivda, Y. Kudenko, J. Kunze, G. Lamanna, C. Lazzeroni, R. Leitner, R. Lenci, M. Lenti, E. Leonardi, P. Lichard, R. Lietava, L. Litov, D. Lomidze, A. Lonardo, N. Lurkin, D. Madigozhin, G. Maire, A. Makarov, I. Mannelli, G. Mannocchi, A. Mapelli, F. Marchetto, P. Massarotti, K. Massri, P. Matak, G. Mazza, E. Menichetti, M. Mirra, M. Misheva, N. Molokanova, J. Morant, M. Morel, M. Moulson, S. Movchan, D. Munday, M. Napolitano, F. Newson, A. Norton, M. Noy, G. Nuessle, V. Obraztsov, S. Padolski, R. Page, V. Palladino, A. Pardons, E. Pedreschi, M. Pepe, F. Perez Gomez, M. Perrin-Terrin, P. Petrov, F. Petrucci, R. Piandani, M. Piccini, D. Pietreanu, J. Pinzino, M. Pivanti, I. Polenkevich, I. Popov, Yu. Potrebenikov, D. Protopopescu, F. Raffaelli, M. Raggi, P. Riedler, A. Romano, P. Rubin, G. Ruggiero, V. Russo, V. Ryjov, A. Salamon, G. Salina, V. Samsonov, E. Santovetti, G. Saracino, F. Sargeni, S. Schifano, V. Semenov, A. Sergi, M. Serra, S. Shkarovskiy, A. Sotnikov, V. Sougonyaev, M. Sozzi, T. Spadaro, F. Spinella, R. Staley, M. Statera, P. Sutcliffe, N. Szilasi, D. Tagnani, M. Valdata-Nappi, P. Valente, M. Vasile, V. Vassilieva, B. Velghe, M. Veltri, S. Venditti, M. Vormstein, H. Wahl, R. Wanke, P. Wertelaers, A. Winhart, R. Winston, B. Wrona, O. Yushchenko, M. Zamkovsky, A. Zinchenko.

Table 1. Current limits on the branching fractions of the main LFV decay modes in charged kaon decays.

Decay mode	Upper limit (at 90% CL)	Experiment	Ref.
$K^+ \to \pi^+\mu^+e^-$	1.3×10^{-11}	BNL E777/865	[1]
$K^+ \to \pi^+\mu^-e^+$	5.2×10^{-10}	BNL E865	[2]
$K^+ \to \pi^-\mu^+e^+$	5.0×10^{-10}	BNL E865	[2]
$K^+ \to \pi^-e^+e^+$	6.4×10^{-10}	BNL E865	[2]
$K^+ \to \pi^-\mu^+\mu^+$	1.1×10^{-9}	NA48/2	[3]
$K^+ \to \mu^-\nu e^+e^-$	2.0×10^{-8}	Geneva-Saclay	[4]
$K^+ \to e^-\nu\mu^+\mu^+$	no data		

2. The NA62 experiment

The main goal of the experiment is to measure the branching ratio of the decay $K^+ \to \pi^+\nu\bar{\nu}$ with ~10% precision [5]. In two years of data taking it is supposed to have about 10^{13} kaon decays inside the fiducial volume [6]. This statistics provides good opportunities for improving the existing limits on the LFV decay modes listed in Table 1. In addition, there will be $\sim 2\times10^{12}$ π^0 decays from the decay mode $K^+ \to \pi^+\pi^0$ (BR $\sim$ 21%) which can be used in other studies.

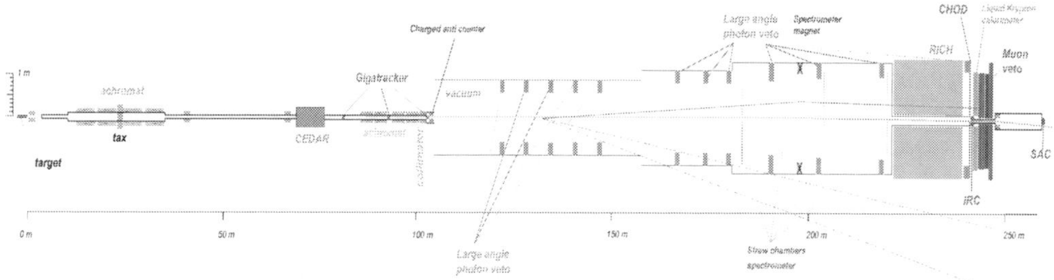

Figure 1. NA62 experimental setup.

The layout of the experimental setup is shown in Fig. 1. The high-intensity (750 MHz) unseparated (~6% of positively charged kaons) hadron beam with a 75 GeV/c momentum is obtained from the SPS primary beam (protons with 400 GeV/c momentum) impinging on a Be target. Kaons are identified by a differential Cerenkov counter (CEDAR). The momentum and direction of beam particles are measured by a silicon pixel beam spectrometer (Gigatracker) with a space resolution ~0.2%. The decay region has a length of ~65 m and is evacuated to 10^{-6} mbar. The momentum of secondary charged particles (produced in kaon decays) is measured by a magnetic spectrometer composed of straw tubes (four stations) and a magnet with p_T ~270 MeV/c. The energy of secondary photons and electrons flying at small angles with respect to the beam axis is measured by a liquid krypton electromagnetic calorimeter, while large angle photons are detected by a system of 12 rings of lead-glass blocks. The identification of secondary particles (muon-pion separation at the level better than 1%) in the momentum range 15–35 GeV/c is done in a RICH detector filled with Ne at atmospheric pressure. A highly segmented muon identification system (Muon veto) is located behind the krypton calorimeter.

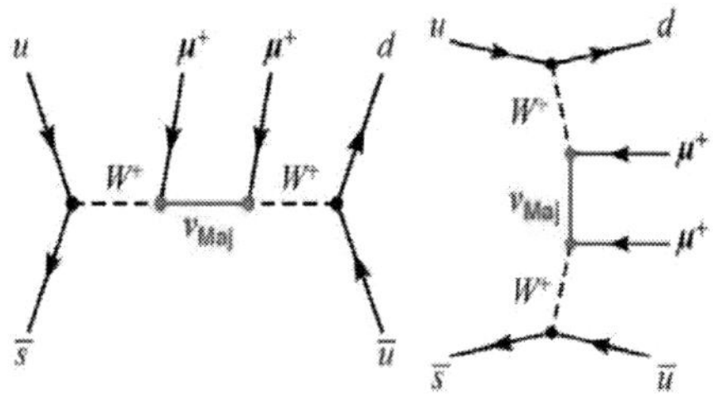

Figure 2. Lowest-order contributions to the decay $K^+ \to \pi^-\mu^+\mu^+$.

3. The decay $K^+ \to \pi^-\mu^+\mu^+$

Among various processes from Table 1 the decay $K^+ \to \pi^-\mu^+\mu^+$ provides a unique possibility to study the neutrinoless double beta decay in the second generation, since it implies the exchange of a virtual Majorana neutrino (Fig. 2).

The best upper limit on the BR of this decay was obtained by the NA48/2 experiment [3]. The result is based on 52 events in the signal region and a background estimation 52.6±19.8 obtained from Monte Carlo simulation. The upper limit was found to be $\mathrm{BR}(K^\pm \to \pi^\mp\mu^\pm\mu^\pm) < 1.1 \times 10^{-9}$(90% CL). The expected NA62 single-event sensitivity can reach 10^{-12}.

4. Search for heavy neutrinos in $K^+ \to \mu^+\nu_h$ decay

Searches for heavy neutrino in $K^+ \to \mu^+\nu_h$ decay can be of two types: production and decay. In the former case the peaks in the muon momentum (or missing mass) distribution are searched for regardless of the neutrino lifetime, while in the latter case the lifetime should be short enough to have a relatively large acceptance for the decay of interest. The current status of heavy neutrino searches (limit on the mixing matrix element squared $|U_{\mu h}|^2$ vs ν_h mass) is shown in Fig. 3.

The ongoing analysis of the 2007 NA62 R_K data [12] shows that in the absence of background the upper limit could be set at the level of 10^{-7} for the mass region $100 < m_h < 400$ MeV/c^2. Preliminary studies including systematics (which is mostly due to the background subtraction) indicate that the limits from Fig. 3 in the high mass region ($300 < m_h < 350$ MeV/c^2) can be improved.

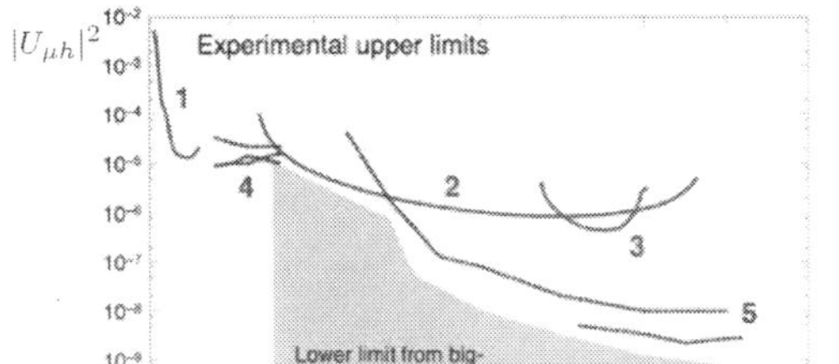

Figure 3. Heavy neutrino search in kaon and pion decays. Production searches are shown in green: 1) $\pi \to \mu\nu_h$, PSI 1981 [7]; 2) $K \to \mu\nu_h$, KEK 1982 [8]; 3)$K \to \mu\nu_h + inv$, LBL 1973 [9]. Decay searches are drawn in red: 4) $K \to \mu\nu_h, \nu_h \to \nu\gamma$ ($\tau_h = 10^{-9}, 10^{-10}, 10^{-11}$ s), ISTRA+ 2012 [10]; 5) $K \to \mu\nu_h, \nu_h \to \mu^\mp e^\pm\nu$ (above) and $\nu_h \to \mu^\mp\pi^\pm$ (below), PS191 1988 [11]. The grey area corresponds to the lower limit from the big-bang nucleosynthesis.

5. Search for "dark photons" in π^0 decays

The U boson (or "dark photon", or A′) is a new light gauge boson that could mediate interactions with dark matter constituents explaining several unexpected astrophysical measurements (see [13], [14]) and the so called $(g-2)_\mu$ anomaly ($> 3\sigma$ discrepancy between measured and predicted values, see [15]). If the U boson mass is smaller than m_{π^0} and greater than $2m_e$ it can be searched for in the decay $\pi^0 \to U\gamma, U \to e^+e^-$ (Fig. 4) with the same final state as the π^0 Dalitz decay. Kaon decays are an abundant source of π^0's tagged with the charged pion in $K^+ \to \pi^+\pi^0$ decay. Fig. 5 illustrates the current status of the U boson search.

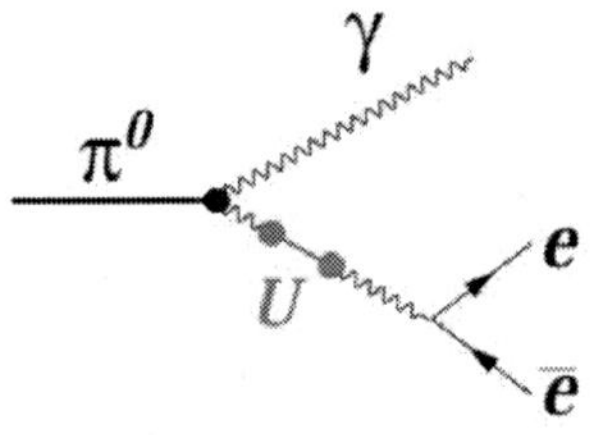

Figure 4. Contribution to the the $\pi^0 \to e^+e^-\gamma$ decay from U boson.

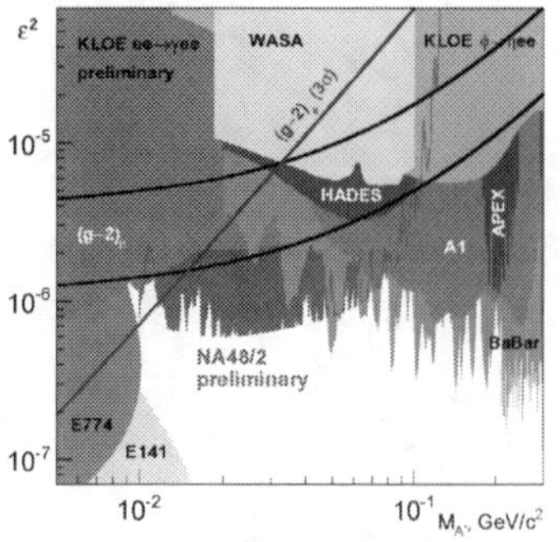

Figure 5. Constraints on the U boson search (ϵ is a mixing parameter). Grey regions are excluded.

The preliminary result of NA48/2 analysis based on $\sim 5\times10^6$ π^0 Dalitz decays improves existing limits in the range 10–60 MeV/c^2. The NA62 experiment can collect ~50 times more π^0 Dalitz decays with a mass resolution of ~1 MeV/c^2 for the e^+e^- pair. The sensitivity on the ϵ^2 in the mass range $m_U < 100$ MeV/c^2 can reach $\sim 10^{-7}$.

6. Conclusions

The NA62 provides a wide spectrum of studies beyond the baseline: LFV kaon decays, π^0 decays. These decays allow to study various phenomena like neutrinoless double beta decay in the second generation, heavy neutrinos, "dark photons" and more.

References

[1] A. Sher et al., Phys. Rev. D72 (2005) 012005.
[2] R. Appel et al., Phys. Rev. Lett. 85 (2000) 2877.
[3] NA48 collaboration, J. Batley et al., Phys. Lett. B 697 (2011) 107.
[4] A. Diamant-Berger et al., Phys. Lett. B62 (1976) 485.
[5] G. Anelli et al., *CERN-SPSC-2005-013, SPSC-P326.*
[6] NA62 Technical Design Document, NA62-10-07; https://cdsweb.cern.ch/record/1404985.
[7] R. Abela, Phys. Lett. B105 (1981) 263.
[8] R. Hayano et al., Phys. Rev. Lett. 49 (1982) 1305.
[9] C. Pang et al., Phys. Rev. D8 (1973) 1989.
[10] V. Duk et al., Phys. Lett. B710 (2012) 307.
[11] G. Bernardi et al., Phys. Lett. B203 (1988) 332.
[12] NA62 collaboration, C. Lazzeroni et al., Phys. Lett. B719 (2013) 326.
[13] PAMELA collaboration, O. Adriani et al., Nature 458 (2009) 607.
[14] DAMA collaboration, R. Bernabei et al., Eur. Phys. J. C56 (2008) 333.
[15] F. Jegerlehner and A. Nyffeler, Phys. Rept. 477 (2009) 1.

Journal of Physics: Conference Series **556** (2014) 012068 doi:10.1088/1742-6596/556/1/012068

Searches for New Physics with the TREK Detector

C Djalali (for the TREK Collaboration)

Department of Physics and Astronomy, University of Iowa, Iowa City, IA 52242, USA

E-mail: chaden-djalali@uiowa.edu

Abstract. The KEK Experiment E246 has provided the best upper limit of the transverse muon polarization in $K_{\mu 3}$ decays to date. The E246 detector is now being upgraded by the TREK collaboration and upcoming experiments with this detector at the J-PARC facility will allow new high-precision measurements in search for physics beyond the Standard Model. The first round of planned experiments include the search for lepton universality violation in a measurement of the ratio of the K_{e2} and $K_{\mu 2}$ decay widths, as well as the search for heavy sterile neutrinos and possible massive gauge boson A' in the 10 to 100 MeV mass range. The search for T violation in kaon decays requires a higher intensity hadron beam and will be carried out later.

1. Introduction

The discovery of the Higgs boson, which until recently was the only missing predicted particle in the Standard Model (SM), has further established the overwhelming success of the SM in explaining the majority of phenomena observed in particle physics. At the same time, it has become clear that the SM cannot be the ultimate theory. Issues such as the neutrino mass problem, the hierarchy problem, the baryogenesis problem and the dark matter and dark energy problem clearly show that the SM is an effective theory for the low energy scale ($\sim$TeV). New physics beyond the SM should lie in the energy region of a few TeV or higher. Among many theories beyond the SM, supersymmetry theory (SUSY), which predicts the existence of supersymmetric partners for all of the SM particles can elegantly explain many of the issues mentioned above. Among the experiments looking for physics beyond the SM, there are several precision measurements of various kaon decay channels. The TREK collaboration is preparing for two of these experiments with high-intensity kaon beams at the Japan Proton Accelerator Research Complex (J-PARC). Using the current J-PARC luminosity, experiment E36 [1] will measure the ratio R_K of the K_{e2} and $K_{\mu 2}$ decay widths of stopped kaons and provide complementary data to previous in-flight-decay experiments. The data from E36 will also allow searching for a heavy sterile neutrino and a massive gauge boson A', "dark photon". Experiment E06 [2] will improve the limit on the magnitude of the T-odd transverse muon polarization, P_T, in $K^+ \to \pi^0 \mu^+ \nu$ decays set by KEK experiment E246 [3] by a factor of 20.

2. The TREK detector

The detector system (Figure 1) will be an upgraded version of the previous KEK-PS E246 experiment [4]. A collimated mixed K^+/π^+ beam of 800 MeV/c will pass through a beam Cherenkov counter — for kaon identification — and a degrader system before being stopped in a segmented, active fiber target. Final-state particles will be either detected in the CsI(Tl) electromagnetic calorimeter or tracked (C1, C2, C3, and C4) and momentum analyzed in the toroidal spectrometer system. Time-of-flight counters (TOF1 and TOF2), an aerogel Cherenkov detector close to the target, and lead glass Cherenkov detectors (PGC) at the exit of the spectrometer will provide for excellent e/μ separation. Additional improvements consist of a new readout system for the CsI(Tl) calorimeter, which allows for a high counting rate, and, new active polarimeters for the muons and decay positrons inside a new muon field magnet.

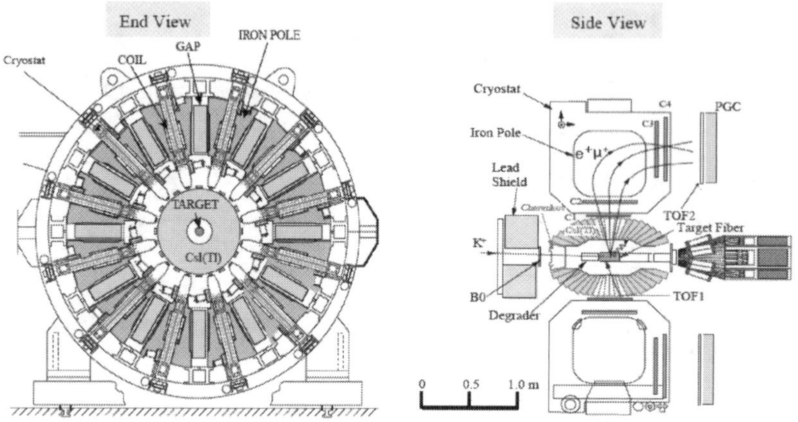

Figure 1. E246 detector upgraded for E06 and E36 J-PARC Experiments.

3. The E36 Experiment

Lepton universality, which requires the identical coupling constant of the three-lepton generations - the electron, muon, and tau - is a basic assumption in the SM. Violation of lepton universality will clearly indicate the existence of physics beyond the SM. Under the assumption of μ-e universality, the ratio of the electric (K_{e2}) and muonic ($K_{\mu2}$) decay modes (R_K) in the SM is known with excellent accuracy [5]:

$$R_K^{SM} = \frac{\Gamma(K^+ \to e^+\nu_e[\gamma])}{\Gamma(K^+ \to \mu^+\nu_\mu[\gamma])} = \frac{m_e^2}{m_\mu^2}\left(\frac{m_K^2 - m_e^2}{m_K^2 - m_\mu^2}\right)^2 (1 + \delta_{QED}) = 2.477 \pm 0.001 \times 10^{-5} \quad (1)$$

Looking for deviations of R_K from R_K^{SM} provides a clean tool to detect or constrain physics beyond the SM such as lepton violating effects. The E36 experiment seeks to measure R_K with an accuracy of $\Delta R_K/R_K$ of 0.2%. The K_{e2} and $K_{\mu2}$ events will be identified by the charged lepton momenta, $p_e = 247$ MeV/c and $p_\mu = 236$ MeV/c, as determined in the spectrometer. Excellent particle identification is key to the measurement. Particle momenta as well as time-of-flight and Cherenkov detectors will be used to decrease the probability of misidentifying a μ^+ as e^+ to less than 10^{-8}. The CsI(Tl) calorimeter will serve as photon detector for the radiative processes. The R_K value is then derived from the detector-acceptance-corrected number of the identified $K_{e2(\gamma)}$ and $K_{\mu2(\gamma)}$ events, including those with the production of internal bremsstrahlung but excluding structure-dependent contributions. Figure 2 shows the KLOE [6] and NA62 [7] results as well as the projected uncertainties for E36.

The Neutrino Minimal Standard Model (νMSM) is an example of a renormalizable extension of the SM that contains three right-handed (sterile) neutrinos with masses below the electroweak

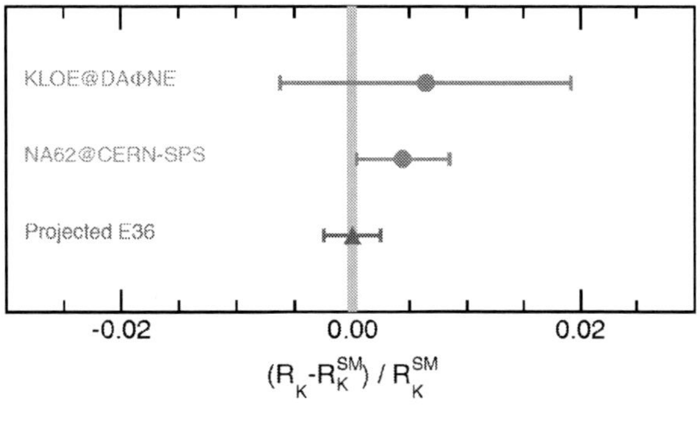

Figure 2. Expected TREK R_K value compared to current measurements. The projected uncertainty of E36 [1] (triangle) is shown at the Standard Model value.

scale [8]. The lightest of the three sterile neutrinos is a dark-matter candidate, the masses of the heavier two sterile neutrinos are degenerate in the νMSM, $M_1 < M_2 \approx M_3$. If the sterile neutrino is lighter than the kaon, it can give rise to leptonic and semi-leptonic decays with relatively large branching ratios. This gives a possibility to confirm or rule out the νMSM by looking for massive sterile neutrinos in kaon decay experiments. The E36 experiment will be looking for heavy sterile neutrinos (N) in the $K^+ \to \mu^+ N$ decay using the μ^+ polarimeter. The νMSM predicts the branching ratio to be up to 10^{-6}, which is accessible by the TREK experimental apparatus. This decay has some important features such as (1) a narrow peak structure in the momentum spectrum; (2) a 100% μ^+ polarization in the direction opposite of the muons from the $K^+ \to \mu^+\nu$ decay. Measuring both the μ^+ momentum and polarization will improve the sensitivity of the branching ratio to better than 10^{-8}.

Furthermore, studying the $K^+ \to \mu^+\nu_\mu A'$ channel can provide constraints on the massive gauge boson A' ("dark photon") in the 100 MeV mass range. The A' particle could be identified in the e^+e^- invariant-mass distribution through its decay, $A' \to e^+e^-$. In the E36 experiment the e^+e^- pair will be detected in the CsI(Tl) calorimeter. Preliminary, simulated invariant-mass distributions for various assumed masses ($M_{A'} = 10$ MeV/c^2 to 100 MeV/c^2) show that A' with mixing parameters as low as $\epsilon^2 \approx 10^{-6}$ can be detected. The observation of the $K^+ \to \pi^+ A'$ could be an alternative signal process in this search. The A' would be identified in the π^+ momentum distribution as measured in the TREK spectrometer.

4. The E06 experiment

The transverse muon polarization (P_T) in K$\to \pi\mu\nu$ ($K_{\mu3}$) decays with T-odd correlation was suggested by Sakurai [9] about 50 years ago to be a clear signature of T violation. Unlike other T-odd channels such as in nuclear beta decays, P_T in $K_{\mu3}$ has the advantage that the final state interactions (FSI), which may mimic T violation by inducing a spurious T-odd effect, are very small ($< 10^{-5}$) [10]. This argument applies most particularly to $K^+_{\mu3}$ decay where the FSI contribution is only from higher order loop levels and is calculable. Another important feature of a P_T study is the fact that the contribution to P_T from the SM is nearly zero ($\sim 10^{-7}$). Therefore, any non zero P_T will be indicative of new physics beyond the SM (Figure 3).

The KEK experiment E246 studied the $K_{\mu3}$ decay of stopped kaons and put the most stringent limit to date on this observable, $|P_T| < 5 \times 10^{-3}$ (90% C.L.) [3]. The new experiment E06 is aiming to improve that limit by a factor of, at least, 20 and to reach a sensitivity of $\approx 2 \times 10^{-4}$ (90% C.L.) [2]. A factor of 20 improvement of the statistical uncertainty requires the high-intensity K^+ beam at J-PARC and a new active muon polarimeter with increased acceptance and sensitivity. The new polarimeter will also help reduce systematic uncertainties compared to what was achievable in E246.

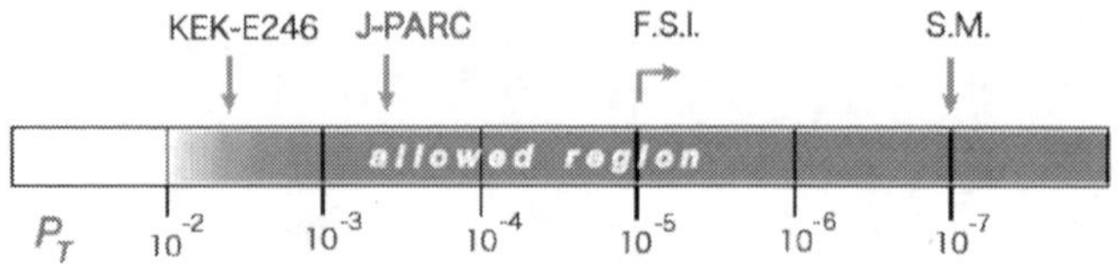

Figure 3. Experimental status of $K_{\mu 3}$ P_T physics relative to the SM prediction. The Allowed region represents the range of possible non zero P_T from non-SM CP violation models.

5. Summary

In summary, the TREK collaboration has a rich experimental program looking for new physics beyond the SM by studying the decay of stopped kaons with two experiments, E36 and E06, at J-PARC. The first experiment, which does not require the high luminosity beam at J-PARC, is scheduled to run in 2015 and will search for lepton universality violation in a measurement of the ratio of the K_{e2} and $K_{\mu 2}$ decay widths, for a heavy sterile neutrino in the muon-momentum distribution from $K^+ \to \mu^+ N$, and for a possible massive "dark photon" in the $K^+ \to \mu^+ \nu A' \to \mu^+ \nu e^+ e^-$ and $K^+ \to \pi^+ A'$ reactions. The second experiment will search for T violation in kaon decays through the measurement of the transverse muon polarization in the $K^+ \to \pi^0 \mu^+ \nu_\mu$ reaction. Experiment E06 requires the active muon polarimeter and maximum kaon-beam intensities and will run when higher intensity beams are achieved at J-PARC.

Acknowledgments

This work was supported in the US in part part by the National Science Foundation under the award NSF-0856010 and the DOE, the NSERC and NRC in Canada, the KAKEN-HI Grants-in-Aid for Scientific Research in Japan, and in Russia by the Ministry of Science and Technology. The Japan Proton Accelerator Research Complex (J-PARC) is jointly and cooperatively run by the Japan Atomic Energy Agency (JAEA) and KEK.

References

[1] Kohl M and Shimizu S 2010 *J-PARC Experiment E36: Measurement of* $\Gamma(K^+ \to e^+\nu)/\Gamma(K^+ \to \mu^+\nu)$ *and Search for heavy sterile neutrinos using the TREK detector system*, `http://j-parc.jp/researcher/Hadron/en/pac_1007/pdf/KEK_J-PARC-PAC2010-04.pdf`
[2] Imazato J and Kohl M 2006 *J-PARC Experiment E06: Measurement of T-violating Transverse Muon Polarization in* $K^+ \to \pi 0 \mu^+ \nu$ *Decays*, `http://trek.kek.jp/proposal/E06_proposal.pdf`
[3] Abe M, Aliev M, Anisimovsky V, Aoki M et al. 2006 *Phys. Rev.* D**73** 072005
[4] Abe M et al. 2003 *Nucl. Instrum. Meth.* A**506** 60, hep-ex/0302001.
[5] Cirigliano V and Rosell I 2007 *Phys. Rev. Lett.* **99** 231801, 0707.3439
[6] Ambrosino F et al. 2009 *Eur. Phys. J. C* **64** 627, 0907.3594
[7] Lazzeroni C et al. 2013 *Phys. Lett.* B**719** 326, 1212.4012
[8] Canetti L, Drewes M, and Shaposhnikov M 2013 *Phys. Rev. Lett.* **110** 061801, 1204.3902.
[9] Sakurai J J 1957, *Phys.Rev.* **109** 980
[10] Efrosinin V, Khriplovich I, Kirilin G and Kudenko Y 2000 *Phys. Lett.* B**493** 293, hep-ph/0008199.

A two-stage approach in the the search for $\mu - e$ conversion with COMET

A Edmonds[1]

[1] Department of Physics and Astronomy, University College London, Gower Street, London WC1E 6BT, UK

E-mail: a.edmonds@ucl.ac.uk

Abstract. The COMET experiment will search for the charged lepton flavour violating process $\mu^- + N(Z,A) \rightarrow e^- + N(Z,A)$ in Al with a single event sensitivity of 3×10^{-17}. This will be a four orders of magnitude improvement on the current limit set by SINDRUM-II in Au. The design, R&D and construction of COMET are all well underway and described here is an update on the current status of the experiment.

1. Introduction to Charged Lepton Flavour Violation

It has been known since the discovery of neutrino oscillations that lepton flavour is not a conserved quantity in nature [1]. However, although neutrino oscillations are evidence of lepton flavour violation in the neutral lepton sector, there has been no observation of this phenemonon among the charged leptons.

In fact, this is what would be expected from the Standard Model (SM) since it predicts these charged lepton flavour violation (CLFV) processes to have tiny branching ratios of $O(10^{-54})$. This is a result of large suppressions in Feynman diagrams such as the one shown in Figure 1, which arise from the fact that the neutrino is oscillating in a virtual loop.

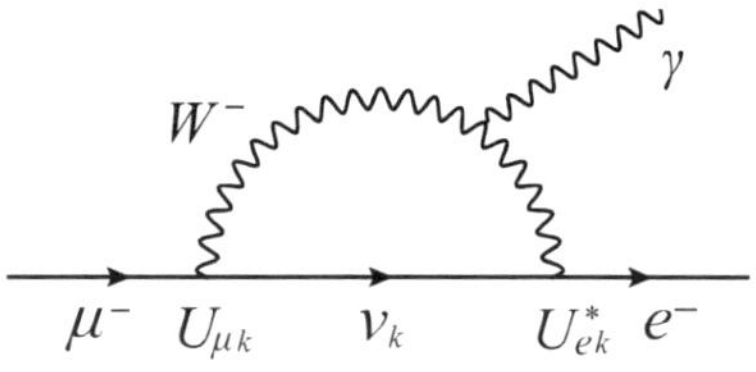

Figure 1. Standard Model Feynman Diagram for the $\mu \rightarrow e\gamma$ process.

Since the SM prediction is so small, any observation of a CLFV process would be clear evidence of beyond the Standard Model physics and so there are many experiments that are looking for such processes.

The CLFV processes that are searched for in the muon sector are given in Table 1 along with their current upper limits.

Table 1. The current limits in muon CLFV searches.

CLFV Process	Current Limit at 90% CL	Experiment	Reference
$\mu^+ \to e^+\gamma$	$< 5.7 \times 10^{-13}$	MEG	[2]
$\mu^+ \to e^+e^-e^+$	$< 1.0 \times 10^{-12}$	SINDRUM	[3]
$\mu^- + \text{Au} \to e^- + \text{Au}$	$< 7 \times 10^{-13}$	SINDRUM-II	[4]

COMET will be searching for the $\mu^- + N(Z,A) \to e^- + N(Z,A)$ process (also known as $\mu - e$ conversion) in aluminium.

The $\mu - e$ conversion process occurs when a muon is stopped by an atom and falls down to the 1s state, forming a muonic atom. The muonic atom then decays with a certain lifetime (864 ns for aluminium [5]), with one of its possible decay processes being an interaction with the quarks in the nucleus to produce a single electron.

The muon's mass defines the energy of the initial state and so the electron in the final state will have a very well-defined energy approximately equal to it. The full equation for the energy of $\mu - e$ conversion electrons is $E_e = m_\mu - E_b - E_{rec}$ where E_b is the binding energy of the 1s state of the muonic atom and E_{rec} is the energy taken away by the recoiling nucleus. For aluminium, the electron energy is 104.9 MeV.

Another advantage of this process is that it remains a simple signal to search for at high muon rates since there is no need to determine coincidences like in $\mu \to e\gamma$ and $\mu \to eee$.

2. The COMET Experiment

The COMET experiment will be built in two phases. Phase-I will begin in 2016 and is seeking to achieve a single event sensitivity (S.E.S) of 3×10^{-15} and Phase-II will begin in 2019 with a S.E.S of 3×10^{-17}. The experimental layouts of Phase-I and Phase-II are shown in Figures 2 and 3 respectively.

As shown by the figures, COMET Phase-II has a distinctive S-shape and Phase-I is a smaller scale version that is identical up to the end of the first 90° bend. The advantage of taking a staged approach is that it will be possible to search for $\mu - e$ conversion at an intermediate sensitivity to the final design goal as well as allowing for the investigation of the backgrounds to be encountered in Phase-II with the actual beamline that will be used.

Searching for $\mu - e$ conversion with such a high sensitivity requires a very intense muon beam and this will be generated by firing protons at a production target and collecting the backwards travelling pions with a high solenoidal magnetic field. This will result in a muon beam intensity of $O(10^9)$ μ^- per second.

After the beam is produced, it will travel around the first C-section, designed to be long enough so that the majority of the pions decay into muons and to select the low momentum particles. There are two reasons for this selection. First, it is only low momentum muons that will stop in the aluminium stopping target and also, any high momentum muons above 77 MeV/c could decay in flight and produce electrons of the same energy as the $\mu - e$ conversion signal which would be a major background to the experiment.

Once the muon beam gets around the first C-section, it will then enter the muon stopping target section where it will be stopped in the stopping target. Muonic atoms will then be formed and, after decaying, the emitted electrons will be guided through the final C-section which acts as an electron spectrometer to select 104.9 MeV/c electrons that will be evidence of the $\mu - e$ conversion process. The momentum and energy of the electrons will then be measured by the detector.

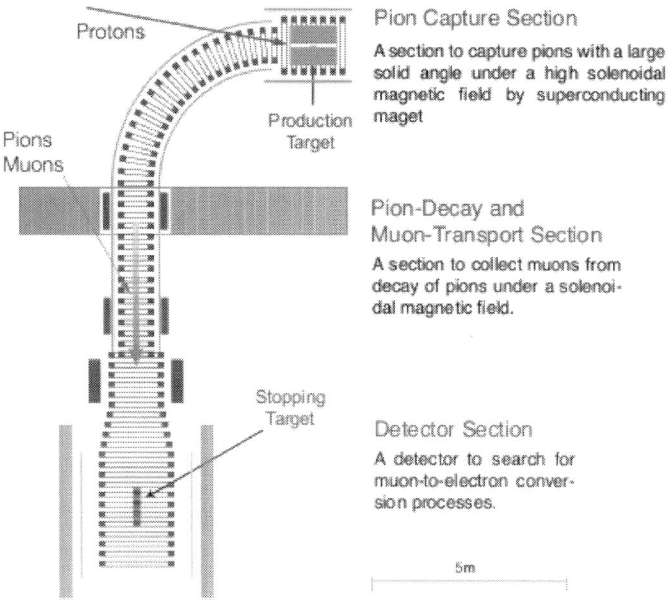

Figure 2. Experimental Layout for COMET Phase-I.

The detector will consist of a straw tube tracker with excellent momentum resolution and an electromagnetic calorimeter (ECAL) that will add both redundancy to the measurement and also act as a trigger for the DAQ. Together these two parts are known as the StrECAL.

For Phase-I, only the first 90° of the experiment will be built and, in order to reach the desired sensitivity, a cylindrical detector (named CyDet) will be used for the $\mu - e$ conversion search instead. This gives an increased geometrical acceptance but is affected more by any other particles emitted from the stopping target such as protons which consequently impact the data rates. In addition to the CyDet, there will also be a prototype StrECAL to be used for background investigations.

3. Proton Beam

COMET will use an 8 GeV, 3.2 kW (Phase-I) or 56 kW (Phase-II), pulsed proton beam at J-PARC. By using a pulsed beam we can utilise a delayed-timing window for our measurement in order to remove prompt backgrounds.

A diagram of the proton time structure is given in Figure 4 and shows the proton pulses (black, diagonal stripes) will have a 1.1 μs separation in time. Once the muon beam reaches the stopping target, there will be an immediate flash of prompt backgrounds (light grey, diagonal stripes) which mostly arise from pion contamination in the beam. The stopped muons then decay slowly (dark grey, vertical stripes) and a time window (dark grey, horizontal stripes) is opened sufficiently long after the prompt backgrounds have passed through the experiment that the search for a signal event (solid black) can start.

Also in this diagram is an illustration of what happens if there is insufficient extinction of the proton beam between pulses (small, striped black pulse in the centre of the plot). Having residual protons between the main pulses means that the corresponding muons arrive at the

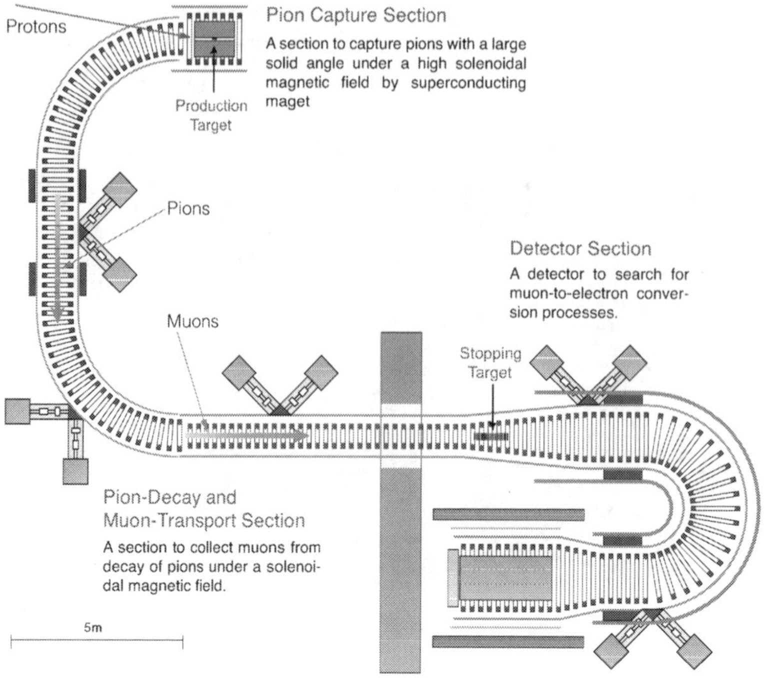

Figure 3. Experimental Layout for COMET Phase-II.

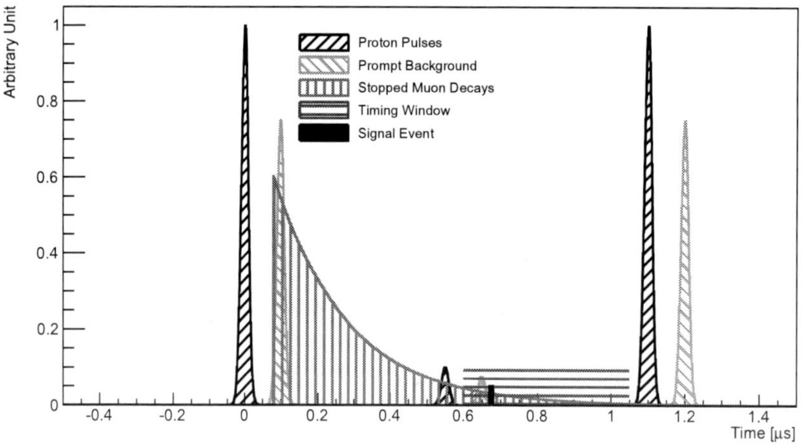

Figure 4. The time structure of the COMET proton beam and associated backgrounds.

stopping target later than they should. This means that a smaller prompt background flash occurs which arrives in the measurement window and so will obscure any search for the $\mu - e$ conversion process. An extinction factor of 3×10^{-11} is necessary to overcome this problem and, in order to get it down to the required level, a novel injection method has been developed and has been successfully demonstrated at J-PARC in May 2014.

4. Background Estimates

As for any rare search measurement, it is very important to know the background processes that could mimic the signal and to estimate how much impact they will have. For COMET Phase-I, 0.03 background events are expected per signal event for a branching ratio of 1×10^{-15} and 30 days running. The important backgrounds are muon decay-in-orbit, which contributes 0.01 background events per signal event and radiative pion capture which also contributes 0.01 background events per signal event.

The decay-in-orbit (DIO) background is the only intrinsic physics background in the experiment and occurs when the stopped muon decays normally into an electron and two neutrinos. The electron energy spectrum for this process is a rapidly falling distribution above 52.8 MeV but, if the neutrinos take very little energy, the electron can have energies very close to the signal region. Although the end-point of this spectrum is below the signal energy, in reality, the signal and background could be smeared into each other due to detector resolution effects.

The radiative pion capture (RPC) background occurs when pion contamination in the beam is captured by the aluminium nucleus in the stopping target. This leaves the nucleus in an excited state and, as it decays back to the ground state, it emits high energy gamma rays that can pair-produce electrons with an energy in the signal region. This is a prompt process and so is suppressed by the extinction factor described in section 3.

5. Cylindrical Detector - CyDet

For COMET Phase-I, a cylindrical detector (CyDet) will be installed to search for the $\mu - e$ conversion process. This increases the geometrical acceptance of the detector but also increases the rate of hits since any particles emitted from the stopping target will enter the detector. The emitted particles mostly consist of protons from nuclear muon capture and so, for Phase-II, this will not be a problem since there will be a final C-section that will charge select the protons away.

The CyDet will be composed of two parts. A cylindrical drift chamber (CDC) to measure the momentum of the electrons and a pair of Cherenkov counters to act as a trigger.

The requirements of the CDC are that it needs to have a gas gain of $> 10^5$, a position resolution in the transverse plane of $< 250\,\mu$m, a position resolution in the z direction of < 2 mm and the gas needs to reduce the amount of multiple scattering in order to minimise the amount of energy loss and achieve the best possible momentum resolution.

The current design for the CDC is based on the design of the CDC that has been built for Belle-II. This was constructed at KEK and so allows the exploitation of tooling and expertise that already exists in the community. The specific parameters of the CyDet geometry are given in Table 2.

There are two important parameters that help in the reduction of the backgrounds. First, the inner radius of the CDC is large enough so that the large flux of low p_T electrons that occur from DIO do not enter the detector and, second, the inner wall is thick enough to remove protons that are emitted from the target but not so thick that any electrons that pass through it lose too much energy and so any signal electrons lose enough energy to be smeared into the background.

Currently, there is a small-scale prototype of the CDC that has undergone cosmic ray tests at both KEK and Osaka and the procurement of the final CyDet is underway.

Table 2. Geometry parameters of the CyDet for COMET Phase-I.

Parameter	Value
CDC Inner Radius	55 cm
CDC Outer Radius	84 cm
CDC Length	150 cm
CDC Drift Gas	He:iC_4H_{10} (90:10)
Trigger Counter Material	Scinitillator and Cherenkov Counter
Trigger Counter Length	300 mm
Trigger Counter Thickness	35 mm
Magnetic Field	1 T

6. Straw Tube Tracker and ECAL - StrECAL

The StrECAL detector will be the main detector for Phase-II but a prototype will also exist for Phase-I to measure backgrounds in preparation for Phase-II. It will consist of a straw tube tracker with five stations (4 layers of straws in each station) and an electromagnetic calorimeter.

The straw tube tracker will be based on the design of the tracker built at JINR for NA62 and so a collaboration has been set up between KEK and JINR to undertake design and R&D work. Currently, a 1:1 scale prototype of a tracker station has been built with fewer channels and is undergoing various tests at KEK.

The requirements for the ECAL are that it needs a resolution of $< 5\%$ at 105 MeV, a low trigger rate, a spatial resolution of $< 1.5\,\text{cm}$ and a quick response of $< 100\,\text{ns}$ so that it can be used as the trigger.

At present, there are two candidate crystals under consideration: GSO and LYSO. Both of these have been tested in a beam test at Tohoku University in March 2014 and a decision on which crystal will be used will be made in the immediate future.

7. Conclusion

COMET will be searching for the charged lepton flavour violating process $\mu^- + N(Z,A) \rightarrow e^- + N(Z,A)$. COMET will have a single event sensitivity of 3×10^{-15} in Phase-I which will run in 2016 and a single event sensitivity of 3×10^{-17} in Phase-II which will begin in 2019.

References

[1] Beringer J *et al.* (Particle Data Group) 2012 *Phys.Rev.* **D86** 010001
[2] Adam J, Bai X, Baldini A, Baracchini E, Bemporad C *et al.* 2013 *Eur.Phys.J.* **C73** 2365 (*Preprint* 1303.2348)
[3] Bellgardt U *et al.* (SINDRUM Collaboration) 1988 *Nucl.Phys.* **B299** 1
[4] Bertl W H *et al.* (SINDRUM II Collaboration) 2006 *Eur.Phys.J.* **C47** 337–346
[5] Measday D 2001 *Phys.Rept.* **354** 243–409

Journal of Physics: Conference Series **556** (2014) 012070 doi:10.1088/1742-6596/556/1/012070

Leptogenesis

Pedro Schwaller

CERN, Theory Division, CH-1211 Geneva 23, Switzerland

E-mail: pedro.schwaller@cern.ch

Abstract. Leptogenesis [1] is one of the leading mechanisms for explaining the observed baryon asymmetry in the universe. In this proceeding I review the basic features of leptogenesis, before discussing some recent developments in the theoretical computation of the resulting baryon asymmetry. Finally I comment on the possible connection with low energy neutrino data and testability of the mechanism.

1. Introduction

The baryon asymmetry of the universe (BAU) has been measured precisely both from observations of the cosmic microwave background (CMB), and from the abundances of light elements in the universe which depend on the baryon density at the time of big bang nucleosynthesis (BBN). Today the entropy normalised baryon asymmetry is known to be [2]

$$Y_{\Delta B} = \frac{n_B - n_{\bar{B}}}{s} = (8.59 \pm 0.13) \times 10^{-11}\,, \tag{1}$$

where the relation to the baryonic energy density is given by $Y_{\Delta B} \approx 3.89 \times 10^{-9}\Omega_B h^2$ [3]. To generate such an asymmetry the three Sakharov conditions have to be satisfied: 1. Baryon number (B) must be violated; 2. C and CP must be violated; 3. The universe must depart from thermal equilibrium at some early time.

Within the standard model (SM) the first two conditions are satisfied, namely B is violated at high temperatures in the early universe by non-perturbative sphaleron processes, and C and CP are violated by the weak interactions and the quark Yukawa couplings. However the electroweak phase transition within the SM is not strong enough to allow departure from equilibrium, and no other strong phase transition is expected at higher temperatures, while at lower temperatures B violation is no longer active.

Leptogenesis is based on the see-saw mechanism, which explains the smallness of neutrino masses by introducing heavy right-handed neutrinos N_i. Yukawa couplings between the left-handed doublets lepton doublets and the right-handed neutrinos then induce masses for the SM neutrinos ν_i which are parametrically of the order $m_\nu \sim y^2v^2/M$, where y is the neutrino Yukawa coupling, v is the Higgs vacuum expectation value and M is the right-handed neutrino mass scale. Sub-eV neutrino masses are then possible with order one couplings provided that the right-handed neutrino masses are of order 10^{9-15} GeV.

When the temperature in the early universe drops below the mass M_1 of the lightest right-handed neutrino, N_1, it will decay into a lepton and a Higgs doublet, or their corresponding anti-particles. Complex phases in the neutrino Yukawa couplings can lead to CP violation in the decay rates, such that more leptons than anti-leptons can be produced in those decays. If the decay rate Γ is slower than the expansion rate of the universe $H(T)$ at temperatures $T \sim M_1$, the right-handed neutrino distributions will depart from thermal equilibrium, such that a net lepton asymmetry can be generated. This asymmetry is then partially transferred to the baryon sector through electroweak sphalerons, thus producing a baryon asymmetry in agreement with Sakharov's conditions. The departure from equilibrium of N_1 depends on the same combination of parameters as the resulting light neutrino masses, and it turns out that the observed neutrino masses are exactly consistent with the requirement $\Gamma/H(T = M_1) \approx \mathcal{O}(1)$.

2. Theory Progress

In order to understand whether the observed neutrino masses and mixing angles are consistent with the observed BAU, one has to be able to compute $Y_{\Delta B}$ as function of the Yukawa couplings and masses. In recent years a lot of progress has been made to formulate the problem using non-equilibrium quantum field theory, see e.g. [4, 5, 6], and references therein for earlier work.

The main advantage of this approach is that it provides a robust framework for deriving and studying quantum effects like CP-violation and flavour oscillations in transport equations. Furthermore it also allows a systematic inclusion of thermal effects on the CP asymmetry and on scattering rates, which can be significant in particular at early times $T \gtrsim M_1$.

In strong washout scenarios, which are characterised by $\Gamma \gtrsim H(T = M_1)$, inverse decays $\ell\phi$, $\bar{\ell}\phi^\dagger \to N_1$ efficiently wash out asymmetries produced at earlier times. The final asymmetry is determined at the time when the inverse decays freeze out, $T \ll M_1$, and does not depend significantly on either the initial conditions or on thermal corrections, which mostly affect the asymmetry at earlier times. Since strong washout is also favoured by our current knowledge of the neutrino mass parameters, we will in the following focus on recent developments that affect the final asymmetry in that regime.

2.1. Flavour Effects

While it was always clear that the lepton asymmetry produced by N_1 decays is produced in a linear combination of flavour eigenstates, it was only realised in [7, 8] that this could have a significant impact on the final asymmetries, if leptogenesis takes place at a time where one or more charged lepton Yukawa couplings are in thermal equilibrium.

Roughly, this can be understood as follows: When the interactions mediated by the tau Yukawa coupling are in equilibrium (i.e. faster than the Hubble rate), the produced asymmetry is projected onto either the tau lepton flavour or an orthogonal linear combination. In general each flavour combination has a different washout rate, while the overall asymmetry production rate stays the same. This can result in a sizeable increase of the final lepton asymmetry in scenarios with moderate to strong washout.

In order to properly understand the transition regime where the flavour interactions enter thermal equilibrium, the quantum kinetic equations of motion for the flavoured lepton number densities, including flavour off diagonal correlations, were derived from non-equilibrium quantum field theory in [9]. In Fig. 1 it can be seen that the total asymmetry smoothly increases form the unflavoured to the flavoured limit (dotted and dashed lines) as the leptogenesis scale decreases, and there is a broad transition regime where neither approximation yields an accurate result.

2.2. Scattering Rates

An important ingredient for a precise treatment of flavour effects is the flavour relaxation rate, i.e. the rate of processes mediated by the charged lepton Yukawas. The leading processes, namely decays and inverse decays of the Higgs boson into left- and right-handed leptons appearing at $\mathcal{O}(y_\tau^2)$, are not allowed kinematically in the early universe where all particles are approximately massless. Instead one has to consistently compute all processes at order $g^2 y_\tau^2$ in order to arrive at

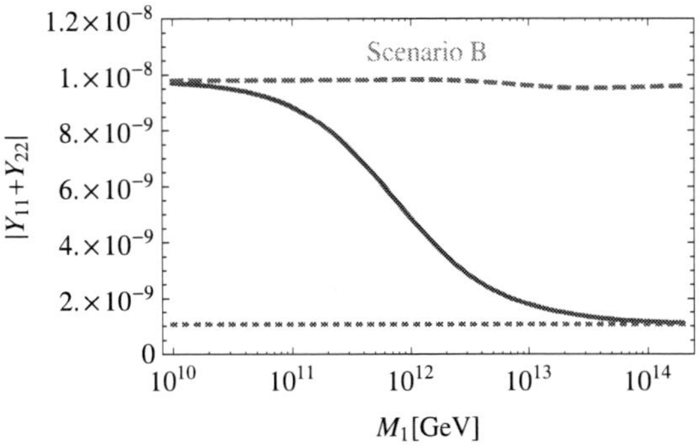

Figure 1. Total lepton asymmetry as a function of the Leptogenesis scale M_1. Shown are the solutions of the exact flavoured equations of motion (solid, blue) as well as the fully flavoured (dashed, green) and unflavoured (dotted, red) approximations. Taken from Ref. [9].

a reliable leading order result. At finite temperature, this not only includes scattering processes involving electroweak gauge bosons and top quarks, but also $1 \rightarrow 2$ decay processes that become allowed once thermal masses are included, and processes with multiple soft scatterings of gauge bosons which have to be resummed [10].

A first calculation of the charged lepton Yukawa equilibration rate was performed in [11]. For the temperature range relevant to leptogenesis, the charged lepton flavour relaxation rate was found to be $\Gamma_\alpha^{\mathrm{fl}} \approx 5 \times 10^{-3} y_\alpha^2 T$, where α labels the different lepton flavours. This rate is dominated by scattering processes of the form $\ell A \leftrightarrow \phi e_R$ and $\ell \bar{e}_R \leftrightarrow A\phi$, where a lepton is exchanged in the t-channel. Here A denotes an electroweak gauge boson, and e_R a right-handed charged lepton.

These processes diverge when the momentum exchange goes to zero, and the divergence is regulated by the thermal mass of the exchanged lepton. Using the 2PI formalism for calculating these rates in non-equilibrium QFT, this regularisation is implemented automatically since the corresponding scattering processes arise from self energy diagrams involving resummed fermion propagators, which include both thermal masses and widths. A calculation of the corresponding neutral lepton Yukawa interactions was presented in [12] using different techniques, and we have explicitly shown that we reproduce the same logarithmic enhancement of the rate, however a discrepancy in the $\mathcal{O}(g^2)$ contribution remains to be resolved.

2.3. Spectator Effects

The washout, and thus the final asymmetry, can also be affected by interactions that do not change the lepton asymmetry themselves, so called *spectator effects*. To understand this it is important to notice that Leptogenesis not only produces an asymmetry in lepton doublets, but also in the Higgs doublets. This asymmetry is then partially transferred into a chiral asymmetry between third generation quark doublets and right-handed top quarks, since top Yukawa interactions are always in thermal equilibrium. This overall redistribution of the asymmetry leads to an effective reduction of the washout rate.

Depending on the Leptogenesis scale, different SM interactions may reach thermal equilibrium during the leptogenesis process. Possible candidates include interactions mediated by third and second generation Yukawa couplings and the strong and electroweak sphaleron processes. The usual approach to deal with this was to assume that for a given scale, a process is either fully equilibrated or inactive. In our discussion of flavour effects above however we have seen that the transition regime for an interaction to reach equilibrium can span more than an order of magnitude, and this is similarly the case for spectator effects.

In [13] we have therefore considered the case of spectators reaching thermal equilibrium during Leptogenesis, by including additional degrees of freedom in the Boltzmann equations. As a simple toy example we have considered the case where the tau Yukawa interaction reaches thermal equilibrium, while neglecting any lepton flavour effects. What happens then is that parts of the asymmetry are transferred to an asymmetry in right-handed leptons, where it is protected from washout.

While we again observed a smooth interpolation between the limiting cases at high and low scales, for strong washout there is an intermediate regime where the final asymmetry is larger than in either limit. The reason for this is that for strong washout, the asymmetry at intermediate stages is much larger than the final freeze out asymmetry. Now parts of this larger asymmetry are transferred to the right-handed leptons, but since the spectator interaction is not fully equilibrated yet, this asymmetry is only transferred back to the lepton doublets after the washout ends, thus protecting some asymmetry from being washed out.

In the toy example this enhancement of the asymmetry can reach up to 60%, while in more realistic examples where either the sphalerons or the bottom Yukawa enter thermal equilibrium it only provides corrections at the 10% to 20% level. Another important consequence here is that the spectator "remembers" the asymmetry at earlier times, therefore it can reintroduce a dependence on initial conditions and on thermal corrections even in the strong washout regime.

3. Thoughts on Testability

Due to the high scales involved, it is unlikely that the right-handed neutrino sector can ever be accessed experimentally. It is therefore reasonable to ask what can be learned from low energy neutrino observations, and to which extend this mechanism can be probed.

One would expect that a measurement of the low energy neutrino CP phase should have some impact on Leptogenesis. In general this is not the case however: In the absence of flavour effects, the lepton asymmetry is independent of the entries in the U_{PMNS} matrix. Including flavour effects reintroduces some dependence, however without further restrictions on the high-scale parameters Leptogenesis can still work for any value of the neutrino mixing parameters.

On the other hand the LHC could have severely disfavoured the Leptogenesis mechanism. The reason is that it relies on the existence of the SM, or a perturbative extension thereof (like the MSSM), up to the Leptogenesis scale. The discovery of a SM-like Higgs boson in a mass range that is consistent with vacuum stability up to very high scales, and nothing else (yet), can therefore be taken as positive evidence that indeed Leptogenesis could be the mechanism responsible for our existence.

In the future a confirmation of the Majorana nature of neutrinos and of a non-vanishing CP phase could provide further evidence for Leptogenesis. In contrast, an observation of large lepton number violating processes at the LHC or elsewhere would be problematic, since the new washout processes induced by such interactions would erase any lepton asymmetry produced by Leptogenesis [14].

References

[1] M. Fukugita and T. Yanagida, Phys. Lett. B **174** (1986) 45.
[2] P. A. R. Ade *et al.* [Planck Collaboration], Astron. Astrophys. (2014) [arXiv:1303.5076 [astro-ph.CO]].
[3] S. Davidson, E. Nardi and Y. Nir, Phys. Rept. **466** (2008) 105 [arXiv:0802.2962 [hep-ph]].
[4] M. Garny, A. Hohenegger, A. Kartavtsev and M. Lindner, Phys. Rev. D **80** (2009) 125027.
[5] M. Beneke, B. Garbrecht, M. Herranen and P. Schwaller, Nucl. Phys. B **838** (2010) 1.
[6] A. Anisimov, W. Buchmller, et. al., Annals Phys. **326** (2011) 1998 [Erratum-ibid. **338** (2011) 376].
[7] A. Abada, S. Davidson, F. X. Josse-Michaux, M. Losada and A. Riotto, JCAP **0604** (2006) 004.
[8] E. Nardi, Y. Nir, E. Roulet and J. Racker, JHEP **0601** (2006) 164.
[9] M. Beneke, B. Garbrecht, C. Fidler, M. Herranen and P. Schwaller, Nucl. Phys. B **843** (2011) 177.
[10] A. Anisimov, D. Besak and D. Bodeker, JCAP **1103** (2011) 042.
[11] B. Garbrecht, F. Glowna and P. Schwaller, Nucl. Phys. B **877** (2013) 1.
[12] D. Besak and D. Bodeker, JCAP **1203** (2012) 029.
[13] B. Garbrecht and P. Schwaller, arXiv:1404.2915 [hep-ph], to appear in JCAP.
[14] F. F. Deppisch, J. Harz and M. Hirsch, Phys. Rev. Lett. **112** (2014) 221601.

Journal of Physics: Conference Series **556** (2014) 012071 doi:10.1088/1742-6596/556/1/012071

CPT and Lorentz symmetry tests with entangled neutral kaons at KLOE/KLOE-2

Michał Silarski on behalf of the KLOE-2 Collaboration

Institute of Physics, Jagiellonian University, PL-30-059 Cracow, Poland

E-mail: michal.silarski@uj.edu.pl

Abstract. The KLOE experiment at the DAΦNE ϕ-factory located at the INFN Laboratory in Frascati, Italy collected data corresponding to 2.5 fb^{-1} of integrated luminosity. Neutral kaon pairs produced in ϕ-meson decays offer a unique possibility to perform tests of fundamental discrete symmetries. The entanglement of the two kaons is exploited to search for possible violation of CPT symmetry and Lorentz invariance in the context of the Standard-Model Extension (SME) framework. A new approach to the analysis of $\phi \to K_S K_L \to \pi^+\pi^-, \pi^+\pi^-$ events has been adopted allowing us to independently measure all four CPT violating parameters Δa_μ appearing for neutral kaons in the SME. The final KLOE results on Δa_μ will be presented. These are presently the most precise measurements in the quark sector of the SME.

1. Introduction

The invariance under CPT operation, the combined action of charge conjugation (C), parity reflection (P), and time-reversal (T), is fundamental in the framework of any quantum field theory. CPT is also one of the best tested symmetries having passed very stringent and diverse experimental tests (eg. $|m_{K^0} - m_{\overline{K^0}}/m_{av}| < 6 \times 10^{-19}$ [1]). As a consequence of Lorenz invariance, unitarity and locality of quantum field theory CPT may be however violated in the presence of quantum gravity effects (QG). In several QG models CPT can be violated via some mechanism which violates also standard Quantum Mechanics (QM) [2, 3], eg. causing decoherence of an entangled quantum state. As a consequence the quantum mechanical operator generating CPT transformations cannot be consistently defined [4]. CPT symmetry may be violated also through spontaneous breaking of Lorentz symmetry [5, 6] or locality [7] of interactions. In these cases the well-defined generator of the CPT symmetry does not commute with the effective Hamiltonian of the system [2]. On phenomenological grounds the CPT and Lorentz symmetries violation induced by quantum gravity effects is expected at energies of the order of $\sim 10^{20}$ GeV, however they might give rise to observable effects also in the low energy regime, particularly in the entangled neutral meson systems [8]. These effects have been searched for in the B mesons system by the BaBar collaboration [9] and in the neutral kaon system by the KLOE experiment operating at the DAΦNE ϕ-factory. KLOE has provided the wolrd's best upper limits for phenomenological parameters describing CPT violation by decoherence of entangled neutral kaons [10] and by violation of the Lorentz symmetry [11] occurring within the framework of Standard Model Extension (SME) as it will be discussed in this article.

2. Test of CPT and Lorentz symmetries with the KLOE detector

The KLOE experiment [12, 13, 14] operates at the DAΦNE e^+e^- collider working at a center of mass energy of about 1020 MeV. Positron and electron beams collide at an angle of π-25 mrad, producing ϕ mesons with non-zero momentum in the horizontal plane, $|p_\phi| \sim 15$ MeV, in the KLOE reference frame [11].
Neutral kaons from ϕ meson decays are produced in a coherent quantum state:

$$|i\rangle = N\left(|K_S, \vec{p_1}\rangle\,|K_L, \vec{p_2}\rangle - |K_L, \vec{p_1}\rangle\,|K_S, \vec{p_2}\rangle\right) , \tag{1}$$

where $\vec{p_1}$ and $\vec{p_2}$ denote the kaon momenta and N is a normalization factor expressed in terms of the CP impurities ϵ_S and ϵ_L: $N = \sqrt{(1+|\epsilon_S|^2)(1+|\epsilon_L|^2)/2(1-\epsilon_S\epsilon_L)}$. Due to the quantum entanglement of the two neutral kaons their decays are correlated. Thus, the kaons double differential rate of decay into two final states $f_1$and f_2 contains a characteristic time interference term [15]:

$$I(f_1, f_2, \Delta t) \sim |\eta_1|^2 e^{-\Gamma_L \Delta t} + |\eta_2|^2 e^{-\Gamma_S \Delta t} - 2|\eta_1|^2|\eta_2|^2 e^{-\frac{(\Gamma_L+\Gamma_S)}{2}\Delta t}\cos(\Delta m \Delta t + \phi_2 - \phi_1) \tag{2}$$

where Δt denotes the proper decay times difference and η_1 and η_2 are the following decay amplitude ratios: $\eta_i = |\eta_i| e^{i\phi_i} = \frac{\langle f_i|T|K_L\rangle}{\langle f_i|T|K_S\rangle}$. Γ_L and Γ_S denote the widths of K_L and K_S meson, respectively. The decay intensities of entangled kaons appear to be very sensitive to any CPT violating effect at the level of the interesting Planck scale region, presently unreachable in other similar systems [15]. If both kaons decay to the same final state (e.g. $\pi^+\pi^-$) the interference pattern in Eq. 2 is very sensitive to any deviation from unity of the ratio η_1/η_2 in the interference region of $\Delta t \approx 0$. These deviations may be present due to the CPT violation which in the framework of SME manifests to lowest order only in the CPT violating parameter δ_K equal to: $\delta_K = (\epsilon_S - \epsilon_L)/2$. This parameter exhibits a dependence on the 4-momentum of the kaon:

$$\delta_K \approx i\sin\phi_{SW} e^{i\phi_{SW}} \gamma_K (\delta a_0 - \vec{\beta_K} \cdot \Delta\vec{a})/\Delta m , \tag{3}$$

where ϕ_{SW} is the superweak phase, γ_K and $\vec{\beta_K}$ are the boost factor and velocity of kaon and Δm denotes the difference between K_L and K_S mass. The four Δa_μ parameters are a measure of the Lorentz symmetry violation. Neglecting higher order contributions to CP violation the two η parameters can be written as:

$$\begin{aligned} \eta_1 &\cong \epsilon_K - \delta_K(\vec{p_1}, t_s) \\ \eta_2 &\cong \epsilon_K - \delta_K(\vec{p_2}, t_s) , \end{aligned} \tag{4}$$

where $\epsilon_K = (\epsilon_S + \epsilon_L)/2$. Since the Earth is rotating the CPT and Lorentz violating parameters are usually considered in the fixed stars reference frame which makes δ_K dependent also on the sidereal time t_s after the transformation from the laboratory frame [15].
At KLOE a search for the Lorentz and CPT symmetries violation have been performed by measurements of the interference pattern for both kaons decaying to $\pi^+\pi^-$ final states. A sample of 1.7 fb^{-1} of integrated luminosity gathered in the 2004-2005 data taking period have been used in this analysis. Moreover, we have generated two Monte Carlo samples: one contains all the ϕ meson decay channels (used for analysis optimization); the other is a MC signal sample which have been used for efficiency and decay time difference resolution determination. The first preselection of data was done requiring two vertices with two tracks of opposite curvature. Assuming the pion hypothesis for every track we have then applied a set of kinematical cuts [11]:

- $|m_{\rm trk} - m_{\rm K}| < 5$ MeV/c^2, where $m_{\rm K}$ denotes the kaon mass and $m_{\rm trk}$ is the invariant mass of the kaon reconstructed from the tracks assuming charged pion mass hypothesis: $\vec{p}_{1(2)} = \vec{p}_{\pi^+_{1(2)}} + \vec{p}_{\pi^-_{1(2)}}$ and $E_{1(2)} = E_{\pi^+_{1(2)}} + E_{\pi^-_{1(2)}}$.

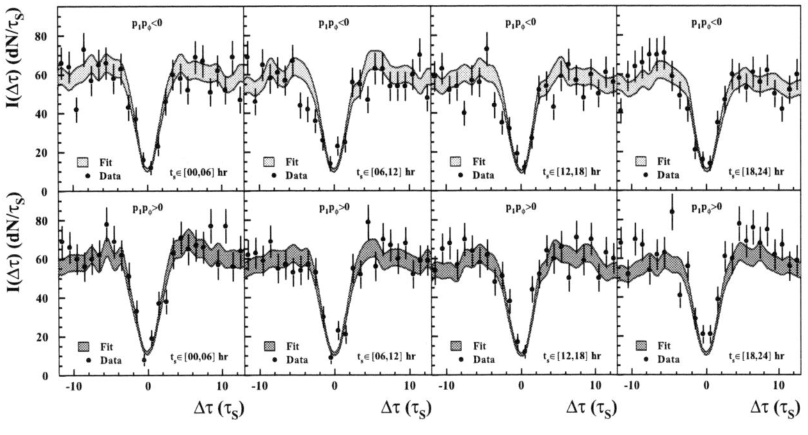

Figure 1. Distributions of the double differential decay rates into $\pi^+\pi^-, \pi^+\pi^-$ final states measured with the KLOE detector. The top panel contains distributions for four bins of sidereal time when kaon emmited in the forward hemisphere ($\cos\theta > 0$) in the KLOE reference frame was moving along with the ϕ momentum (i.e. $\vec{p}_{1(2)} \cdot \vec{p}_\phi > 0$). The lower panel shows distributions for a subset of data where kaon emmited in the forward hemisphere was moving in the opposite direction (i.e. $\vec{p}_{1(2)} \cdot \vec{p}_\phi < 0$). The results of the fit is indicated together with uncertainties by colored bands (figure taken from [11].)

- $\sqrt{E^2_{\text{miss}} + |\vec{p}_{\text{miss}}|^2} < 10$ MeV, where $\vec{p}_{\text{miss}} = \vec{p}_\phi - \vec{p}_1 - \vec{p}_2$ and $E_{\text{miss}} = E_\phi - E_1 - E_2$;
- -50 MeV $^2/\text{c}^4 < m^2_{\text{miss}} < 10$ MeV$^2/\text{c}^4$, where $m^2_{\text{miss}} = E^2_{\text{miss}} - |\vec{p}_{\text{miss}}|^2$
- $|p^*_{1(2)} - p^*_0| < 10$ MeV/c, where $p^*_{1(2)}$ is the momentum of the kaon, as derived from tracks, in the ϕ meson reference frame, while $p^*_0 = \sqrt{s/4 - m^2_{\text{K}}}$ and $\sqrt{s}$ is the center-of-mass energy.

According to MC simulations the remaining data sample contains a small fraction of irreducible background mainly due to kaon regeneration in the thin beryllium cylindrical foil inside the interaction region ($\sim 2\%$) and due to the $e^+e^- \to 4\pi$ reaction. In order to improve the accuracy of decay time measurements a dedicated global fit procedure has been performed [11]. Since the resolution of decay time difference deteriorates at large pion opening angles $\theta_\pi$we have rejected all events with $\cos\theta_\pi < 0.975$.

The data sample was divided into two subsets: one containing events where kaon emmited in the forward hemisphere ($\cos\theta > 0$) in the KLOE reference frame was moving along with the ϕ momentum (i.e. $\vec{p}_{1(2)} \cdot \vec{p}_\phi > 0$), and the second one with events where kaon emmited in the forward hemisphere was moving in the opposite direction (i.e. $\vec{p}_{1(2)} \cdot \vec{p}_\phi < 0$). Moreover, the dataset was divided into four bins of sidereal time. For every subset of data the decay intensity distribution as a function of Δt was determined. The distributions were fitted simultaneously with the interference pattern taken from Eq. 2 and assuming possible modulation effects induced by the CPT-violating parameter δ_K as in Eq. 3. The resulting eight distributions are presented in Fig. 1, where the difference of decay times Δt is expressed in the units of K_S lifetime. The colored bands represent the fit uncertainty due to finite MC simulation statistics and efficiency correction.

The values of Δa_μ parameters determined from the fit are:

$$\Delta a_0 = (-6.0 \pm 7.7_{stat} \pm 3.1_{syst}) \times 10^{-18} \text{ GeV},$$
$$\Delta a_x = (\ 0.9 \pm 1.5_{stat} \pm 0.6_{syst}) \times 10^{-18} \text{ GeV},$$

$$\Delta a_y = (-2.0 \pm 1.5_{stat} \pm 0.5_{syst}) \times 10^{-18} \text{ GeV},$$
$$\Delta a_z = (\ 3.1 \pm 1.7_{stat} \pm 0.5_{syst}) \times 10^{-18} \text{ GeV}.$$

These results do not reveal any CPT or Lorentz symmetries violation and constitute first measurement of these quantities in the kaon sector. The detailed description of the systematic uncertainties can be foud in [11].

3. Future prospects

The CPT and Lorentz symmetries violation are one of the main physics topics of the KLOE-2 experiment which is continuing and extending the physics program of its predecessor [16]. For the forthcoming run the KLOE performance have been improved by adding new subdetector systems: the tagger system for the $\gamma\gamma$ physics, the Inner Tracker based on the Cylindrical GEM technology and two calorimeters in the final focusing region [17]. The new inner detector will increase the tracking and vertexing capabilities of KLOE allowing for significant reduction of systematic uncertainties which together with the higher statistics expected by KLOE-2 will significantly improve the present result.

Acknowledgments

This work was supported in part by the EU Integrated Infrastructure Initiative Hadron Physics Project under contract number RII3-CT- 2004-506078; by the European Commission under the 7th Framework Programme through the 'Research Infrastructures' action of the 'Capacities' Programme, Call: FP7-INFRASTRUCTURES-2008-1, Grant Agreement No. 227431; by the Polish National Science Centre through the Grants No. DEC-2011/03/N/ST2/02641, 2011/01/D/ST2/00748, 2011/03/N/ST2/02652, 2013/08/M/ST2/00323, 2013/11/B/ST2/04245, and by the Foundation for Polish Science through the MPD programme and the project HOMING PLUS BIS/2011-4/3.

References

[1] Olive K A et al. (Particle Data Group) 2014 *Chin. Phys. C* **38** 090001
[2] Bernabeu J, Ellis J, Mavromatos N E, Nanopoulos D V and Papavassiliou J 2006 *Nuclear Physics B* **744** 180
[3] Ellis J, Hagelin J S , Nanopoulos D V, Srednicki M, 1984 *Nucl. Phys. B* **241** 381
[4] Wald R, 1980 *Phys. Rev. D* **21** 2742
[5] Kostelecky V A, Samuel S, 1989 *Phys. Rev. D* **39** 683
[6] Kostelecky V A, Potting R, 1991 *Nucl. Phys. B* **359** 545
[7] Barenboim G, Lykken J, 2003 *Phys. Lett. B* **554** 73
[8] Huet P, Peskin M, 1995 *Nucl. Phys. B* **434** 3
[9] Lees J P et al., 2012 *Phys. Rev. Lett.* **109** 211801
[10] Ambrosino F et al., 2006 *Phys. Lett. B* **642** 315 Di Domenico A, 2009 *J. Phys. Conf. Ser.* **171** 012008
[11] Babusci D et al., 2014 *Phys. Lett. B* **730** 89
[12] Adinolfi M et al., 2002 *Nucl. Instrum. Methods Phys. Res. A* **488** 51
[13] Bossi F et al., 2008 *Riv. Nuovo Cim.* **31** 531
[14] Adinolfi M et al., 2002 *Nucl. Instrum. Methods Phys. Res. A* **482** 363
[15] Di Domenico A (Ed.), 2007 *Handbook on Neutral Kaon Interferometry at a ϕ-Factory* (Frascati Phys. Ser., vol 43)
[16] Amelino-Camelia G *et al.* 2010 *Eur. Phys. J. C* **68** 619
[17] Moricciani D 2011 *PoS* **EPS–HEP2011** 198

Journal of Physics: Conference Series **556** (2014) 012072 doi:10.1088/1742-6596/556/1/012072

The *Belle II* Experiment

Giulia Casarosa[1]

INFN - Sezione di Pisa, Largo B. Pontecorvo 3, Pisa, Italy

E-mail: giulia.casarosa@pi.infn.it

Abstract. The *Belle II* experiment will analyze the products of the e^+e^- collisions at the center-of-mass energy corresponding to the mass of the $\Upsilon(4S)$ produced at the high-luminosity B-Factory SuperKEKB. The experiment will collect 50 ab^{-1}, allowing to probe the Standard Model at a never-reached level, and to put very stringent constraints in the space of parameters of many New Physics models. I report here the expected precision that *Belle II* will reach with the full dataset on a selection of observables in the charm and B physics sectors.

1. Introduction

The most important result achieved by the B-Factories BELLE and *BABAR* was the confirmation of the CKM mechanism as the one responsible for CP violation (CPV) in the Standard Model (SM), which earned Kobayashi and Maskawa the Nobel prize in 2008. This measurement was only one of those performed with the 1.5 ab^{-1} of data that they collected together. Most of the measurements have confirmed the SM predictions, other measurements are in tension with the predicted value, but the size of the disagreement is not large enough to open to a particular New Physics (NP) scenario. Moreover, there are a certain number of observations that are not well accommodated in the SM, e.g. the size of the asymmetry between matter and antimatter. In order to find an explanation for these observations, a significantly larger data sample is needed. *Belle II* sits on the intensity frontier path to NP, with its data it will be possible to search for NP beyond the TeV scale.

2. The High-Luminosity B-Factory SuperKEKB and The Detector *Belle II*

In order to increase the data sample by almost two orders of magnitude, a new accelerator concept is needed. SuperKEKB is an upgrade of the KEKB accelerator, based on the nano-beam scheme [1], and it is designed to reach the highest luminosity ever reached of $8 \times 10^{35}\text{cm}^{-2}\text{s}^{-1}$. Some of the new beam features will have a direct impact on the physics analyses. The size of the beams on the transverse plane at the interaction point will be reduced by a factor 20 with respect to KEKB. The luminous region is used as a constraint in the decay and therefore physics analyses will benefit from this reduction. The center-of-mass boost is reduced by roughly a factor two with respect to the previous B-Factories, increasing the detector hermiticity, that is crucial in missing-energy analyses. The consequent reduction of the vertices separation (Δz) deteriorates the resolution of the Δt[2] that is recovered by a completely new vertex detector. Finally, the factor 40 higher instantaneous luminosity will imply higher background rates (and consequently

[1] for the *Belle II* Collaboration.
[2] the difference of decaying time of two particles, e.g.: B and $\bar{B}$, or mother and daughter particles.

higher detector occupancy, radiation damage, fake hits) and higher event rate (demanding very high DAQ and computing performances).

The *Belle II* detector is a general purpose detector and it is described in detail elsewhere [2]. The tracking system is composed by a 6-layer silicon Vertex Detector (VXD) and the Central Drift Chamber. The particle identification is performed by the AEROGEL Cherenkov ring imaging and the time-of-flight sub-detectors. The K_L^0 and muon detector is installed outside the electromagnetic calorimeter. The detector has been developed in order to reach similar or better performances that the former BELLE one. Improvements are expected, in particular in the interaction point and secondary vertex resolution and in K_S^0 and π^0 reconstruction.

3. Expectations on B Physics

The B-Factories are perfect environments to study B physics: in the e^+e^- collision a boosted $Y(4S)$ is produced and immediately decays in a couple of entangled B mesons. Depending on the signal signature, the *other* B (the tag-side B) is fully or inclusively reconstructed. The full reconstruction is used with weak signal signatures (e.g. two neutrinos in the final state), and it is cleaner but less efficient than the inclusive reconstruction. A reconstructed B meson sample with high-efficiency flavour/charge tagging is available for the analyses. With the full reconstruction tool missing-energy analyses are straightforward.

3.1. The Unitary Triangle and CP Violating Phases

One of the great achievements of the B-Factories is the measurement of the angles and sides of the Unitary Triangle (UT). *Belle II* will continue improving the constraints on the UT. In particular, the expected precisions with the full statistics on the three angles α/ϕ_2, β/ϕ_1, γ/ϕ_3 are $1°$, $0.4°, 1.5°$, respectively. Improvements on the sides measurements are also expected: between 1.2% and 1.4% for $|V_{cb}|$ and 2.4% and 3% for $|V_{ub}|$.

The size of CPV in the SM is orders of magnitude smaller that the one needed to explain the asymmetry between matter and antimatter. One of the missions of *Belle II* will be to search for new CP violating phases. One place to look for additional CP violating phases are B^0 decays through penguin processes in ϕK_S^0, $\eta' K_S^0$ and $K_S^0 K_S^0 K_S^0$. The expected sensitivity on the parameter S[3] as a function of the integrated luminosity is shown in Figure 1. The expected

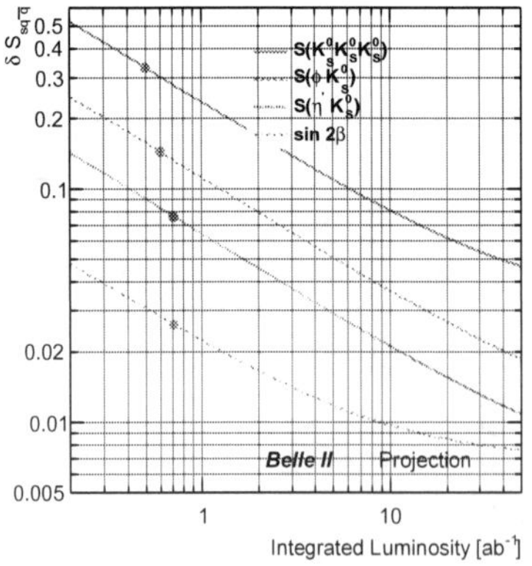

Figure 1. Expected error on the parameter S as a function of the integrated luminosity for different final states.

[3] the equivalent of $\sin 2\beta/\sin 2\phi_1$.

precision with the full dataset should be good enough to distinguish different theory models [3].

3.2. Right-Handed Currents and Rare B decays

In certain New Physics scenarios, helicity-suppressed amplitudes (e.g. $b \to s\gamma_R$) can be enhanced. The sensitive observable is represented by the parameter S mentioned above. Analysing the full dataset *Belle II* should be able to either exclude or confirm this enhancement. For example, the experimental precision on S for the $K^0_S\pi^0\gamma$ final state is expected to be around 0.035, while certain theory models predict a value of 0.5 [3].

Rare decays as $B \to h^{(*)}\nu\nu$, with $h = K, \pi$ are very clean from the theoretical point of view, and this makes them golden places where to look for New Physics contributions. Standard Model expectations on the branching ratio for channels with pion in the final state are orders of magnitude below the expected upper limit reachable with the full *Belle II* statistics (around 10^{-5}) and therefore any significant signal would be a sign of NP. The SM branching ratios for channels with kaons in the final states are higher, e.g. *Belle II* should be able to measure the branching ratio of the decay $B \to K^{*0}\nu\nu$ with an error of 30% with the full dataset.

3.3. Charged Higgs

The 2-Higgs doublet model can be tested in $B^+ \to \tau^+\nu$ and $B^0 \to D^{(*)-}\tau^+\nu$ decays. In the case of the $\tau^+\nu$ final state, the analysis consists in fully reconstructing the tag-side B, and then measure the branching ratio of the signal channel. This measurement put constraints in the $(\tan\beta, m_{H^\pm})$ space, and with the 50ab^{-1} of *Belle II*, the allowed region will considerably shrink. In the $D^{(*)}\tau\nu$ case, the observable is a ratio of branching ratios:

$$R = \frac{Br(B \to D^{(*)}\tau\nu)}{Br(B \to D^{(*)}\ell\nu)} \quad (1)$$

The current measurements of R are in tension with the SM showing an excess of around 3σ. *Belle II* will be able to confirm the excess already with 5 ab^{-1}. The expected precision with the full dataset is expected to be 2% (3.5%) for the $D^*\tau\nu$ ($D^*\tau\nu$) final state.

4. Charm Physics Expectations

B-Factories are excellent places to study charm physics since the production of charm is at the same level as the production of bottom. The analyses techniques are quite different since there is not a coherent production of $D^0 - \bar{D}^0$: flavour tagging is done selecting D^0 from D^{*+} decays and there is no access to relative strong phases. As for B physics, time-dependent analysis are possible, as well as D meson full reconstruction for neutrinos and inclusive analyses.

4.1. Mixing and Indirect CPV

One of the most interesting things in charm is the study of mixing and CP violation. The reason for this is that the possible new physics must couple to up-type quarks, opposed to what should happens for all the other mixing systems. This means that the information on NP that lies in the charm sector is complementary to what we can find in the B and kaon systems.

Belle II will be essential in the direct measurement of the mixing parameters x and y through the time-dependent analysis on the Dalitz Plot of the $D^0 \to K^0_S\pi^+\pi^-$ decays. The sensitivity on the mixing parameters should reach the level of 0.05% for y and 0.08% for x. In this analysis the CP violating parameters $|q/p|$ and $\phi = \arg(q/p)$ will also be directly measured. This analysis will be limited by the systematic error associated to the modelling of the decays on the Dalitz Plot.

In Figure 2 the expected allowed region for the mixing and CP violating parameters after the combination of all measurements including the expectations for *Belle II* full dataset results by the Heavy Flavour Average Group. The allowed regions are much smaller than the current ones, attesting the significant contribution of *Belle II* to the charm sector.

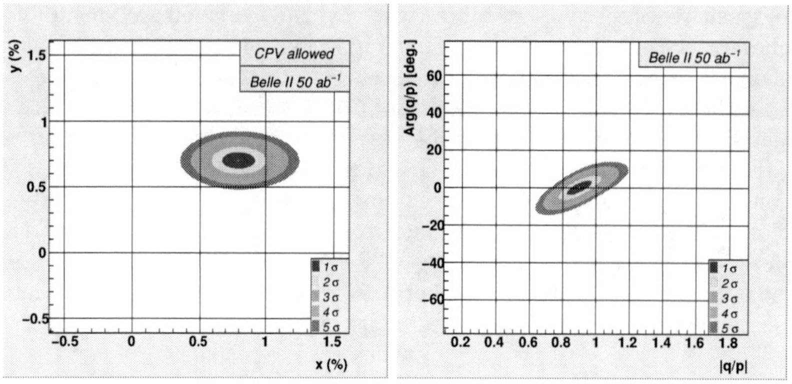

Figure 2. Expected allowed regions for the mixing (left) and the CP violating (right) parameters with *Belle II* full statistics. The central values are $(x, y) = (0.8, 0.7)$ and $(|q/p|, \phi|) = (0.9, 0)$.

4.2. Time-Integrated CPV

The search for CP violation in charm decays will be also performed in the measurement of the A^f_{CP} asymmetries defined as:

$$A^f_{CP} = \frac{\Gamma(D \to f) - \Gamma(\bar{D} \to \bar{f})}{\Gamma(D \to f) + \Gamma(\bar{D} \to \bar{f})} \qquad (2)$$

The most important contribution of *Belle II* will be on channels with neutrals in the final state such as $D^0 \to K^0_S\pi^0, \pi^0\pi^0, K^0_S\eta$ and $D^+ \to \pi^+\pi^0$. The expected sensitivity will reach 0.03% (for the $K^0_S\pi^0$ final state).According some models [4] direct CPV in radiative decays ($D^0 \to \phi\gamma, \rho\gamma$) can be enhanced above 1%, corresponding to the level of the expected experimental sensitivity. It would therefore be possible to measure it with *Belle II* full dataset.

5. Conclusions

Flavour physics will continue to play a fundamental role in the process of understanding Nature in the next decade. The *Belle II* physics program is very rich, it can open a many NP scenarios. The complementarity with LHCb physics is established, both experiments are needed in case something unexpected will be found. Moreover, the existence of observables with comparable precision will be an important check for both experiments. The construction of the accelerator is expected to be completed by mid 2015, while *Belle II* construction is ongoing. We expect the first physics run in 2017, with the full dataset of 50 ab^{-1} collected in the next 6 years.

References

[1] P. Raimondi, A. Seryi, "A Novel Final Focus Design for High Energy Linear Colliders", Jun 2000 Published in 'Vienna 2000, EPAC 00' 492-494
[2] *Belle II* Technical Design Report, KEK Report 2010-1, arXiv:1011.0352
[3] Physics At Super B Factory, KEK Report 2009-12, arXiv:1002.5012
[4] G.Isidori and J.F. Kamenik "Shedding Light on CP Violation in the Charm System via $D \to V\gamma$ Decays", Phys. Rev. Lett. 109, 171801

Journal of Physics: Conference Series **556** (2014) 012073 doi:10.1088/1742-6596/556/1/012073

The LHCb Upgrade

S Akar on behalf of the LHCb Collaboration
CPPM, Aix-Marseille Université, CNRS/IN2P3, Marseille, France
E-mail: simon.akar@cppm.in2p3.fr

Abstract. The LHCb experiment is a high-precision spectrometer searching for New Physics via measurements of CP violation and rare decays in the b- and c-quark sector. The quality of the results obtained from the data collected during the first Run of the LHC has demonstrated the excellent performance and the robustness of the detector. In order to significantly increase the statistical precision on theoretically clean observables in the heavy flavor sector, the level of collected data by the LHCb detector must be increased much beyond $1\,\text{fb}^{-1}$ per year. Therefore, it is planned to upgrade the detector, which will allow a 40 MHz readout with a much more flexible software-based triggering system and redesigned sub-detectors.

1. Introduction

The LHCb experiment [1],[2] is a general-purpose forward spectrometer operating at the LHC accelerator. The detector covers a pseudo-rapidity range of $2 < \eta < 5$, which is complementary to the other LHC detector ranges. During Run 1 (2010-2012), the detector demonstrated excellent operation performance, with about 99% of working detector channels, allowing to record more than 90% of the data delivered by the LHC adding up to a total of about $3\,\text{fb}^{-1}$ of data saved to disk over the period 2010-2012. In 2011, the detector was already running at the design luminosity and number of visible interactions per bunch crossing, respectively of $\mathcal{L} = 2 \times 10^{32}\,\text{cm}^{-2}\text{s}^{-1}$ and $\mu = 0.4$. In 2012, the LHC accelerator improved its performance and both $\mathcal{L}$ and μ increased above these design values. In order to cope with the delivered luminosity, the LHCb detector started to adjust constantly the beam overlap to maintain a constant instantaneous luminosity of twice the design one at $\mathcal{L} = 4 \times 10^{32}\,\text{cm}^{-2}\text{s}^{-1}$ during a run, and with $\mu = 1.5$.

During LHC Run 3, the objective is to increase the instantaneous luminosity to $\mathcal{L} = 2 \times 10^{33}\,\text{cm}^{-2}\text{s}^{-1}$. However, in this condition the current design of the LHCb detector does not allow to further improve the data rate as well as the efficiency. Indeed, it is limited in terms of data bandwidth to 1 MHz, instead of the LHC bunch crossing frequency of 40 MHz, due to the implementation of a first-level hardware trigger (L0). In particular, with an increased luminosity, the architecture of the L0 trigger does not allow for efficient triggering on purely hadronic channels. As shown on Figure 1, while the physics yield for muonic channels increases with the luminosities, hadronic ones saturate above a luminosity of $3 \times 10^{32}\,\text{cm}^{-2}\text{s}^{-1}$ and is essentially half of the one for muonic channels at $\mathcal{L} = 4 \times 10^{32}\,\text{cm}^{-2}\text{s}^{-1}$. In consequence, it is mandatory to upgrade the current detector in order to increase the data rate as well as the efficiency to reach the high level of experimental precisions that is required in the flavor sector.

The upgrade of the detector is planned during the second Long Shutdown (LS2) of the accelerator in 2018-2019, and should allow LHCb to increase the yields in the muonic and

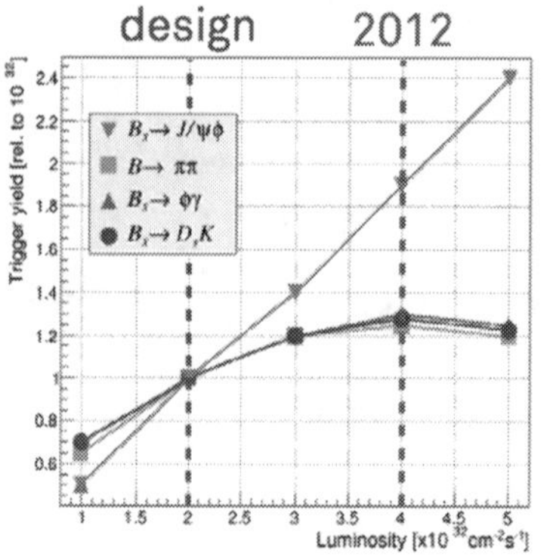

Figure 1: Trigger yield for different B decays as function of the instantaneous luminosity.

hadronic channels by a factor five and ten, respectively. The objective is to operate at a constant instantaneous luminosity of $2 \times 10^{33}\mathrm{cm}^{-2}\mathrm{s}^{-1}$ during a period of ten years, corresponding to about $50\,\mathrm{fb}^{-1}$ of collected data. These running conditions correspond to ten times the current design luminosity with increased pileup of a factor 5. To reach the desired performance, all the front-end electronics as well as the detector components with embedded electronics need to be replaced. In addition, some of the detector components, in particular those closest to the interaction point, need to be redesigned to cope with the higher expected occupancies and radiation.

The plans for the LHCb upgrade are documented in the Letter of Intent [3] and in the Framework Technical Design Report of the LHCb Upgrade [4], and for each subsystem in a collection of dedicated Technical Design Reports [5],[6],[7],[8].

2. Physics motivation

Up to now, the results in heavy flavour physics are fully consistent with the CKM mechanism. However, a new source of CP violation beyond the Standard Model (SM) is needed to explain, for example, the amount of matter in the universe.

Extensive searches for evidence of New Physics (NP) beyond the SM at the TeV scale are currently being performed at the LHC. The absence of positive results coming from direct searches of NP effects performed in high-energy colliders, indicates that if such NP exist, corresponding effects would arise beyond the electroweak scale. The LHCb experiment is well designed to perform searches of NP effects via indirect searches, by performing precise measurements of specific decay processes involving loop diagrams. In the presence of NP, new heavy particles could enter the loop and provide new amplitudes and phases that could dramatically modify the observables [9],[10]. The current results in B and charm physics obtained at the LHC are, so far, consistent with the SM predictions, but the theoretical precision on flavour observables in many decay processes are still well below the experimental uncertainties [4]. The aim of the upgrade of the LHCb detector, with $50\,\mathrm{fb}^{-1}$ of collected data, is to significantly decrease the experimental uncertainties to reach values comparable to or below the theoretical precision. Some key channels are briefly presented here.

A good example of LHCb's potential in the search for NP are the rare decays $B_s^0 \to \mu\mu$ and $B^0 \to \mu\mu$. LHCb's current uncertainty on branching ratio $\mathcal{B}(B_s^0 \to \mu\mu)$ is 1.5×10^9 [11], while the expected statistical sensitivity with the upgraded detector reaches 0.15×10^9 , whereas the theoretical precision is 0.3×10^9. Rare decays involving $b \to s$ transitions can also be considered as key channels for the upgrade of the LHCb detector. Indeed, in $B^0 \to K^*(K)\mu\mu$ decays, the expected statistical sensitivity on many observables such as the forward-backward asymmetry, others also related to the angular distributions, as well as the isospin asymmetry are expected to reach the theoretical uncertainties. The excepted statistical sensitivity on CP violation parameters, especially in the B_s system, should also get close to the theoretical precision, as for example the mixing induced phase ϕ_s in $B_s^0 \to J/\psi\phi$ or $B_s^0 \to J/\psi f_0(980)$ decays.

3. The LHCb upgrade strategy

3.1. The trigger system

With the current detector, the main bottleneck for increasing the data rate as well as the efficiency at a luminosity of $2 \times 10^{33}\,\mathrm{cm^{-2}s^{-1}}$ is the existing readout architecture, shown in Figure 2. The strategy for the upgrade of the LHCb experiment is to remove the L0 trigger implemented in the hardware of the individual sub-detectors. In the L0 frame, the rate of accepted events is reduced from the LHC bunch crossing frequency of 40 MHz to about 1 MHz based on the transverse energy and transverse momentum measured by the calorimeters and the muon system, respectively. In a second step, events are built in the software based Higher Level Trigger (HLT), using the information from the entire detector. The final output rate to storage on disk corresponds to 5 kHz.

In the detector upgrade the new readout architecture will be implemented as a trigger-less system [5], where the full detector information at the inelastic event rate of 30 MHz will be used in a fully-flexible software based trigger, as shown in Figure 3. This full software trigger should apply selections as similar as possible to those applied in offline analyses to maximize trigger efficiencies and to minimize systematic uncertainties. The full software trigger will write to tape with a rate of 20 to 100 kHz.

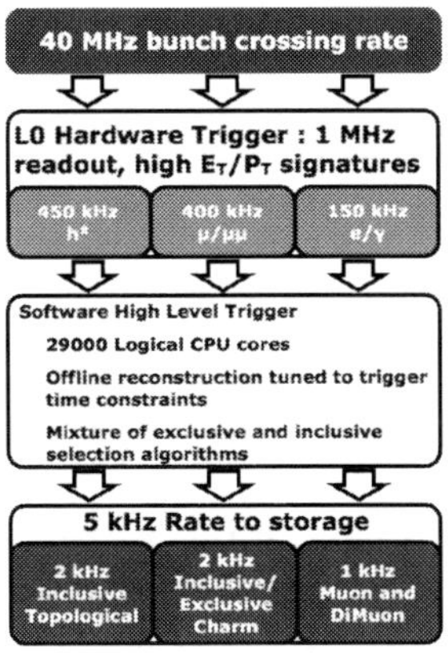

Figure 2: Schematic view of the LHCb trigger system, as implemented until 2012.

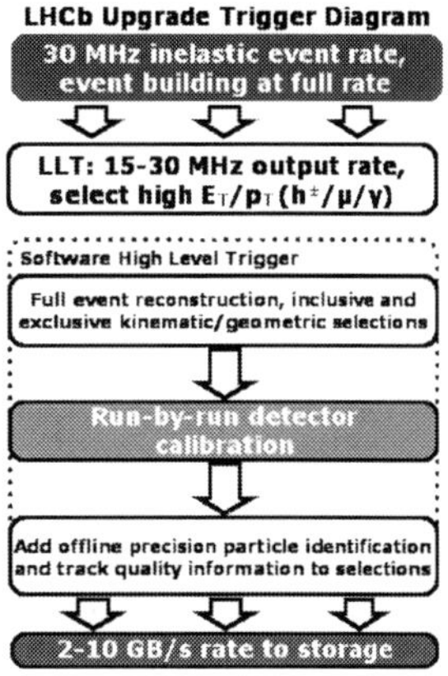

Figure 3: Schematic view of the full-software trigger scheme for the LHCb upgrade.

3.2. Detector Upgrade

In addition to the new readout system briefly discussed in Section 3.1, many sub-detectors must be redesigned to cope with the harsher conditions in which the LHCb upgrade detector will be running. The most significant changes concern the tracking system, including the vertex detector, which will be completely redesigned.

The vertex detector (VELO) upgrade [6], which is essential for vertices identification and track reconstruction, will benefit from a new pixel detector with smaller cells of $55 \times 55\,\mu\mathrm{m}^2$. In order to minimize the production of secondary particles, the thickness of the sensors will be reduced, down to $200\,\mu$m. Additionally, to improve the impact parameter resolution and to increase the track acceptance, the distance between the first sensors and the beam will be reduced from 5.5 to 3.5 mm in the fully closed configuration of the VELO. Simulation studies, presented in Reference [6], show that the expected performance of the upgraded VELO should be maintained, if not improved, compared to the already excellent performance obtained with the current VELO.

The rest of the current tracker system is composed of the Trigger Tracker (TT) and the T-stations, placed upstream and downstream of the magnet, respectively. The TT will be replaced by the Upstream Tracker (UT), using the same technology as in the TT, i.e. silicon micro-strips, but with finer granularity, especially in the inner part, in order to reduce the occupancy and ghost tracks rate. The material budget will also be lowered, with sensors of reduced thickness: 300 μm compared to 500 μm in the TT. The T-stations, currently using silicon strips in the inner-most part and straw tubes in the outer part will be replaced by a single technology using layers of scintillating fibres with a diameter of 250 μm. The readout, made of silicon photo-multipliers, will be placed outside the acceptance at the edge of each of the three $5 \times 6\,\mathrm{m}^2$ panels. Details and expected performance of both the UT and the T-stations upgrade can be found in Reference [7].

The particle identification (PID) system, which is composed of two Ring Imaging Cherenkov (RICH) detectors, one upstream and the other downstream the magnet, the electromagnetic and hadronic calorimeters as well as five muon chambers, will also undergo modifications during the upgrade. Indeed, in the case of the RICH placed upstream the magnet, the optics will be redesigned and the aerogel will be replaced by a CF_4 radiator, in order to cope with the higher expected occupancies. For the same reason and due to the removal of the L0 trigger, the Scintillating Pad Detector and the Preshower detectors, as well as the first muon chamber, all placed in front of the electromagnetic calorimeter, will be removed. Details and expected performance of the upgraded PID system can be found in Reference [8].

4. Conclusion

During the first Run of data taking at the LHC, the LHCb detector demonstrated excellent performance, leading to impressive results with a high-precision level on many heavy flavor observables. An upgrade of the LHCb detector is proposed, with a new readout system and a full software trigger, in order to enhance the physics yields, especially in the purely hadronic channels, which will increase by a factor ten with respect to the foreseen yields in Run 2. A total dataset of at least $50\,\mathrm{fb}^{-1}$ will be collected during a period of ten years. This will allow to significantly decrease the statistical uncertainties on many key measurements, some of which will be reaching the current theoretical precisions, which is required in order to investigate small deviations from the SM. All the documents concerning the LHCb upgrade have now been released by the LHCb collaboration and approved by the CERN Research Board. The road-map is on schedule for an installation of the LHCb upgrade that will start with the LS2 in 2018.

References

[1] LHCb Collaboration 2003 *LHCb technical design report: Reoptimized detector design and performance* CERN-LHCC-2003-030
[2] Alves A A *et al* (LHCb Collaboration) 2008 The LHCb Detector at the LHC *J. of Instrumentation* JINST3(2008)S08005
[3] LHCb Collaboration 2011 *Letter of Intent for the LHCb Upgrade* CERN-LHCC-2011-001
[4] LHCb Collaboration 2012 *Framework TDR for the LHCb Upgrade: Technical Design Report* CERN-LHCC-2012-007
[5] LHCb Collaboration 2014 *LHCb Trigger and Online Upgrade Technical Design Report* CERN-LHCC-2014-016
[6] LHCb Collaboration 2013 *LHCb VELO Upgrade Technical Design Report* CERN-LHCC-2013-021
[7] LHCb Collaboration 2014 *LHCb Tracker Upgrade Technical Design Report* CERN-LHCC-2014-001
[8] LHCb Collaboration 2013 *LHCb PID Upgrade Technical Design Report* CERN-LHCC-2013-022
[9] Grossman Y and Worah M P 1997 *Phys. Rev.* B. **395** 241-9
[10] Bigi I Y and Sanda A I 2000 *CP Violation* (Cambridge: Cambridge University Press)
[11] Aaij R *et al* (LHCb Collaboration) 2013 *Phys. Rev. Lett.* **110** 021801

The NA62 run control

N. Lurkin

On behalf of the NA62 collaboration[1]
School of Physics and Astronomy, University of Birmingham, Birmingham B15 2TT, UK

E-mail: ncl@hep.ph.bham.ac.uk

Abstract. The NA62 experiment at CERN SPS will soon start taking data to measure the ultra-rare decay $K^+ \to \pi^+\nu\bar{\nu}$. The requirement to reach a rejection factor of 10^{12} can be obtained with a large number of VETO and particle identification detectors as well as momentum measurement spectrometers. All components are included in the Trigger and/or Data Acquisition system. The RunControl has been designed to centrally link them to achieve their control and monitoring while still allowing non-experts shifters to run the experiment easily.

1. Introduction

The NA62 experiment at the CERN SPS aims to collect about 100 events of the ultra-rare kaon decay $K^+ \to \pi^+\nu\bar{\nu}$ with less than 10% background over 2 years of data taking. This process is forbidden at tree level and can occur only through a Flavour Changing Neutral Current loop. It is theoretically very clean, the short distance contribution dominates and the hadronic matrix element can be related to other well measured processes. The prediction of the branching fraction by the Standard Model has been computed to an exceptionally high degree of precision[1]: $BR(K^+ \to \pi^+\nu\bar{\nu})_{th} = (7.81 \pm 0.75 \pm 0.29) \times 10^{-11}$.

This channel is therefore an excellent probe of the standard model as any small deviation from this value would indicate new physics. The combined experimental value for this branching ratio is $BR(K^+ \to \pi^+\nu\bar{\nu})_{exp} = (1.73^{+1.15}_{-1.05}) \times 10^{-10}$, extracted from the 7 detected events in the E787 and E949 experiments at Brookhaven National Laboratory[2] but its precision is not sufficient to be a significant test of new physics.

To achieve this ambitious goal, NA62 needs to consider at least 10^{13} K^+ decays from the secondary 75 GeV/c unseparated hadron beam with $\sim$6% K^+. The in-flight decay technique sets the scale of the detector which has a longitudinal extention exceeding 200 m, enclosing the 65 m long fiducial decay region.

2. The NA62 RunControl

The RunControl is a centralised control and monitoring application whose purpose is to allow a non-expert shifter to supervise the data taking easily while still giving specialists the possibility to achieve a high level of control of their own equipment.

[1] Participating Institutes: Birmingham, Bratislava, Bristol, Bucharest, CERN, Dubna, Fairfax, Ferrara, Florence, Frascati, Glasgow, Liverpool, Louvain, Mainz, Merced, Moscow, Naples, Perugia, Pisa, Prague, Protvino, Rome I, Rome II, San Luis Potos, Stanford, Sofia, Turin

2.1. Technologies and architecture

In order to have a coherent system along with the other central control systems of the experiment (DCS, Gas System, Cryogenics) the industrial software "WinCC Open Architecture" was a natural choice. This software is the central part of two frameworks - JCOP and UNICOS - already widely used at CERN for the LHC experiments. Some of the components provided by these developer toolkits are of special relevance for NA62:

- DIM (Distributed Interface Management) as the communication layer between the RunControl and the equipment spread across the experiment[3].
- The FSM toolkit managing finite state machines with SMI++ processes[4] according to their definition (states, transition rules, actions).
- The configuration database tool to define and apply sets of parameters from database.
- The Farm Monitoring and Control for the management of the PC farm.

The software is implemented as a hierarchical tree of Finite State Machine (FSM) where the state of each node is either defined by a set of rules summarizing the state of the children nodes or by the evaluation of logical expressions involving parameters provided by the devices.

In this tree, each external node implements a FSM model that represents a specific hardware or software devices (boards, computer, crates and control software programs). The state of this node is computed according to the value of parameters received from the corresponding piece of equipment. The internal nodes group their own children nodes into logical subsystems (subdetectors, PC farm). Finally the top node represents the global state of the experiment.

A change of state is always propagated upwards from the device where this change originated to the top node. Conversely, a command can be issued at any node and is always propagated downwards to all the children until it reaches a device node. At this point, the command is generated and transmitted through the network using the DIM protocol. This command is received by the device control software implementing a standardized interface and encapsulating the knowledge of the device, which will then execute the proper sequence of actions on the hardware.

2.2. Infrastructure

The RunControl is a distributed system spread across different machines in the experiment. The core of the system is located on a dedicated WinCC OA data server on the technical network. A machine bridging the networks hosts the DIM managers which are responsible for loosely binding all the devices in the experiment and the RunControl. The DAQ of the experiment will continue to run correctly by itself in case of a connectivity problem or if this node is cut from one or the other network. As soon as communication is re-established, all the clients will reconnect to the DIM servers and the RunControl will resume control over the experiment. The bridge also receives external information such as the start and end of burst signals or information about the current run from the accelerator and beam line. The user interface itself runs on a different computer in the control room and is remotely connected to the main system.

References

[1] Brod J, Gorbahn M and Stamou E 2011 *Phys. Rev. D* **83**(3) 034030
[2] Artamonov A V *et al.* (E949 Collaboration) 2009 *Phys. Rev. D* **79**(9) 092004
[3] Gaspar C, Donszelmann M and Charpentier P 2001 *Computer Physics Communications* **140** 102 – 109
[4] Franek B and Gaspar C 2004 *Nuclear Science Symposium Conference Record, 2004 IEEE* vol 3 pp 1831–1835

Kaon Identification at NA62

Francis Newson[1]
School of Physics and Astronomy, University of Birmingham, B15 2TT
E-mail: fon@hep.ph.bham.ac.uk

Abstract. The NA62 experiment at the CERN SPS aims to measure the branching ratio of the very rare kaon decay $K^+ \to \pi^+\nu\bar{\nu}$, collecting $\mathcal{O}(100)$ events with a 10% background in 2 years of data taking. A modified CEDAR (Cherenkov Detector with Achromatic Ring focus), named KTAG (Kaon TAGger), will be used to positively identify kaons in the beam, to reduce the background from scattered pions.

1 The NA62 Experiment

The NA62 experiment is a kaon decay in flight experiment at CERN which will begin taking data in October 2014. The primary aim is to measure the branching ratio of the very rare decay $K^+ \to \pi^+\nu\bar{\nu}$ which, in the Standard Model, is predicted to be $BR(K^+ \to \pi^+\nu\bar{\nu}\ ) = (7.81 \pm 0.75 \pm 0.29) \times 10^{-11}$ [1]. High momentum (75 GeV/c) kaons will be obtained by impinging protons from the CERN SPS accelerator onto a target to produce an unseparated beam of protons (22%), pions (72%), and kaons (6%) [2], which will be allowed to decay in vacuum. Since the experimental signature of the signal decay is a single pion and no other detected particles, there is a potential background from beam pions interacting with the residual gas in the vacuum tank. NA62 will use an upgraded CEDAR detector, called KTAG, to positively identify kaons in the beam, relaxing the requirements on the vacuum purity which would otherwise be required.

The specifications of KTAG are driven by the properties of the beam and the background rejection requirements of the NA62 experimental strategy. KTAG must reject pions with a mistagging rate below 10^{-4}, while maintaining a kaon tagging efficiency above 95%. Since the total rate of beam particles is 750 MHz, the kaon crossing time must be measured with a resolution better than 100 ps.

2 Operation of CEDAR and KTAG

When particles travel through a gas at a velocity greater than the local speed of light, they emit Cherenkov radiation at an angle which is a function only of the particle velocity and the refractive index of the medium. In the NA62 beam which has a fixed momentum, a particle's velocity depends only on its mass, and at fixed temperature, the refractive index of a gas depends only on the pressure and the wavelength of light passing through it. The CEDAR detector (Cherenkov Detector with Achromatic Ring focus [3]), uses a chromatic corrector lens to compensate for the wavelength dependence, so that, at a given pressure, light from particles passing through it can be focussed on a ring whose opening angle is determined almost completely by the particle mass.

[1] On behalf of the NA62 Collaboration: Birmingham, Bratislava, Bristol, Bucharest, CERN, Dubna, Fairfax, Ferrara, Florence, Frascati, Glasgow, Liverpool, Louvain, Mainz, Merced, Moscow, Naples, Perugia, Pisa, Prague, Protvino, Rome I, Rome II, San Luis Potosí, Stanford, Sofia, Turin

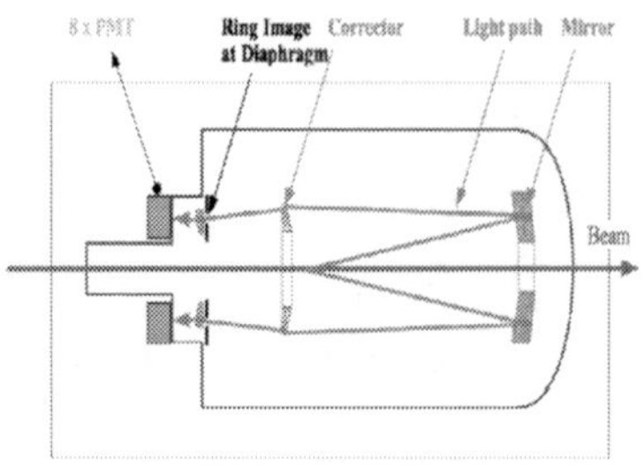

Figure 1: Light path through the CEDAR detector

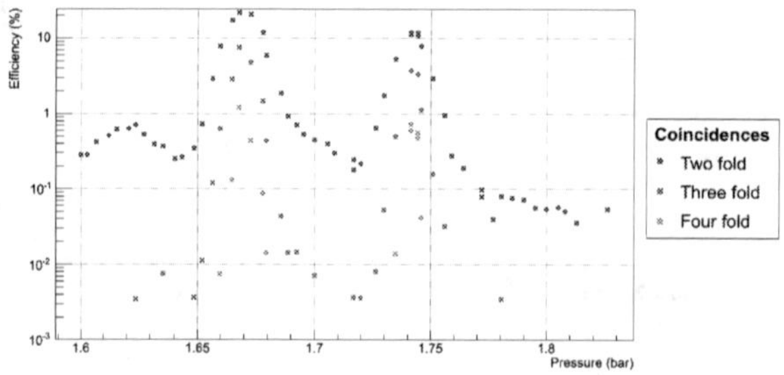

Figure 2: Pressure scan from the 2012 Technical Run showing peaks due to pions (left) and kaons (right)

Figure 1 shows the layout of the CEDAR detector. The beam passes through a steel vessel, 4.5 m long and filled with compressed nitrogen. Cherenkov photons emitted in the gas are reflected at the downstream end of the vessel before passing through a chromatic corrector. The light is focussed on a ring at the upstream end of the detector, whose width (due to chromatic dispersion) is $\Delta r = 2\,\text{mm}$. A diaphragm of variable annular aperture allows selection of light at a fixed radius – corresponding to particles of a fixed mass, determined by the gas pressure.

In the original CEDAR design, Chenrenkov light was detected by 8 photomultiplier tubes (PMTs) at the upstream end of the detector. The high rate expected at NA62 (50 MHz of kaons) would correspond to an unmanageably high flux of 250 MHz of photons per PMT, so a new system has been designed using more PMTs. Light from each of the 8 windows is reflected radially outward by a spherical mirror, spreading the photons evenly over an array of 48 PMTs. The rate of photons on an individual PMT is then reduced to 5 MHz, which can be read out and digitized with a custom electronics chain.

The partially instrumented KTAG was tested in a Technical Run in 2012 [4] and performed as expected. Figure 2 shows the results of scanning the gas pressure to identify light from both pions and kaons.

3 Conclusion

The kaon tagging requirements of NA62 have been met by upgrading the CEDAR detector to perform in a high rate environment. The detector is expected to meet all performance requirements and will begin taking data in October 2014.

4 References

[1] Brod J, Gorbahn M and Stamou E 2011 *Phys. Rev.* D **83**(3) 034030

[2] Hahn F *et al.* (NA62 Collaboration) 2010 *NA62: Technical Design Document* NA62-10-07 `https://cds.cern.ch/record/1404985`

[3] Bovet *et al.* 1982 *The CEDAR counters for particle identification in the SPS secondary beams: a description and an operation manual* CERN-1982-013

[4] NA62 Collaboration *2013 NA62 Status report to the CERN SPSC* CERN-SPSC-2013-009 (SPSC-SR-115)

Journal of Physics: Conference Series **556** (2014) 012076 doi:10.1088/1742-6596/556/1/012076

HERAFitter - an open source QCD fit framework

Andrey Sapronov
on behalf of the `HERAFitter` team

Joint Institute for Nuclear Research, 141980 Russia, Dubna, Moscow Region, Joliot-Curie 6.

E-mail: `sapronov@cern.ch`

Abstract. The `HERAFitter` project was established to increase applicability of the QCD analysis in the hadron collider experiments. The framework may be used to extract parton density functions from a variety of experimental measurements and to assess the impact of new data on the parton density determination. It may also be employed to perform data consistency checks and to test theoretical models. This short article covers a poster contribution on the BEACH2014 conference held at the Birmingham University.

In the era of hadron colliders the proton parton density functions (PDF) have been gaining an increasing attention. While the parton interactions within a hadron cannot be duly described using perturbative approximation, the factorization theorem helps isolating the hard scattering cross section components which can be determined empirically. Improving precision of PDFs is important as it limits our knowledge of cross sections in Standard Model (SM) or beyond SM calculations. The `HERAFitter` [1] is an open source project created to facilitate QCD analysis of experimental data and to increase the accuracy of PDF determination.

A detailed review and comparison of PDF analyses can be found in [2, 3]. The technique of PDF extraction in QCD analysis is based on fitting the experimental data with theoretical predictions. The predictions are obtained with parametrized partonic distributions at a starting scale and then are evolved to the measurement scale using DGLAP evolution [4, 5, 6, 7, 8]. The objective is to utilize as much information from precision data as possible, taking care regarding treatment of the uncertainties and accounting for possible correlations. The means of data fitting, theoretical evaluation, QCD evolution and elaborate uncertainty treatment are all implemented in the `HERAFitter` framework.

There are several particularities related to PDF extraction procedure. One of them is that different processes have different sensitivity to certain quark densities. This allows a broad range of studies for available experimental data in order to estimate their impact on the determination of PDFs. The second is the universality of partonic distributions which allows to perform valuable cross checks and validation of correlated data analysis.

The `HERAFitter` framework was built and extended to benefit from these two aspects. The analysis method relies on the χ^2 estimate of how well the theoretical predictions describe experimental data for a given PDF parametrisation. The program allows tests of various ansatz on the PDF parametric form and QCD evolution approaches (DGLAP, dipole models, unintegrated PDFs). Different forms of χ^2 based on the use of nuisance parameters or on the full covariance matrix can be chosen in the `HERAFitter` for the minimisation procedure.

The instrumentation to calculate theoretical predictions for a list of significant ep, pp and $p\bar{p}$

processes is provided with a number of options. The DIS processes have several heavy quark flavor schemes implemented in different approximations including the fixed-flavour (FFN) and variable flavour number (VFN) schemes. VFN schemes with various treatments for the heavy quark thresholds include the Thorne Roberts (TR) scheme at LO, NLO and NNLO [9, 10] as provided by the MSTW group, the ACOT scheme at LO and NLO as provided by the CTEQ group. The QCDNUM also provides the calculations of the DIS structure functions in the zero-mass VFN and FFN schemes. The Drell-Yan and jet production processes calculations at fixed order are accessible via fast cross section evaluation tools such as APPLGrid [11] and FastNLO [12]. The data on top-quark production can be matched with Monte Carlo calculations by HATHOR [13]

A number of analysis results obtained using the `HERAFitter` package have been published recently. These include: NLO QCD fit based on all published H1 inclusive cross section data [14]; beauty quark mass measurement using ZEUS data [15]; QCD analysis of combined HERA charm data [16]; ATLAS Drell-Yan measurements [17, 18]; PDF determination using CMS and ATLAS inclusive jet data [19, 20]; strange quark density measurements in ATLAS and CMS [21, 22]; Dipole model analysis and TMD gluon density with HERA data [23, 24].

A paper by the `HERAFitter` developers' team devoted to the study of (N)LO PDFs with correlated uncertainties [25] has been accepted for publication and a general description of the package is prepared for journal submission. A complete list of results as well as the newest releases of the `HERAFitter` program can be accessed at the official web-page www.herafitter.org.

References

[1] HERAFitter https://www.herafitter.org
[2] De Roeck A and Thorne R 2011 *Prog.Part.Nucl.Phys.* **66** 727–781 (*Preprint* `1103.0555`)
[3] Watt G 2011 *JHEP* **1109** 069 (*Preprint* `1106.5788`)
[4] Gribov V N and Lipatov L N 1972 *Sov. J. Nucl. Phys.* **15** 438
[5] Gribov V N and Lipatov L N 1972 *Sov. J. Nucl. Phys.* **15** 675
[6] Lipatov L N 1975 *Sov. J. Nucl. Phys.* **20** 94
[7] Dokshitzer Y L 1977 *Sov. Phys. JETP* **46** 641
[8] Altarelli G and Parisi G 1977 *Nucl. Phys. B* **126** 298
[9] Thorne R S and Roberts R G 1998 *Phys. Rev. D* **57** 6871 (*Preprint* `hep-ph/9709442`)
[10] Thorne R S 2006 *Phys. Rev.* **D73** 054019 (*Preprint* `hep-ph/0601245`)
[11] Carli T *et al.* 2010 *Eur. Phys. J.* **C66** 503–524 (*Preprint* `arXiv:0911.2985`)
[12] Kluge T, Rabbertz K and Wobisch M 2006 483–486 (*Preprint* `hep-ph/0609285`)
[13] Aliev M, Lacker H, Langenfeld U, Moch S, Uwer P *et al.* 2011 *Comput.Phys.Commun.* **182** 1034–1046 (*Preprint* `arXiv:1007.1327`)
[14] Aaron F *et al.* (H1 Collaboration) 2012 *JHEP* **1209** 061 (*Preprint* `arXiv:1206.7007`)
[15] Abramowicz H *et al.* (ZEUS Collaboration) 2014 (*Preprint* `1405.6915`)
[16] Abramowicz H *et al.* (H1 and ZEUS Collaborations) 2013 *Eur. Phys. J.* **C73** 2311 (*Preprint* `arXiv:1211.1182`)
[17] Aad G *et al.* (ATLAS Collaboration) 2014 *JHEP* **1406** 112 (*Preprint* `1404.1212`)
[18] Aad G *et al.* (ATLAS Collaboration) 2013 *Phys.Lett.* **B725** 223–242 (*Preprint* `1305.4192`)
[19] Chatrchyan S *et al.* (CMS Collaboration) 2014 *CMS PAS* **SMP-12-028**
[20] Aad G *et al.* (ATLAS Collaboration) 2013 *Eur.Phys.J.* **73** 2509 (*Preprint* `arXiv:1304:4739`)
[21] Aad G *et al.* (ATLAS Collaboration) 2012 *Phys. Rev. Lett.* **109** 012001 (*Preprint* `arXiv:1203.4051`)
[22] Chatrchyan S *et al.* (CMS Collaboration) 2014 *submitted to Phys. Rev.* **D** (*Preprint* `arXiv:1312.6283`)
[23] Hautmann F and Jung H 2014 *Nuclear Physics B* **883** 1 – 19 (*Preprint* `1312.7875`)
[24] Luszczak A and Kowalski H 2014 *Phys.Rev.* **D89** 074051 (*Preprint* `1312.4060`)
[25] HERAFitter Developers Team and Lisovyi M 2014 (*Preprint* `arXiv:1404.4234`)

Journal of Physics: Conference Series **556** (2014) 012077 doi:10.1088/1742-6596/556/1/012077

BEACH 2014 Theory Summary

Sebastian Jäger

Department of Physics and Astronomy, University of Sussex, Falmer, Brighton BN1 9QH, UK

Abstract. I summarize key aspects of the quest for physics beyond the Standard Model in flavour physics as discussed at the BEACH 2014 conference in Birmingham.

1. Introduction

I thank the organisers for inviting me to BEACH 2014 in Birmingham and bestowing on me the honour of giving the theory summary. The many excellent theoretical talks, many of them review talks in their own right, make both for a rich source of topics and a challenge of selection; I apologise at the outset for omissions. The conference started with the low-energy QCD effects in spectroscopy and Kaon physics and ended with the early universe, reflecting appropriately the wide range of energy scales relevant to, and probed through, flavour physics (Figure 1). As indicated in the figure, flavour and heavy-quark physics has seen the birth and use of

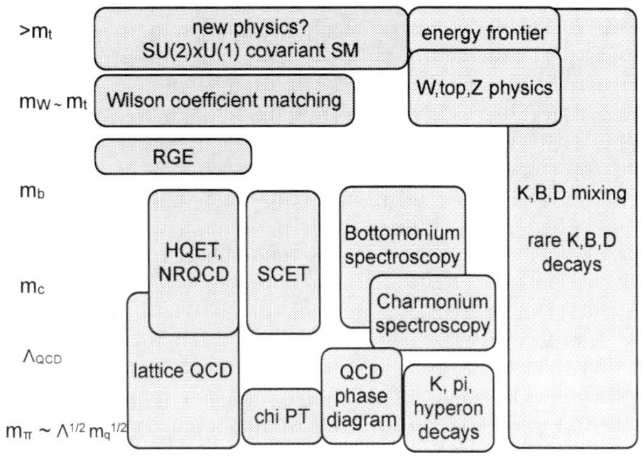

Figure 1. Energy scales, observables, and techniques relevant to flavour physics.

many fruitful theoretical concepts, including the weak Hamiltonian and its renormalisation-group evolution, heavy-quark expansions/HQET/NRQCD and soft-collinear effective theory, and provides the impetus for many developments in lattice QCD.

2. Flavour physics and the BSM landscape

Flavour physics played a key role in constructing the Standard Model (SM), including the invention of Cabibbo mixing to address a 2σ tension in weak decay data [1], the resolution of a naturalness problem in the $K_L - K_S$ mass difference based on the then hypothetical charm quark [2], and the prediction of a third generation by Kobayashi and Maskawa [3] to accommodate CP violation in K_L decay (ϵ_K).

With the discovery of a scalar particle consistent, so far, with the SM Higgs boson, another naturalness problem looms larger than ever in the hierarchy between the electroweak scale and any fundamental scale at which new degrees of freedom appear, including $M_{\rm Planck}$, $M_{\rm GUT}$, $M_{\rm seesaw}$. The known natural beyond-SM (BSM) scenarios all involve, so far hypothetical, particles at the TeV scale and were reviewed in the opening talk by Kamenik [4]. They all bring in new sources of flavour violation. Will flavour physics again be the lead in constructing the next Standard Model? Nobody knows, but it is entirely possible. Indeed, there are several interesting puzzles in the data, some new and some older but persistent, that have been discussed at this meeting. No BSM particles have been identified in the high-p_T experiments yet; this certainly raises the importance of virtual probes of the kind discussed at this conference. Even without a BSM discovery so far, the wealth of data has led to a great deal of bottom-up phenomenology in the last few years, largely phrased in terms of ad-hoc models or fits to effective field theory parameters, at the expense of top-down model building; indeed there was not much discussion of model-specifics at this meeting.

The discovery power of flavour physics arises as follows. A generic FCNC amplitude behaves as

$$\mathcal{A} = \mathcal{A}_{\rm SM} + \mathcal{A}_{\rm BSM}. \tag{1}$$

A typical observable, say a rare decay rate, provides a (schematic) constraint

$$|\mathcal{A}_{\rm BSM}|^2 + 2\,{\rm Re}\,\mathcal{A}^*_{\rm BSM}\mathcal{A}_{\rm SM} \;=\; \Gamma_{\rm exp}(1 \pm \Delta^{(\rm exp)}) - |\mathcal{A}_{\rm SM}|^2(1 \pm \Delta^{(\rm SM)}). \tag{2}$$

Hence, although new physics decouples as $M_Z^2/M_{\rm NP}^2$, flavour observables can probe well beyond the energy frontier provided sufficient statistics and theoretical precision are available. It is perhaps worth comparing the situation in flavour physics with that in precision Higgs physics. Assuming a "little hierarchy" $M_Z \ll M_{\rm NP}$, the leading BSM physics can be parameterized in terms of a large number of dimension-six, $SU(2)_W \times U(1)_Y$ invariant operators [5, 6]. B decays alone probe more than 100 operators even when assuming lepton flavour conservation, far more than in the Higgs case. (Disentangling them is a formidable but not impossible task, see below.)

Figure 2 [4] shows the energy scales probed by the four types of meson-antimeson mixing. The most stringent constraint comes from ϵ_K, pointing either to (approximately) CP-conserving new physics, to a substantial "little hierarchy" between the weak scale and the new physics, or to some kind of flavour suppression mechanism resulting in small mixing angles suppressing strangeness-changing neutral-current transitions. CP-conserving observables provide a less stringent constraint, especially for Kaons, but still apply in the absence of new sources of CP violation. It is worth noting that this situation can change qualitatively in the future, in particular with progress in lattice QCD that seems imminent (see below). Once a deviation is seen in one or more experiments, flavour physics may guide us toward an understanding of the origin of flavour, perhaps in the form of a dynamical theory, perhaps based on symmetry principles.

From the phenomenological point of view, matching the experimental precision comprises disentangling perturbative short-distance physics from long-distance nonperturbative QCD

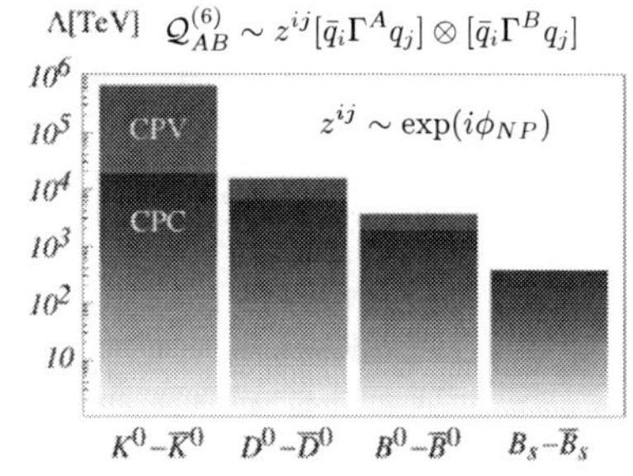

Figure 2. Generic bounds on the scale of CP violating and CP-conserving new physics from meson-antimeson mixing as shown by Kamenik [4].

effects, and calculating the latter or determining them from data was the subject of a number of presentrations at this conference.

3. Kaons

CP violation in $K-\bar{K}$ mixing (ϵ_K) is the prototypical flavour precision observable and generically provides the most stringent constraints on physics beyond the SM [4]. It originates from $\Delta F = 2$ box diagrams with W and u, c, t quarks. The dominant contribution, involving internal top quarks, needs to be complemented by lattice calculations of the nonperturbative normalisation B_K, which is given in terms of the matrix element of a local operator,

$$B_K \propto \langle \bar{K}^0 | (\bar{s}_L \gamma^\mu d_L)(\bar{s}_L \gamma_\mu d_L) | K^0 \rangle.$$

Including perturbative calculations of short-distance charm and up contributions, one has [7]

$$|\epsilon_K| = (1.81 \pm 0.28) \times 10^{-3}, \qquad \Delta M_K^{\rm SD} = (3.1 \pm 1.2) \times 10^{-15}\ {\rm GeV}, \tag{3}$$

while experiment gives [8]

$$|\epsilon_K|^{\rm exp} = (2.228 \pm 0.011) \times 10^{-3}, \qquad \Delta M_K^{\rm exp} = (3.484 \pm 0.006) \times 10^{-15}\ {\rm GeV}. \tag{4}$$

Comparing the measured value of ϵ_K to the theory prediction constrains the CKM unitarity triangle and allows to put constraints on BSM physics, which can involve additional local operators. The constraint from ΔM_K is less stringent due to the larger error and the unknown long-distance contribution, which could interfere destructively with a possible BSM contribution.

In fact, as discussed by Christ [9], already for ϵ_K accuracies have now reached the level where both refined lattice renormalisation schemes (or perhaps a direct calculation of the RG-invariant bag parameter $\hat{B}_K$) and precision determinations of the charm and up loop contributions become important. The latter two correspond to non-local matrix elements of the sort

$$\langle \bar{K}^0 | T \left\{ [(\bar{s}_L \gamma^\mu d_L)(\bar{c}_L \gamma_\mu c_L)](x) [(\bar{s}_L \gamma^\mu d_L)(\bar{c}_L \gamma_\mu c_L)](0) \right\} | K^0 \rangle.$$

These include a perturbative part from scales $\sim m_c$ (although $\alpha_s(m_c)$ is relatively close to the strong-coupling regime), which matches onto the same local operator, and further contributions from scales $\sim \Lambda_{\rm QCD}$. Lattice QCD calculations with dynamical charm quarks are starting to access the required correlation functions directly. These developments also raise the exciting prospect of extending the energy reach of ΔM_K as a probe of CP-*conserving* new physics by an order of magnitude in the near future.

Beyond mixing, the super-rare decays $BR(K^+ \to \pi^+\nu\bar{\nu})$ and $BR(K_L \to \pi^0\nu\bar{\nu})$ provide two further very clean probes of physics beyond the Standard Model, with SM theory errors at the 10% level. Though not discussed in any detail from a theory point of view at this conference, experiment is on the move: The goal at NA62 (K^+, CERN, now running), where the Birmingham group is heavily involved, is a 10% measurement (assuming the SM branching ratio), while KOTO (J-PARC, restarting in 2015) expects to observe the K_L mode, which violates CP. Both will provide stringent constraints on, or measurements of, the sdZ vertices, very sensitive to $SU(2)$-breaking BSM effects such as stop-scharm mixing originating from flavour-violating A-terms (eg [10, 11, 12]). Exciting lattice prospects were discussed [9] for a holy grail of Kaon theory, the two two-body $K \to \pi\pi$ decay amplitudes ($I = 0$ and $I = 2$). From a BSM point of view, the experimental value of the direct-CP violation observable $\epsilon'/\epsilon = (1.66 \pm 0.23) \times 10^{-3}$ [8] is complementary to mixing and rare K decay in that it conveys information about a variety of short-distance couplings, including the sdZ vertex and the chromo-dipole couplings $\bar{s}\sigma^{\mu\nu}P_{L,R}dG_{\mu\nu}$. From an SM/QCD point of view, there is also the question why the decay amplitude into the $I = 0$ final state is more than 20 times larger than that into the $I = 2$ state (the $\Delta I = 1/2$ rule), which predates the advent of the SM. Both questions require the computation of the complex amplitude ratio $A(I = 2)/A(I = 0)$. There is a long history of sophisticated and rather successful calculations based on chiral perturbation theory and the large-N_c limit, as reviewed by Pich [13]. The full-QCD, $N_c = 3$ results have been elusive so far. It is all the more remarkable that the RBC/UKQCD collaboration now appears on track for a fully realistic QCD calculation, with the prospect of a 20% calculation of ϵ'/ϵ. There are many more uses of Kaons, such as constraining the CKM matrix element V_{us}, for which I refer to [9, 13].

4. Heavy quarks

The three heavy quarks benefit, to varying degree, from a hierarchy $m \gg \Lambda$, which gives rise to heavy-quark spin symmetries and is the basis for various effective field theories. Brambilla reviewed the theory and applications of quarkonia spectra and furthermore discussed the nature of heavy-quark bound states close to or above threshold, where, unlike for states well below threshold, the heavy-quark expansion (formalised through NRQCD and pNRQCD) no longer leads to a calculable framework in terms of potentials (which could be determined perturbatively or via lattice QCD simulations) [14]. These include in particular states such as the $X(3872)$ that appear to behave as a molecular state made of two heavy-light systems. The nature of the $X(3872)$, which within experimental accuracies sits right on top of the $D^{0*}\bar{D}^0$ threshold, was discussed in more detail by Nieves [15]. If this picture is correct, it predicts a variety of other nearby states, and we may hope that more data will shed light on this very difficult multi-scale QCD problem. The fate of charmed hadrons in the quark gluon plasma, and other current topics in high-temperature QCD, were reviewed by Steinheimer [16]. Heavy and multiply-heavy baryons were the subject of talks by Azizi [17] and by Erkol [18].

5. Charm

Charm mixing and decay provide sensitivity to short-distance physics similarly to the Kaon case. Unfortunately, the benefits of having a (slightly) heavy quark are outweighed by the loss of chiral perturbation theory together with an adverse CKM/GIM structure, which leads, in the SM, to a complete long-distance dominance for mixing as well as decays, including rare ones. Petrov discussed $D - \bar{D}$ mixing and CP violation in hadronic charm decays [19]. If one assumes the absence of large cancellations between the SM and new-physics contributions, the measured mass difference bounds the new physics scale, but less so than in the Kaon case (Figure 2). The non-observation of CP violation in mixing gives a (stronger) constraint (on CP-violating new physics); in this case the long-distance dominance in the SM leads to a suppression: the three-generation, CP-violating SM is approximately reduced to a two-generation, CP-conserving theory.

The situation is quite different for the difference of direct CP asymmetries, $\Delta a_{\rm CP} = a_{\rm CP,KK} - a_{\rm CP,\pi\pi}$. Previously measured to be 3σ away from zero to much excitement, $\Delta a_{\rm CP}$ is no longer significant as of 2013. In order to turn this into a constraint on the new physics scale, one would require knowledge of (or make estimates about) the strong phases of the SM contributions.

Fajfer discussed a large number of rare semileptonic and radiative D decay rates and asymmetries [20]. All of them suffer from large and rather uncertain, long-distance-dominated SM contributions, but there are a few examples where the experimental bound still significantly exceeds the SM estimates and new physics could give a dominant contributions, such as $D^0 \to \mu^+\mu^-$.

Overall, the power of charm to detect new physics still appears relatively bleak, with the exception of mixing. Eventually, lattice simulations with dynamical charm quarks may open up charm to precision phenomenology, at least for the $\Delta C = 2$ case (and perhaps beyond [19]).

6. Beauty and Truth

B physics comprises the lion's share of observables in flavour physics, many of them sensitive to SM parameters and/or new physics. The advent of the LHCb experiment has brought a wealth of measurements of new decay modes and unprecedented precision for old ones, though some of the most relevant constraints, notably $BR(B \to X_s\gamma)$, still come from the B-factories. Besides the mass splittings and CP violation in $B^0 - \bar{B}^0$ and $B_s - \bar{B}_s$ mixing, leptonic, rare semileptonic, and two-body hadronic decays were discussed at this conference. Important theoretical tools are lattice QCD and expansions in Λ/m_b. More specifically,

- Lattice QCD, reviewed by Jüttner [21], can provide B decay constants, comprising the nonperturbative input to $B \to \tau\nu$ and $B_q \to \mu^+\mu^-$ decay, for dynamical, light quarks at their physical masses, with uncertainties of a few percent. Matrix elements of $\Delta B = 2$ operators needed for mixing can also be calculated with uncertainties of order 5%, and the same is true for form factors for a decay into a stable particle (with regard to QCD) such as a pion or a K, as long as the momentum transfer q^2 is close to maximal (the energy of the light particle is small). One needs to treat the b-quark in an effective theory (HQET or NRQCD), which can however be systematically implemented on a lattice. First calculations of $B \to K^*$ form factors have also appeared, but only for unphysical quark masses for which the K^* is stable. The realistic case of, for instance, $B \to K^*[\to K\pi]\mu^+\mu^-$ would require a $B \to K\pi$ form factor calculation, which is at an early stage conceptually [22]. A complementary method, light-cone sum rules, was discussed by Khodjamirian; and can be used to obtain form factors at small q^2 (as well as other nonperturbative quantities quantities). It also has the ability to describe a $B \to K\pi$ (or similar) form factor. Form factor calculations in a quark model were the topic of a talk by Hernandez [23].
- Observables related to mixing, the topic of a talk by Bobeth [24], comprise the mass difference, the lifetime difference, time-dependent CP violation in several modes, and CP violation in mixing (the flavour-specific or semileptonic CP asymmetries). Data show a good overall degree of consistency with the SM (Figure 3). In the case of B_d mixing, there is a mild tension. In the case of B_s mixing, the best fit is right on top of the SM expectation. One should note however that the consistency between the different inputs is not good, in particular, the D0 measurement of the dimuon charge asymmetry, which depends on

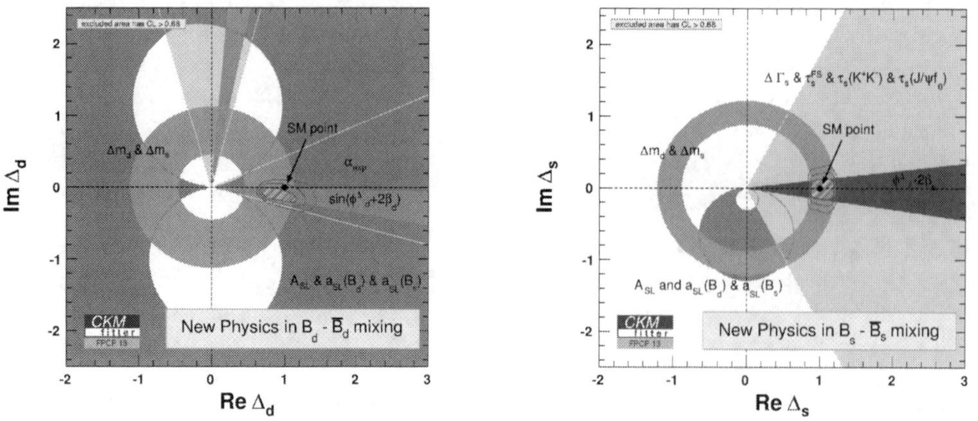

Figure 3. Combined constraints on $B_d - \bar{B}_d$ mixing (left) and $B_s - \bar{B}_s$ mixing (right) by the CKMfitter group (as of FCPC 2013) [25].

both the semileptonic CP asymmetry and on the lifetime differences, is about 3σ away from the SM prediction. Possible interpretations in terms of new decay modes of B_d were discussed by Bobeth [24]. In particular, it appears possible in a model-independent fashion to explain the discrepancy between the D0 result and other data with extra contributions to the lifetime difference $\Delta\Gamma_d$ (as opposed to the semileptonic CP asymmetry) from either four-quark operators or semileptonic operators with τ fields, without getting disagreement with exclusive B decay data.

Another issue that will become important with the increased accuracy with which LHCb measures time-dependent CP violation in B_d and B_s decays is the "penguin pollution" in the various decay modes, discussed by Fleischer at this conference [26] together with U-spin-symmetry-based methods for controlling these effects, which are not yet taken into account in official projected sensitivities.

Two-body charmless B decays exhibit QCD factorisation in the heavy quark limit, i.e., they factorise into perturbatively calculable kernels and nonperturbative objects such as form factors. Bell reported on the theory status including ongoing computations which will allow to compute direct CP asymmetries to next-to-leading order [27].

- The combined CMS-LHCb results on leptonic decay $B_s \to \mu^+\mu^-$ and $B_d \to \mu^+\mu^-$ read [28]

$$BR(B_s \to \mu^+\mu^-)^{\rm exp} = 2.8^{+0.8}_{-0.6} \times 10^{-9}, \qquad BR(B_d \to \mu^+\mu^-)^{\rm exp} = 3.9^{+1.6}_{-1.4} \times 10^{-10},$$

and are consistent with SM expectations of

$$BR(B_s \to \mu^+\mu^-) = (3.65 \pm 0.23) \times 10^{-9}, \qquad BR(B_d \to \mu^+\mu^-) = (1.06 \pm 0.09) \times 10^{10},$$

at the 1.2σ and 2.2σ levels, respectively, also showing a certain tension (which is 2.3σ in the ratio [28]). I find it extremely impressive that both rates are already measured to better than 40% accuracy, in spite of their extreme smallness, which shows the potential to find even small deviations from the SM, such as through a modified Zbs vertex. The theoretical expectations given above are the outcome of an NNLO-QCD and NLO-electroweak calculation, subject of a talk by Steinhauser [29], which also investigated issues such as the fate of the helicity suppression underlying the smallness of the rate under soft photon emission. The theory errors, which now are below 10%, are dominated by parametric uncertainties from f_B and f_{B_s}, which will further reduce with progress on the lattice.

- The most intriguing picture presents itself when one goes to rare semileptonic $b \to sll$ decays, reviewed by [30], which entail several anomalies. The bulk of the information is provided by the angular analysis of $\bar{B} \to \bar{K}^* \mu^+ \mu^-$, which involves 12 separate angular observables, each of which a function of the dilepton invariant mass. A global fit to coefficients in the BSM effective Lagrangian indicates a discrepancy with the SM (Figure 4 (left); see also [31]). The effect can be attributed to a sizable negative BSM shift of the Wilson coefficient

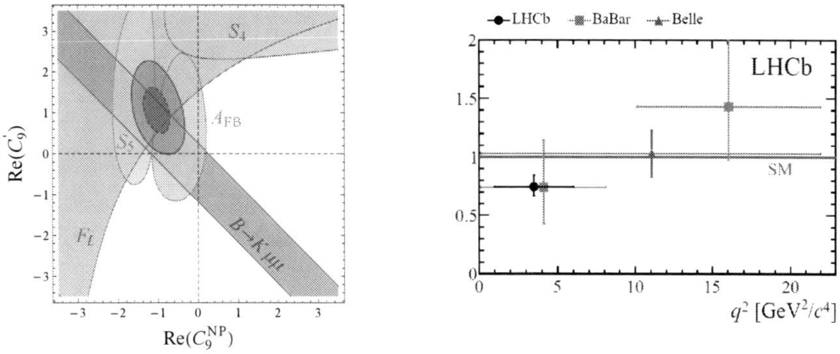

Figure 4. Left: Fit of $b \to sll$ data to two Wilson coefficients in the effective BSM hamiltonian [32]. Right: Lepton-number-(non)universality ratio $R_K = \Gamma(B^+ \to K^+\mu^+\mu^-)/\Gamma(B^+ \to K^+e^+e^-)$ [33, 34].

C_9, multiplying the operator $Q_9 \propto (\bar{s}_L \gamma^\mu b_L)(\bar{\mu}\gamma_\mu \mu)$. The devil may, however, be in the detail. The angular observables are quadratic functions in helicity amplitudes, the most complicated of which are the vector amplitudes, schematically

$$H_V(\lambda) \propto \tilde{V}_\lambda(q^2) C_9 + \frac{2\, m_b m_B}{q^2} \tilde{T}_\lambda(q^2) C_7 - \frac{16\, \pi^2 m_B^2}{q^2} h_\lambda(q^2).$$

They involve interference of three different terms, including two nonperturbative local form factors and a further, nonlocal nonperturbative term. In order to claim an effect of the stated significance, the necessary precision on the form factors at small dilepton mass, more specifically on deviations from their relations in the heavy-quark limit, is on the order of 5% (Figure 5 left), and in my view it is doubtful that existing calculations (based mainly on light-cone sum rules) can be trusted to such a high level of precision. A conservative treatment of the power corrections [35] (see also [36]) results in a lower significance of the effect (Figure 5 right), of about 1σ only.

On the other hand, the picture of a negative correction to the SM Wilson coefficient C_9 is consistent with the relatively low branching ratio in the decay $B^+ \to K^+\ell^+\ell^-$. Also here the conclusion depends somewhat on nonperturbative form factor normalisations. The latter largely drop out of the ratio $R_K = \Gamma(B^+ \to K^+\mu^+\mu^-)/\Gamma(B^+ \to K^+e^+e^-) = 0.745^{+0.090}_{-0.074}$ [34] (in the bin $1\,\mathrm{GeV}^2 < q^2 < 6\,\mathrm{GeV}^2$), which points at lepton universality violation. If this result stands, it could indeed be accommodated by a muon-specific negative shift to C_9, consistent with what is inferred from the $B \to K^*\mu^+\mu^-$ angular analysis, but in the muonic operator only (as opposed to that containing the electronic vector current). Such scenarios are possible in UV-complete models, see e.g. [37]. In this context an interesting new development first shown at this conference is the systematic use of $SU(2) \times U(1)_Y$-invariant effective theory, which appears justified by the likely hierarchy $M_{\rm NP} \gg M_Z$. This rules out tensor operators as an explanation of any of the anomalies and allows to correlate the R_K measurement with $B_s \to \mu^+\mu^-$ decay, and to rule out any role of scalar or pseudoscalar operators, leaving only the semileptonic current-current operators which includes the operator Q_9 mentioned above. Finally, the same decay $B^+ \to K^+\mu^+\mu^-$

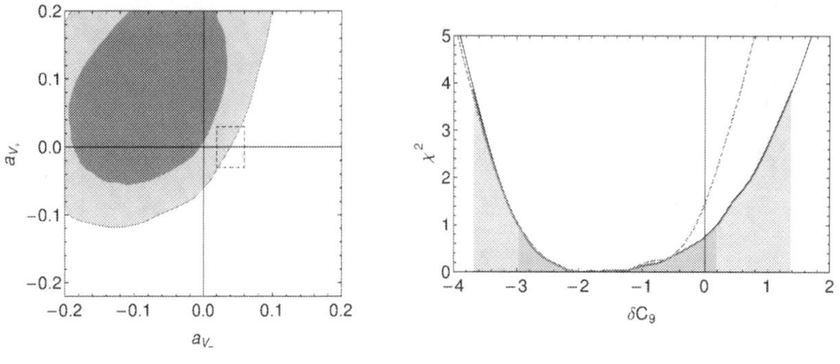

Figure 5. Left: profile χ^2 of the angular observable P_5' in the [1..6] GeV^2 bin, as a function of two power correction parameters. Right: profile χ^2 for the Wilson coefficient $C_9^{\rm NP}$ [30, 38].

also shows a rather pronounced resonant structure above the open charm threshold, the significance of which is unclear at present.

The intriguing situation on several fronts makes B-physics a definite place to watch for upcoming developments, such as updates of the angular analysis in $B \to K^*\mu^+\mu^-$ with the full run I dataset, the first angular analysis in $B \to K^*e^+e^-$, and high-statistics results based on the upcoming LHC run II and, before the decade is out, from the Belle 2 super flavour factory.

The top quark was also discussed at this meeting, in talks by Pecjak [39] and by Zhang [40]. Due to the unprecedented top samples at the LHC, it has started to become the subject of precision studies, among them searches for BSM effects in flavour-changing top decays. As with the lighter flavours, assuming a little hierarchy the BSM contributions to top production and decay can be described in terms of effective field theory and the data used to put model-independent constraints on its operator (Wilson) coefficients [40].

7. Leptons

In the lepton sector, flavour is known to be violated by neutrino oscillations. It is widely expected that lepton number is also violated, as happens in the seesaw models, which would give rise to the unique (and lepton-number-violating) dimension-5 operator beyond the Standard Model. All data is presently consistent with this picture, with exception of the well-known LSND anomaly. Status and prospects were reviewed by Pakvasa [41] and by Palazzo [42]. After the establishment of $\theta_{13} \approx 9^\circ \neq 0$, the main questions are: (i) do neutrino masses have a lepton-number-violating origin (search for neutrinoless double beta decay), (ii) is CP violated in the lepton/neutrino sector, (iii) is the neutrino mass hierarchy normal or inverted, (iv) are there sterile neutrinos (and if so at what scale)? There are indications in the global fit for a nonzero ("Dirac", lepton-flavour-violating) CP-violating phase $\delta \in (\pi, 2\pi)$, although the significance depends on the mass hierarchy and is never above 2σ. This and the other questions will, however, take a long time to resolve definitely. If lepton number is indeed violated, one can also have CP violation through the "Majorana" phases in the neutrino mass matrix.

The fact that lepton flavour violation is observed implies that charged lepton-flavour-violating processes must also occur, such as $\mu \to e\gamma$. They were the subject of a talk by Paradisi [43]. In the SM (supplemented by neutrino masses), the tiny masses of the neutrinos (when compared to the W mass) imply a near-perfect GIM cancellation, with no hope to see a signal. The situation can be very different beyond the SM. SUSY seesaw models, for example, communicate the flavour violation above the seesaw scale through to low-energy scales via renormalisable terms, such as off-diagonal elements of the slepton mass matrices. The resultant SUSY contributions can saturate the current bounds on many lepton-flavour-violating decay modes. An interesting aspect are relations within the general MSSM between the radiative lepton-flavour-violating decays and the flavour-conserving but BSM-sensitive anomalous magnetic moment $(g-2)_\mu$ of the muon. The interest derives from the fact that the latter has shown a 3σ deviation from the SM prediction for a number of years now. The correlation and the theory of $(g-2)_\mu$ was reviewed in more detail by Velasco-Sevilla [44].

8. Origin of matter

The most important aspect of CP violation may be that our existence may be thanks to it. More specifically, the observed excess of matter over antimatter in the universe can be explained by out-of-equilibrium CP-violating processes (together with violation of baryon number and charge conjugation symmetry) in the early universe. There are many models and mechanisms providing a successful baryogenesis, and it is impossible to conclude from the observation of the baryon asymmetry alone which one is correct. Some of the more compelling ones are, however, connected to CP-violating flavour physics at the TeV scale. Electroweak baryogenesis in a Two-Higgs-doublet model was discussed by Huber [45]. Here new degrees of freedom near the weak scale effect a strong first-order electroweak phase transition; the CP violating processes occuring in the course of this are found to be correlated with rare B decay and $B_s - \bar{B}_s$ mixing, as well as electric dipole moments of elementary particles. Schwaller discussed leptogenesis [46], whereby heavy SM singlets in a seesaw model decay to Higgs + lepton in a CP-asymmetric fashion; the necessary baryon number violation (which converts part of the lepton number into baryon number) takes place at much lower energies/temperatures via sphaleron processes. An important topic are thermal corrections to CP violation and a proper account of the tau Yukawa coupling.

9. Outlook

In conclusion, while it can be at times disappointing that the eagerly anticipated signatures of new physics below the TeV scale have not materialised, we are in the lucky circumstance of being showered with new results on a regular basis. While, for the time being, most of them imply ever tighter constraints on BSM physics, we should note that the arguments motivating physics at or near the TeV have not diminished. Any of the several tensions mentioned can, with more data and/or better theory, turn into significant falsifications of the Standard Model; and they may in fact already be pointing at a consistent picture. That the search for new physics would be a laborious affair was, by the way, predicted here in Bimingham, as the university motto attests (Figure 6 left). Thanks to our experimental colleagues, there is no shortage of work for flavour theorists these days. While, as in any ambitious project, success is not certain, I am confident that, in the end, Hilbert will be proven right (Figure 6 right).

Figure 6. Left: The University of Birmingham's motto, "Through struggle to high energies." Right: Hilbert's tomb [Source: Wikimedia Commons].

Acknowledgment
I would like to thank the organisers warmly for such an exciting and perfectly organised conference. I am grateful to the many theory speakers for their clear presenations, which made my task easier than it could have been, and to Alexey Petrov for photographs used on my slides (not reproduced here). Support from the UK STFC under grants ST/J000477/1 and ST/L000504/1 is acknowledged.

References
[1] Gell-Mann M and Levy M 1960 *Nuovo Cim.* **16** 705
[2] Gaillard M and Lee B W 1974 *Phys.Rev.* **D10** 897
[3] Kobayashi M and Maskawa T 1973 *Prog.Theor.Phys.* **49** 652–657
[4] Kamenik J *these proceedings, and references therein*
[5] Buchmuller W and Wyler D 1986 *Nucl.Phys.* **B268** 621–653
[6] Grzadkowski B, Iskrzynski M, Misiak M and Rosiek J 2010 *JHEP* **1010** 085 (*Preprint* 1008.4884)
[7] Brod J and Gorbahn M 2012 *Phys.Rev.Lett.* **108** 121801 (*Preprint* 1108.2036)
[8] Olive K *et al.* (Particle Data Group) 2014 *Chin.Phys.* **C38** 090001
[9] Christ N *these proceedings, and references therein*
[10] Colangelo G and Isidori G 1998 *JHEP* **9809** 009 (*Preprint* hep-ph/9808487)
[11] Buras A J, Ewerth T, Jager S and Rosiek J 2005 *Nucl.Phys.* **B714** 103–136 (*Preprint* hep-ph/0408142)
[12] Isidori G, Mescia F, Paradisi P, Smith C and Trine S 2006 *JHEP* **0608** 064 (*Preprint* hep-ph/0604074)
[13] Pich A *these proceedings, and references therein* (*Preprint* 1410.3783)
[14] Brambilla N *talk at this conference (unpublished)*
[15] Nieves J *these proceedings, and references therein* (*Preprint* 1409.4390)
[16] Steinheimer J *these proceedings, and references therein*
[17] Azizi N *these proceedings, and references therein*
[18] Erkol G *these proceedings, and references theirein*
[19] Petrov A *these proceedings, and references theirein*
[20] Fajfer S *these proceedings, and references therein*
[21] Jüttner A *talk at this conference (unpublished)*
[22] Briceo R A, Hansen M T and Walker-Loud A 2014 (*Preprint* 1406.5965)
[23] Hernandez E *these proceedings, and references therein* (*Preprint* 1409.4207)
[24] Bobeth C *these proceedings, and references therein*
[25] Charles J *et al.* (CKMfitter Group) 2005 *Eur.Phys.J.* **C41** 1–131 (*Preprint* hep-ph/0406184)
[26] Fleischer R *these proceedings, and references therein* (*Preprint* 1408.2497)
[27] Bell G *talk at this conference (unpublished)*
[28] Tolk S *Talk at Implications of LHCb measurements and Future Prospects, CERN, 15 October 2014*
[29] Steinhauser M *these proceedings, and references therein*
[30] Martin Camalich J *talk at this conference (unpublished)*
[31] Descotes-Genon S, Matias J and Virto J 2013 *Phys.Rev.* **D88** 074002 (*Preprint* 1307.5683)
[32] Altmannshofer W and Straub D M 2013 *Eur.Phys.J.* **C73** 2646 (*Preprint* 1308.1501)

[33] Shires A *talk at Flavour of New Physics in $b \to s$ transitions, Institut Henri Poincar'e, 3 June 2014*
[34] Aaij R *et al.* (LHCb collaboration) 2014 *Phys.Rev.Lett.* **113** 151601 (*Preprint* 1406.6482)
[35] Jäger S and Martin Camalich J 2013 *JHEP* **1305** 043 (*Preprint* 1212.2263)
[36] Beaujean F, Bobeth C and van Dyk D 2014 *Eur.Phys.J.* **C74** 2897 (*Preprint* 1310.2478)
[37] Altmannshofer W, Gori S, Pospelov M and Yavin I 2014 *Phys.Rev.* **D89** 095033 (*Preprint* 1403.1269)
[38] Jäger S and Martin Camalich J *in preparation*
[39] Pecjak B *talk at this conference (unpublished)*
[40] Zhang C *these proceedings, and references therein* (*Preprint* 1410.2825)
[41] Pakvasa S *these proceedings, and references therein*
[42] Palazzo A *these proceedings, and references therein*
[43] Paradisi P *talk at this conference (unpublished)*
[44] Velasco-Sevilla L *these proceedings, and references therein*
[45] Huber S *talk at this conference (unpublished)*
[46] Schwaller L *these proceedings, and references therein*

AUTHOR INDEX

AUTHOR INDEX